THE CASSINI-HUYGENS MISSION

THE CASSINI-HUYGENS MISSION

Orbiter In Situ Investigations

Volume 2

Edited by

CHRISTOPHER T. RUSSELL
University of California, California, U.S.A.

KLUWER ACADEMIC PUBLISHERS
DORDRECHT / BOSTON / LONDON

A.C.I.P. Catalogue record for this book is available
from the Library of Congress

ISBN 978-94-015-7069-5 ISBN 978-1-4020-2774-1 (eBook)
DOI 10.1007/978-1-4020-2774-1

Published by Kluwer Academic Publishers
P.O. Box 17, 3300 AA Dordrecht, The Netherlands

Sold and distributed in North, Central and South America
by Kluwer Academic Publishers,
101 Philip Drive, Norwell, MA 02061, U.S.A.

In all other countries, sold and distributed
by Kluwer Academic Publishers,
P.O. Box 322, 3300 AH Dordrecht, The Netherlands

Printed on acid-free paper

Cover illustration: Courtesy of ESA

Softcover reprint of the hardcover 1st edition 2004

TABLE OF CONTENTS

Foreword

Planetary exploration is generally conducted with one of five different techniques. In order of increasing complexity these are: remote sensing from 1 Astronomical Unit (AU), observations from a spacecraft flying by the planet, orbital observations, atmospheric and/or landing probes, and sample return missions. For the Moon, Mars and Vesta for which meteorites are available, samples literally fall in our laps, but for most of the planets, and in particular for the Saturnian system, we must visit them in order to obtain detailed understanding.

Despite its great distance from the Sun, almost 10 AU, Saturn's great size and extended ring system, has enabled Saturn to be studied for centuries using the first of these techniques. Nevertheless Saturn's atmosphere, unlike Jupiter's has a veil of haze that cloaks the planet, keeping much of the dynamics of that atmosphere hidden from view. While scientists have learned much about the Saturnian system from remote sensing from the vicinity of Earth, this knowledge has generally fueled our fascination, rather than deepened our understanding.

The first glimmering of understanding arose over two decades ago from technique number two with three flyby missions to Saturn in quick succession: Pioneer 11, Voyager 1 and Voyager 2. However, the time spent at Saturn by these three missions is numbered only in days. The time needed to obtain understanding of the dynamics of the atmosphere, the structure in the rings, the properties of the moons, the harmonics of the magnetic and gravitational fields and the dynamics of the magnetospheric plasma requires years of observation, not days. An orbiting spacecraft is required to provide long-term, in situ observations. The Cassini orbiter answers that need.

As described in Volume 1 of this three-volume set, Cassini is a comprehensive attack on the nature of the Saturnian system, using techniques three and four on our list, measurements with the Cassini orbiter and the Huygens atmospheric probe/lander. Volume 1 described the Huygens probe in detail, the overall scientific objectives of the mission, and the mission and spacecraft design. In this volume we describe the in situ investigations carried by the orbiter: plasma and neutral particles, energetic electrons, ions and neutrals, the magnetic field, radio and plasma waves, and dust particles. Volume 3 will be devoted to the remote sensing instruments.

As with Volume 1, this compilation is due to the efforts of many individuals, especially the referees and authors who have contributed to producing a very readable and complete description of the investigations. We especially wish to

Space Science Reviews **114:** vii–viii, 2004.

thank Anne McGlynn who assisted me in the initial stages of the assembly of this volume and Marjorie Sowmendran who completed the effort upon Anne's retirement.

C. T. Russell
University of California Los Angeles
January, 2004

CASSINI PLASMA SPECTROMETER INVESTIGATION

D. T. YOUNG[1], J. J. BERTHELIER[3], M. BLANC[4], J. L. BURCH[1], A. J. COATES[5], R. GOLDSTEIN[1], M. GRANDE[7], T. W. HILL[8], R. E. JOHNSON[10], V. KELHA[11], D. J. MCCOMAS[1], E. C. SITTLER[9], K. R. SVENES[12], K. SZEGÖ[13], P. TANSKANEN[14], K. AHOLA[16], D. ANDERSON[1], S. BAKSHI[9], R. A. BARAGIOLA[10], B. L. BARRACLOUGH[2], R. K. BLACK[1], S. BOLTON[6], T. BOOKER[1], R. BOWMAN[1], P. CASEY[1], F.J. CRARY[1], D. DELAPP[2], G. DIRKS[1], N. EAKER[1], H. FUNSTEN[2], J. D. FURMAN[1], J. T. GOSLING[2], H. HANNULA[11], C. HOLMLUND[11], H. HUOMO[15], J. M. ILLIANO[3], P. JENSEN[1], M. A. JOHNSON[9], D. R. LINDER[5], T. LUNTAMA[11], S. MAURICE[4], K. P. MCCABE[2], K. MURSULA[14], B. T. NARHEIM[12], J.E. NORDHOLT[2], A. PREECE[7], J. RUDZKI[1], A. RUITBERG[9], K. SMITH[1], S. SZALAI[13], M.F. THOMSEN[2], K. VIHERKANTO[11], J. VILPPOLA[14], T. VOLLMER[9], T. E. WAHL[6], M. WÜEST[1], T. YLIKORPI[11] AND C. ZINSMEYER[1]

[1]*Southwest Research Institute, San Antonio, TX, U.S.A.*
[2]*Los Alamos National Laboratory, Los Alamos, NM, U.S.A.*
[3]*Centre d'étude des Environnements Terrestre et Planetaires, CNRS, St. Maur, France*
[4]*Observatoire Midi-Pyrenées, Toulouse, France*
[5]*Mullard Space Science Laboratory, University College London, Surrey, England*
[6]*Jet Propulsion Laboratory, Pasadena, CA, U.S.A.*
[7]*Rutherford Appleton Laboratory, Oxfordshire, England*
[8]*Rice University, Houston, TX, U.S.A.*
[9]*Goddard Space Flight Center, Greenbelt, MD, U.S.A.*
[10]*University of Virginia, Charlottesville, VA, U.S.A.e*
[11]*VTT Information Technology, Espoo, Finland*
[12]*Norwegian Defense Research Establishment, Kjeller, Norway*
[13]*KFKI Research Institute for Particle and Nuclear Physics, Budapest, Hungary*
[14]*University of Oulu, Oulu, Finland*
[15]*Nokia Corporation, Helsinki, Finland*
[16]*TEKES Technology Development Centre, Helsinki, Finland*

(Received 8 February 1998; Accepted in final form 9 January 2004)

Abstract. The Cassini Plasma Spectrometer (CAPS) will make comprehensive three-dimensional mass-resolved measurements of the full variety of plasma phenomena found in Saturn's magnetosphere. Our fundamental scientific goals are to understand the nature of saturnian plasmas primarily their sources of ionization, and the means by which they are accelerated, transported, and lost. In so doing the CAPS investigation will contribute to understanding Saturn's magnetosphere and its complex interactions with Titan, the icy satellites and rings, Saturn's ionosphere and aurora, and the solar wind. Our design approach meets these goals by emphasizing two complementary types of measurements: high-time resolution velocity distributions of electrons and all major ion species; and lower-time resolution, high-mass resolution spectra of all ion species. The CAPS instrument is made up of three sensors: the Electron Spectrometer (ELS), the Ion Beam Spectrometer (IBS), and the Ion Mass Spectrometer (IMS). The ELS measures the velocity distribution of electrons from 0.6 eV to 28,250 keV, a range that permits coverage of thermal electrons found at Titan and near the ring plane as well as more energetic trapped electrons and auroral particles. The IBS measures ion velocity distributions with very high angular and energy resolution from 1 eV to 49,800 keV. It is specially designed

Space Science Reviews **114:** 1–112, 2004.

to measure sharply defined ion beams expected in the solar wind at 9.5 AU, highly directional rammed ion fluxes encountered in Titan's ionosphere, and anticipated field-aligned auroral fluxes. The IMS is designed to measure the composition of hot, diffuse magnetospheric plasmas and low-concentration ion species 1 eV to 50,280 eV with an atomic resolution M/ΔM ~70 and, for certain molecules, (such as N_2^+ and CO^+), effective resolution as high as ~2500. The three sensors are mounted on a motor-driven actuator that rotates the entire instrument over approximately one-half of the sky every 3 min.

Keywords: Saturn, Titan, magnetosphere, space plasma, ion composition

1. Introduction

Saturn's magnetosphere comprises a unique plasma environment very different from that of the Earth or other planets. Although it shares a rapidly rotating magnetic field with the other giant planets, Saturn is distinguished by internal plasma sources such as Titan's atmosphere and ionosphere and the rings. The existence and location of Titan also permits the Cassini orbiter to execute a 4-year tour unlike that of any planetary magnetosphere to date including the Earth's. Over 40 close flybys of Titan and another dozen or so of the icy satellites present many opportunities for studies of magnetospheric interactions with planetary atmospheres and surfaces. Highly inclined passes near or through the auroral zone, and a wide sampling of both magnetic local time and latitude, constitute an unparalleled opportunity for comprehensive studies of the morphology and dynamics of the magnetosphere. Moreover, nearly all of the macroscopic phenomena mentioned here are associated with microphysical processes such as wave–particle interactions that can be studied to great advantage during the tour.

In order to be fully responsive to mission science objectives, the CAPS instrument is designed to make the most comprehensive possible suite of plasma measurements within constraints imposed by the mission and spacecraft itself. The balancing of measurement requirements, available technologies, and resource constraints has led to many tradeoffs that shaped the final execution of the CAPS design. A single plasma sensor cannot carry out the wide range of measurement objectives presented by the mission and therefore CAPS is made up of three sensors (see detailed descriptions in Sections 4 through 6). The Electron Spectrometer (ELS) measures differential electron velocity distributions making detailed studies of secondary electron fluxes that contribute to ionization and chemical processes taking place at Titan and elsewhere. At tens of keV ELS is expected to contribute to studies of trapped electrons and those associated with saturnian aurora. Throughout its energy range ELS will provide a global survey of plasma density, temperature and electron pitch angle distributions that are needed to derive a comprehensive view of plasma dynamics within the magnetosphere and, for roughly 50% of the mission, in the solar wind and magnetosheath.

At 9.5 AU the solar wind has cooled to become highly supersonic (Mach numbers ~10 to >40). The large amount of time that Cassini spends in the solar wind presents a rare opportunity to study both its intrinsic characteristics and its interactions with the magnetosphere of Saturn and possibly, the comet-like magnetosphere of Titan. The Ion Beam Spectrometer (IBS) is capable of the very high energy and angular resolution necessary not only for solar wind measurements, but also for observing ion ram fluxes at Titan and any auroral ion beams that might exist.

Many of the key questions of plasma origins and processes can only be answered through knowledge of plasma ion composition. Saturn's magnetosphere is known from Voyager to contain a wide variety of ion distributions that derive primarily from the icy surfaces of the satellites and rings, Titan's atmosphere, and, to a lesser extent, from Saturn's atmosphere and solar wind. Unfortunately very little is known about the composition of these plasmas and how the composition might affect magnetospheric phenomena. The Ion Mass Spectrometer (IMS), a high sensitivity, high-resolution mass spectrometer, is designed to provide comprehensive measurements in all regions of the magnetosphere. IMS relies on time-focused optics combined with carbon foil technology. It is designed to separate atomic species with high resolution, and to identify isobaric molecular species such as CH_4^+, NH_2^+, and O^+ (all with $M/Q = 16$, where M/Q is the mass/charge ratio) or N_2^+ and CO^+ ($M/Q = 28$) that would otherwise require a very large conventional instrument to achieve. Because of its broad energy range IMS can be used to study the composition of Titan's ionosphere at a few eV, complementing the Ion and Neutral Mass Spectrometer (Waite *et al.*, 2004), or to study energetic trapped ions, complementing the MIMI/CHEMS investigation (Krimigis *et al.*, 2004).

We first present a discussion of investigation science objectives and measurement requirements (Section 2). We then describe the design, development, calibration, and operation of the CAPS instrument in some detail (Sections 3 through 7) and conclude with a discussion of instrument operations and modes (Section 8) and examples of performance data from the Cassini encounters with Earth and Jupiter (Section 9). A table of acronyms can be found in the Appendix.

2. Scientific Objectives

Cassini's broad scientific mission to study in depth the entire saturnian system, including its magnetosphere, admits of an equally broad range of scientific objectives for the CAPS investigation. These are primarily the saturnian magnetosphere and aurora, Titan's ionosphere and magnetosphere and the tenuous ionospheres of the rings and icy satellites. In the remainder of this section we discuss the CAPS scientific objectives in more detail as a way of providing motivation and background for measurement requirements and instrument design that are the focus of this paper.

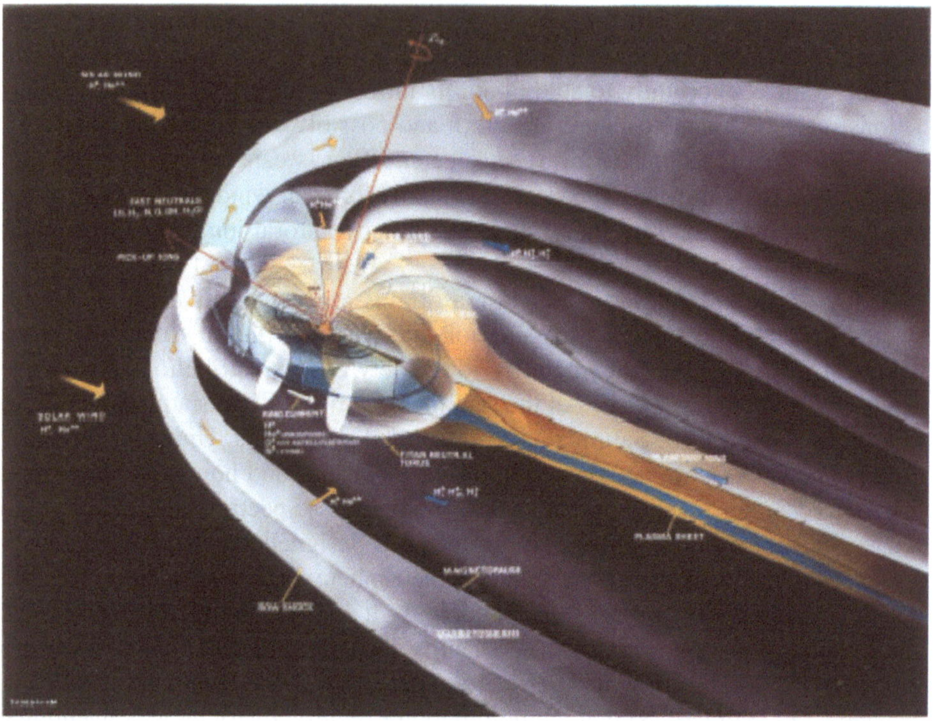

Figure 1. Three-dimensional rendering of the Saturnian magnetosphere showing solar wind streaming from the right, the donut-shaped torus shed by Titan, the inner region of material shed by icy satellites, and the orange-colored plasma sheet stretching out into the magnetotail. (Painting courtesy of J. Tubb, Los Alamos National Laboratory.)

2.1. Saturn's Magnetosphere and Aurora

Saturn's magnetosphere (Figure 1) envelops the extensive ring system and the collection of large icy satellites. It also encloses the orbit of Titan during average solar-wind conditions, although a strong solar-wind gust can push the magnetopause inside Titan's orbit temporarily on the day-side. The satellites and rings provide sources and sinks of plasma, thereby affecting the dynamics as well as the composition of the magnetospheric plasma. The plasma, in turn, affects the evolution of satellite surfaces and even the motion of the smallest particulates, providing a natural laboratory for *in-situ* study of dust–plasma interactions that have implications for solar-system evolution and many other astrophysical applications.

2.1.1. *Plasma Sources and Sinks*

Saturn's magnetosphere has three distinct internal sources of plasma: the upper atmosphere of Saturn (dominated by atomic hydrogen), the icy-satellite/ring system and associated neutral-gas cloud (dominated by water products), and Titan (dominated by atomic nitrogen and perhaps atomic hydrogen). These are illustrated schematically in Figure 1. Data from Pioneer 11 and Voyagers 1 and 2 suggest that all three sources are effective, but their relative importance in various regions remains controversial (cf. Shemansky *et al.*, 1985; Richardson *et al.*, 1986; Richardson and Eviatar, 1987; Blanc *et al.*, 2004). External sources (solar wind and interstellar gas) are less evident in Voyager data but present in principle, and no less important to detect if present. Apart from its possible role as an external particle source for the magnetosphere, solar wind at 9.5 AU is expected to have high Mach numbers, providing a unique opportunity to study bow shock dynamics at very high-sonic Mach numbers. Also, the weaker interplanetary magnetic field at 9.5 AU will make shock layers thicker than at 1 AU and allow CAPS, with 2-s time resolution, to spatially resolve shocks. During the length of the tour we expect a very large number of bow shock crossings. We will look for reflected- and diffuse-ion populations within the foreshock (Thomsen, 1985), accelerated electrons in the foreshock region (Klimas, 1985), and the leakage of magnetospheric plasma into the solar wind.

We will exploit four techniques to distinguish the source of resident plasma. One is to monitor atomic and molecular ion composition as a function of position in the magnetosphere. A wide variety of anticipated ion species must be resolved, including H^+, H_2^+, H_3^+, He^{2+}, O^+, O^{2+}, OH^+, H_2O^+, H_3O^+, N^+, and N_2^+. It is particularly important to resolve *M*/*Q* 16 (O^+ from the icy satellite/ring system) from *M*/*Q* 14 (N^+ from Titan), which CAPS will easily do. Information on ion composition is often necessary but not always sufficient to distinguish the source of the plasma. For example, O^+ and other water products can originate from any of the icy satellites or E ring particles, whereas H^+ can originate from any of the several anticipated sources. A second clue to source location is the ion pitch-angle distribution (Figure 2). For example, a beam distribution centered on the particle source cone signifies an ionospheric source (Figure 2a), a pick-up ring distribution signifies a nearby equatorial source (Figures 2b and 2c), a shell or highly anisotropic "pancake" (isotropized ring) distribution signifies a more remote equatorial source (Figure 2d), and a quasi-Maxwellian distribution probably reflects a distant (e.g., solar-wind) source. (The precise shapes of the distributions will depend on the relative location of the source, the particle transit times and the rate of scattering due to wave–particle interactions.) The pick-up ring distribution may evolve into a shell, as observed near comets (Coates, 2003 and references therein), or into an anisotropic "pancake" distribution, depending on the ratio of the pickup and ion cyclotron wave velocities (Crary and Bagenal, 2000). Both regimes are expected to occur within Saturn's magnetosphere. A third clue to source location is provided by a radial map of flux-shell plasma content, which tends to peak in source regions and to dip in loss regions. This quantity is obtained by integration of the ambipolar equilibrium equation along magnetic field lines, the accuracy of that depends on accurate measurements of the ion mass spectrum as well as the ion and electron temperatures and anisotropies. The fourth clue to source location

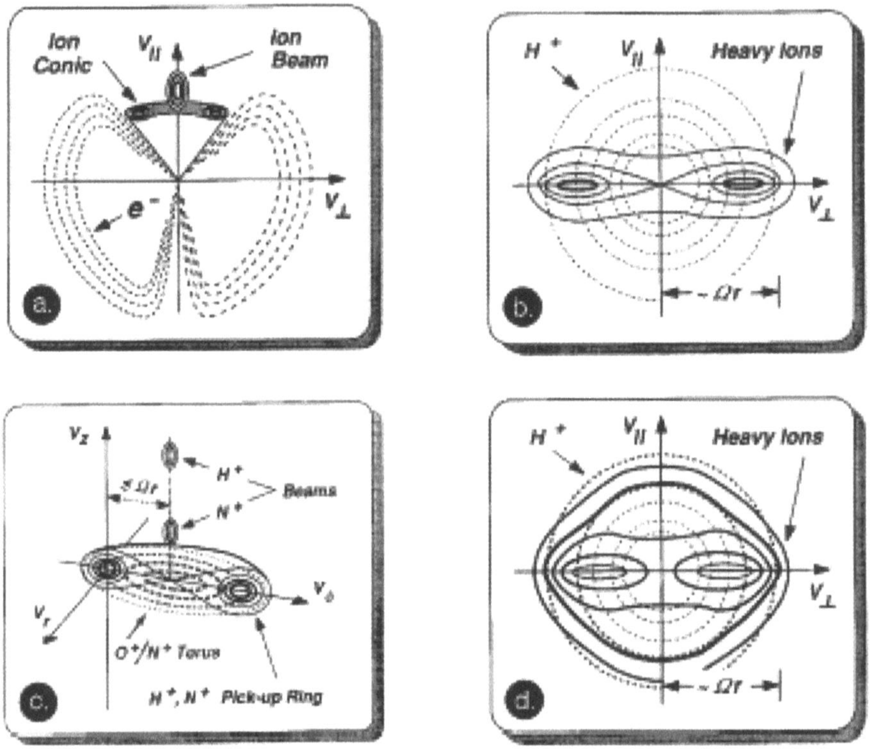

Figure 2. Representative ion velocity space distributions in the vicinity of Saturn.

is the corotation lag of magnetospheric plasma: the large-scale radial variation of the partial corotation speed provides information about the global outward mass transport rate (Hill, 1979), whereas localized departures from the general radial trend provide information on the local mass-loading rate due to ionization and charge exchange (Pontius and Hill, 1982). These last two signatures (flux-shell content variations and corotation lags) are apparent in Voyager data (Richardson, 1986; Eviatar and Richardson, 1986), and clearly contain a wealth of information that can be extracted with the greatly enhanced space/time coverage afforded by Cassini.

Plasma sinks (i.e., loss mechanisms) are also important as determinants of magnetospheric dynamics. These include absorption by solid bodies (icy satellites and E ring in particular), precipitation into Saturn's atmosphere, recombination (particularly dissociative, e.g., $H_2O^+ + e^- \rightarrow H + OH$), radial convective transport, and charge exchange (which is the primary loss mechanism). Each mechanism leaves a characteristic signature in plasma composition, energy, and/or pitch angle. These in turn can be exploited to determine their relative importance as a function of time and location in the magnetosphere.

2.1.2. *Plasma Transport*

The internal plasma sources described above all produce unstable particle distributions, which, among other effects, almost certainly drive radial convective ($\mathbf{E} \times \mathbf{B}$) transport. The solar-wind interaction may also drive an Earth-like convection system, particularly on the night side where Saturn's ionosphere virtually disappears (Kaiser *et al.*, 1984a, b; Connerney and Waite, 1984), thereby reducing the coupling between Saturn and its magnetosphere. The spatial and temporal organization of the resulting flow is, however, completely unknown, and such information is critical to our understanding of the dynamics of Saturn's magnetosphere.

For example, does the convection consist primarily of small-scale eddy circulations that can be described in terms of a radial diffusion coefficient (e.g., Hood, 1985), or is there a persistent global-scale pattern? If a global pattern is present, is it organized with respect to saturnian longitude (indicating rotational control and an intrinsic magnetic-field asymmetry (e.g., Hill *et al.*, 1981)) or with respect to local time (indicating ionospheric or solar-wind control)? Saturn's kilometric radiation (SKR), for example, shows evidence of both rotational (Warwick *et al.*, 1981) and solar-wind (Desch and Rucker, 1983) control. Is plasma lost from the magnetosphere primarily through the formation of magnetotail plasmoids? If so, do plasmoids form as the result of the solar wind interaction as at Earth (e.g., Hones, 1979 and references therein) or as the result of planetary rotation as probably occurs at Jupiter (e.g., Vasyliunas, 1983 and references therein)? What is the origin of the high-density inclusions detected by the Voyagers in the outer dayside magnetosphere – are they vestiges of a Titan plume wrapped around Saturn (Eviatar *et al.*, 1982), or blobs of the central plasma sheet slung off by centrifugal force (Goertz, 1983), or something else? To address these and other questions we require not only accurate determination of the ion bulk flow speed, but also high spatial (hence temporal) resolution of boundaries between different flow regimes. Spatial precision is needed for accurate mapping of convection boundaries along the magnetic field to the ionosphere for comparison with auroral emission features observed by the UVIS instrument (Esposito *et al.*, 2004) and SKR emissions observed by the RPWS instrument (Gurnett *et al.*, 2004).

2.1.3. *Auroral Processes*

Voyager observations of SKR and UV emissions suggest strongly that parallel (magnetic-field-aligned) voltage drops enhance auroral precipitation at Saturn, as they do at Earth. Parallel voltages arise to maintain continuity of Birkeland currents along the converging magnetic field. They develop somewhere above the ionosphere where the velocity of current-carrying particles is a maximum, i.e., where the ratio of magnetic-field strength to plasma number density is greatest. This probably occurs below the minimum altitude reached by the Cassini orbiter, which would preclude direct observation of the accelerated electron beam below the voltage drop. There are, however, two distinct signatures of a parallel voltage drop that are discernible at higher altitudes: (1) the presence of upward ion beams (Figure 2a), either strictly field-aligned beams resulting from direct parallel acceleration, or "conics" resulting from transverse acceleration followed by diamagnetic repulsion (e.g., Gorney *et al.*, 1981); and (2) the enlargement of the electron loss cone resulting from the tendency of the parallel electric field beneath the spacecraft to counteract the magnetic mirror force on the electrons (e.g., Mizera *et al.*, 1981). These two high-altitude signatures have been observed simultaneously from the DE-1 spacecraft above Earth's aurora, and intercalibrated both with each other and with the classical low-altitude electron beam signature observed simultaneously by the DE-2 spacecraft orbiting beneath the acceleration region on the same field lines (Reiff *et al.*, 1988). Thus we have powerful analytical tools, tested in Earth orbit, for assessing the parallel voltage distribution along high-latitude field lines traversed by Cassini. It is also possible, although unlikely, that the Cassini orbiter will cross the aurora acceleration region. The location of the acceleration region depends on the electron density at high latitudes, which is poorly constrained by existing data. If sufficiently low densities place the acceleration region above $3R_S$ from body center, then Cassini would pass through this region.

To apply these tools it is essential to resolve the atmospheric loss cones (or source cones) of both ions and electrons. It is also important to bear in mind that the terrestrial signatures cited above refer to an upward parallel electric field, which covers most terrestrial cases but not necessarily most saturnian cases. Significant parallel electric fields are generally upward at Earth because the flux of current-carrying electrons available from the ionosphere typically exceeds that available from the magnetosphere by a wide margin (Knight, 1973), a condition that may not apply at Saturn, particularly on the night side.

2.2. Titan and Its Interaction with Saturn's Magnetosphere

The interaction of Titan with Saturn's magnetosphere provides an opportunity to study a unique regime of the parameter space relevant to the interaction of magnetized plasma with a non-magnetized body. Here, we briefly assess our current understanding of four important facets of Titan's interaction with the magnetosphere of Saturn (see Ip, 1992, for more details) that are directly relevant to the CAPS investigation.

Ionosphere. Ionization of Titan's atmosphere above 700 km results from the action of solar EUV, impact ionization caused by the incoming corotating flow (~20% of that caused by EUV, Keller *et al.*, 1994a, b; Luna *et al.*, 2003), and precipitation of magnetospheric electrons. Because the corotation direction and the direction of solar radiation differ around its orbit, any part of Titan's ionosphere may have different contributors to ionization at any one time (Nagy and Cravens, 1998; Figure 2). The best direct evidence for the ionosphere is the Voyager 1 observations of a plasma and magnetic wake behind Titan's trailing hemisphere

(Hartle *et al.*, 1982a,b). Quantitatively, only upper limits on the ionospheric electron density ($\sim$2400 $\pm$ 1100 cm^{-3} at the terminators) could be derived from radio occultations (Bird *et al.*, 1997). Models of Titan's ionosphere suffer from a lack of observational constraints and from difficulties inherent in Titan's environment: complex ion chemistry coupled with the neutral N_2–CH_4 atmosphere, and intricate boundary conditions set by the interaction with Saturn's magnetic field. Elaborated models of Titan's ionosphere (e.g., Ip, 1990; Keller and Cravens, 1994; Keller *et al.*, 1994; Cravens *et al.*, 1998, Ledvina, *et al.*, 1998), which predict an electron density peak near 1200 km (the exobase is at 1500 km), show that solar photons are presumably the principal agent of ionization, and that chemistry is initiated by the formation of N^+ and N_2^+. Such species can lead to many more complex molecules including H_2CN^+, CH_5^+, $C_2H_5^+$, $C_3H_m^+$.

Magnetic Field Interactions. The magnetospheric flow past Titan is expected to be sub-magnetosonic (M_S and $M_A \sim 0.5$; Ness *et al.*, 1982 a,b) over most of Titan's orbit. MHD simulations (Hansen *et. al*, 2001) suggest that the flow may be weakly super-magnetosonic when Titan is on the dusk side of the magnetosphere. Thus, the Titan/magnetosphere interaction is distinct from both the Venus-Mars/solar-wind interaction ($M_A > 1$, $M_S > 1$), and the Io-torus/magnetosphere interaction ($M_A < 1$, $M_S > 1$). No fast upstream shock is expected, except in those rare instances when Titan is in the upstream solar wind or when Titan is in the dusk side of the magnetosphere. Either mass-loading or Titan's ionospheric Pedersen conductivity can cause the magnetospheric flux tubes to slow down, drape around Titan, and form an ionospheric wake downstream (Luhmann, 1996).

The wake resembles an induced magnetotail with the northern and southern lobes comprising oppositely directed field lines (Ness *et al.*, 1982a,b). The draping of the field lines at Titan, whose magnetotail diameter is $\sim 2R_T$, is more extreme than at Venus (diameter $\sim 3R_V$), or Mars (diameter $\sim 5R_M$) (Luhmann *et al.*, 1991). Brecht *et al.* (2000) have obtained initial results of global hybrid numerical simulations of Titan's magnetic interactions that reveal the complexity of the interaction caused by ion kinetics (Ledvina *et al.*, 2000). The spatial scale of the interaction, which is determined by the heavy ion gyroradius or inertial scale length, depends significantly on mass loading of the flow. Recent models indicate that Titan's ionosphere supports currents that exclude the magnetospheric field from altitudes below about 1000 km (Lindgren *et al.*, 1997).

Escape of Charged Particles. At the interface with magnetospheric plasma flow, charged particles are removed continuously from Titan's ionosphere, and some neutrals above the exobase are ionized. Newly created particles are accelerated to the local plasma corotation rate, implying exchange of momentum and energy between Saturn's magnetosphere and Titan's atmosphere. The draping of the magnetic field lines around Titan is associated with this momentum transfer. In a three-dimensional multi-species model, Nagy *et al.* (2001) have found that tailward escape of heavy ions creates a flux $\sim 6.5 \times 10^{24}$ ions/s. Kopp and Ip (2001) and others have argued that mass loading is asymmetric at Titan because the ion gyroradius is of the order of Titan's radius. Because MHD models neglect gyroradius effects, this asymmetry emphasizes the need for a kinetic model. There is no evidence of electron acceleration at Titan. Instead, a bite-out of electrons with energies >800 eV was observed (Bridge *et al.*, 1981), which is suggestive of magnetospheric electron absorption by Titan (Hartle *et al.*, 1982). Because corotation still prevails at $20R_S$, the charged particles escaping Titan tend to form a torus around Saturn. However, frequent motions of the magnetopause, as well as effects of convection, displace particles radially from their original position. This may result in a multitude of dense, cool plasma "blobs" or "plumes" in the outer magnetosphere, which is otherwise filled by a hot, tenuous plasma. Plumes were observed in the neighborhood of Titan's orbit by Eviatar *et al.* (1982), who interpreted them as a plume wrapped around Saturn. Goertz (1983) proposed that the plumes were instead detached from Saturn's inner plasma-sheet and centrifugally transported outward.

Escape of Neutrals. Diffuse neutral gas dominates the particle environment of Saturn: the neutral to plasma density ratio is typically about 10 (Richardson, 1998). Neutrals may also eventually escape Titan's upper atmosphere (e.g., Shematovich *et al.*, 2003). The anticipated species are H, H_2 and N. The existence of a hydrogen cloud has been confirmed by Voyager measurements of Lyman-alpha emission. This cloud probably connects to the extended hydrogen corona of Saturn (Broadfoot *et al.*, 1981; Shemansky and Hall, 1992) and to hydrogen-rich icy surfaces in the inner magnetosphere. Molecular hydrogen, H_2, may result from the photolysis of CH_4 but the existence of an H_2-cloud remains speculative. Nor has a cloud of neutral nitrogen been observed. Monte Carlo model calculations show that neutrals escape by non-thermal processes initiated by UV photons, and precipitating electrons and ions (e.g., Ip, 1992; Keller and Cravens, 1994; Keller *et al.*, 1994; Shematovich *et al.*, 2003), processes often lumped together as atmospheric sputtering (Johnson, 1994). This loss rate has recently been shown to be sensitive to the slowing and deflection of co-rotating ions and to the flux of locally produced pick-up ions (Brecht *et al.*, 2000; Shematovich *et al.*, 2003). Therefore, the measurement of plasma ion energies and fluxes near Titan by CAPS will be critical in modeling neutral interactions with the atmosphere.

Neutrals that escape Titan become distributed in a torus as Titan orbits Saturn. Charge-exchange between the neutral torus and magnetospheric ions, as well as electron impact dissociation, direct photoionization, and electron-impact ionization, are appreciable sources of H^+ and N^+, as well as molecular ions, in the magnetospheric plasma. Studying Titan's interaction with Saturn's magnetosphere will enable us to set important constraints for our general understanding of Titan's upper atmosphere and ionosphere. Progress on the four key problems cited above requires high-temporal resolution to identify spatial boundaries, high-angular resolution to track plasma acceleration, and high-mass resolution to separate and identify neighboring ion species. We anticipate that CAPS performance will allow us to achieve these objectives. Many close encounters with Titan during the tour will be essential because the outer magnetosphere of Saturn is highly variable and also because unique information will be provided by variations in the local time geometry of each fly-by.

2.3. Icy Satellites and Ring Particles

Scenarios for the formation of the icy satellites all assume that volatiles other than water were part of the initial composition (Stevenson, 1982). However, until recently the only volatile clearly seen by Pioneer, Voyager and Earth-based observers is water. On the other hand, atoms and molecules are ejected from surfaces by a number of processes. Because the energies of the ejected atoms and molecules are too small to escape from Saturn, this material either recondenses or is ionized and picked up by the corotating magnetic field. Plasma in the inner magnetosphere has been shown to come from satellites and ring particles (see discussion below) therefore it should be possible to use CAPS data to determine their surface compositions (Johnson and Sittler, 1990).

A principal process for ejection of neutrals from the surfaces of the satellites and ring particles is sputtering by the plasma itself, in which case the plasma is self-sustained (Huang and Siscoe, 1987). Noll *et al.* (1997) reported an observation suggestive of O_3 primarily on the trailing hemispheres of Dione and Rhea, and the possible presence of O_3 requires that O_2 exist in the ice (Johnson and Jesser, 1996). The observation of O_3 is important for two reasons. First, it is a clear indication that magnetospheric plasma ions impact the surfaces of Dione and Rhea (Johnson and Quickenden, 1997) and, second, it confirms that these ions produce new chemical species from the surface materials (Johnson, 1990; Johnson *et al.*, 1997; Delitsky and Lane, 1997). Therefore, this observation strengthens the suggestion that the plasma in Saturn's magnetosphere is a product of sputtering of ring particle and satellite surfaces by energetic ion impact. This bombardment also complicates analysis and understanding of the surface composition because reactive nitrogen ions that diffuse inward from Titan's torus are implanted into the icy surfaces.

Earlier telescopic and spacecraft observations were also suggestive of plasma bombardment and modification of the surfaces of the icy satellites. Differences in reflectance, particularly at short wavelengths, between the leading and trailing hemispheres were suggestive of radiation damage and sputtering of ice by the plasma. Differences in weak IR water bands between the hemispheres also were suggestive of plasma erosion and modification. Finally, preliminary modeling of the composition and spatial distribution of the plasma appear to confirm its self-sustained production.

As noted above, even the primary composition of the plasma was uncertain until recently. The lack of mass resolution on the Voyager PLS and LECP instruments allowed the hypothesis that N and H from Titan could be the dominant source of plasma in the inner magnetosphere rather than H_2O from the satellites and ring particles. This issue was decisively settled by the observations of gas-phase OH co-existing with the plasma (Shemansky *et al.*, 1993). These observations combined with modeling (e.g., Ip, 1995; Jurac *et al.*, 2002) confirmed that the icy satellites and ring particles were the principal source of plasma in the inner saturnian magnetosphere. However, the recent estimates of source rates for nitrogen from Titan (Shematovich *et al.*, 2003) are such that N^+ diffusing inward form Titan's torus may be a significant, possibly dominant component of the *energetic* heavy ions in the inner magnetosphere.

The amount of OH seen by Voyager was larger than that initially predicted due to plasma bombardment alone (Johnson *et al.*, 1989), suggesting that additional processes cause the loss of surface material. Recent modeling has confirmed that within $\sim 4.5 R_S$ sources other than the sputtering of satellite surfaces contribute (Jurac *et al.*, 2001, 2002), whereas at larger distances satellite sputtering dominates (Shi *et al.*, 1995a, b).

Eventually, inner magnetospheric sources of plasma begin to overlap Titan's neutral out-flow. The other possible neutral and plasma sources are sputtering of E-ring grains (Morfill *et al.*, 1993), self-erosion of E-ring grains by collisions (Horanyi *et al.*, 1992), particulate bombardment of the icy satellites (Haff *et al.*, 1983; Burns and Mathews, 1986), micro-meteorite erosion of the main rings (Ip, 1984), and plasma sweeping of the ambient gas emanating from the main rings (Ip, 1995). Jurac *et al.* (2001, 2002) have shown that there is likely to be "unseen" material orbiting in the vicinity of Mimas and Enceladus, acting as sources of the plasma and the neutral OH cloud.

In addition to determining the composition and dynamics of the plasma in the inner saturnian magnetosphere, the CAPS data, combined with modeling of ambient neutrals and plasma, will be able to definitively unravel the physical processes determining various source strengths. However, what may be more important to planetary science is the following. Because the rings and satellites are sources, and different objects will dominate the local source at different distances from Saturn, CAPS can obtain compositional information on individual objects. In addition to the spatial distribution of composition, which can be masked by diffusion, measurement of the energy and pitch-angle distributions of a particular species will distinguish freshly produced ions from those that have diffused inwards or outwards. Such measurements would be particularly advantageous on a close pass by a satellite or in regions where ring sources dominate.

Observations of the icy satellites and rings have a long history but there is very little definitive composition information on these objects. Although Cassini will carry imaging spectrometers that in principle will be able to identify species other than H_2O, the plasma bombardment itself can make such identifications difficult. First, sputtering depletes surface layers of the most volatile species, causing enhanced diffusion from depth and loss of volatiles. Moreover, bombardment also chemically alters the materials. For example, H_2O is decomposed to H_2 and O_2, NH_3 can be converted into N_2H_4 and N_2, and carbon bearing species into volatiles such as CO, CO_2 and refractory carbon chains. These species are often difficult to observe spectroscopically.

By contrast, the sputtering process acts as a natural mass spectrometer with atomic and molecular ions produced roughly stoichiometrically (Johnson and Sittler, 1990). CAPS IMS can identify the presence of intrinsic parent species and is capable of separating important atomic and molecular ions that have nearly the same molecular weight. The target list of separable sputter products includes (in order of increasing mass-to-charge ratio): C^+ ($M/Q = 12$); CH^+ ($M/Q = 13$); N^+, CH_2^+ ($M/Q = 14$); NH^+, CH_3^+ ($M/Q = 15$); O^+, NH_2^+, CH_4^+ ($M/Q = 16$); OH^+, NH_3^+ ($M/Q = 17$); H_2O^+ ($M/Q = 18$); and H_3O^+ ($M/Q = 19$). Other, heavier molecules of importance in the sputtering process that can be separated include: C_2^+, ($M/Q =$ 24), C_2H^+ ($M/Q = 25$); CN^+, $C_2H_2^+$ ($M/Q = 26$); and CO^+, N_2^+ ($M/Q = 28$). The CAPS data in conjunction with modeling can therefore provide definitive analysis of satellite and ring surfaces for the planetary community.

3. Design Approach

The stated goal of the Cassini mission is to achieve measurements "at least 10-times better than that of Voyager." In order to meet this goal, Voyager observations (Sittler *et al.*, 1983) and models of the saturnian plasma environment (Richardson and Sittler, 1990; Richardson, 1995; Richardson *et al.*, 1998) were used to develop CAPS science objectives, performance guidelines, and sensor requirements. A second important goal relative to Voyager performance is that CAPS should provide complete coverage of electron and ion velocity distributions, eliminating gaps between Voyager PLS (Bridge *et al.*, 1977) and LECP (Krimigis *et al.* 1977) instruments. Those gaps were <10 eV for electrons and ions; 6 to 14 keV for electrons, 6 to 30 keV for protons, and 6 to $\sim$70 keV for O^+. A third key goal is to measure unambiguously the composition of all major atomic and molecular ions from 1 to 50,000 eV.

With reference to Figure 3 and Table I, our measurement goals call for ELS to obtain medium-resolution electron energy-angle spectra; the IBS to obtain high-resolution energy-angle spectra; and the IMS to obtain high-mass resolution and medium energy-angle resolution of ions. Figure 3 is a simplified overview of the CAPS instrument layout and particle optics. All three sensors have in common that they are based on charged particle motion in electrostatic fields. After entering the sensors through wedge-shaped fields-of-view, particle trajectories are dispersed in electric fields and then measured using electron-multiplier detectors. The ELS and IBS optics separate electrons and ions respectively by energy/charge (E/Q) ratio and by elevation angle of arrival (out of the plane of Figure 3). The second angle, azimuth, is obtained by sweeping the sensor fields-of-view using a motor-driven actuator. From knowledge of detector counting rates as a function of energy and two angles, particle velocity distributions can be deduced. The IMS optics also

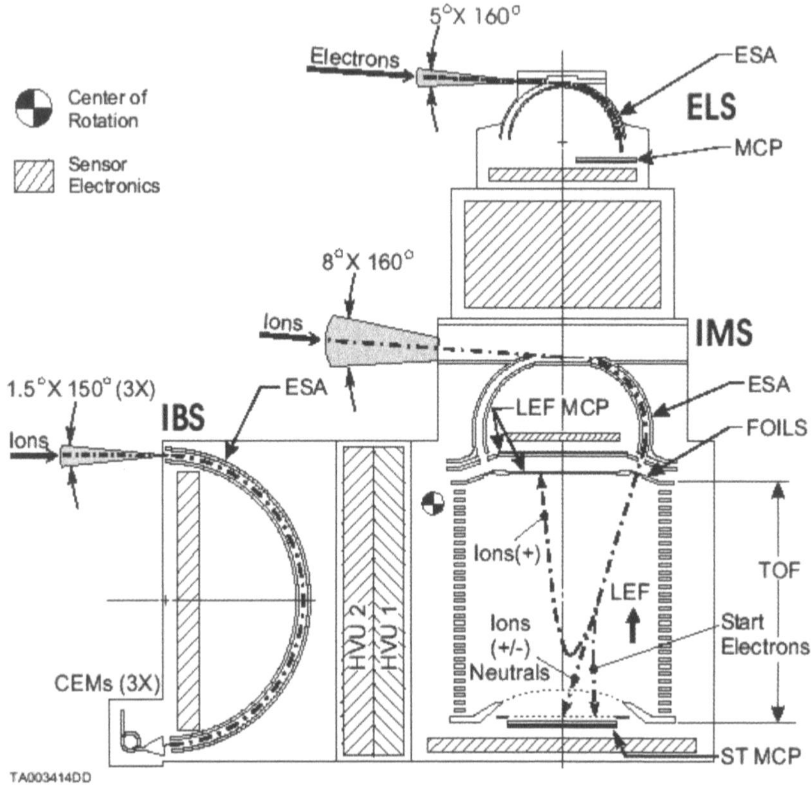

Figure 3. Optical layout, fields-of-view, and key sensor elements of CAPS shown in the *X–Y* (azimuthal) plane of the spacecraft (see Figure 4). Cross-hatched areas Figure 3 indicate sensor electronics subsystems. Heavy dashed lines suggest the general shape of particle trajectories.

separate ions by *E*/*Q* and angle of arrival, but then in addition disperse them by time-of-flight (TOF) in a novel high-resolution mass spectrometer. IMS is capable of separating major ion species to $\sim$1% of the total flux, and minor ion species to $\sim$0.1% or better of the total flux.

An important design consideration is the location of CAPS on the spacecraft (Figure 4). Of particular concern was obtaining good separation from the main Cassini engines and thrusters (potential sources of chemical contamination), separation from the radioisotope thermoelectric generators (potential source of penetrating background radiation), and separation from any sources of electrostatic charging. With all these considerations in mind, the best location for CAPS turned out to be on the underside of the fields-and-particles pallet (Figure 4) adjacent to the MIMI/CHEMS instrument (Krimigis *et al.*, 2004) and just below the INMS (Waite *et al.*, 2004). Although meeting all of the above criteria for location, CAPS still did not have an acceptable field-of view because it was fixed to the spacecraft body and thus could only view in directions constrained by spacecraft orientation. In order to counteract this limitation, the CAPS sensors were mounted on a rotating

TABLE I
CAPS sensor performance summary.

Parameter	IMS Med. Res.	IMS High Res.	ELS	IBS
Energy/charge response				
Range (eV/e)	1–50,280		0.6–28,750	1–49,800
Resolution $(\Delta E/E)_{\mathrm{FWHM}}$	0.17		0.17	0.014
Angular response				
Elevation sectors (number)	8		8	3
Instantaneous FOV $(\mathrm{AZ} \times \mathrm{EL})_{\mathrm{FWHM}}$	$8.3° \times 160°$		$5.2° \times 160°$	$1.4° \times 150°$
Angular resolution $(\mathrm{AZ} \times \mathrm{EL})_{\mathrm{FWHM}}$	$8.3° \times 20°$		$5.2° \times 20°$	$1.4° \times 1.5°$
Mass/charge response				
Range (amu/e)	$1 \sim 400$	$1 \sim 100$	–	–
Resolution $(M/\Delta M)_{\mathrm{FWHM}}$	8	60	–	–
Energy-geometric factor* (cm^2 sr eV/eV)	5×10^{-3}	5×10^{-4}	1.4×10^{-2}	4.7×10^{-5}
Temporal response				
Per sample (s)	6.25×10^{-2}		3.125×10^{-2}	7.813×10^{-3}
Energy-elevation (s)	4.0		2.0	2.0
Energy-elevation-azimuth (s)			180	

*Applies to total field-of-view and includes efficiency factors.

platform driven by a motor actuator capable of sweeping the CAPS instrument by $\sim 180°$ around an axis parallel to the spacecraft Z-axis (Figure 4). In this way nearly 2π sr of sky can be swept approximately every 3 min regardless of spacecraft motion or lack thereof. Although not ideal for plasma measurements under all circumstances (e.g., when the spacecraft body blocks the direction looking into a plasma flow), careful design of observing periods permits effective performance under most conditions.

Although adding a rotating platform provides a means of turning the instrument, the spacecraft nonetheless occludes parts of the FOV as shown in Figure 5. At about $+80°$ azimuth parts of the fields and particles pallet (FPP), the neighboring LEMMS instrument, and RTG shielding obscure the CAPS FOV. Encroachments are actually

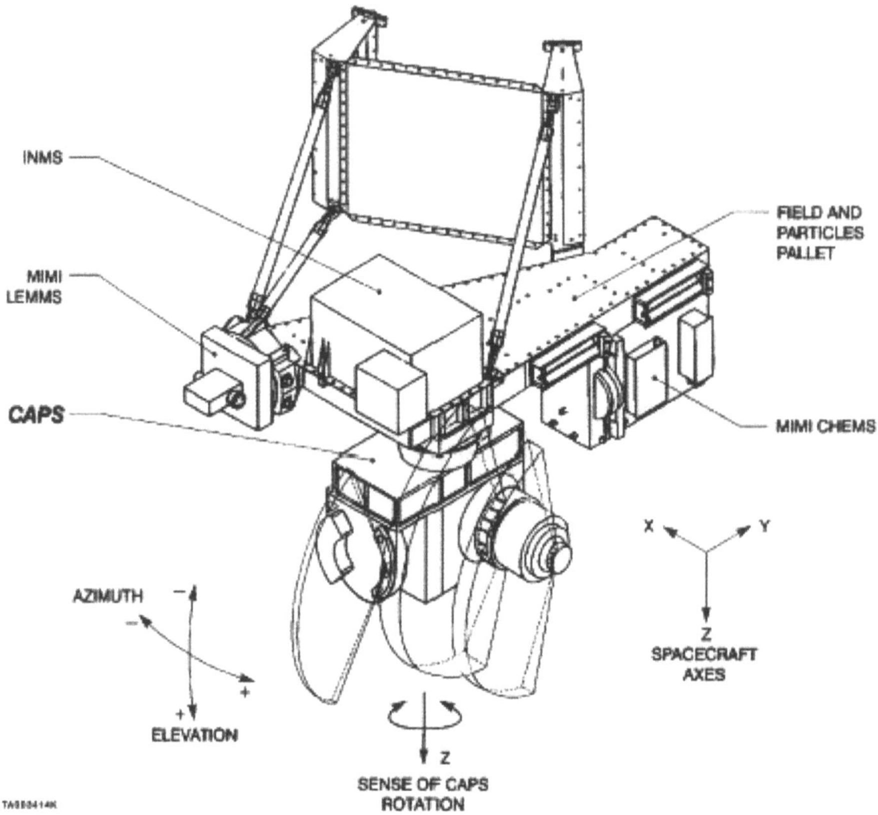

Figure 4. Location and orientation of CAPS on the Fields and Particle Pallet. Note the definitions of azimuth (in the spacecraft X–Y plane) and elevation (parallel to the spacecraft Z-axis) angles. These will be used throughout the paper to describe instrument orientations and fields-of-view (FOV).

larger than shown here because of multi-layer thermal insulation blankets that stand off from all spacecraft surfaces by ~5 cm.

4. Electron Spectrometer

4.1. Principles of Operation

The ELS sensor (Figure 6) is a hemispherical top-hat electrostatic analyzer (ESA) similar to that described by Carlson *et al.* (1983). Its implementation is based closely on the High-Energy Electron Analyzer (HEEA), part of the Cluster Plasma Electron and Current Experiment (PEACE) (Coates *et al.*, 1992; Johnstone *et al.*, 1997). The ELS energy range and angular field-of-view (FOV) overlap considerably with the MIMI/LEMMS solid-state electron detectors (Krimigis *et al.*, 2004), producing complete coverage on Cassini from 1 eV to ~250 keV with no gaps.

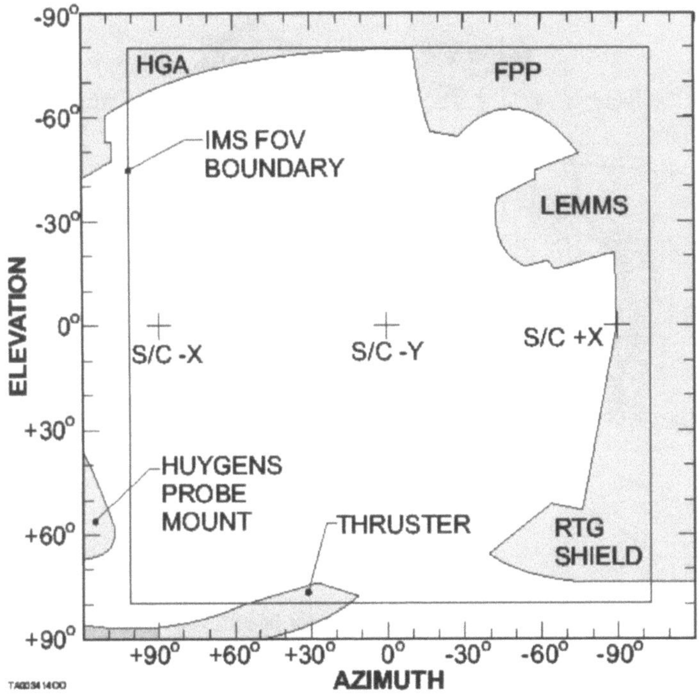

Figure 5. All-sky projection of the CAPS IMS field-of-view. Encroachment on the CAPS FOV are caused by surrounding spacecraft structures (shaded areas). Similar encroachments occur for IBS and ELS sensors.

Electrons enter the sensor via a grounded baffle (Figure 6) and then pass between concentric hemispherical electrostatic analyzer (ESA) plates before impacting on an annular micro-channel plate (MCP) detector. Angular and energy resolution of the ELS are determined by the relative spacing between the two concentric hemispheres, $R_0/\Delta R$. In addition, the analyzer energy acceptance is proportional to $R_0/\Delta R$ times the voltage applied to the inner hemispherical plate. An energy spectrum is obtained by changing the voltage on the inner hemisphere in discrete, programmable steps. Electron direction of arrival in elevation is determined from the position at which it strikes the detector, recognized by the anode positioned behind the MCP (Figure 7). A number of innovative aspects from PEACE have been incorporated in the design of the ELS analyzer, including reduction of photoelectron susceptibility (Alsop *et al.*, 1998) and high-relative mechanical accuracy (Woodliffe and Johnstone, 1998) that minimizes errors in electron energy measurements.

When operating, the ELS executes consecutive energy sweeps in which the selected energy (voltage) is held for a fixed accumulation time (31.25 ms) and then stepped down to the next level. One quarter of the accumulation interval is dead time that permits readout of the detector counters and settling of the sweep high voltage.

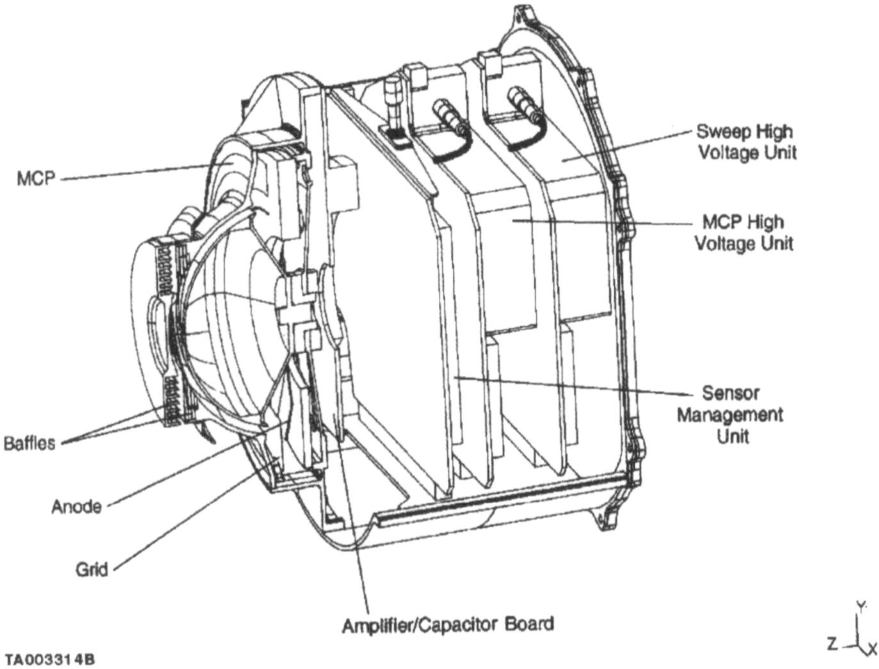

Figure 6. Cutaway drawing of the ELS sensor and electronics unit.

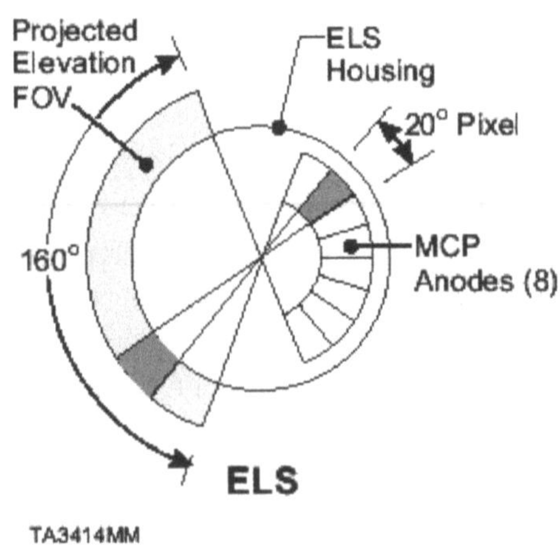

Figure 7. ELS field-of-view in the elevation plane showing its mapping to detector pixels.

In normal operation, a 64-level logarithmic energy spectrum is scanned between 0.6 and 28,000 eV in 2 s. The sequence is repeated until ELS is commanded to do otherwise. Three high-voltage step tables are stored in the ELS. Sweep Table-AI, the default mode, contains 64 values log-spaced over the energy range 0.6 eV–28.75 keV separated by 16% decrements this will likely be the workhorse mode of ELS in the saturnian magnetosphere. Energy separation in this mode is matched to the analyzer pass-band to ensure contiguous energy coverage. Alternatively, 32 values out of the 64 available can be selected by setting the starting point of the energy sweep to any of the top 32 steps. Sweep Table B contains 32 values with 25% decrements. This mode scans over a range of 1–1000 eV and is tailored to solar wind measurements. Voltage Table C consists of 32 values with 36% decrements over an energy range of 1.8–22,000 eV. It is designed to provide faster time resolution (1 s/sweep) over most of the available energy range. A fixed-step mode is also available to facilitate ground calibration and to enable high-time resolution measurements at a fixed energy if needed.

4.2. Electron Optics

Studies by Carlson *et al.*, (1983) indicate that a bending angle of 75° is an optimal tradeoff between resolution and sensitivity for a top-hat ESA. Once the shape and alignment of the hemispheres was selected, secondary electron and UV rejection became major optical design considerations. In order to minimize their effects, the input collimator aperture incorporates a saw-tooth baffle structure designed to reduce particle and solar UV scattering. The central baffle section has a spherical profile that maintains the desired electric field in the ESA. A series of concentric ring-shaped baffles on the top inner surface of the outer hemisphere forms a second line of defense against stray UV and photoelectrons. The combination of these two features ensures that there is no direct line-of-sight from the aperture to the hemispherical solid surfaces.

Potential effects of sunlight in the sensor were further reduced by application of a highly absorbent, diffusely reflecting surface layer of copper oxide crystals (grown using the Ebanol-C process (Alsop *et al.*, 1998)) deposited on all internal surfaces. The film is electrically conducting, has good adhesion, and is sufficiently thin (less than 8 μm) and uniform to maintain the analyzer's mechanical accuracy requirements. During operation, the inner ELS hemisphere is set to one of a programmable series of positive voltage steps (the outer hemisphere is grounded). Stepping this voltage shifts the narrow band of electron energies transmitted by the ESA. Electrons emerging after a field-defining grid reach the MCP detector. In order to maintain a satisfactory analyzer bending-angle and also to prevent high-voltage breakdown, the MCP could not be located at the optimum focus position behind the ESA. Instead, a grid was placed at the focus directly below the analyzer exit, with the MCP positioned below this and 90° away from the analyzer entrance. (In any case the coarse anode pattern does not require very good focusing).

The grid between the analyzer exit and the MCP defines the 160°-wide elevation FOV of the sensor. The grid is made from Laser-cut phosphor bronze plated in gold. An optimum design thickness of 125 μm was obtained by considering electric field definition requirements versus mechanical strength. By biasing the grid at -8 V

TABLE II
ELS key sensor data and dimensions.

Parameter	Value
ESA type	Spherical top-hat
Mean radius	4.15 cm
Plate spacing	0.30 cm
Analyzer constant	6.3
Plate bending angle	75°
Top-hat set-back angle	19.0°
Top-hat aperture radius	1.35 cm
Detector	Chevron MCP
Detector anode inner radius	3.95 cm
Detector anode outer radius	4.35 cm

(normally at 0 V) to repel electrons and by setting the plate voltage to its minimum (0.1 V), the background count-rate due to penetrating radiation can be measured. The 160° annular segment of the grid is divided into tapered windows at 2° intervals and has a calculated transparency in excess of 80%.

4.3. Ray-Tracing and Modeling Results

The electron spectrometer has been extensively studied by numerical simulation. Optical design, UV susceptibility and total electron fluence during the mission were all simulated. The resulting design is similar to that of the HEEA (Johnstone *et al.*, 1997), except that an analyzer bending-angle of 75° was chosen. Table II gives key sensor data and dimensions for the ELS.

The electrostatic modeling performed for PEACE has been described elsewhere (Woodliffe, 1991). The potential distribution in a three-dimensional electrostatic model of the instrument was solved using the Laplace equation and spline interpolations between the grid points. Analyzer response was calculated in three ways using electron ray-tracing based on: (1) a regular starting grid, (2) a Monte Carlo approach, and (3) tracing of the outside edge of the response function, i.e., the extreme limiting trajectories. The latter technique was a quick way of determining instrument response and establishing the major design parameters. Then Monte Carlo particle tracing was used to study detailed analyzer response and for comparisons with calibration.

The results of electron optical modeling are summarized in Table III. The acceptance space of the analyzer can be thought of as three-dimensional in energy, elevation angle and azimuthal angle. In the simulation, electrons are started at a range of angles and energies using the second technique above. For each dimension,

TABLE III

Comparison of simulated and measured ELS analyzer characteristics.

Parameter	Simulated value	Measured value in 125 eV calibration	Measured value in 960 eV calibration
Elevation FWHM (°)	20.0	20.20 ± 0.23	20.26 ± 0.27
Azimuth FWHM (°)	5.24	6.45 ± 0.06	5.68 ± 0.04
Azimuth midpoint (°)	–	0.41 ± 0.07	0.06 ± 0.04
Energy FWHM/eV	–	3.46 ± 0.01	26.18 ± 0.08
Energy midpoint/eV	–	20.34 ± 0.04	152.08 ± 0.23
$\Delta E/E$ (%) FWHM	16.75	17.02 ± 0.05	17.21 ± 0.06
Analyzer constant	6.35	6.16 ± 0.01	6.31 ± 0.01
Geometric factor (20° anode, 100% efficiency) (cm^2 sr eV/eV)	1.7×10^{-3}	$8 \pm 1 \times 10^{-4}$	$8 \pm 1 \times 10^{-4}$

TABLE IV

Simulated ELS peak count rates per 20° anode for typical Saturn magnetospheric conditions.

Location	Temperature (eV)	Density (cm^{-3})	Counts (s^{-1})
Solar wind	1	0.1	427
Magnetosheath	50	0.1	2137
Plasma sheet	100	30	854700
Magnetosphere	300	0.1	5128

the response is summed over the other two to produce the full width at half maximum in each dimension. A simulated geometric factor is also tabulated.

The susceptibility of ELS to background from solar UV was also simulated. Assuming a cosine law for reflection and a reflectivity of 0.5%, we found a rejection ratio of $\sim 10^{-8}$. A more sophisticated model based on the HEEA sensor was reported by Alsop *et al.* (1998), including the effect of shims introduced into ELS to reduce susceptibility to UV.

Calculations were performed to estimate the number of counts anticipated for particular plasma environments during the mission. Maxwellian distributions of specified density, temperature and bulk velocity formed the input to the ELS detector simulation program, which calculates the number of counts to be expected in each angular and energy bin. Table IV shows the count rates per 20° anode at the expected peak of some typical distributions. Note that for a Maxwellian distribution the peak count rate occurs at twice the temperature in eV.

4.4. Detailed Design

4.4.1. *Mechanical*

The sensor head assembly, which is generally cylindrical in cross-section, consists of an entrance collimator and baffles, ESA hemispheres, and MCP detector and anode. The sensor, mounted integrally with the ELS electronics compartment, is attached to the top of the IMS collimator assembly (Figure 6). This arrangement places the ELS aperture as far as possible from the surface of the Cassini spacecraft. Two flat side panels carry card guides for four electronics boards. Connectors on the board edges mate with a motherboard in the lower part of the compartment providing ease of access. Flexible circuit cables link the motherboard to the CAPS DPU interface connector, mounted on one of the flat side panels, and to the capacitor/amplifier board mounted behind the MCP anode. Pins on the back of the anode plug into sockets on the capacitor/amplifier board when the anode is installed.

The sensor head design incorporates very accurate relative positioning of the hemispheres (design goal 1%, equating to a total tolerance of 30 μm; Woodliffe *et al.*, 1998), which ensures accurate knowledge of the selected electron energy at all positions around the detector. Aluminum alloy milled to a wall thickness of 1.6 mm forms the outer shell of the instrument. An additional 3 mm of aluminum located directly above the MCP provides radiation shielding.

4.4.2. *Detectors*

After leaving the ESA, electrons incident on the front face of the detector each cause an amplified cloud of charge collected by an anode at the rear of the detector (Figure 7). The detector consists of a chevron MCP pair with a gold-coated copper spacer 66-μm thick positioned between the two plates. The purpose of the spacer is to lower the voltage required for a particular gain, hence allowing more scope for increasing MCP bias voltage as required over the mission. The effect of the spacer is also to improve gain uniformity over the whole detector. At operating voltage, the measured FWHM pulse height distribution is 130%. The resistivity of the glass in the MCP is low enough to allow the plate to respond to count-rates up to 1×10^4 mm^{-2} s^{-1} or approximately 10^6 electrons per anode per second, without saturation causing significant gain degradation. The MCP high voltage can be varied from 0 to +3.5 kV in steps of approximately 60 V. This allows the MCP bias to be increased throughout the mission to recover possible gain loss. The bias voltage at the input to the MCP is maintained at +150 V to ensure all electrons have sufficient energy to be detected. During calibration, the operational voltage on the MCP was approximately +2.4 kV.

Electrons leaving the rear of the MCP traverse a gap of 500 μm before striking the anode. A voltage of +82 V applied between the anode surface and the back surface of the MCP optimizes spreading of the charge cloud leaving the MCP. The anode has eight discrete 20°-wide electrodes separated by 150 μm. The active anode area is formed by 10 μm thick gold on a Deranox 975 Alumina substrate.

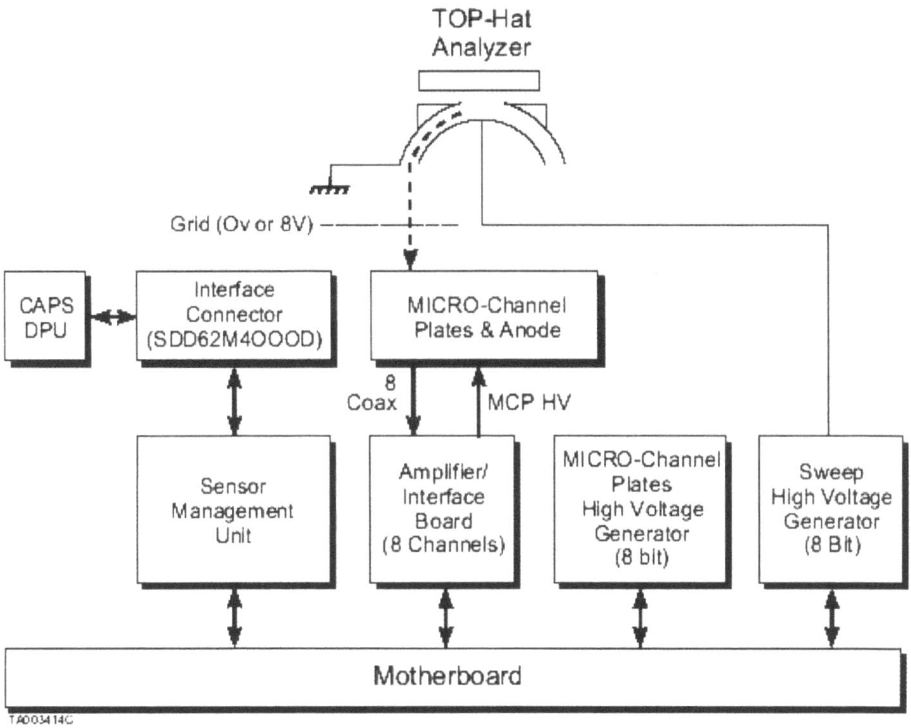

Figure 8. Schematic ELS electrical block diagram.

The area of the separator contacting the MCP is coated with 10 μm of gold. A signal ground plane incorporated into the bottom layer of the multilayer ceramic provides electromagnetic screening of the anode from the analyzer structure.

4.4.3. *Sensor Electronics*

A functional block diagram of ELS is shown in Figure 8. The electronics are accommodated on four circuit boards integrated to a single motherboard consisting of flexible and rigid sections. This design eliminates the need for an internal cable harness, and at the same time couples ELS to the CAPS Data Processing Unit (DPU) interface connector.

Amplifier/Capacitor Board. MCP pulses collected on eight anodes are passed to an equal number of Amptek® A111F charge amplifier/discriminators that convert raw signals above a predetermined threshold into 5 V, 300 ns logic pulses. Thresholds were set in hardware to 3.4×10^5 electrons, which yields an equivalent level of 25 mV (into 2.3 pF), giving good rejection of electronic noise. A decrease in the threshold level by 2.5 mV increases spurious electronic noise counts by a factor of 10. (This relationship holds over a wide range of thresholds. MCP dark counts and penetrating radiation are the main remaining contributors to background.).

A further consideration in threshold selection was cross-talk that might couple MCP signals from one anode to the next. ELS anode cross-talk is below 3%, which is not enough to induce a signal on its own, but could induce spurious counts when added to electronic noise. Convolving the two noise spectra (electronic and cross-talk) provided a check that showed that the chosen threshold was set correctly.

The A111F amplifiers show variation in deadtime with input pulse amplitude, especially within a factor of 2 of threshold, as well as a variation of output pulse width with input pulse amplitude, all of which were characterized during calibration. Front and rear MCP bias voltages are provided by Zener diodes, which require filtering at these low currents (around 10 μA). The MCP anodes are biased at high voltage (Figure 8) so signal pulses must be decoupled by high-voltage capacitors before the signal goes to the amplifiers that share the same circuit board with the HV bias/anode coupling circuitry. The HV section was carefully designed and laid out to support a maximum field of 800 V/mm.

Sensor Management Unit (SMU). The SMU receives and interprets sensor commands sent by the CAPS DPU and accumulates and transmits ELS data back to the DPU. It stores the sequence of high-voltage steps to be applied to the analyzer, the grid voltage setting, and the MCP voltage table. SMU circuitry supplies stimulation test pulses of variable amplitude and frequency to the amplifier/discriminator channels. Under control of the CAPS DPU, the SMU clock speed can be successively halved to lengthen the data acquisition period from 31.25 to 1000 ms/step, creating progressively longer energy sweeps. Furthermore, the sample deadtime can be varied between 25 and 12.5% of the sample period to increase counting rate capability at high rates.

High Voltage Supplies. The ELS contains two high-voltage supplies. A low noise supply biases the MCP at voltages up to +3.7 kV at 25 μA with 6-bit resolution. A second supply powers the ESA with 64 or 32 stepped voltage levels between +4200 and +0.1 V. This wide dynamic range meant that great care had to be taken at low output levels to avoid external noise affecting the pulse-width modulator that sets the voltage levels. A 12-bit digital-to-analog converter (DAC) controls the ESA output voltage using an "expanding DAC" technique to reach 16-bit resolution at low energies, thus achieving voltage accuracy of 1% or 0.1 V, whichever is greater. The supply steps at a minimum interval of 31.25 ms and settles in 8 ms. The entire HV converter section of the circuit board is shielded to protect low-voltage, low-noise circuitry from interference or possible breakdown.

4.5. Calibration

The ELS was calibrated in the Mullard Space Science Laboratory (MSSL) electron calibration facility developed for Cluster (Johnstone *et al.*, 1997). A photograph of ELS in the calibration system appears in Figure 9. A mercury lamp generates UV that strikes a gold layer deposited on a quartz disk. From this photoelectrons are extracted by applying a bias potential to the gold surface, creating an electron beam

Figure 9. Photograph of the ELS flight unit in the MSSL calibration chamber. Gold-plated foil was used to prevent unwanted electrostatic charging in the calibration chamber.

15 cm in diameter with divergence less than 1° (at 1 keV) and good uniformity over the ELS aperture. During calibration ELS was mounted on a two-axis rotary table and turned to allow electrons from defined directions to enter (a short discussion of calibration theory can be found in Section 6.2.2). A μ-metal shield inside the vacuum chamber shielded the calibration volume by reducing the residual magnetic field to less than 10% that of the Earth. Electron beams with energies above ~30 eV showed minimal directional deviation.

Beam current measurements that provide absolute calibration were made with a faraday cup and picoammeter. During calibration sequences beam stability was monitored with a CEM. A tritium source provided a cross check after each sensor re-configuration to maintain consistency during calibration.

Calibration of the ELS engineering model has been described elsewhere (Linder *et al.*, 1998). Calibration of the flight model was made at ten electron energies between 2.3 and 16,260 eV. At each energy step a matrix of approximately 500 $\times$ 10$\times$ 10 aximuthal $\times$ elevation $\times$ ESA voltage sweeps were taken (the actual number varied with energy step). Two basic types of data were taken: First a finely stepped elevation angular scan was made at constant energy and beam azimuth angle. Second, a full three-dimensional calibration (energy, elevation, azimuth) was obtained at defined resolutions in the three dimensions. The most detailed calibrations were made at 125 and 960 eV (Figure 10a and b). Each plot shows the ELS response as a function of elevation angle, summed over the other two dimensions. In each case

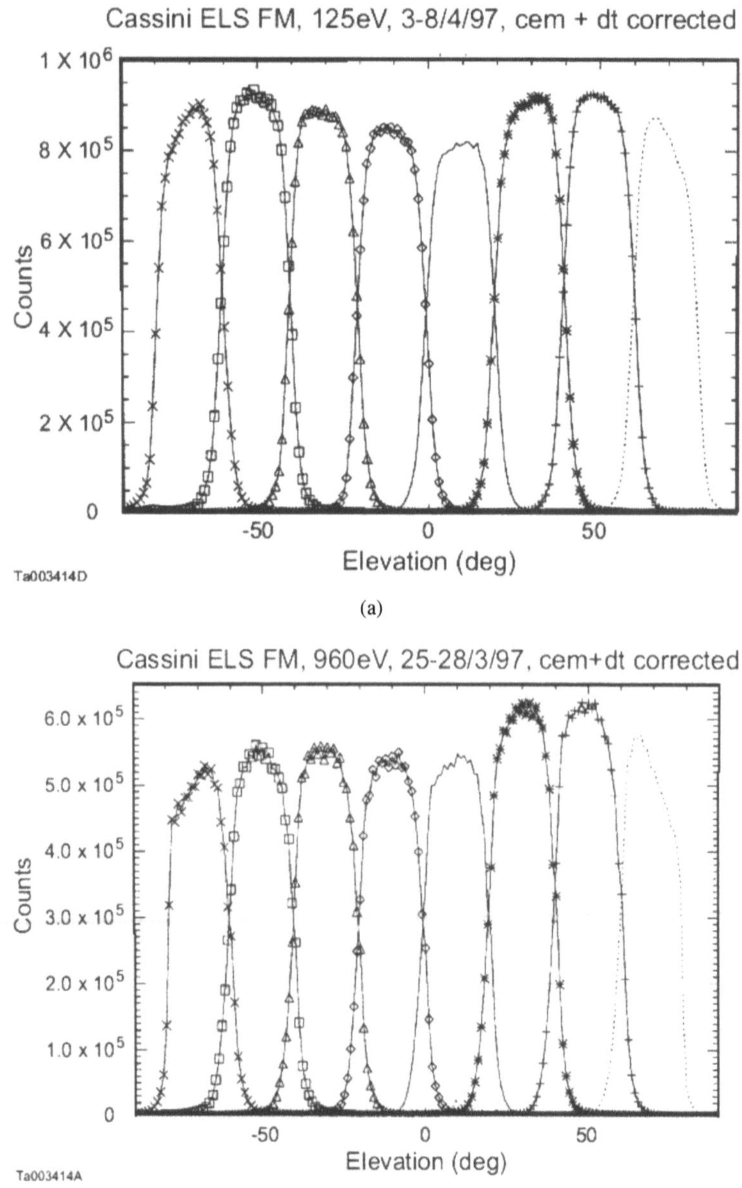

Figure 10. (a) ELS calibration data showing elevation response of the eight anodes at 125 eV. (b) ELS calibration data showing elevation response of the eight anodes at 960 eV.

some 150,000 data points, corrected for dead time and beam monitor readings, were summed to produce a plot. The eight anodes show a nearly uniform response with some loss of transmission at the two end anodes. This is to be expected because the grid holder cuts off incident trajectories at ±80° elevation. A summary of 125 and

960 eV calibration data is included in Table III. Energy-angle scans with a 125 eV electron beam were made at the azimuthal center of each of the eight anodes. These are plotted in spectrogram format in Figure 11. Taken together, Figures 10 and 11 show that analyzer performance in three-dimensions is consistent from one anode to the next and deviates little from instrument simulations.

4.6. Performance

Calibration results in the previous section show that the mechanical construction accuracy of the analyzer (see Johnstone *et al.*, 1997), and therefore the anticipated scientific performance of the instrument, is excellent (Table V). Analyzer response widths agree with simulations and are close to those originally proposed. The geometric factor is based on a nominal MCP voltage setting.

Response of ELS to solar UV also was measured during calibration. In common with Alsop *et al.* (1998) we find energy-dependent rejection efficiency. Setting the grid potential to -8 V and grounding the inner hemisphere made it possible to distinguish between photons themselves and photoelectrons reaching the MCP. The results showed an excellent rejection ratio (i.e. ratio of dark current background to background measured with UV entering the aperture) of $\sim 10^{-10}$ at high electron energy and a worst case of $\sim 10^{-8}$ at low energy. The intensity of Lyman α at Saturn is approximately 2.4×10^9 cm^{-2} s^{-1} so the solar UV background at Saturn should be negligible. Using tritium or an electron beam as a source, end-to-end tests showed that secondary electron production inside ELS is minimal. ELS performance is summarized in Table V.

Section 9 of this paper contains examples of ELS data taken during Cassini's swingby of the Earth in August 1999 and its encounter with Jupiter in December 2000 to January 2001. Beginning with the jovian encounter, ELS (and CAPS as a whole) has been operating continuously and successfully when mission plans permit.

5. Ion Beam Spectrometer

The IBS is specifically designed to provide high resolution, 3-D measurements of the energy and angular distribution of any beamed ion populations encountered during the course of the mission. This instrument, based on an earlier design by Bame *et al.* (1978), has four principal measurement objectives: (1) afford context for saturnian magnetospheric studies by providing solar wind and bow shock measurements, (2) search for ion beams in the saturnian magnetosphere and study high-latitude source/loss cones in the cusp and auroral regions, (3) analyze thermal plasma distributions during transits through Titan's upper atmosphere, and (4) provide solar wind science data when the opportunity arises during the mission.

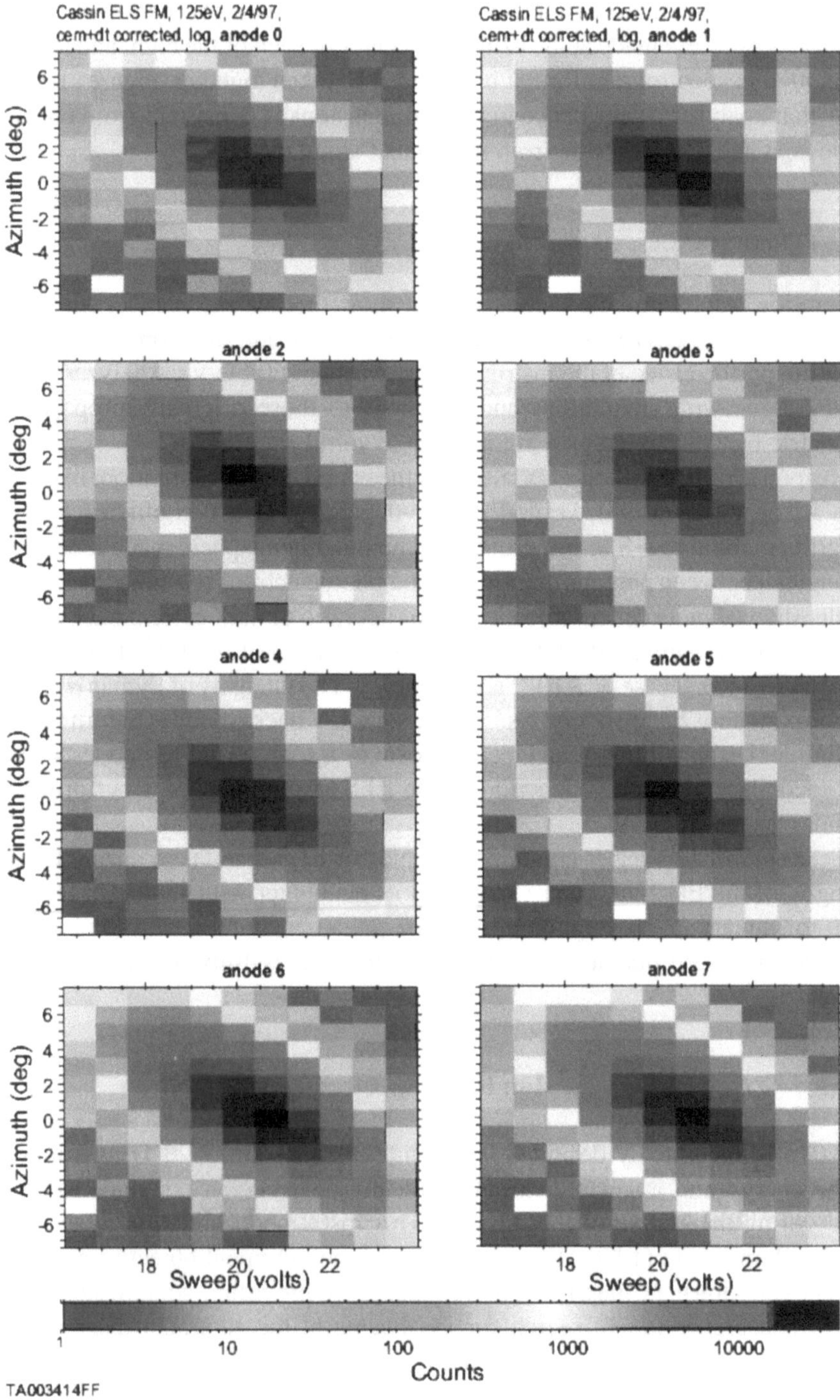

Figure 11. Spectrograms of ELS response in azimuth versus energy. Each spectrogram corresponds to an elevation passband shown in Figure 10a (125 eV beam).

TABLE V
ELS detailed performance summary.

Parameter	Value
Energy range (eV)	0.6–28,250
Resolution $\Delta E/E$ (%)	16.75[a]
Field of view (°)	5.24[a] × 160
Angular resolution (°)	5.24[a] × 20
Analyzer constant measured on FM at 960 eV (eV/V)	6.31
Geometric factor[b] (cm^2 sr eV/eV)	
(1) per 20° anode	8×10^{-4}
(2) per complete FOV	6.4×10^{-3}

[a]Value from simulation.
[b]Based on nominal MCP voltage setting.

5.1. Principles of Operation

Similar to the ELS, the IBS is based on the principles of a curved-electrode electrostatic analyzer. The primary differences, aside from its larger radius, is that the spherical IBS electrodes extend 178° from the entrance aperture to channel-electron multiplier detectors located at the exit. Positively charged ions enter the spectrometer through one of three flat, grounded apertures. They then acquire trajectories that are parts of conic sections due to the central electric force field present between the spherical electrodes (Figure 12). The inner plate has a variable (stepped) negative potential applied to it whereas the outer plate is at ground potential. Only those ions with a very small range of entrance energies and angles will transit the narrow gap between the nested hemispheres and be counted by particle detectors located at the ESA exit. Ions with too large energy or with angles of arrival more than $\sim +1°$ from the aperture normal will not be bent sufficiently by the electric field and will be lost on impact with the outer analyzer plate. Those with too little energy or with angles within less than $\sim -1°$ of the aperture normal will be pulled into the inner plate and also lost to the system. By sequentially stepping the potential on the inner plate and counting particles that transit the ESA, the energy spectrum of the ambient ion population can be readily determined.

A unique aspect of the IBS is the method used to obtain high-angular resolution 3-D velocity space measurements. On the basis of the crossed-fan FOV concept employed in an earlier solar wind ion instrument (Bame *et al.*, 1978), it is possible to obtain the required angular resolution by tilting the acceptance fans of each aperture 30° relative to the others (Figure 12). Because each of the three fans requires only a single non-imaging detector, it is possible to measure the velocity distribution of

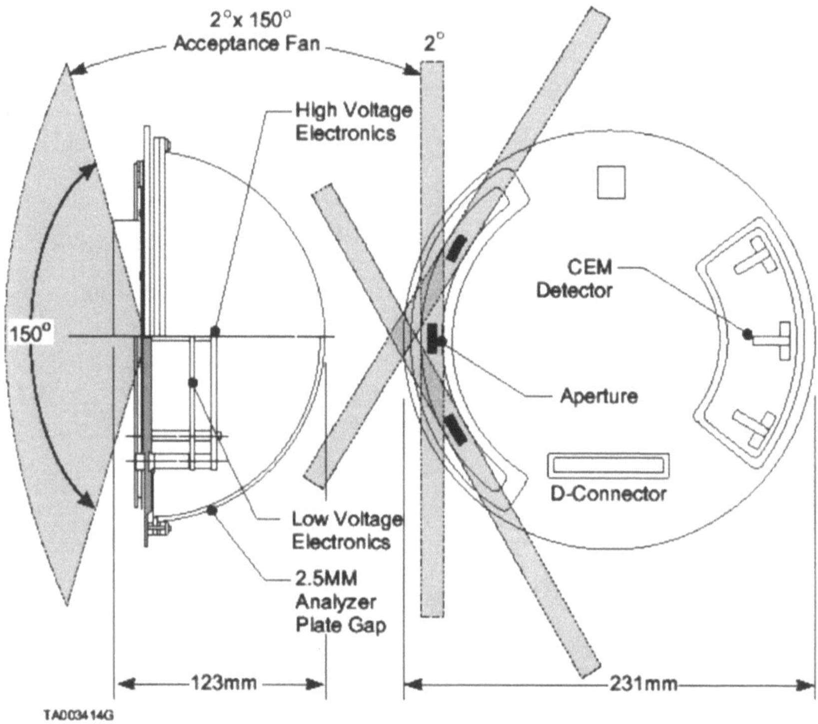

Figure 12. Elevation and plan views of the IBS sensor. Note that the three FOV fans are tilted by 30° and offset from one another in the vertical direction. The cutaway drawing at left shows power supplies mounted within the IBS analyzer dome.

the ambient plasma with a minimum of complexity and resources. Information on the instantaneous viewing direction of each of the fans combined with the energy analysis provided by the ESA provides a nearly complete energy-angle distribution of the ambient plasma ion population.

5.2. Detailed Description

For a variety of reasons, the IBS was allocated minimal resources of approximately 1 kg and 1 W and optimized to near those values (Table XI). These constraints, plus a requirement to be able to search almost the entire unit sphere for ion beams and analyze them with adequate energy and spatial resolution, dictated the overall IBS design.

Key IBS sensor data and dimensions are given in Table VI. The choices were dictated by the requirement for high angular and energy resolution while simultaneously obtaining sufficient particle throughput needed to measure narrow solar wind beams expected at 9.5 AU. To minimize weight but still provide adequate radiation shielding, the inner ESA hemisphere (made from aluminum) was optimized to a

TABLE VI
IBS key sensor data and dimensions.

Parameter	Value
ESA type	Hemispherical
Mean radius	10.00 cm
Plate spacing	0.25 cm
Analyzer constant	19.0
Plate bending angle	178°
Aperture (curved)	0.25 cm × 1.50 cm
Aperture radius of curvature	10.0 cm
Detector	CEM

thickness of 0.7 mm and that of the outer to 0.8 mm. The tolerances for manufacturing the hemispheres and aligning them relative to one another were quite stringent. Vilppola *et al.* (1993) have shown through simulations that inaccuracies of ~ few tens of μm are important.

Normally, the interior surfaces of the hemispheres would have been grooved and blackened to suppress UV photons that could scatter into the detectors, but grooving of the large thin plates is impractical. Originally, we avoided blackening the interior analyzer gap because the large ESA bending angle of 178° requires numerous bounces before photons could be transmitted to the detectors, which should greatly suppress unwanted background. In addition, we expect to encounter very low-energy ions (1 to a few eV) in Titan's ionosphere. Transport of ~1 eV ions through the relatively long path length of the IBS ESA (~315 mm) necessitates a highly uniform field between the plates. We therefore wished to avoid using the usual Ebanol-C black coating used on ELS because of the possibility of introducing surface potential "patchiness" in the analyzer gap. (For 1 eV ions the inner analyzer plate potential is only −50 mV and a variation of only a few mV in surface potential along the path would be unacceptable). Therefore, the ESA hemispheres were not grooved or blackened but were instead carefully coated with pure gold, a poor reflector of UV. However, during early testing it was found that UV transmission through the ESA was much higher than that expected at high-polar angles, apparently due to "channeling" along minute machining grooves that remained in the hemispheres. As a consequence, in the end the hemispheres were blackened using the Ebanol-C process and UV rejection fell to $\sim 10^{-10}$. None of the anticipated problems associated with Ebanol actually occurred.

As mentioned above, the unique aspect of the IBS is its method of determining 3-D plasma velocity space distributions by means of crossed-fan geometry. There are three curved 2.5 × 15 mm apertures in the IBS faceplate (Figure 12), each with a nominal acceptance fan of ±1.5° FWFM in azimuth (set by the ESA characteristics) and ±75° FWFM in elevation angle (set by the apertures) from the normal to the

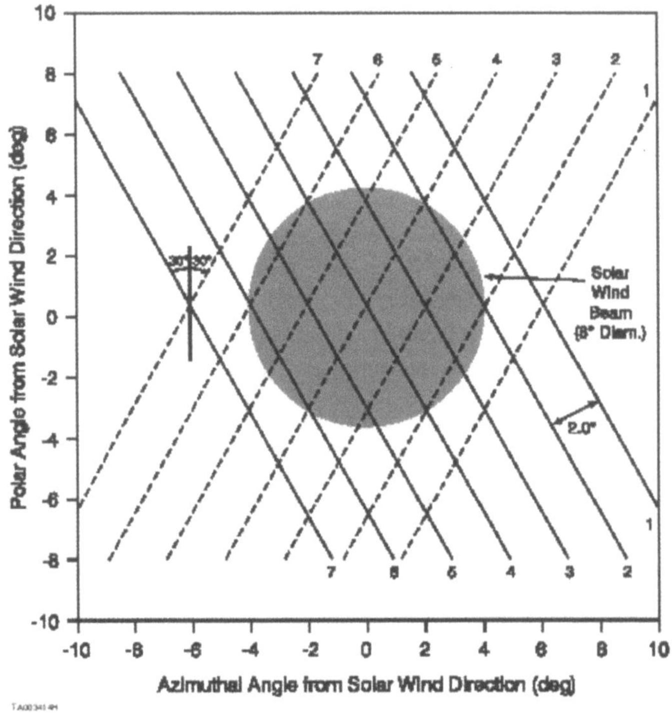

Figure 13. Velocity space coverage of the IBS crossed fan electrostatic analyzer. The solid and dashed lines represent the centers of the two slanted fans. The central aperture fan is omitted for clarity.

plane of the aperture. If we define the middle aperture as being along the 0° radius from the center of the instrument faceplate, the other two apertures are located at ±30° relative to it. There are three CEM detectors located 180° around the faceplate from each of the apertures, i.e. in the position where ions entering the apertures from any transmitted direction come to a focus. The FOV of the middle aperture is oriented such that its long (polar) dimension is parallel to the azimuthal (Z) axis of the CAPS actuator. The FOV of the other two apertures are therefore “crossed” with inclinations of ±30° with respect to that of the middle aperture.

Ions transmitted through each of the apertures are detected by the corresponding CEM, giving an instantaneous 1-D view of an ion distribution. A 3-D measurement of the plasma velocity distribution can be built up from each aperture by simultaneously sweeping the energy passband of the instrument and rotating the actuator (ACT) and/or the spacecraft itself. Figure 13 illustrates how spatial coverage is obtained. Data from the three individual acceptance fans are combined (Figure 14) and used to obtain a 3-D distribution measurement with angular resolution as high as 1.4° × 1.4° (FWHM). The maximum angular resolution falls off gradually with increasing polar (elevation) angle due to the natural broadening of the azimuthal acceptance angle as polar angle increases from the aperture normal (see Figure 17 below).

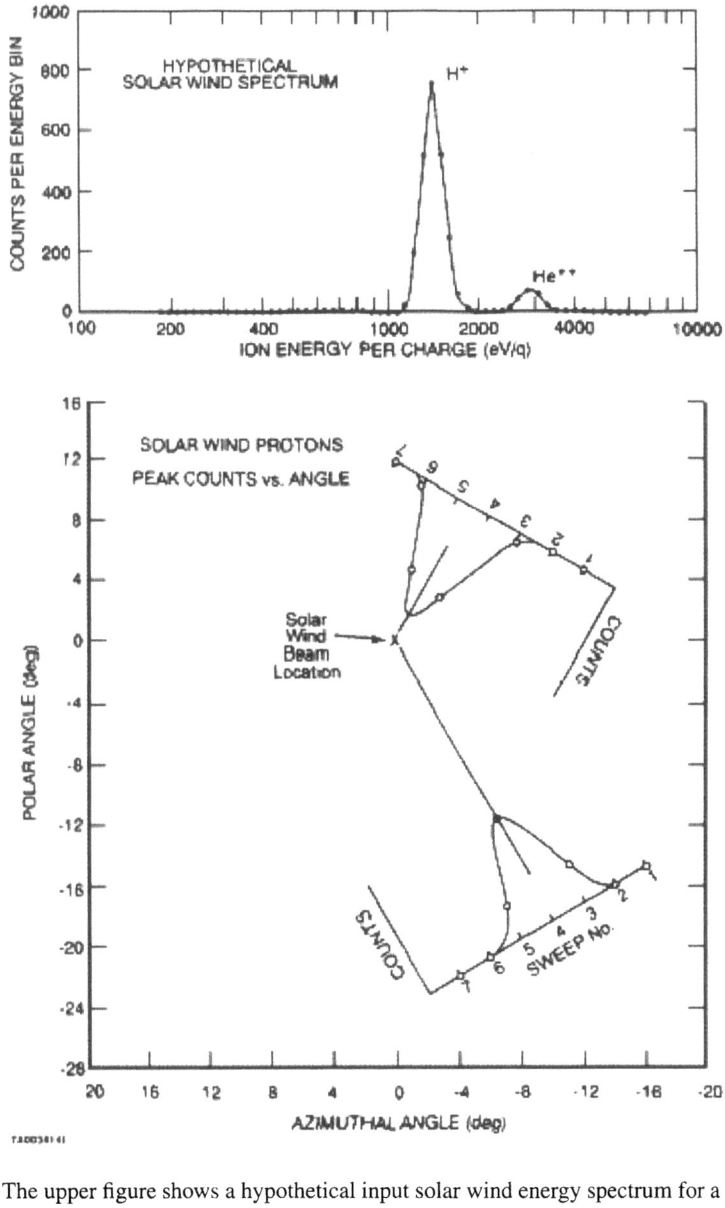

Figure 14. The upper figure shows a hypothetical input solar wind energy spectrum for a single fan. The lower figure illustrates the method for determining ion beam position in velocity space from crossed-fan measurements (after Bame *et al.*, 1978).

It can be seen that two fans would be sufficient to obtain the required measurements. However IBS incorporates three fans for a number of reasons. One is redundancy: if one of the three CEMs fails, it is still possible to obtain fully 3-D results. If two channels fail, useful 2-D data can still be obtained. Second, given an unobscured FOV, IBS can potentially obtain 3-D coverage over ~80% of space.

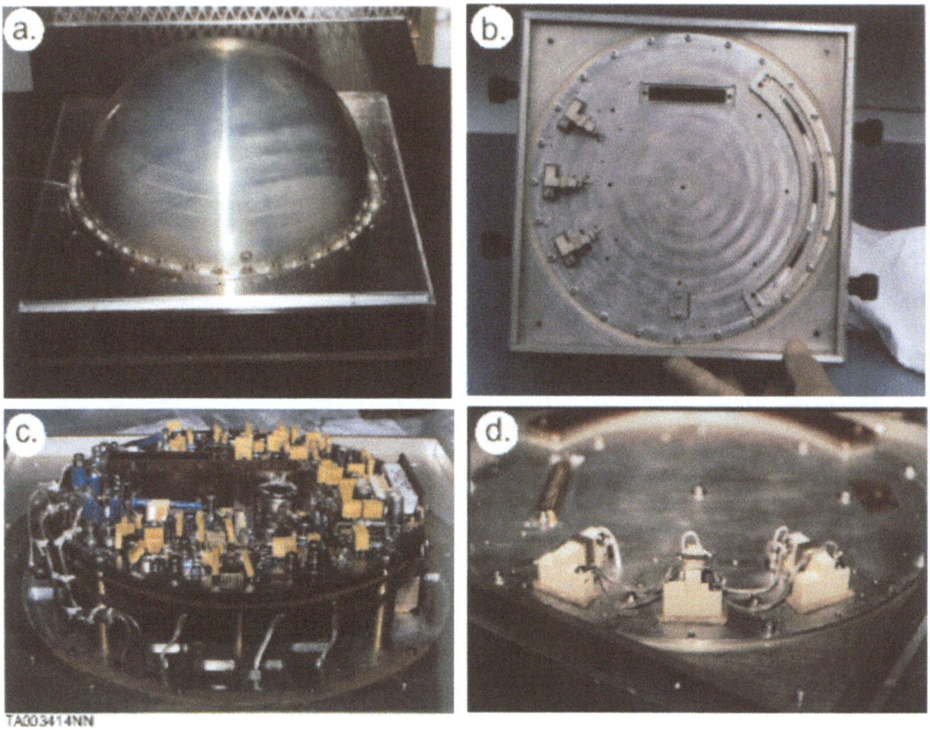

Figure 15. Views of the IBS flight unit during assembly. (a) The completed unit seen from the rear showing the outer hemispherical plate. (b) View of the inside of the front plate of IBS showing the three CEM detectors on the left and aperture holes on the right. (c) Details of IBS high-voltage supplies that are mounted inside the ESA inner dome. (d) Details of the ceramic CEM detectors.

The three matched detectors used in IBS (Figures 12 and 15d) are custom-built CEMs chosen for their rugged construction out of solid ceramic, their high-gain characteristics (typical plateau $\sim 10^8$), and their availability in nearly any desired form factor (the manufacturer is Dr Sjuts Optotechnik GmBH of Germany). The input funnel interior dimensions are 5×20 mm, which is more than adequate to intercept all of the ions converging at each of the sensor's three focal points. Ninety percent transmission grids are stood off from the front of the CEM funnels and biased negatively with respect to the funnels to minimize secondary electron loss. The input assembly (grid and funnel) is biased at negative high voltage so that all transmitted ions receive post-acceleration equivalent to the CEM bias. A CEM responds to an incoming ion by emitting an electron charge pulse at its exit, which is biased at about -100 V. Electrons are attracted to the anode, which is at signal ground, and are then fed to an Amptek A111F hybrid preamplifier/discriminator. The resulting output pulses are routed to counters located in the CAPS DPU where samples are compressed and inserted into the telemetry stream by the CAPS Central Processing Unit 1 (CPU1).

5.3. Ray-Tracing and Modeling Results

Initial modeling of the IBS was chiefly concerned with determining manufacturing tolerances required to achieve the desired energy-angle resolution and transmission efficiency of the ESA (Vilppola *et al.*, 1993). Analysis of simulation results showed that it was necessary to align the two ESA hemispheres relative to one another with an accuracy of better than $\sim 25\mu$ m to obtain the desired energy resolution of $\Delta E/E = 0.015$. Furthermore, deviations of the ESA plate surfaces could not depart from perfect sphericity by more than 300 μm (0.3% of plate radius) if >90% transmission efficiency was to be maintained. Guided by this analysis, fabrication, assembly and metrology techniques were developed to help ensure that individual IBS hemispheres were within specification. Unfortunately, it was not possible to directly measure the inner surfaces of the analyzer gap once the ESA was fully assembled. Thus unexpected irregularities that might be introduced during assembly potentially could go unnoticed. Laboratory calibrations were employed to determine whether specifications had been maintained and sensor response conformed to expectations.

5.4. Electronics

A schematic block diagram of the IBS electronics is presented in Figure 16. All of the electronics are mounted on two circular circuit boards housed in the ESA hemispherical cavity (Figure 15c). The upper board contains both the ESA and the CEM high-voltage power supplies, whereas the low-voltage amplifier/discriminators, test pulser, and digital-to-analog converters (DACs) are contained on the lower board. The electrical interface from the IBS to the CAPS DPU is a single 62-pin connector. The weight of the internal IBS electronics is 42% of the total weight of the IBS sensor, or 490 g.

The CEM HV supply, which provides a negative bias voltage to the three IBS detectors in parallel, is commanded by an 8-bit word over a range of 0 to −4.0 kV. The ESA supply biases the inner ESA hemisphere with a stepped negative potential that determines the instantaneous energy of transmitted ions. For IBS to be able to function as desired, the ESA HVPS must cover the voltage range −0.05 to −2600 V in 1.284% increments with a time between steps of 7.8125 ms. In order to have sufficient resolution over the entire voltage range, the ESA supply was designed with three ranges (−0.05 to −1.85 V, −1.85 to −69.6 V, and −69.6 to −2600 V) each of which is 12-bit controllable. This results in a total of 12,288 possible voltage steps although only 852 logarithmically spaced steps are required for full coverage of the voltage range with 1.284% spacing. During operation a 14-bit data word controlling the ESA stepper supply is written to a latch in the IBS electronics every 7.8125 ms (one-eighth of the IMS stepping interval of 62.5 ms). The basic operational mode consists of a 255-step voltage scan (corresponding to an

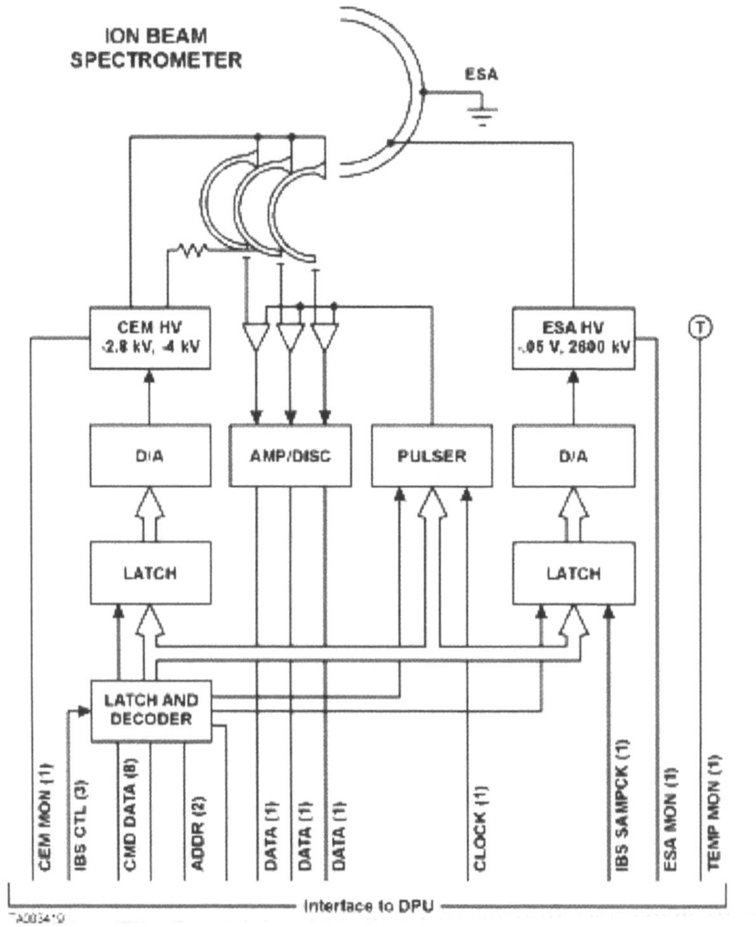

Figure 16. Schematic IBS electrical block diagram.

energy scan) lasting 2 s. In planned operations this may consist of a single, 255-step scan, two 127-step scans, or half of a 510-step, interleaved scan. Data values for generating the energy sweep are stored in CPU1 memory. Energy sweep steps are verified by returning an analog monitor voltage that is digitized and entered into the CAPS housekeeping data.

5.5. Calibration

Initial calibration results showed that IBS responds as expected (Figure 17) except for the presence of a few small but unusual features. Of these, a double-bend in the energy-polar angle response (Figure 17c) is the most notable. To address these findings, a more accurate simulation of the sensor was developed and used

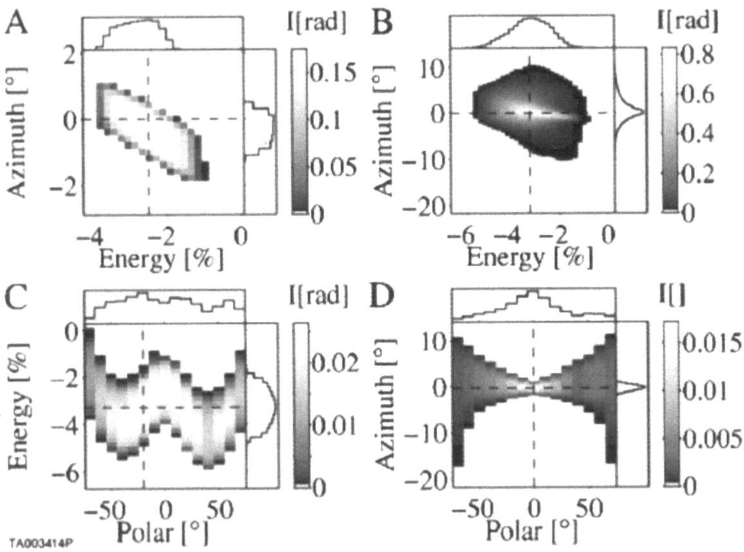

Figure 17. Spectrograms summarizing detailed calibration of the IBS.

to investigate IBS response in more detail (Vilppola *et al.*, 1996). Improvements made to the model included addition of curved apertures (Figure 15b) at the correct standoff distance from the ESA plates, introduction of fringing fields, and a realistic description of the ion beam that matched that used in calibration. These upgrades to the model did not, however, account for the bend feature of Figure 17c. A further refinement of the model allowed the introduction of slight asymmetries in ESA plate geometry (Vilppola *et al.*, 2001). The resulting simulations with asymmetric hemispheres and a slight (few tens of microns) misalignment of the two hemispheres produced good agreement with laboratory results. Although the flight sensor's ESA plates may indeed be slightly misaligned as suggested by simulations, the response of the engineering model IBS is almost identical to the flight model, which is surprising if a random misalignment occurred. Moreover, the responses of the three individual fan apertures in both IBS models are also very similar. This suggests that the unexpected calibration response is due to a small systematic error in alignment or is inherent in the overall electro-optic design and not a function of alignment accuracy. Our latest simulations suggest that a small systematic manufacturing fault might be to blame.

Calibration of IBS took place in ion beam facilities at Los Alamos and SwRI (the same facilities that are used to calibrate IMS). The work at Los Alamos concentrated on angle–angle and angle–energy responses while absolute energy and sensitivity calibrations took place at SwRI. At Los Alamos ions were produced in a radio-frequency discharge ion source and then accelerated down a 3-m flight tube into the calibration chamber where IBS was located. Both external supplies and the IBS internal power supplies were used in calibration. A nitrogen beam was typically used which was accelerated to between 0.3 and 60 keV. Beam location and uniformity were measured but not absolute ion current. Typical operating pressures were in the low 10^{-8} Torr range.

The ion current extracted from the Los Alamos source was quite stable but not easily varied over a wide dynamic range. Therefore a series of slits were employed to adjust the current delivered to the target chamber. IBS was mounted on a stand whose orientation could be adjusted in one translational axis (across the beam) and two rotational axes. Thus the incident ion beam could be made to impinge on the IBS aperture at any desired combination of elevation and azimuthal angles. All of the diagnostic, motion control and data acquisition systems were computer-controlled. During a typical calibration run, three separate 1-D scans in energy, elevation and azimuthal angle would be taken across the center of the response function. Three central 2-D cuts through the response function would then be taken, followed by a series of energy-azimuth cuts along the elevation axis at 10° intervals. Finally, data were corrected for deadtime losses and variation in the beam current during the calibration run. Angular data were transformed from laboratory to spacecraft coordinates. Figure 17 shows an example of the IBS energy–angle response to a 10.0 keV beam of H^+ ions.

After calibration at Los Alamos, the IBS was integrated with the CAPS flight instrument and underwent checks of calibration in the ion beam at SwRI. The latter, which is similar to that at Los Alamos, is described in Section 6. Typical beam spread at SwRI was $\Delta\theta \approx 0.2°$ and $\Delta E/E \approx 0.005$. One difference in the two calibration systems was important: the integrated CAPS unit was positioned to calibrate mainly IMS. Thus the rotation axes of the calibration goniometer were centered on the IMS FOV. Since the plane of the IBS apertures is offset 32.0 cm from the central axis of the IMS and ELS sensors (Section 7.1), the IBS aperture plane was neither co-planar nor co-aligned with the other two sensors: Any rotation of CAPS tended to move the IBS FOV out of the ion beam. The displacement of the aperture through the small IBS maximum azimuthal angular acceptance of $\approx \pm 2°$ was less than 1 cm, well within the diameter of the calibration beam as seen from IBS. Thus only calibrations of the energy response at azimuth = elevation = 0° and of absolute sensitivity (see the discussion in Section 6.6 below) were possible.

5.6. Performance

The detailed results of IBS absolute and elevation calibrations with an N^+ beam are summarized in Table VII. There are slight but consistent trends in the analyzer constant K and in the energy and angle resolutions. Taken at face value, the trend in K suggests that the analyzer plates appear to be closer together at higher voltages, or, alternatively that the voltages produced by the IBS ESA supply are too low by 0.26% at 0.25 keV and too high by 0.47% at 25 keV. The latter would be within the allowed error range of the supplies. Trends in energy and azimuthal angular resolution are toward slightly higher resolution at higher energies. By extrapolating

TABLE VII
IBS detailed performance summary.

Parameter	Value
Energy range (eV/e)	0.95 to 49,800
Energy resolution $(\Delta E/E)_{FWHM}$	0.014
Field-of-view $(^\circ)_{FWHM}$	1.4×150
Angular resolution (°)	1.4×1.4
Analyzer constant (average)	19.0
Geometric factor (cm^2 sr eV/eV)	4.7×10^{-5}
Effective area (cm^2)	0.081
Dead time (μs)	0.86

data taken between 0.25 and 30.0 keV we have estimated performance for all but the lowest ion energies.

Initial absolute efficiency measurements made at SwRI found an efficiency of about 30% of the calculated value. It is possible that this disagreement arises in the beam system's ability to measure total beam current for a tightly collimated FOV such as IBS's. The beam current monitor is located 1-m upstream from the IBS aperture and 4 m from the ion source. The monitor's acceptance in angle, energy, and spatial coordinates are much larger than that of IBS, particularly in the elevation direction. This could cause the monitor to overestimate the amount of beam actually filling the IBS acceptance phase space.

IBS has performed extremely well since initial operations began during the Earth encounter in August 1999. Examples of IBS data taken at Earth and the encounter with Jupiter in December 2000 to January 2001 are presented in Section 9. IBS has been operating continuously since late 2000, with the exception of an 8-month gap in 2003, when the CAPS instrument was off while spacecraft and instrument flight software were being tested.

6. Ion Mass Spectrometer

The IMS is designed to carry out three primary functions in response to CAPS scientific objectives. In describing and discussing the IMS it helps to rank these functions in rough order of time resolution: (1) Obtain rapid measurements of the velocity distributions of major ion species up to 50 keV. These data will be used to determine magnetospheric boundaries, structures in the region of Titan and icy satellites, and particle acceleration regions. (2) Measure the atomic, molecular, and, where possible, isotopic composition of saturnian plasmas in order to determine chemical and physical processes leading to the creation, transport, and loss of plasma populations. (3) Carry out a thorough survey of general characteristics of the magnetosphere including its composition.

6.1. Measurement Requirements and Approach

6.1.1. *Requirements*

Our approach to designing the IMS was taken largely from the three requirements given above. The first centers on measurement of rapid changes in the distribution functions of the dominant ion species (>1% of the total), for example, H^+, H_3^+, He^{++}, O^+, OH^+, H_2O^+, or N_2^+ depending on location in the magnetosphere (see discussion in Section 2). Characteristic scales are, for example, on the order of the time needed to cross plasma boundaries (~few seconds). It is important to realize, however, that characterization of rapid plasma changes is limited by the performance of the ACT and by spacecraft motions (nominal rotation rates of 1°/s and <0.15°/s, respectively). Thus the highest time resolution for two-dimensional snapshots in energy-elevation is a few seconds, while for 3-D scans the resolution is ~3 min. A comparable time base requirement (~minutes) is set by the combination of spacecraft relative velocity and target body radius. In order to fulfil the second requirement IMS must return energy-resolved high-resolution *M*/*Q* spectra from which minor ion species (<1% of the total) can be identified and their relative abundances measured. This function is more nearly like that of a traditional mass spectrometer. The typical time scale for this measurement should be roughly that of the 3-D measurement (a few minutes) in order to sample the entire region of phase space available at any one time. In addition to survey products, the high-resolution *M*/*Q* versus *E*/*Q* data will serve as a ground truth check of the high time-resolution ion data.

A complete understanding of the saturnian magnetosphere requires knowledge of its average characteristics as a function of both time and position. Because of the large number of candidate plasma sources (see Section 2), ion composition can be expected to be a strong function of location. This leads to the third primary requirement placed on IMS, namely an accurate, contiguous survey of as many ion species and their energy distributions as possible. This survey will be carried out under the aegis of the Cassini magnetosphere and plasma science (MAPS) working group to which CAPS contributes. Although limited in some ways by telemetry rate considerations, we have developed a special survey mode to optimize CAPS contributions to this survey (see Section 8.1).

6.1.2. *Approach*

There is a well-known tradeoff in conventional mass spectrometry between sensitivity and mass resolution. In order to build a space flight spectrometer several issues must be addressed: (1) Magnetic mass spectrometers do not usually provide high resolution at ion energies above ~3 keV because the size of the spectrometer grows proportionately to ion velocity. Quadrupole spectrometers (e.g., Waite *et al.*, 2004), another alternative, function well only up to energies of ~0.1 keV. (2) Although energetic particle telescopes are able to distinguish ion mass and mass/charge, they typically are insensitive to heavy ions such as O^+ below ~50 keV/e (e.g., Krimigis *et al.*, 2004). (3) Sensitivity is proportional to aperture acceptance (solid angle and area), so it is essential to develop a sensor having both a large FOV and a large acceptance area. Neither of these factors is conducive to good resolution in conventional mass spectrometers.

Given these considerations, our approach was to choose a novel technique based on isochronous (time-focusing) mass spectrometry based on carbon foils. Using carbon foils as a means of generating timing signals in TOF space instruments is well known (Gloeckler and Hsieh, 1976; Hamilton *et al.*, 1990; Young *et al.*, 1991). What is novel about the IMS design is its capability for isochronous performance over a FOV as large 360° (McComas *et al.*, 1990; McComas and Nordholt, 1990). The combination of carbon foil and isochronous methods solved several measurement problems. In the first place, all other things being equal, aperture size determines mass resolution in conventional mass spectrometry. Here, however aperture size is determined by carbon foil size, which in principle can be made as large as desired. Secondly, the foils can be arranged in an axisymmetric geometry that permits the mass spectrometer to view 360° if desired, thus increasing solid angle of acceptance and FOV appreciably. Finally, a third feature of the IMS TOF optics is that the combination of a high-resolution time-focusing linear electric field (LEF) with time-dispersive optics yields not only high-mass resolution but also high sensitivity because all daughter products leaving the carbon foil can be detected.

6.2. Principles of Operation

6.2.1. *Overview*

To achieve IMS measurement goals we have matched the properties of a toroidal ESA to those of the TOF analyzer. The combination of an ESA with carbon-foil based TOF measurements has been previously developed and flown in space by Gloeckler and colleagues (see Gloeckler and Hsieh, 1976; Gloeckler *et al.* (1995, 1998); Hamilton *et al.*, 1990), Hovestadt *et al.*, 1995, and by Young *et al.* (1991, 1998). Early development of the LEF concept by Gloeckler *et al.* (1995) resulted in a high-resolution 1-D (in energy) isochronous instrument used to measure the collimated flow of the solar wind. However, McComas *et al.* (1990) and McComas and Nordholt (1990) were the first to describe the principles and operation of a cylindrically symmetric LEF analyzer capable of making 2-D energy-angle measurements. McComas *et al.* (1998) and Nordholt *et al.* (1998) further describe the LEF application to IMS.

Ions initially enter the top-hat portion of the ESA through a grounded collimator (Figure 18). A flat circular plate truncates the inner ESA toroid at the point where it joins the top-hat section. There are two advantages to toroidal geometry: (1) Toroids have two radii of curvature that independently control ion focusing within the TOF section (in spheres the two radii are degenerate and equal, in cylinders one radius

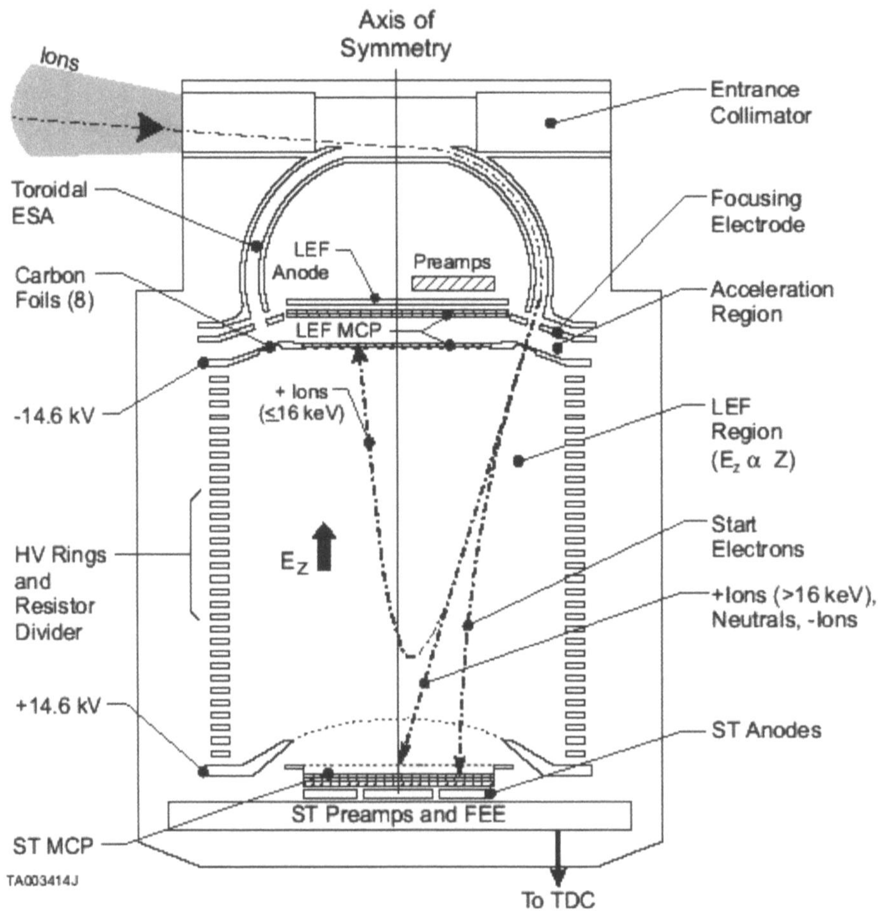

Figure 18. Schematic diagram of the IMS optics, detectors, and placement of high-speed FEE electronics.

is infinite). (2) The top-hat entrance aperture of a toroid like the one used in IMS is larger per unit of ESA plate surface area than for a comparably sized spherical top-hat, giving higher sensitivity per unit of sensor weight (Wollnik, 1971; Young *et al.*, 1988). In the elevation plane (orthogonal to that of Figure 18), ions are formed into a beam by a combination of vanes in the entrance collimator and field-correcting slits (one per 20° pixel) at the exit of the ESA. Figure 19 illustrates the collimated FOV and mapping to the two IMS detectors.

A voltage applied to the inner ESA electrode creates an electric field in the top-hat that deflects ions into the toroid. As with ELS and IBS, only particles within a particular range of E/Q and direction of arrival are transmitted through the ESA to the TOF analyzer. Ions that successfully exit the ESA are accelerated by -14.56 kV into one of eight ultra-thin ($1\sim 1.5\ \mu\text{g/cm}^2$) carbon foils distributed

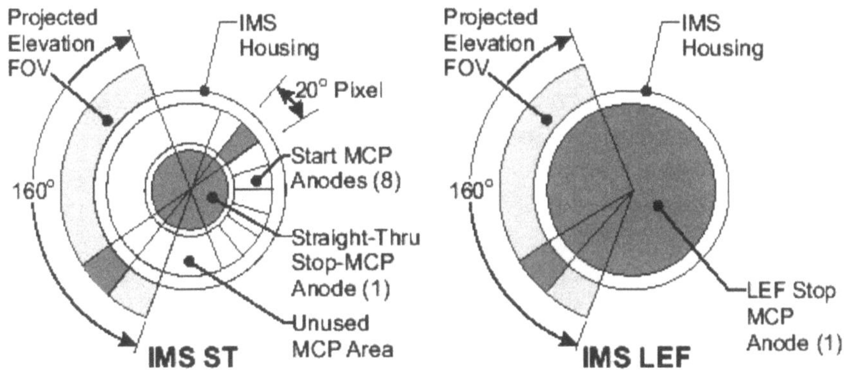

Figure 19. The IMS field of view in the elevation plane showing the mapping of angular pixels to the IMS detectors.

around the entrance to the TOF analyzer (one foil per angular "pixel"). Ions fall through the accelerating field, gaining sufficient velocity perpendicular to the foil so that they penetrate the foils even at the lowest external energies (~1eV). Higher post-acceleration voltages would naturally be desirable, but would have driven up instrument size, mass, and power in addition to risk. (The IMS supplies are capable of delivering −16.0 kV and the optics are rated at −15.0 kV but are tested much higher. However for operational reasons the IMS voltages are set at ±14.56 kV.)

6.2.2. *Instrument Theory*

Figure 20 shows the equipotential distribution in the LEF along with typical particle trajectories. Ions enter from below, normal to the foil surface (scattering in the foil is neglected in this simple presentation). From the figure it is clear that the shape of the LEF near the foil area pushes electrons outward toward the outer edge of the "straight-through" (non-mirroring) ST detector where "start" timing events are recorded. Negative ions have higher momentum and are pushed only slightly outwards in the upper part of the field. Neutrals of course follow a rectilinear path through the field, while positive ions below ~15 keV (external energy) are reflected in the field. The following first-order theory of instrument performance applies only to characteristic trajectories traveling through the central portion of the LEF.

Ion Mass Identification. In the simplest case of an atomic species that does not break up in a foil and that can be identified by a single TOF peak, the parent ion corresponds directly to an identifiable ion of mass/charge (M/Q). The distinction of mass per *charge* is important because IMS is not able to identify mass or charge separately, only the ratio. Identification of parent molecular species (consisting of two or more atoms) is more complicated and follows a multi-step process. First, individual peaks are identified in the TOF spectrum on the basis of calibrated properties such as peak centroid and shape. Second, patterns or combinations of

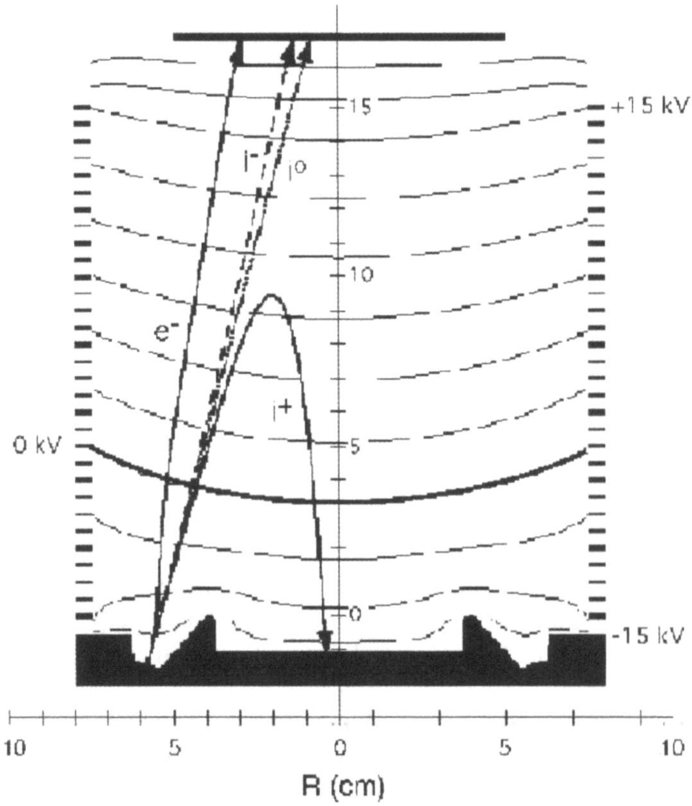

Figure 20. Characteristic trajectories in the LEF obtained by ray-tracing.

daughter peaks are recognized and the daughter peak identified (e.g., a C^+ peak that might originate in any number of carbon compounds). Finally, the relative abundances of any daughter species are used to identify the parent molecule. (It may be possible with experience to go directly from the first step to the last.)

Ions with total energy E and charge Q at infinity are accelerated by a potential U_A, giving them a total energy $E + QU_A$ or, more conveniently, $E/Q + U_A$ since the ESA measures E/Q rather than E. The ions strike the carbon foil, exiting with energy "per charge" $\sigma(E/Q + U_A)$ (for neutral species the total energy is used). The factor σ accounts for energy losses due to collisions. Although molecular species will break up in the foil, the daughter products all travel with approximately the same center of mass velocity, allowing us to write the following equation for any peak measured at time T:

$$M/Q = (\sigma/k_1^2(E/Q + U_A)T^2L^{-2} \qquad (1)$$

where $k_1 = 22.85$, E/Q and U_A are measured in kV, M/Q is in amu/charge: L is the particle path length in cm, and T is in units of ns (10^{-9} s). The acceleration potential

U_A has a nominal value of -14.56 kV but can be varied if necessary although doing so would affect calibration.

Mass resolution $M/\Delta M$ is determined primarily by the time resolution of the LEF optics. Because the LEF portion of the IMS is time focusing resolution can be quite high (~60 at $M/Q = 20$) including all effects of energy and angle straggling. LEF spectra are used in the interpretation of ST spectra. There, resolution is degraded by path length variations, which are caused by the range of trajectory angles leaving the ESA, combined with angle energy straggling in the foil. Writing the uncertainties as ΔT, ΔL and ΔE the mass resolution is:

$$M/\Delta M = [(T/\Delta T)^2 + (L/\Delta L)^2 + (E/\Delta E)^2]^{1/2} \tag{2}$$

Positively Charged Ions with E_0/Q_0 <~15.5 kV. An ion leaving the carbon foil and satisfying the condition $\sigma(E_0 + Q_0 U_A)\cos^2\theta < 14.56$ keV, where $\theta = 14°$ is the angle between the foil normal and the central IMS axis of symmetry, will execute simple harmonic motion before striking the LEF detector (see Figure 20). The motion of ions in the LEF is analogous to simple harmonic motion, depending only on ion M/Q and independent of energy or angle of entry into the field to first order:

$$T = \pi(M/Qk_2)^{1/2} \tag{3}$$

The effective "spring constant" k_2 has been determined from calibration to be 2.327×10^6 amu/(C s^2). Positively charged daughter fragments from molecular breakup have energies per charge proportional to their fraction of the parent molecule's mass. The LEF mirror field has been purposely "detuned" slightly from strict linearity to take advantage of this fact and optimize IMS resolution for molecular fragments. The square root relation in (3) holds very well despite detuning (see below). Other departures from the ideal field such as inhomogenieties cause ions to take different paths in the field, introducing broadening into mass peaks (Equation 2).

Positively Charged Ions with Internal Energy per Charge E_0/Q >~15.5 kV. Ions in this energy range are retarded by the linear electric field, but eventually reach the "straight-through" (ST) detector located at the far end of the LEF, opposite the foils. Their TOF is given by:

$$T = (M/k_2 Q)^{1/2} \sin^{-1}\{[L(k_2/2)^{1/2}]/[E/Q]^{1/2}\} \tag{4}$$

Negatively Charged Ions. These see a spatially varying field as they travel through the TOF region. Hence their flight times must be determined by integrating their velocity over the path.

$$T = (M/2k_2 Q)^{1/2} \quad \ln\left\{\left[k_2^{1/2}\mathrm{L} + (k_2 L^2 + E/Q)^{1/2}\right]/(E/Q)^{1/2}\right\} \tag{5}$$

Neutral Particles. Most particles leaving the foils are charge neutral. Their TOF is given simply by their path length and velocity.

$$T = k_1 L[M/E]^{1/2} \tag{6}$$

Sensitivity. In addition to identification of ion species, the other most important quantity of physical interest is the three-dimensional velocity distribution function for a particular ion species at infinity, $f_i(\mathbf{v})$. The quantity that IMS measures during a sampling interval Δt is the counting rate C. Assuming that the part of the distribution function measured at any one time is constant and uniform within the IMS range of acceptance, then, following the work of Johnstone *et al.* (1987) and others,

$$f_i(\mathbf{v}) = \mathbf{n} C M_i^2 / 4E^2 G_i \Delta t \tag{7}$$

where $\mathbf{n}$ is a unit vector indicating the direction of $\mathbf{v}$, and G is the differential IMS "geometric" factor:

$$G_i = A\varepsilon_i \tau_i \langle \Delta\Omega \Delta E/E \rangle \tag{8}$$

where A is the entrance aperture area, $\varepsilon_i(E, M)$ is the energy and mass dependent efficiency of carbon foils and detectors as well as the effects of and circuit deadtimes, $\tau_i(E, M)$ is the transmission dependence of foils, grids and optics, and $\langle \Delta\Omega \Delta E/E \rangle$ is the solid angle and energy acceptance averaged over the entrance aperture. G has units of cm^2 sr eV/eV. The efficiency factor in (8) can be thought of as converting counts/s into ions/s.

Similar equations also apply to the ELS and IBS. Although ELS will observe only electrons, IBS is capable of observing and separating several ion species in a cold, flowing plasma as will occur during low-altitude Titan flybys. In that case its mass-dependent efficiency will have to be taken into account to obtain the correct relative abundance of ion species.

6.3. Instrument Description

Unlike ELS and IBS, which contain their own power supplies and data acquisition electronics, the IMS consists of a separate sensor unit and six separate electronics modules: a fast front-end electronics (FEE) performing signal acquisition, a time-to-digital converter (TDC), a spectrum analyzer module (SAM), a dedicated central processing unit (CPU2), and two high-voltage units (HVU1 and HVU2). The FEE is located in the sensor itself (Figure 18), the TDC, SAM and CPU2 are inside the DPU, and the HVU's are outside near the sensor (Figure 3). Table VIII summarizes key IMS optical parameters.

Ions entering the IMS collimator are subject to an electric field set by a negative voltage on the inner electrode of the ESA. The value of this voltage is given by $V_{\mathrm{ESA}} = KE_0$ where $K \approx R_0/2\Delta R \approx 6.25$ for IMS, R_0 is the mean major radius of the ESA, and ΔR is the electrode spacing. The ESA high voltage (HV) supply is relatively complex. It generates stepped voltages at a 62.5 ms cadence, covering 63 voltage steps plus one "flyback" step every 4 s. The broad range of energies covered by IMS ($E_{\mathrm{MAX}}/E_{\mathrm{MIN}} = V_{\mathrm{MAX}}/V_{\mathrm{MIN}} = 5 \times 10^4$) requires three separate

TABLE VIII
IMS key sensor properties.

Analyzer	Parameter	Value/type
Electrostatic analyzer	–	Toroidal top-hat
	Mean major radius	6.50 cm
	Mean minor radius	1.30 cm
	Plate spacing	0.40 cm
	Analyzer constant	6.25
	Analyzer bending angle	93.0°
	Top-hat set-back angle	11.0°
	Top-hat aperture radius	2.29 cm
Mass spectrometer	**LEF**	Isochronous time-of-flight
	Spring constant	2.327×10^6 amu/(Cs^2)
	Detector	Z-stack Galileo MCP
	MCP outer radius	4.33 cm
	MCP thickness	0.10 cm
	Channel length/diameter	40:1
	STOP anode radius	3.75 cm
	ST	Linear time-of-flight
	Slant path length foil to detector	18.8 cm
	Detector	Z-stack Galileo MCP
	MCP outer radius	4.33 cm
	MCP thickness	0.10 cm
	Channel length/diameter	40:1
	START anode annulus radii	2.10, 3.80 cm
	STOP anode radius	2.00 cm

digital to analog converters that together yield 3×2^{12} "micro" steps. Normally a logarithmic voltage scan based on 63 log-spaced "macro" steps chosen from among the micro-steps is stored in tables used to command the ESA voltage. In principle the voltage table can be set to scan the ESA in a wide range of patterns. The other HVU2 supplies service the IMS microchannel plate (MCP) detectors, whose gains can be changed independently.

After leaving the ESA, ions encounter a guard electrode set at a distance of 5 mm from the ESA exit plane. The guard electrode is biased at approximately one-half the voltage on the inner ESA toroid to help shield the smaller ESA field (as low as a few mV/m) from the TOF acceleration field ($\sim$1 kV/mm). From the vicinity of the guard electrode ions are accelerated by -14.56 kV through 1.1 ± 0.1 μg/cm^2 thick "carbon" foils and into the LEF analyzer. The linear electric field is produced by voltages of -14.56 kV applied to the foil end of the LEF and $+14.56$ kV to the opposite (ST) end (Figures 18 and 20). Independent positive and negative high-voltage converters located in the HVU1 supply these voltages. Each supply is controlled by an 8-bit command word. A 10^{10} Ω resistor-divider network applied to 30 co-axial rings creates a series of potentials that generate the LEF. One guard ring near the center of the ring stack is grounded to create a zero potential surface (Figure 20). In addition to its vertical, linear component, radial fields arise as a natural consequence of the solution to Laplace's equation in cylindrical coordinates. These act to push positive ions outward from the central axis and focus electrons toward it. By choosing an ESA bending angle of 14°, the inward radial velocity associated with ion bounce motion is balanced against acceleration caused by outward radial field components, thus optimizing ion transmission.

Because the carbon foils covering the entrance to the TOF are very thin and relatively large (7 mm × 18 mm at the center line), they are mounted on stretched 333 lines-per-inch, 70% transmissive nickel mesh. As ions pass through the foils they come into an equilibrium charge state that depends on the incident species, its speed, and the foil material (Funsten *et al.*, 1993a, b, 1994, 1995), but is fundamentally independent of the incident charge state. To illustrate this, the probabilities of hydrogen, helium, carbon, and oxygen ions exiting a foil as singly charged positive or negative ions are shown as a function of energy in Figure 21. Above 15 keV (the lowest energy ions seen in the TOF) the ratio of neutral to positive to negatively charged hydrogen is ~35:15:1 changing to ~60:40:1 at 30 keV. The same ratios for carbon, on the other hand, are ~11:2:1 and ~9:2:1, respectively, which indicates a higher proportion of negative ions relative to hydrogen (typically ~5:1) and corresponding lower proportions of positive ions. For oxygen the comparable ratios are ~15:1:4 and ~7:1:2, respectively, indicating the preponderance of negatively charged oxygen caused by its large electron affinity. It is interesting that for all three species the fraction of ions exiting as neutrals remains at ~70% regardless of energy over this range. Helium has a positive ion yield that is quantitatively similar to oxygen's but the exit charge fraction of its metastable negative ion is $<3\times10^{-4}$ (Funsten *et al.*, 2001). Other ions have different positive, negative, and neutral yields, which generally fall within the range between that of hydrogen and helium. Typically, the fraction of ions exiting the foil in charge state $Q = +2$ is less than 2%. Molecules incident on the foil are fragmented and, depending on the same parameters as for atoms, may exit the foil in a variety of daughter products and charge-states (Funsten *et al.*, 1994).

As ions exit a foil, they eject secondary electrons with a yield of two to three per incident ion on average (Ritzau and Baragiola, 1998). Secondaries are steered by nearby electrodes as they accelerate toward the outer annulus of the ST MCP. Electron transit times are ~1 ns with negligible spread in times of arrival. An electron striking the ST MCP generates a signal on one of eight annular anodes located at the exit of the MCP (anodes and foils are aligned in elevation). The MCP signal is amplified and split into two components in the FEE. One identifies the elevation position of an event, the other starts the TOF measurement. In the nominal case of a single atomic species, a daughter product strikes either the central "stop" region of the ST MCP or the LEF MCP.

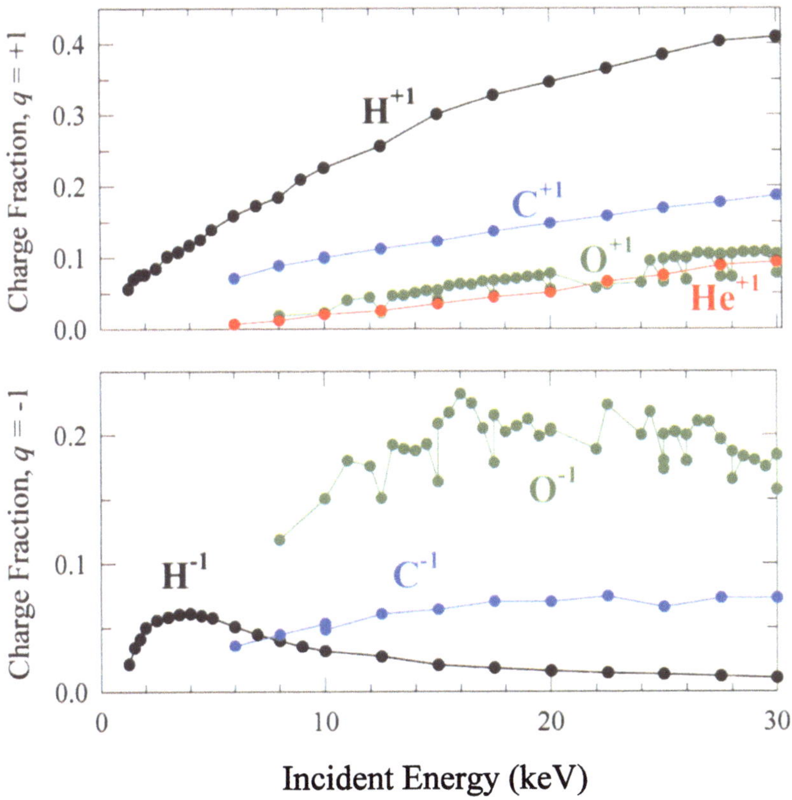

Figure 21. Equilibrium exit charge state fractions for charge states $Q = +1$ (upper panel) and $Q = -1$ (lower panel) for H, He, C, and O incident on a nominal 0.5 μg cm^{-2} carbon foil as a function of incident energy. Data are taken from Funsten, *et al.* (1993a, b, 1994, and 1995).

The ST MCP consists of a so-called Z-stack of three individual plates mounted in proximity to each other. A typical operating gain is ~1 to 3 × 10^6 although this is an adjustable parameter controlled by changing the MCP bias voltage. The LEF MCP has a different configuration from the ST detector. Instead of a three plates in proximity, the MCP plate facing into the ion beam is located on the top of the TOF ring stack (Figures 18 and 23d). The remaining two plates are mounted 12 mm away from the TOF stack near ground potential. This feature conveniently decouples the LEF MCP signal from negative high voltage. Electron charge clouds from an MCP are collected on one of the anodes and then amplified in the FEE. Nearly all daughter products leaving the foils (except for those scattered out of the LEF region) have the potential to be detected. Particles striking the ring electrodes can cause secondary electrons that are detected as background.

Either a LEF or a ST stop event halts the timing measurement. In order to be counted as a valid TOF event, the TDC logic requires that start and stop events both be valid as well as coincident. In the case that "molecular" settings are chosen in the TDC logic, a valid start and a least one stop must be coincident. The timing and outcome of all events are of course probabilistic, which must be taken into account during analysis.

In addition to detecting monatomic species, the IMS design also has fundamental advantages for measuring molecular ions. As noted earlier, nearly all polyatomic species break up during passage through foils (Funsten *et al.*, 1994). Molecular fragments exit with approximately the same center-of-mass velocity as the incident ion so their kinetic energy is proportional to their mass fraction. In order to take advantage of this fact, the part of the LEF field near the TOF foils (where low-energy daughter fragments are reflected) was "de-tuned" from the exact LEF solution by strengthening it slightly (effectively increasing the local spring constant). The effect on positively charged fragments (such as H^+ from protonated molecules) is to make them return slightly faster than their counterparts entering as H^+ parent ions. Trial and error determined the initial field estimate, after which Monte Carlo simulations were employed to find a solution with as few overlapping peaks as possible between atoms and molecules of interest. Details of this process are found in Nordholt *et al.* (1998).

Either the LEF or ST MCP or both may detect molecular fragments. The TDC identifies the later correlated event with high probability as a molecular event. This information, along with LEF and ST times-of-flight, are passed to the spectrum analyzer module (SAM) where a look-up table is used to identify the molecule. Triatomic and larger polyatomic molecules may produce more than two stop signals, but only two are processed. However, detailed TOF spectra can be examined on the ground to identify these ions. As an example of IMS capabilities, the TOF system can separate N, C, and O molecular fragments from N_2^+ and CO^+, giving it an effective mass resolution of 1 part in ~2500 (Figure 33 below).

TOF spectra obtained from the LEF have several properties worth noting. The first is that, because ions with $E_0 < 15.5$ keV are time-focused regardless of energy, TOF spectra change very little with energy, making spectral deconvolution relatively easy (see Figure 34 below). The second property is that ST spectra are dispersed in TOF, which is both an advantage and disadvantage in the deconvolution process. It is an advantage because the electric field disperses, and therefore helps distinguish, negative from neutral from positive daughter products (listed in order of increasing TOF). It is a disadvantage because the lack of time-focusing means that peaks are broader (lower resolution), and there are many more chances for peak overlap.

6.4. Mechanical Design

The IMS collimator is located at the top of the unit (Figure 22). Seven equally spaced vanes divide the 160° FOV into eight elevation channels each 20° wide. The remaining vanes are closed to particles but are needed to support mechanical loads created by the ELS. All entrance collimation is coated with copper sulfide black to

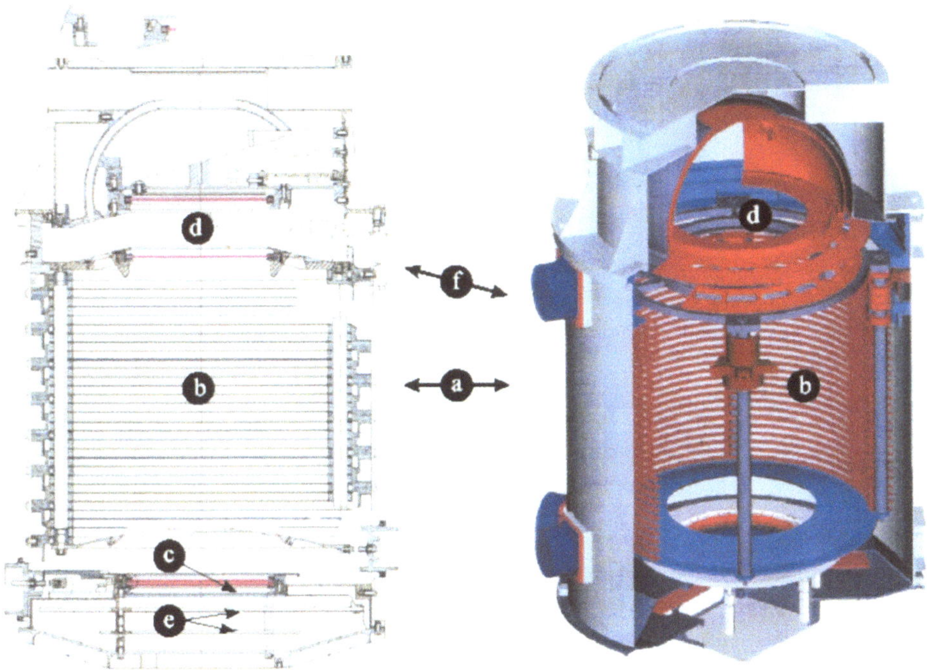

Figure 22. Elevation view of the IMS sensor shown as a detailed engineering drawing at left and as a cutaway perspective drawing at right. The letters a, b, c, etc. in the drawing refer to details visible in Figure 21.

suppress UV and particle scattering. During ground handling and through launch, the IMS aperture was covered by a thin sheet of metallized Kapton designed to help maintain chemical cleanliness and protect the many fragile sensor components (carbon foils, MCP detectors, and blackened optical surfaces) from mechanical damage. Once in space, the IMS cover was released and retracted by a mechanism similar to a roll-up window shade.

The inner dome of the toroidal analyzer is mounted concentrically with the outer by a cantilever located in the blind portion of the IMS FOV (just above (d) in Figure 22). The outer electrode is part of the IMS housing and is held at ground potential. ESA surfaces are also chemically blackened to suppress UV and particle reflections. All materials used inside the IMS are either polished aluminum or ceramic in order to maintain high levels of vacuum cleanliness and high-mechanical precision.

Figure 23b shows the LEF 30-ring assembly. Four ceramic spacers separate each ring from the next by 5.0 mm. One spacer on each ring is designed to hold one of the 30-resistor divider network used to generate the potentials for the LEF. Prior to assembly, electrical contact with the rings is made through gold-plated copper

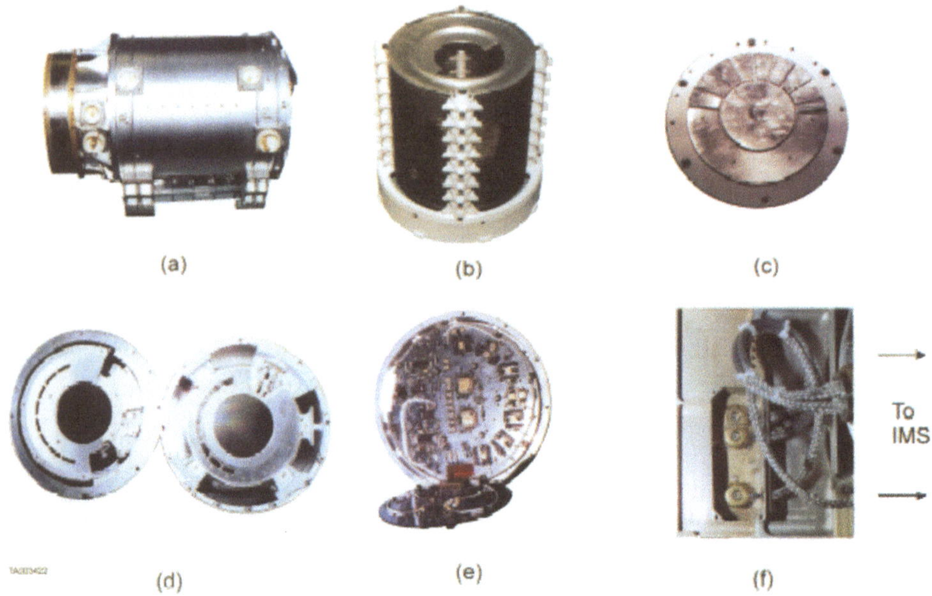

Figure 23. Details of the IMS assembly keyed to Figure 20: (a) IMS housing showing aperture at left and ceramic high-voltage connectors (small white cylinders on the main body). (b) Assembled stack of 30 high-voltage rings used to generate the LEF. Four stacks of ceramic "vertebrae" that position the field rings and hold the resistor divider chain are visible. (c) ST MCP anodes showing eight individual start anodes at the top and the single ST STOP anode in the center. (d) A view looking along the central optical axis of the LEF into the opened IMS. The eight ESA field correction slits can be seen on the left in a semicircular arrangement. The corresponding foils are seen on the right. The dark areas in the center of both pieces are the upper and lower microchannel plates that make up the LEF MCP stack. (e) View of the FEE electronics. The output from the LEF MCP is sent via cabling external to the IMS housing to the FEE board. (f) View looking down on the HVU1 at terminals that connect to cables going to the –15 kV (top) and +15 kV electrodes on the TOF unit.

foils soldered to the ends of each resistor. Four long metal rods pass through holes in the ceramic spacers between the rings, holding the entire stack in compression. To provide the required stiffness of the ring stack, each spacer is attached to the outer IMS housing cylinder (Figures 22a and 23a). All six HV cables are terminated by ceramic feed-throughs that are protected by mechanical covers designed to allow the volumes around the terminals to outgas freely (Figure 23f). Protection from system noise required mounting the sensitive (~few mV signals) FEE inside the IMS housing.

The entrance to the LEF region can be seen most easily to the left in Figure 22d. The eight rectangular windows seen in this view are frames that carry the carbon foils. In the center of the assembly is the LEF MCP. Large openings between the foil assembly and the edge of the ceramic mounting ring provide ample gas conductance during launch and afterwards. The IMS cylindrical housing is approximately 22 cm in diameter × 34 cm long.

6.5. Electrical Design

A number of key science and resource requirements drove the electrical design of IMS.

1. Resolve TOF spectra to 0.75 ns over a range set by the fastest ions (H^+ at 65 keV internal energy) and the slowest (heavy ions at 15 kV internal energies). For the ST portion of the instrument the central path length is 18.8 cm, giving a H^+ flight time of 53 ns. Using 1600 ns as the upper limit on TOF we find that even with an energy loss of, say, 90% in the foils and an energy straggling tail of 300%, a limiting observable mass of ~400 is achievable. For the LEF portion we can use Equation (3) to find the limiting observable mass to be ~200 amu, although in practice it is doubtful that masses much above a few hundred amu will make it through the foil.
2. Compress the raw IMS data stream from (2048 TOF × 2 (LEF and or ST) × 8 elevation anodes × 16 bits)/0.0625 s = 8.4 Mbits/s to a more manageable ~10 kbits/s. As discussed below, this problem is solved using on-board deconvolution of TOF spectra.
3. Obtain mass-resolved energy spectra continuously over 1 to 50,000 eV in 4.0 s, requiring a high voltage slew rate of 150 kV/s.
4. Generate ± ~15 kV high voltages for the TOF optics.
5. Keep mass, power and complexity to a minimum.

In the following section we describe the IMS electronics subsystems by tracing the order in which signals are acquired and processed, namely FEE to TDC to SAM, and then to CPU2.

6.5.1. *Front-End Electronics*

Once charge from an MCP detector is collected on an anode, it enters a series of amplifiers, discriminators, and logic that identifies the event (see Figure 24) and begins processing. In this discussion the signals are identified as "START" and "STOP." Both event types are handled in essentially identical ways. When a START event is detected the signal is split into two parts. One is used to identify the location of the event in one of eight elevation angles or one of the two stop channels (LEF or ST). This part of the signal (termed the identification or ID signal) is processed through fast (0.2 μs deadtime) preamp and level-discriminators and can be used to correct for deadtime in the slower (2.2 μs) circuits. The other "half" of the START signal is sent to the TDC where it triggers the timing circuitry. Similar processes apply to the STOP signals.

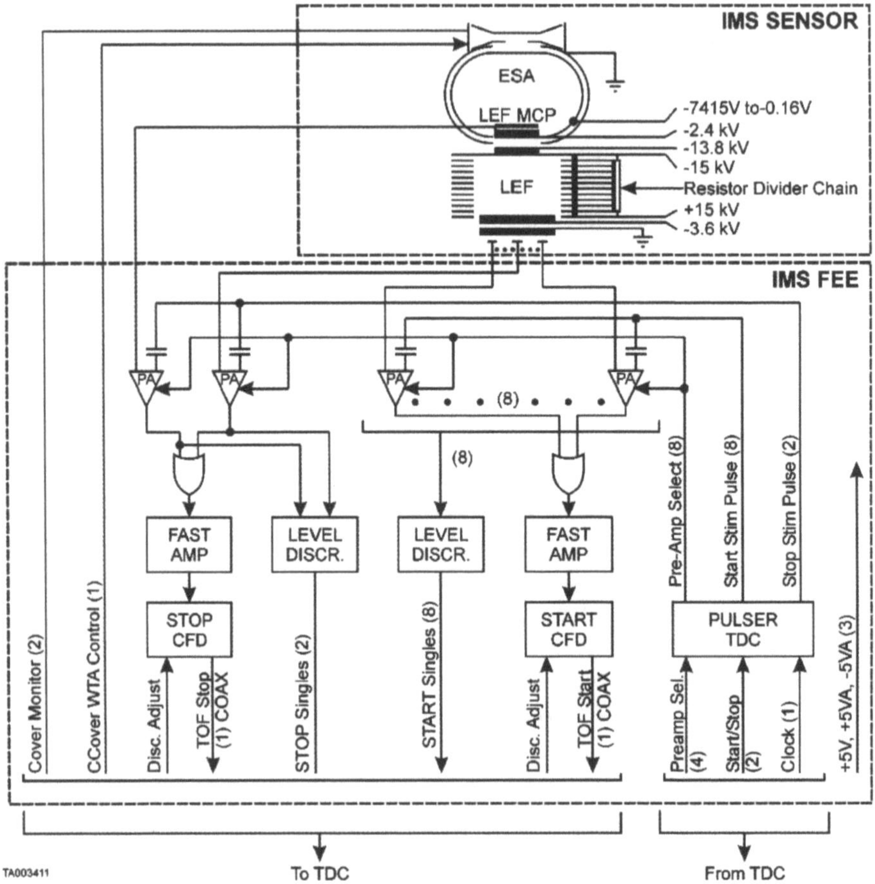

Figure 24. Schematic block diagram of the IMS front end electronics (FEE) and high-voltage connections.

6.5.2. *Time-to-Digital Converter*

Once a START pulse triggers the TDC, the state of the ID discriminators is captured giving the identity of the event origin (Figure 25). Subsequent START events are ignored for timing purposes but are counted as ID events. Once initiated, the TDC converts time directly into a digital word using a clock and vernier technique in the following manner. A valid START enables a gated 80 MHz clock that increments a counter. A subsequent STOP that occurs later than 40 ns (the pulse pair resolution of the TDC) after the START event, inhibits that counter, thus providing a coarse time measurement with a resolution of 12.5 ns. If no STOP event is recorded the counter times out at 1600 ns, corresponding to the longest time required for the heaviest, lowest energy ions to cross the TOF optics as discussed above. To obtain finer time resolution, the phase of the clock relative to the clock edge is measured at the instant the STOP is received by means of the delay line vernier (Figure 25)

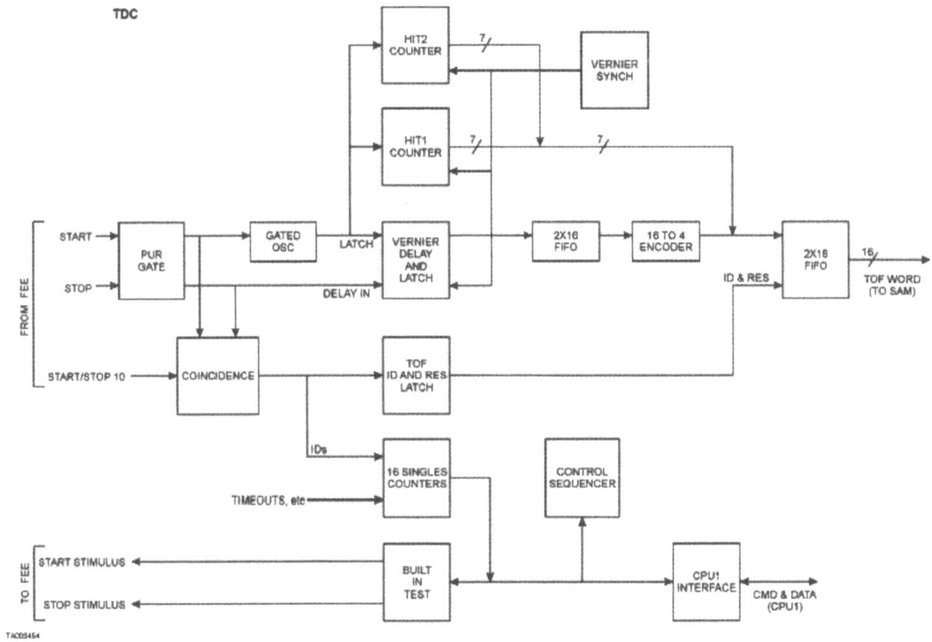

Figure 25. Schematic block diagram of the IMS time-to-digital converter (TDC).

implemented using a 12.5 ns delay line with 16 taps. As the STOP pulse propagates down the delay line, a changing pattern of 1's and 0's appears on the delay line taps. This pattern is latched at the end of every clock period. When a STOP occurs the resulting digital pattern represents the time of the STOP signal relative to the clock edge. The FEE logic then encodes it into a binary word that becomes the fine time measurement. Thus the TDC resolution is effectively 12.5 ns/16 = 0.781 ns, corresponding to 1600 ns/0.781 ns = 2048 channels. If the TDC is instructed to identify molecular events then the gated clock is fed to two independent counters. The first STOP causes one counter to be inhibited and the vernier pattern to be read out and stored. The second STOP inhibits the second counter and causes a second vernier pattern to be read out.

There are several possible outcomes of events detected by the TDC logic in addition to the ideal outcome. For example, a "real" ion can cause a start event followed by a second ion causing a stop event. This could come about in two ways, each leading to a false result: (1) the start event could be a real start followed by a stop caused by the "wrong" ion, or (2) the start event could be caused by an ion scattered to the start electrodes (see Figure 20) followed by a stop caused by a second ion. Another example of types of false coincidences is that between real start or stop events and background in the detectors, e.g., that caused by penetrating radiation. The rate of accidental coincidence events is dependent in a complicated non-linear way on the total event rates. False events and background must be removed from IMS data before they are usable.

After a TOF measurement is completed, the TDC encodes an 11-bit TOF value, a 3-bit elevation sector value, a resolution bit signaling ST or LEF data, and a "continuation" bit indicating a molecular event. Each 16-bit TOF word is then sent to a first-in-first-out (FIFO) buffer read out by the spectrum analyzer module (see Figure 25). The TDC is then reset in preparation for the next measurement. Among its other functions the TDC logic checks for inconsistent control states, channel ID's, or time measurements in order to minimize systematic errors.

During a single IMS sampling period (62.5 ms) the TDC also accumulates the number of ID events corresponding to the eight start anodes, two stop anodes, two timing discriminator channels, number of times when no coincident stop TOF event was detected, total number of TOF measurements completed, and the number of events recorded in two other configurable channels. The configurable channels are used to diagnose TDC operation and to correct errors that may occur at high-counting rates. They can be set to read out the total number of the following events recorded during a 62.5 ms measurement period: (1) STARTs, (2) STOPs, (3) acquisition errors, (4) TDC fixed deadtime reached (regardless of whether or not a measurement results in a valid TOF measurement), (5) monatomic events, (6) molecular events, (7) TDC resets at the end of a measurement cycle, (8) errors caused by indeterminate channel ID values, and (9) TOF measurements less than the pulse-pair resolution time of 40 ns.

To achieve the best possible performance at high-event rates, the TDC operates as a non-paralyzeable counter with a fixed deadtime of 2.187 μs for TOF measurements. ID measurements (collected in "SINGLES" counters), on the other hand, have a 0.2 μs deadtime making them suitable for correcting the slower timing measurement rates. Finally, the FEE/TDC system has built-in test pulsers that can stimulate any START or STOP pair with one of 24 selectable time delays at periodic event rates up to 1 MHz.

6.5.3. *Spectrum Analyzer Module*

The SAM processing method allows a high level of TOF data compression ($\sim$1000:1) to be achieved with little sacrifice in accuracy. The key to carrying out this process at high-event rates and within the short IMS sampling interval (62.5 ms) is a high-speed, deterministic deconvolution technique that we will refer to as the SAM algorithm. The hardware that supports this algorithm has a distributed, pipelined architecture based on four loosely coupled, reconfigurable modules (Figure 26). Once SAM processes a single data sample (i.e., data gathered in 62.5 ms) they are transferred to the CPU2 processor and the cycle repeated. Although data transfers from FEE to TDC and TDC to SAM occur at random rates, control of all other processing, data transfers and command and control functions are synchronized, taking place during the final 12.5% (7.81 ms) of the sampling interval.

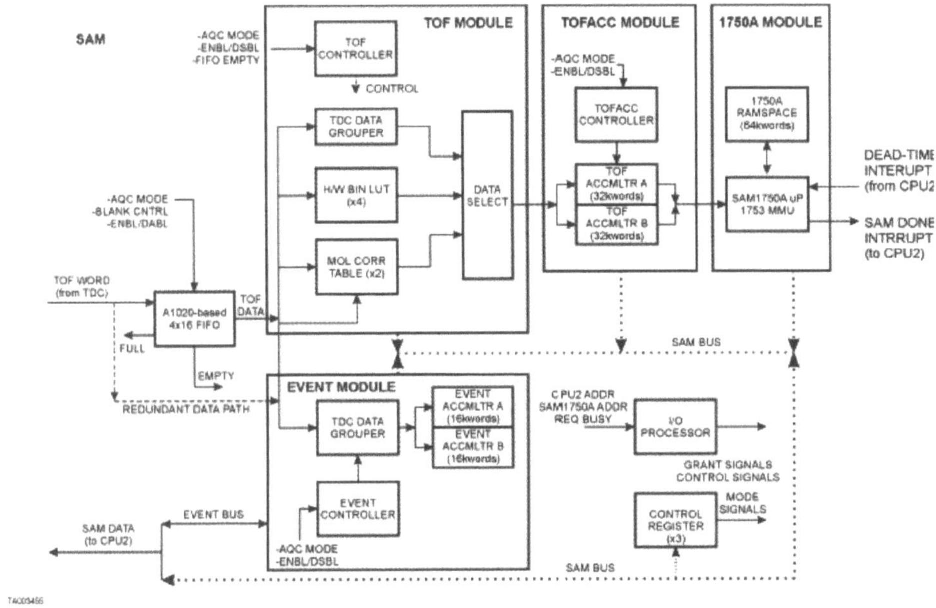

Figure 26. Schematic block diagram of the Spectrum Analyzer Module (SAM).

During this time the ESA voltage is stepped, data registers are read out and reset, and new commands are executed if required.

As a consequence of the primary measurement objectives discussed in Section 6.1.1, two different types of data and two differently timed acquisition cycles form the basis of IMS operation. The data types consist of ion data (ION) identified on-board at the highest possible rate using the SAM algorithm. The second is TOF spectra at the highest possible mass resolution but at a much lower data rate. ION energy spectra have a time resolution of 4.0 s for a maximum of 7 ion species over the full energy range. TOF data on the other hand, also cover the energy spectrum but at lower time resolution (256 s).

The operational cycles associated with data acquisition are defined as A- and B-cycles. The A-cycle lasts 32.0 s. During this time the ESA is scanned through eight, 64-step energy sweeps of 4.0 s each. The slower B-cycle is timed to correspond to eight A-cycles and to the acquisition and transmission primarily of TOF spectra (during both A- and B-cycles, data of both types are being collected). Dual memory banks allow raw data acquisition to proceed while CPU2 and SAM processors both access data from the previous 4.0 s acquisition interval. At the end of each interval, spectra that were processed during the previous interval are read from SAM by CPU2. In addition, new, selectively binned TOF spectra, are read into the SAM processor. There, during the next measurement interval, the SAM algorithm deconvolves TOF spectra, extracting selected ion *M*/*Q* values. Once analysis is

TABLE IX

SAM operational modes.

Mode	Description
1. SNGL_DUAL_STOP	Single-stop or dual stop mode of operation
2. TOF_MODE	
Angle_Collapse_En	Accumulate TOF, collapse Polar Angle
Alt_Angle_Collapse	Accumulate TOF, collapse Alternate Polar Angles
Seq_Event_En	Enable Sequential Event Mode
3. BINNING MODE	
HW_Bin_En	Enable Hardware Binning Mode
SW_Bin_En	Enable Software Binning Mode
4. HW Bin LUT Sel	Select 1 of 4 hardware binning look-up tables
5. Mol LUT Sel	Select 1 of 2 molecular look-up tables
6. ACCUM_EV1_2_EN	Accumulate 1 event or both events from a dual stop event
7. TOF_CMPRS	Enable or disable summing: 4 adjacent TOF ST bins; 2 adjacent LEF bins
8. TOF_DISABLE	Enable or disable TOF and TOFACC Modules

complete, SAM interrupts CPU2, which then reads the *M/Q*-sorted data. SAM is then ready for the next acquisition and processing cycle.

SAM operates as a CPU2 slave. The analysis algorithm is coded in the Ada language and executed in the 1750A microprocessor module (Figure 26). Its address space is accessible by CPU2, and program and local data are loaded into the module by CPU2. However, SAM itself cannot read or write to CPU2 memory. The three remaining modules (TOF, TOFACC, and EVENT) are each autonomously controlled by separate gate arrays. In order to speed up TOF data processing, dedicated paths support concurrent data flow from the input FIFO to the appropriate processing modules (Figure 24). In addition, each module manages its own local processing and external access to data. The SAM operational modes are given in Table IX.

SAM acquires data from the TDC via a FIFO buffer at a maximum rate of 5×10^5 periodic events/s. The FIFO can be configured by command to enable dual-stop operation (for molecules) and data acquisition from selected elevation ID channels. Disabling the FIFO causes data to be routed by a redundant path (Figure 26) to the EVENT and TOF modules.

The TOF module organizes incoming data from the TDC into groupings that correspond to different ion species. The way in which incoming data are grouped depends on parameters in the SAM ION group table, which contains a targeted list of species and their binning parameters (see Table IX). In normal operation, a software-binning mode (option 3 in Table IX) will be used. In this mode 16 TOF bins

× 8 elevation sectors × 2 channels (ST and LEF) are stored during one 0.0625 s accumulation period. During the next acquisition period the SAM algorithm is executed beginning with the first new data stored in the double-buffered memory. If the hardware-binning mode is selected instead, the same quantity of data are selected but then are grouped according to one of four command-selectable look-up-tables (LUTs). In this mode, data are grouped in TOF channels according to their assigned mass peak locations and exported to CPU2 (Figure 26). In general, the software-binning mode requires more analysis time than the hardware-based approach.

Deconvolution of TOF spectra is more complicated in the dual STOP mode used for molecular ions. The reason is that a molecular event (i.e., two valid coincident stops) consists of two daughter products, both of which must be identified and then their identities correlated in order to specify the parent molecule. In this mode incoming TOF events are classified into one of four two-dimensional correlation tables depending on which of the following events occurs: (1) the ST MCP records the first stop, (2) the ST MCP records the second stop, (3) the LEF MCP records the first stop, or (4) the LEF MCP records the second stop. Each of the four tables is 2048 TOF channels deep and contains identification codes for 32 different molecular species. Once a molecular event is identified the SAM algorithm is used to identify the particular parent molecule.

After SAM algorithm processing is completed, all valid TOF events have been sorted into a two-dimensional array consisting of eight elevation sectors by seven ION (atomic or molecular) species. These 56 data products are sent to CPU1/CPU2 shared memory once per energy step. Several data compression modes are available based on summing adjacent energies or angles (Table IX), or by selecting fewer than seven ion species. Yet another mode, used primarily for engineering, stores 32,000 TDC words with each word corresponding to a concatenation of TOF (11-bits) + elevation (3-bits) + TOF resolution (1-bit) + atom/molecule (1-bit). Energy and azimuthal data for a particular set of events are tracked separately.

We would now like to provide some insight into the SAM algorithm (Sittler, 1993). The algorithm's advantages are that it reduces the inversion problem of deconvolving TOF spectra to a deterministic number of multiply and add operations, thus making processing time predictable and the algorithm efficient. Figure 27 contains two examples of TOF spectra. Summing counts within an interval centered on a given mass peak is a straightforward way of obtaining ion peaks from a LEF spectrum (top panel). The bottom panel, however, shows a simulated spectrum obtained from the ST detector in which it is apparent that TOF peaks from different parent species overlap significantly, and simple binning will not give the correct answer. The SAM algorithm responds to this need by providing accurate deconvolution of complex spectra without introducing systematic errors.

Taking advantage of the fact that TOF peak profiles for different species can be determined from calibration, we have developed a relatively simple algorithm that

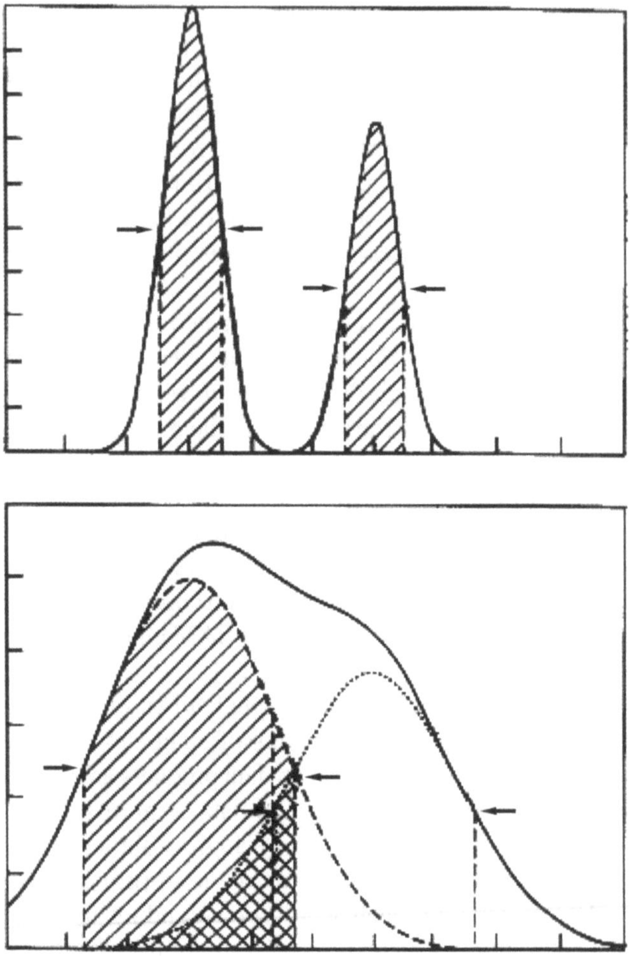

Figure 27. Schematic example of a high-resolution (LEF) TOF spectrum (upper panel) and a low-resolution (ST) TOF spectrum (lower panel).

can be expressed as a set of n-coupled linear equations with constant coefficients. This is shown in Equation 9.

$$
\begin{aligned}
N_1\alpha_1 \sum_{k=1}^{m} {}^1\gamma_1^k + N_2\alpha_2 \sum_{k=1}^{m} {}^1\gamma_2^k + \cdots\cdots + N_n\alpha_n \sum_{k=1}^{m} {}^1\gamma_n^k &= B_1 \\
N_1\alpha_1 \sum_{k=1}^{m} {}^2\gamma_1^k + N_2\alpha_2 \sum_{k=1}^{m} {}^2\gamma_2^k + \cdots\cdots + N_n\alpha_n \sum_{k=1}^{m} {}^2\gamma_n^k &= B_2 \\
N_1\alpha_1 \sum_{k=1}^{m} {}^n\gamma_1^k + N_2\alpha_2 \sum_{k=1}^{m} {}^n\gamma_2^k + \cdots\cdots + N_n\alpha_n \sum_{k=1}^{m} {}^n\gamma_n^k &= B_3
\end{aligned}
\qquad (9)
$$

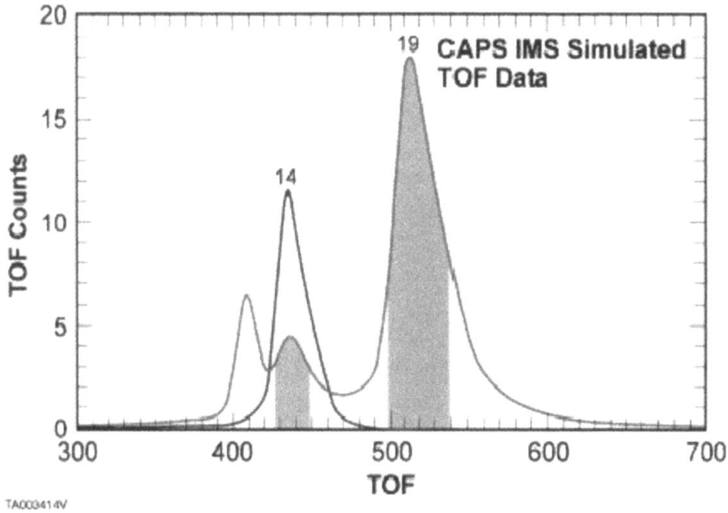

Figure 28. Simulated TOF spectrum resulting from mass 14 and the (fictitious) daughter fragments of mass 19 (in red).

Here the B_j are binned counts for the *j*th binning interval where j is the index indicating ion species. The N_j are the true counts corresponding to the *j*th ion species, the α_j renormalize the N_j so that the product $\alpha_j N_j$ is equal to the total number of TOF events due to ion species j, and ${}^n\gamma_n^k$ gives the fraction of the total counts that the *k*th fragment of ion j contributes to the binning interval of ion i. Basically each row in Equation (9) consists of terms that represent contributions from each ion species and its daughter fragments to the binned counts for a particular binned interval. An application of this algorithm can be seen in Figure 28 where we show a simulated TOF spectrum containing only $M/Q = 14$ and 19 (the mass 19 fragments are fictitious and for illustration purposes only). A binning interval is defined as the inclusive range of TOF channels between the FWHM points measured at a particular peak (roughly the blue areas in Figure 28). The binning interval for the largest peak is the shaded region on the right containing N_j counts. The total under all peaks attributable to mass 19 is $\alpha_j N_j$. The shaded region on the left is associated with a daughter peak of mass 19, showing the contribution of the *k*th fragment of ion j (i.e., of mass 19) to ion i (i.e., mass 14), i.e., the factor ${}^n\gamma_n^k$.

The set of equations in (9) can be written in matrix notation as follows:

$$\begin{bmatrix} a_{11}\ a_{12}\ a_{13} \ \cdots\cdots\ a_{1n} \\ a_{13}\ a_{22}\ a_{23} \ \cdots\cdots\ a_{2n} \\ \vdots \\ a_{n1}\ a_{n2}\ a_{n3} \ \cdots\cdots\ a_{nn} \end{bmatrix} \begin{bmatrix} N_1 \\ N_2 \\ \vdots \\ N_n \end{bmatrix} = \begin{bmatrix} B_1 \\ B_2 \\ \vdots \\ B_n \end{bmatrix} \tag{10}$$

where the total coupling coefficient is given by $a_{ij} = a_j \sum_{k=1}^{m} i\gamma_j^k$. By taking the inverse of Equation (10), we obtain the following

$$\begin{bmatrix} N_1 \\ N_2 \\ \vdots \\ N_n \end{bmatrix} = \begin{bmatrix} g_{11} & g_{12} & g_{13} & \cdots\cdots & g_{1n} \\ g_{21} & g_{22} & g_{23} & \cdots\cdots & g_{2n} \\ \vdots & & & & \\ g_{n1} & g_{n2} & g_{n3} & \cdots\cdots & g_{nn} \end{bmatrix} \begin{bmatrix} B_1 \\ B_2 \\ \vdots \\ B_n \end{bmatrix} \tag{11}$$

In Equation (11) the summed counts N_j are seen to be a weighted sum of the binned data B_j for which the g_{ij} are the weighting coefficients. (Note that $C_j = \alpha_j N_j$ represents the total number of counts originating from ion j, which is what SAM provides to CPU2.) Because ST spectra change with ion energy (recall that ST TOF $\sim$1/velocity $\sim E^{-1/2}$), ST spectral binning intervals and inverse coupling coefficient matrices also change. The SAM ION group tables contain binning intervals and matrices for 16 energy steps covering the full energy range then operating. The values used for a given energy step are computed in SAM by linear interpolation at the end of the preceding energy step. Finally, SAM uses the total number of TOF events processed plus the total number of timeout events to make dead time corrections.

The major advantage of the SAM approach is that the majority of calculations needed to deconvolve ion species, such as peak fitting, can be performed on the ground using calibration data. Only the results, in the form of coefficients that make up the inverse matrix G and the binning intervals, are needed in flight. The SAM method works very well but depends on accurate calibration, inclusion of all observed species in the group tables, and subsequent analysis of contributors to calibration mass peaks.

Algorithm Modeling. Multiple simulations have been performed to test the SAM algorithm. Observed ion peak shapes of parent and daughter particles are modeled by combined gaussian and lorentzian fitting functions. Peaks from daughter species are normalized to unity under the total line profile (Figure 29). Then the sum of the counts in each peak is weighted such that the number of counts for a particular ion, j, are equal to C_j. The procedure is then repeated for all ions making up a particular group.

The data shown in Figure 29 represent simulated spectra from ions with masses 14 through 19. Although the six TOF spectra are simulated, it is fairly obvious that it would be impossible to interpret the summed spectrum (red spectrum labeled "14–19") without some means of deconvolving individual peaks shown below it. The process followed to generate coefficients needed by the SAM algorithm begins by defining a binning interval for each dominant ion peak (in this case the peaks labeled 14–19 in Figure 29). Once binning intervals are specified, it is straightforward to compute the inverse coupling coefficient matrix. Then a simulated spectrum can be generated and data binned and multiplied by the inverse coupling coefficient

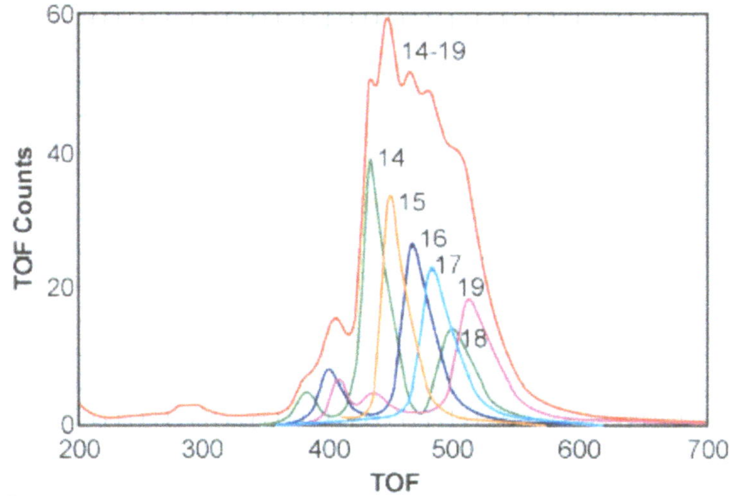

Figure 29. A simulated example of daughter product peaks (M/Q = 14 through 19) and the sum of all daughter-product counting rates (outer envelope in red) based on calibration data. Note daughter fragments from several molecular species such as the water group are present (masses 16-19). Each species is normalized to 1000 counts.

matrix G. In order to estimate the accuracy of this method, poisson uncertainties are introduced randomly and the process repeated 200 times. In this particular example the water group ions (16–19 in Figure 29), are broader than mass 14 and 15 peaks because of greater overlap of the 16–19 peaks in this region ST mass resolution is relatively low (~8).

As an example of the power of the SAM technique, we have considered a case in which the relative distribution of total counts from each ion species in Figure 29 is held the same, but the total number of counts in the entire TOF spectrum is varied from 10 to 10,000 thus varying counting statistics and, one would think, the accuracy of the method. Figure 30 is a graph of the square root of the number of counts assigned to each peak divided by the measured standard deviation from the simulation. The profiles in Figure 30 are relatively flat and independent of count rate. As expected, the best precision is obtained for the most abundant species (14 and 15) with the least precision for masses 16–19. Note, however, that although the relative precision for the least abundant species is low, the average results are correct. The constant values for the errors result from the requirement that the relative abundances are held constant.

6.5.4. *High-Voltage System*

Figure 31 is a schematic of the IMS high voltage (HV) system made up of five supplies controlled independently by CPU1 (see Table X for detailed supply characteristics). As explained earlier, the supplies are contained in two separate units: nominally ±15 kV in HVU1 and the two MCP supplies and ESA stepping supply

TABLE X
IMS high-voltage supply specifications.

Supply	Range (kV)	Resolution (bits)	Precision	Ripple (p–p)
HVU1/N	0 to +16.0/0 to −14.56*	8	±1%	0.1%
HVU1/P	0 to +16.0/0 to + 14.56*	8	±1%	0.1%
HVU2/ST MCP	−0.5 to −3.6	8	±2%	0.5%
HVU2/LEF MCP	−0.5 to −2.4	8	±2%	0.5%
HVU2/ESA	−0.00015 to −7.415	3 ranges × 12	±0.5% or 1 LSB	0.25%

*First range is maximum capability. Second range is normal operating range.

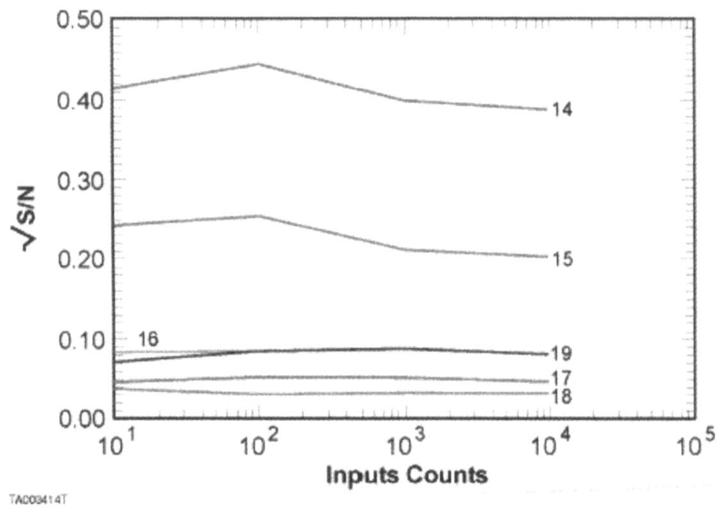

Figure 30. The ratio of the square root of the signal to measured standard deviation of estimates produced by the SAM algorithm as a function of the number of counts in the simulation. Data are from 200 runs of the algorithm using the data in Figure 29.

in HVU2. All supplies feature current-limiting and breakdown protection, as well as analog monitoring (digitized in the data stream). HVU1 is powered by filtered +30 V from spacecraft primary power whereas the HVU2 supplies use +15 V generated by CAPS. Stringent hardware, software and operational precautions were instituted to prevent HV from being applied at inappropriate times. Once in flight, separate commands have to be sent to enable and arm a supply. In addition, high-voltage commands (to any sensor) are checked by CPU1 to ensure they do not violate pre-set constraints on upper HV limits or HV step size. High-voltage cables rated at ≥2.5 times the respective supply voltages connect HVU1 and HVU2 to IMS sensor electrodes (40 kV cables are used for HVU1 connections). All HV connectors were custom designed and fabricated from low-void ceramic to prevent high-electric field concentrations that might lead to HV breakdown over the course

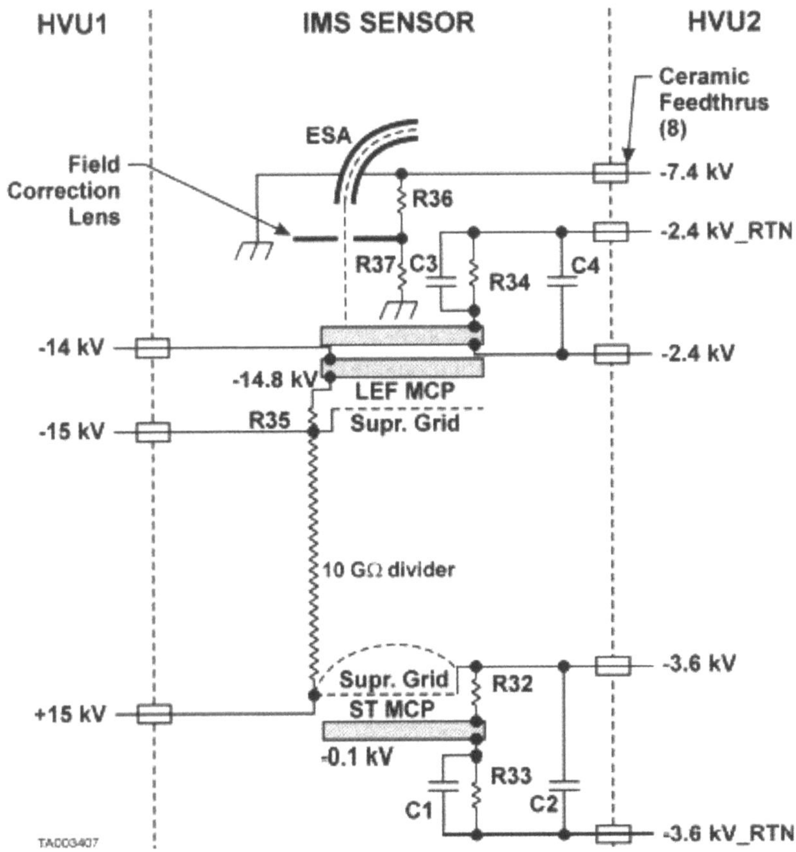

Figure 31. Schematic of the IMS high-voltage system.

of the mission. Metal covers placed over the HV terminals reduce electric stresses and provide mechanical as well as contamination protection.

HVU1. The HVU1 supply is composed of independent positive and negative voltage converters whose output is maintained at $\pm1\%$ over all line and temperature variations (Table X) with very low ripple. The primary purpose of HVU1 is to provide the negative and positive high voltages that together create the LEF by correctly biasing a string of 30-high ohmic resistors running the length of the TOF ring stack (Figures 24 and 31). HVU1 also generates regulated 1200 V referenced to -15 kV that is used to bias the first plate in the LEF MCP stack. Because power for the LEF MCP regulator is derived from the -15 kV supply the latter must reach a minimum of ≈ -12 kV in order for the LEF MCP to reach ~950 V and operate correctly. A voltage drop of -100 V is placed on a grid in front of the MCP to return secondary electrons to the MCP surface, thus maintaining high efficiency.

To make a compact but reliable unit, electric fields in the HVU1 supplies were kept below 1 kV/mm relative to the grounded housing. Transformer and high voltage assemblies were not encapsulated in order to avoid multi-material composite structures that could introduce mechanical stresses that might in turn lead to dielectric faults and eventual breakdown. Instead, after fabrication and cleaning, the interior surfaces and components were coated with Parylene-C, a tough, low-outgasing polymer with high-dielectric strength.

HVU2. Both of the MCP supplies are located in the HVU2 as is the ESA supply. The MCP supplies are virtually identical in construction and operation, relying on a pulse-width modulator control circuit and a resonant converter operating at a switching frequency of 100 kHz. The capacitor networks shown in Figure 31 provide rf filtering. A resistor divider (R36 and R37 in Figure 31) biases the field-correcting aperture located at the ESA exit to approximately one-half the ESA plate potential. Another function of the MCP supplies, in addition to powering the two detectors, is supplying a suppression bias voltage to high-transmission grids to return secondary electrons to the MCP (the same function provided by HVU1 for the LEF MCP). Similar potentials at the back of the third stages of both detectors accelerate electron charge clouds toward their respective anodes.

The ESA programmable power supply presented several challenges that make it different from the MCP supplies. The primary requirements were that it step rapidly and accurately over a wide range of voltages set by the energy range of the instrument (0.16 – 7415 V, taking into account the finite width of the IMS passband). Thus a commandable resolution of 14 bits was required to obtain the precision given in Table X. With this resolution it is possible to cover a wide dynamic range very accurately with, in theory, any type of scan desired (linear, logarithmic, etc.). The supply is designed to step every 62.5 ms with a slew period of 12.5% or 7.81 s, requiring a voltage slew of 150 kV/s.

6.6. Calibration

Returning to the equation for "geometric" factor, we can use Equation (8) to find a way to measure this factor in the laboratory. As discussed by Johnstone *et al.* (1987) and others, Equation (8) can be rewritten:

$$G_i = (Q/J_0)\varepsilon(E_j, M_i)\tau_i(E_j, M_i)\sum_i [\Delta\theta \sin\theta \Delta\varphi(\Delta E/2E)]_{\rm i} C_{\rm i} S_{\rm i} \qquad (12)$$

In this equation J_0 is the beam current density its charge so that their ratio has units of flux. The efficiency and transmission are functions of energy and mass. The summation term is taken over a series of small volumes in energy space centered on $(\theta, \varphi, E)_{\rm i}$ weighted by the counting rate at each location. The summation extends over the entire acceptance of IMS. Finally, the weighting factor $S_{\rm i}$ accounts for variations in ion current over the time that the individual samples are taken.

Calibration was performed with a Von Ardenne plasma discharge source at SwRI and a radio frequency source at LANL (Alton, 1993). Beam intensity was fairly stable over short time periods (~1 h) but could be controlled only by pressure feedback in the source itself. The beam used for calibration at SwRI was monochromatic and plane parallel to a high degree. It was verified to have an angular spread of 0.2 °, about 2.5% of the width of IMS acceptance in angle. Beam width in energy was $\Delta E/E = 0.005$, about 3% ov the IMS passband. Only the mass resolution of the beam separator was below expectations $(M/\Delta M)_{\text{BEAM}} \approx 40$.

During calibration a known ion species was selected, the energy of the beam set, and J_0 measured. Because it is far simpler than varying the energy of the ion beam, the voltage on the ESA was micro-stepped at increments of ~0.3% of the passband. The IMS was placed on a three-axis goniometer (two angles plus translation) and rotated about the center of the IMS FOV to obtain angular measurements at a resolution of ~1° × 2° (AZ).

On the other hand, calibration of the IMS response to mass, energy, and J_0 could be achieved relatively quickly. A wide range of energies was calibrated using N^+ as a standard. Similarly, a wide range of ion species was calibrated using a few specific ESA energies as standards. At every opportunity the value of J_0 was measured to quantify absolute calibration.

Correctly associating the center of TOF peaks with corresponding M/Q values for known gases established TOF calibration, the correction factor $\sigma_i(E, M)$ for energy losses in the foil, and k_2 in Equation 3.

The mass-analyzed ion beam at LANL was used initially to test IMS energy-angle response and TOF characteristics. From the latter, initial coefficients for the SAM algorithm could be calculated. During these tests we discovered cross-talk between elevation-angle channels caused by internal particle scattering. Coating the inner plate of the ESA with copper sulfide black, an anti-scattering agent, eliminated most cross-talk but some remains.

Final calibration of the IMS in flight configuration was carried out at SwRI. Only about 30% of planned activities were completed. Twenty energy steps between 0.256 and 27.56 keV, corresponding to the centers of the IMS energy passbands, were covered. The calibration species included H^+, He^{2+}, He^+, C^+, CH^+, N^+, CH_2^+, NH^+, O^+, CH_4^+, NH_2^+, OH^+, H_2O^+, H_3O^+, $^{20}Ne^+$, $^{22}Ne^+$, N_2^+ and Ar^+. The primary concentration of calibration effort covering nearly all of these species at a single energy took place at 1.024, 9.74 and 27.56 keV. At all other energies calibration concentrated on N^+ and $M/Q = 16$.

Further calibrations of IMS are being carried out in the laboratory with a prototype model refurbished to be identical to the flight model in all essential characteristics. The purpose of this effort is two-fold: First to understand features of the instrument, primarily certain TOF spectra, which were not well characterized during flight unit calibration. Second, we plan to introduce a wider variety of ion species that are more characteristic of what might be expected in the ionosphere of Titan. All calibration data from both the flight and prototype units have been placed in a common format and assembled into an archive kept with flight data.

6.7. Performance

Figure 32 illustrates the characteristic energy-angle response of the IMS ESA for a 16.39 keV Ne^+ beam, plotted as ion count contours in the voltage (energy) elevation angle plane. The slightly asymmetric appearance of the contours is consistent with the simulated response of the toroidal top-hat analyzer (Young *et al.*, 1988). The sawtooth nature of the contour edges is an artifact of goniometer motion. Note that this figure corresponds to Figures 11 and 17a for ELS and IBS, respectively.

A set of TOF spectra taken simultaneously with the ST and LEF anodes for $M/Q = 16$ at 1.024 keV is shown in Figure 33 (a) and (b). The actual ion flight times in Figure 33 are obtained by multiplying the indicated channel numbers by 0.78 ns. As can be seen in the LEF spectrum in Figure 33 (a), H^+, O^{2+}, C^+, N^+ and O^+ can be identified as the daughter products of O^+, CH_4^+ and NH_2^+ all of which were present in the same beam at $M/Q = 16.00$. The mixture can be inferred on the basis of the history of gases leaked into the source and on the makeup of filament materials. Although the calibration beam was not mass separated, it is clear that the IMS resolve beam components. The H^+ peak is proportional to the amount of CH_4^+, and NH_2^+, while the amount of $[O^+ + O^{2+}]$, N^+, and C^+ are proportional to the parent molecular species.

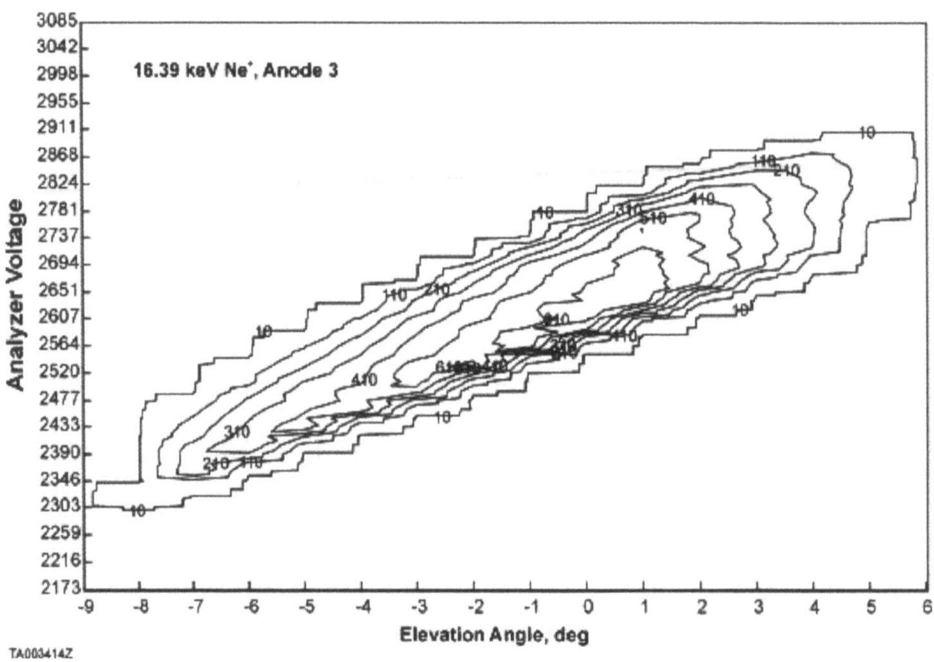

Figure 32. Ion energy (proportional to analyzer voltage) versus azimuthal entrance angle for a single elevation anode of the IMS sensor.

The corresponding ST spectrum shown in Figure 38 (b) illustrates that generally the most significant peak in the ST response is due to neutrals. Since, to first order, all neutral daughter products from an incident molecule travel with the same center-of-mass speed $v_{\rm int} = (2E_{\rm int}/M)^{1/2}$ through the TOF analyzer, a single neutral peak is produced. Energy straggling and scattering in the foil tend to broaden this peak. At 1.024 keV, no positive ions appear in the ST spectrum because they are all reflected by the LEF, appearing exclusively in Figure 33(a). The two peaks labeled "*E*" are the result of de-focused ions that strike the LEF MCP suppression grid and grid holder. The errant ions produce secondary electrons, which are then accelerated by the LEF to the ST detector, resulting in ST peaks. Note that these two "echoes" coincide with the LEF N^+ and O^+ peaks. The LEF field accelerates stable negative ions leaving the foils and thus they appear at earlier times than the neutrals. This accounts for H^-, C^-, and O^- peaks in the example. Nitrogen does not have a stable negative ion and thus does not appear in the spectrum. The noise background in these spectra is primarily due to internal scattering of particles exiting the foil and to accidental coincidences of particle events with each other (for example a START coincident with a non-related STOP event.

Data from calibration spectra have been analyzed and compared to models of IMS behavior given in Equations (3) through (6). For example, in a perfectly linear electric field the TOF should be proportional to the square root of the ion mass/charge, independent of energy. Figure 34 shows a plot of this relation for ion M/Q vales of 7 and 14 (N^{++} and N^+, respectively) over a wide range of energies. Although the electric field has been purposely detuned from a perfectly linear case, the square root relation still holds quite well and can be used to identify peaks in LEF spectra. The notable exceptions are low-energy ions, which straggle badly and tend to reflect in the most non-linear portions of the field. The third curve in Figure 34 is a ghost peak caused by internal scattering (see Figure 33 (a) at TOF channel 170).

The neutral species present in ST spectra experience no electric field and hence their times-of-flight are given by effective path length divided by exit speed (Equation 6). Figure 35 (upper panel) shows a plot of measured neutral particle TOF for a wide range of incident ion M/Q and energies. The square root relation holds quite well and here again it is useful for identifying species in the TOF spectra. Energy straggling and scattering caused by the foil produce some of the scatter in this plot because of the difficulty encountered in fitting the peaks and hence identifying their centroids.

In contrast to neutral species, however, ST ions that overcome the LEF potentials experience a varying force along their paths, so modeling in this case requires integration along the trajectory (Equation 4). As an example, Figure 36 shows a comparison between the measured TOF of C^- and O^- ST ions for several energies in a simple model assuming a perfectly linear electric field. The agreement is not as good as with the LEF and neutral models, but is better than about 10%. We believe that the disparity is partly due to the non-linearity of the electric field, as well as

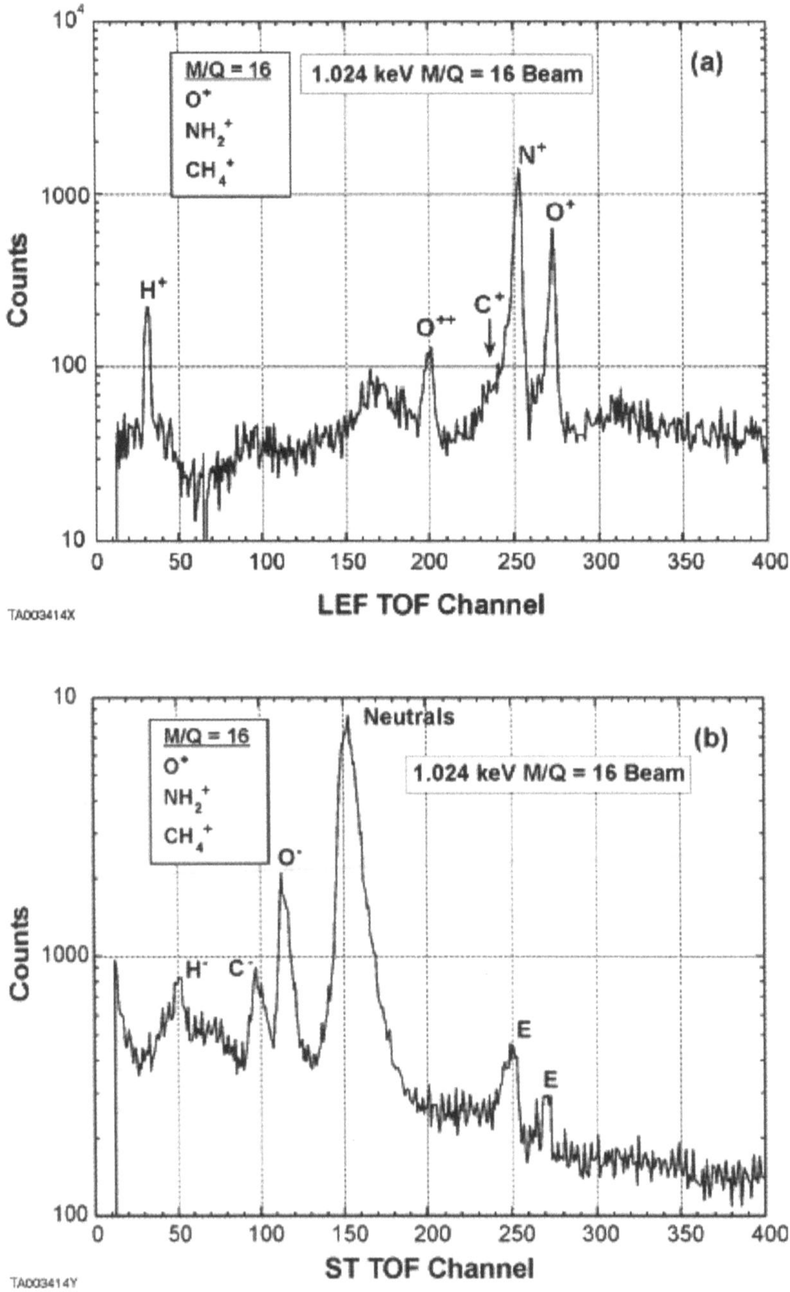

Figure 33. (a) An LEF time-of-flight spectrum for an incident 1.024 keV ion beam made up of mass/charge 16 ions including CH_4^+, NH_2^+, and O^+. (b) The ST TOF spectrum corresponding to the LEF spectrum shown in Figure 38 (a).

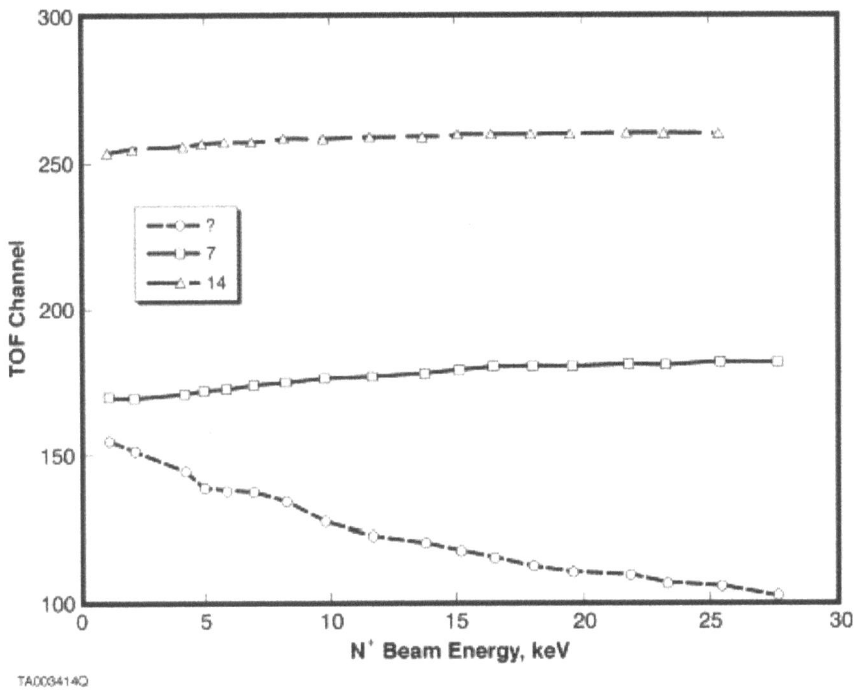

Figure 34. Location of the center of the N^+ and N^{2+} LEF TOF peaks as a function of incident ion energy demonstrating the energy focusing capabilities of the LEF optics. The lower curve corresponds to behavior of the ghost peak evident in Figure 33 (b) at 170 TOF channels.

the mass and energy-dependent straggling and scattering caused by the foil. The foil effects are more severe for lower energies and result in a longer than predicted TOF. In addition, when a molecule breaks up in the foil the resulting components divide their energies proportionally to their masses so each has a lower energy than the original. This effect shifts the TOF of positive molecular fragments from their expected positions had they the same energy as the incident parent molecule. Thus it is possible to distinguish between an ion formed in a molecular breakup and an atomic ion of the same M/Q. Comparisons with the LEF and ST spectra (Figure 36) allows us to distinguish between molecular and atomic species, which would otherwise require a conventional mass spectrometer with much higher resolution.

7. Instrument System Design

This section provides a complete description of the system aspects of the CAPS instrument.

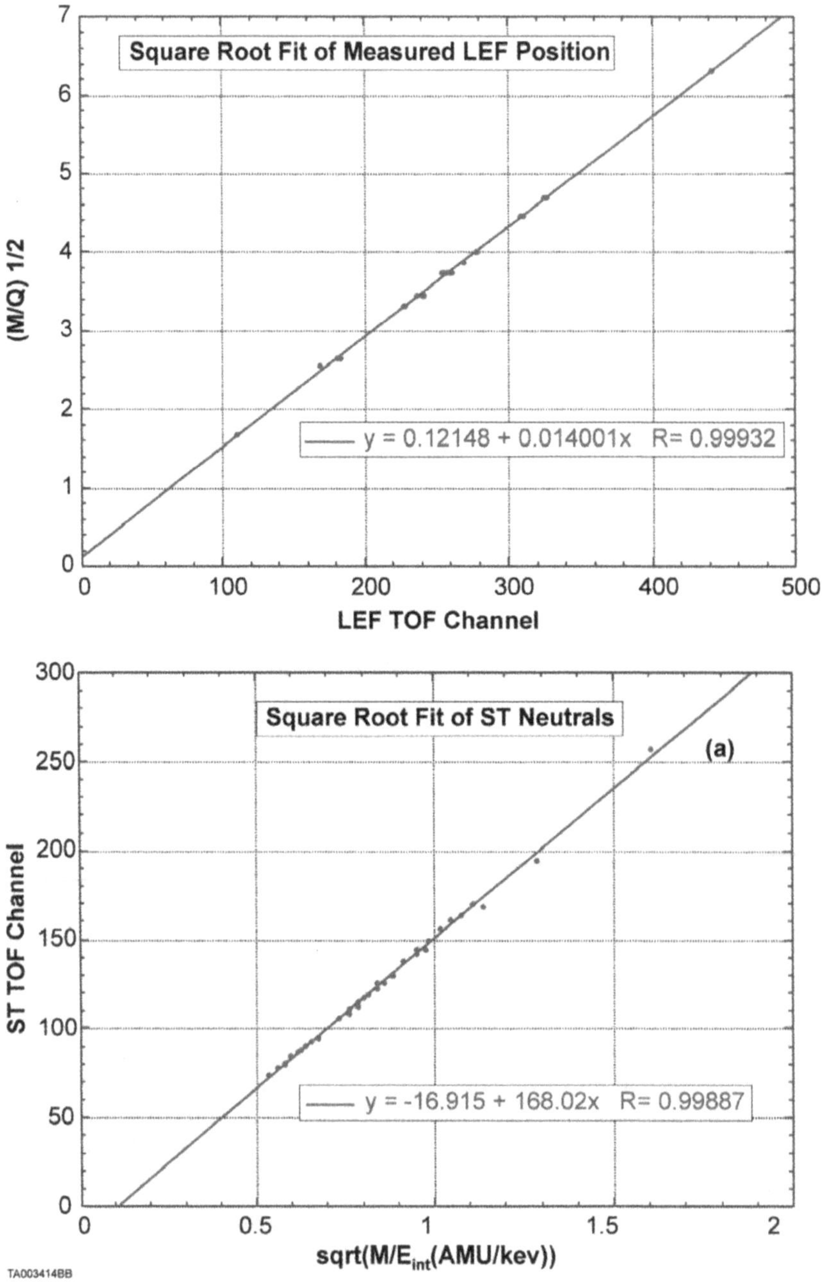

Figure 35. (a) Locations of the centers of TOF peaks as a function of ion M/Q for the ST detector. (b) Locations of the centers of TOF peaks as a function of ion M/Q for the LEF detector. Spectra were obtained at energies between 1 and 27.6 keV/e.

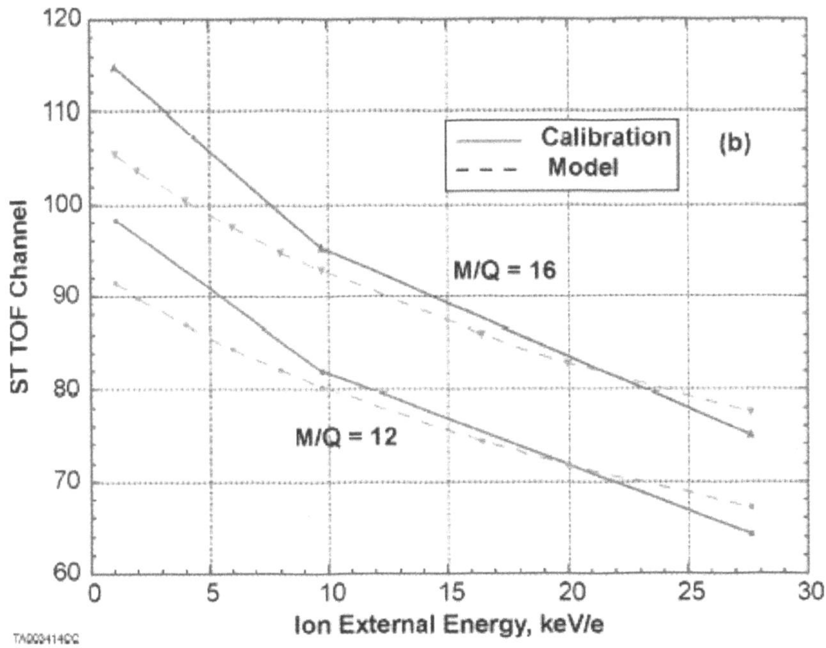

Figure 36. Comparison of the TOF peak location of negatively charged mass 12 and 16 ions as a function of ion energy external to the TOF system. The relatively simple model does not include all details of the IMS LEF fields and energy straggling.

7.1. Mechanical Design

Figure 37 shows the actuator (AC). In addition to the 3 features to note in the photograph in Figure 37 are the high-voltage cables running from the IMS high-voltage units 1 and 2 to the IMS sensor (the cylindrical shape in the background) and the micrometeoroid shielding placed on the outside of the DPU. Table XI summarizes mass and power properties of the CAPS subsystems.

To have compatible electron and ion measurements, the fields-of-view of all three sensors are co-planar in elevation (Figures 4 and 38a). In addition the centers of the IMS and ELS fields-of-view are co-axial whereas that of the IBS is offset by 32.0 cm to the right in Figure 38a.

The data processing unit (DPU) houses the greater part of the CAPS electronic circuitry, and forms the central mechanical platform of the instrument. The large number and complexity of circuits needed to service three sensors, the ACT, and spacecraft interfaces determined the physical size of the DPU. To give a simple example of the electronic complexity of CAPS, consider that each of the eight DPU circuit boards is 19.0 cm $\times$ 19.0 cm square making the total board area used for electronics $\sim$0.5 m^2. An additional $\sim$0.3 m^2 of electronics is located in the individual units for a total of 0.8 m^2.

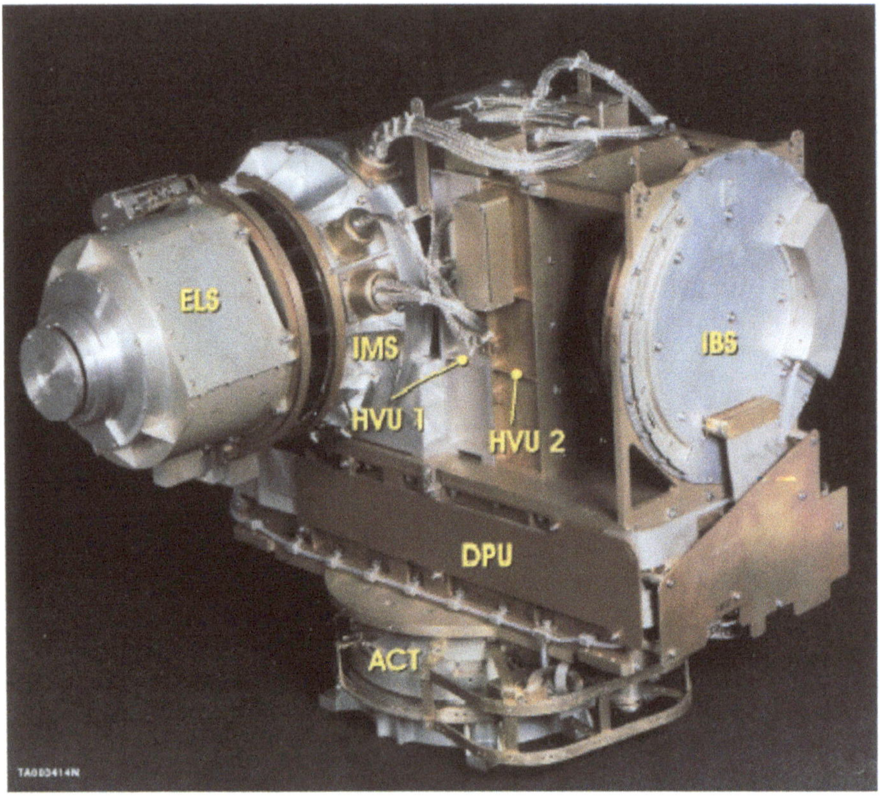

Figure 37. Photograph of the CAPS flight model prior to delivery to the Cassini spacecraft.

The presence of the ACT complicates thermal design because its bearings thermally isolate CAPS, which dissipates a maximum of 21 W, from the FPP. Therefore, the CAPS temperature is controlled independently of the spacecraft through a dedicated radiator (on the $+X$ surface of the DPU) and electrical heaters in the instrument. The CAPS radiator is sized so that operations during the mission result in average instrument temperatures between -20 C and $+30$ C.

The bottom of the DPU structure is attached to a mechanical transition plate that provides attachment points connecting the DPU to the ACT assembly (Section 7.4). Six mounting feet on the base of the ACT secure CAPS to the FPP. During all test and launch activities, whenever the ACT was operated, it was done so in an upside-down configuration (see Figure 3). After final mechanical properties were measured, the transition plate was drilled such that the center of mass of the rotating portion of CAPS was positioned at the center of rotation of the ACT. This technique made it relatively easy to meet spacecraft static and dynamic balance requirements (<0.090 kg m and <0.060 kg m^2, respectively), permitting CAPS to actuate during periods when optical remote sensing (ORS) observations and

TABLE XI
CAPS subsystem characteristics.

Subsystem	Mass (kg)	Power (W)	
		Peak	Average
IBS sensor	1.27	1.09	0.93
ELS sensor	1.40	1.90	1.50
IMS sensor	7.06	7.21	4.77
DPU	6.70	8.50	6.90
Cable and shielding	2.64	–	–
ACT	3.96	2.30	2.30
Totals	23.03 kg	21.00 W	16.40 W

spacecraft communications with Earth are taking place. Tests conducted during the cruise phase of the mission have shown that CAPS does not interfere with spacecraft pointing in any measurable way.

With six complex subsystems located externally to the DPU, 11 interconnecting cables comprising about 350 individual wires were needed outside the DPU structure. Conventional wire cables and associated connectors would have been too complicated, heavy, and bulky to be practical. In addition they would be susceptible to micrometeoroid damage and electromagnetic radiation leakage. Our approach was to design and manufacture a special set of cables made of flexible multi-layer printed circuits that were terminated by imbedded connectors. These so-called "flat" cables, which carried an outer protective shielding layer of copper, were easily bent around corners and secured to the DPU. This also made instrument integration simpler and safer than would have been possible with conventional cables. Once flat cables were implemented, a re-examination of the micrometeoroid hazard by the Cassini project indicated that cable shielding was too thin to prevent penetration by interplanetary or ring particles. Thus impact shielding was increased by introducing 1.5-mm thick aluminum "bumper" shields placed on the $-X$, $-Y$ and $+Y$ faces of the DPU (Figure 37). As an additional precaution, the Cassini project added two layers of quartz "beta-cloth" fabric beneath the spacecraft multi-layer insulation blankets that covered CAPS. With these precautions the probability of the cables or DPU being damaged by micrometeoroid impact was estimated to be less than 4% (most of this risk occurred during successful passage through the asteroid belt).

To avoid contamination by spacecraft photoelectrons, the ELS was mounted at the far end of the IMS collimator, which, in combination with CAPS location on the FPP, placed the ELS FOV as far from nearby spacecraft surfaces as possible. However, this mounting arrangement created a cantilevered inertial load on

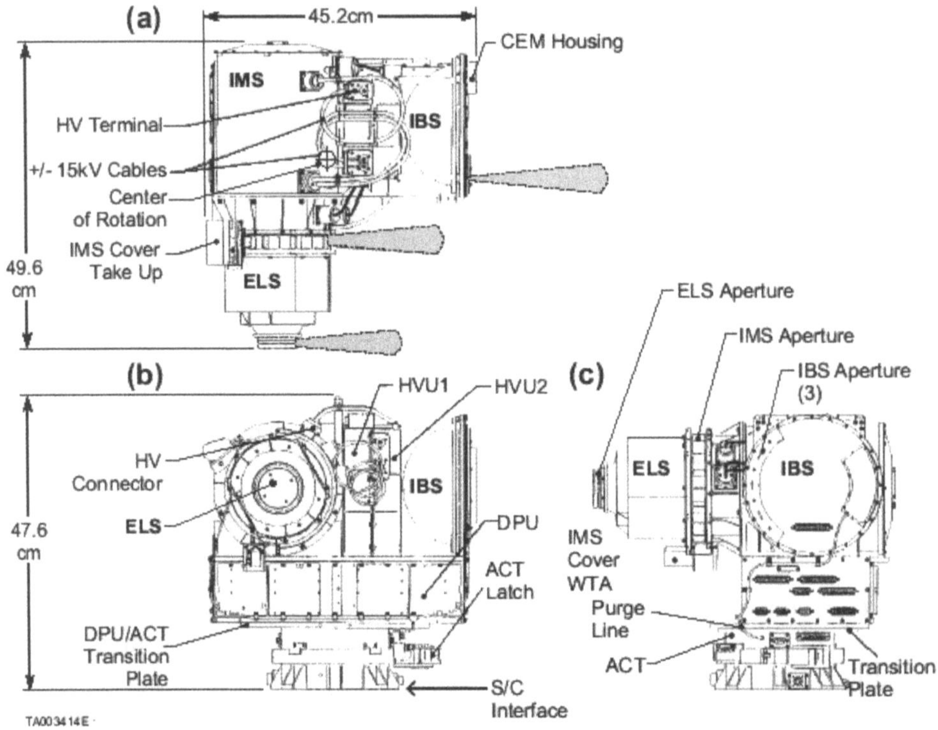

Figure 38. Orthogonal projection drawings of the CAPS flight unit showing mechanical details. Projections are in the spacecraft (a) *X–Y*, (b) *Y–Z*, and (c) *X–Z* planes assuming that CAPS is in the nominal "safe" position shown in Figure 3.

the cylindrical IMS housing. Careful structural analysis was necessary to avoid mechanical resonances.

During its nominal 11-year mission, electronic components will absorb an estimated cumulative radiation dose of 100 krads behind 1.0 mm of aluminum shielding or its equivalent. Shielding calculations performed with a factor of 2 design margin showed that the walls of electronics housings required a minimum thickness of at least 1.5 mm. Although most of the electronic components in CAPS are radiation hardened to the 100 krad requirement, some are not. Care was taken to place sensitive components near the centers of individual circuit boards and then locate the boards in a way that that made optimum use of the enclosures and surrounding spacecraft volumes to provide shielding. As a result, electronic components in CAPS meet the radiation design goal with margins of a factor of two or more.

Final integration of the CAPS Flight Unit on the Cassini orbiter took place at Kennedy Space Center in July 1997. At the time of integration the interface between CAPS and the spacecraft MLI blankets was problematic, especially in areas near the junction of the moving parts of the ACT and the surface of the FPP. MLI blankets were fabricated very precisely in order to avoid snagging CAPS on the FPP. During one phase of flight operations, however, there is some evidence that CAPS did snag on the MLI. For that reason the motion of CAPS has been restricted to $+104^\circ$ to -80° in azimuth, rather than the planned $+104^\circ$ to -104° range.

Electrostatic Cleanliness. Differential spacecraft charging is a serious concern because all three CAPS sensors are capable of measuring particles with energies slightly below 1 eV. To ensure their performance, sensor apertures were coated with conducting black materials (copper oxide or copper sulfide deposited electrochemically), which, because of their microscopically rough surfaces, help reduce both particle and UV scattering by factors of 100 to 1,000. External aperture surfaces are bare aluminum, a sufficiently good conductor to allow adequate charge dissipation. During spacecraft integration concerns over the conductivity and uniformity of indium-tin-oxide MLI surfaces on the orbiter exterior prompted the CAPS team to request that carbon-impregnated Kapton®, similar to that used on Voyager and Galileo, be placed around the instrument. As an additional precaution, MLI blankets near the CAPS apertures were stitched on with thin stainless steel wires grounded through contact with blanket material.

Contamination Control. Many sensor components (detectors, carbon foils, high surface resistance insulators and chemically blackened surfaces) are sensitive to chemical contamination and particulate debris that might be introduced during ground handling, testing, or launch. The CAPS detectors are all electron multipliers that depend on the electron yield of a thin layer of reduced lead-silicate glass. Surface chemical contamination by, for example, compounds found in circuit board coatings, reduces multiplier efficiency and gain. Water and other condensable molecules are easily adsorbed on multiplier surfaces only to outgas once in vacuum, potentially leading to HV breakdown. Carbon foils in IMS interact with incoming ions to emit secondary electrons in much the same way that multipliers do. The foils are hygroscopic and easily form van der Waals bonds with polar compounds such as water (e.g., during launch). High-voltage insulators and high-ohmic resistors ($\sim$10 G Ω) are also concerns. Thus prolonged exposure of internal sensor surfaces to water vapor and other volatiles can create large internal gas loads that need to be avoided. For this reason, and because the carbon foils are sensitive to acoustic vibration during launch, the IMS was fitted with a retractable cover released several months after launch. Operation since then has shown no signs of contamination or foil loss.

7.2. Electrical Design

Figure 39 is a schematic overview of the major components of the CAPS electrical system and its command, control and data flow aspects. The system is made up of

20 individual circuit boards located in the DPU (8), IBS sensor (2), ELS sensor (4), IMS sensor (3), IMS HVU1 and HVU2 (4) and the ACT (1). Functionally the different boards provide each sensor with independent command and control; high voltages (9 independent supplies), high-voltage safe/arm protection and voltage monitoring; data acquisition, processing and compression; and data formatting and transmission to the spacecraft. Control of the entire CAPS instrument depends on a hierarchical network involving three microprocessors, all of which are radiation hardened PACE MIL-STD 1750A′s. CPU1 handles all CAPS command, control and data interface traffic with the spacecraft via a 1553 interface protocol. It also controls IBS, ELS and ACT subsystems and the CPU2. The latter in turn controls IMS data acquisition and processing functions carried out by the TDC and SAM, respectively. Table XII lists the eight DPU electronics boards and their functions.

Spacecraft Interface. The spacecraft data and command interfaces are primarily through the bus interface unit (BIU). The BIU generates real-time interrupt (RTI) signals sent to CPU1 telling it to poll for messages from the spacecraft and to transmit any waiting telemetry packets.

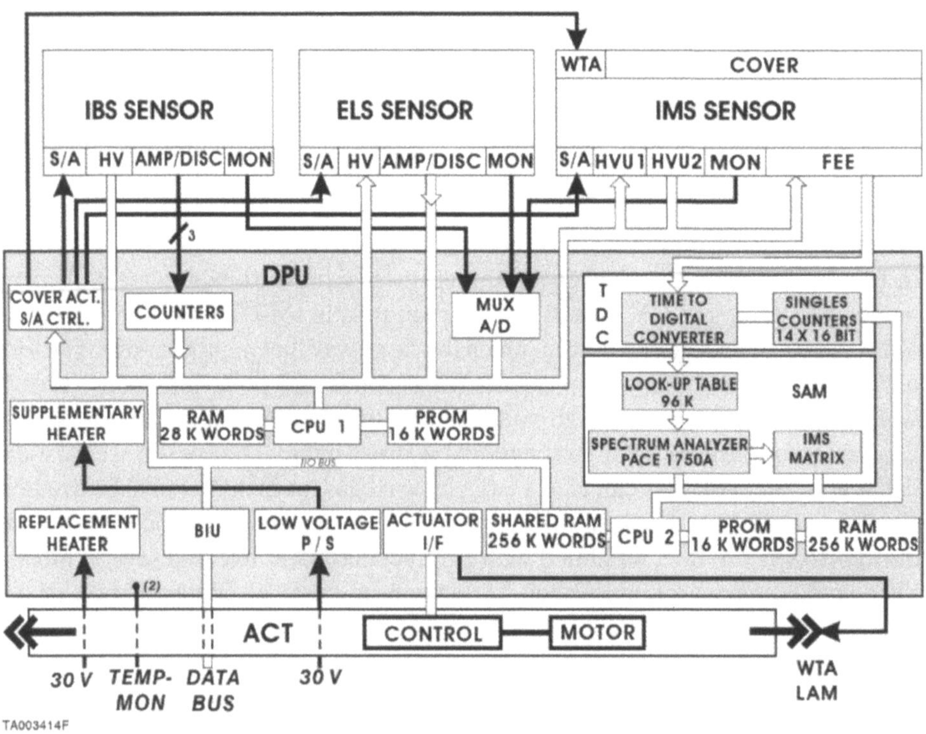

Figure 39. Schematic high-level CAPS electrical block diagram. Sensors are represented by the three boxes in the upper part of the figure, the DPU by the central shaded box, and the ACT by the thin box at the bottom.

TABLE XII
Primary functions of DPU circuit boards.

Board Name	Functions
1. Spacecraft Interface	Bus Interface Unit, ACT interface, IMS cover release, ACT release, heater control
2. Low Voltage Supply	Low voltage converters, +30 V filter, housekeeping
3. IBS/ELS Interface	Master timing, IBS and ELS command and data interfaces, IBS HV control, HV safe/arm
4. IMS Interface	HV control and safe/arm, analog data, power synchronization
5. TDC	TOF data acquisition, TDC and coincidence logic, FEE control, FIFO interface with SAM
6. SAM	Pace 1750A processor, TOF, TOF accumulator, and EVENT modules, local memory for SAM LUT's
7. CPU1	Pace 1750A processor, memory management unit; 256 kwords local memory; 16 kwords PROM; ELS, IBS, HVU1, HVU2, CPU2, ACT control functions
8. CPU2	Pace 1750A processor, memory management unit, 256 kwords local memory, 256 kwords shared memory, 8 kwords boot PROM, SAM contol

CPU1 and CPU2 Microprocessors. The CPU1 and CPU2 microprocessors are located on separate circuit boards together with associated memories. With reference to Figure 39, the CPU1 is the primary CAPS control and data processing unit. Its address and data control bus extend to the IBS/ELS interface board, the IMS high-voltage control board, and the BIU. The CPU2 is dedicated entirely to IMS data acquisition and processing as well as to control of the TDC and SAM. Both CPU1 and CPU2 can access 10^6 16-bit words of memory address space via individual memory management units. CPU2 contains an additional 2,56,000 words of shared memory accessible by CPU1.

Low Voltage Power System. The LVPS provides sensor analog and digital electronics with a variety of regulated voltages including ELS (± 8 and $+5$ V), IBS ($+5$ and -15 V) and HVU2 ($+5$ and ± 15 V). Each high-voltage supply is powered through individual current-limited $+15$ V lines controlled by separate solid-state power switches. The $+15$V lines are fused to prevent failure in any one circuit from damaging others.

IBS/ELS Interface Board. This board contains the master CAPS timing generator used for subsystem synchronization and data acquisition cycles. The timing signal is divided down to three basic periods: 7.8125 ms (the basis for IBS sampling, ACT micro-stepping, and timing interrupts to CPU1), 31.25 ms (ELS sampling interval), and 62.50 ms (IMS sampling interval). The master timing generator allows CPU1 software to synchronize CAPS operations to the scheduled repeat cycle of the spacecraft. This is accomplished when the software sets a bit in hardware to start the master timing circuit after the appropriate real-time interrupt signal has been received from the BIU. The master clock has the capability to be divided down still further to lengthen sensor sampling periods by factors of 2 to 8.

The ELS/IBS board also contains three 16-bit counters making up the IBS data acquisition interface. An 8-bit bus controls high-voltage supplies for the IBS detector and ESA. The ELS data acquisition interface consists of a 16-bit one-way bus to the DPU that reads out eight registers in which detector counts are stored (corresponding to the eight elevation sectors). In addition an 8-channel 8-bit analog multiplexer in ELS sends housekeeping status to CPU1. A similar bus operating in the opposite direction allows CPU1 to command ELS. In addition to these functions, once power is applied to the ACT, CPU1 sends a 16-bit word every 7.8125 ms (equal to the IBS stepping cadence) to control acceleration, scan, deceleration, and stop modes.

IMS HV Interface Board. As mentioned earlier, CPU1 controls operation of the HV supplies located in HVU1 and HVU2. The HVU1, which produces $\pm$14.56 kV for the TOF, is powered directly by the spacecraft +30 V bus through a redundant, optically isolated, current-limited switch controlled by CPU1. HVU1 and HVU2 are controlled independently by 8-bit command words. The HVU2 produces ESA and two MCP detector high voltages in much the same way as the IBS and ELS high-voltage supplies. In normal operation CPU1 writes a 14-bit word to the ESA stepping supply every 62.5 ms to select 63 pre-programmed ESA voltage steps. CPU1 also controls the LEF and ST MCP detector high voltages using separate 8-bit command words.

Two 16-channel analog multiplexers process analog monitors that go into the housekeeping telemetry data. Also on this board are separate SAFE and DISABLE control lines for all nine of the CAPS high-voltage supplies. The lines are routed to a SAFE/ARM connector (removed before launch) located on the board and accessible from the outside of the instrument. Its purpose was to provide visual assurance during ground operations that CAPS sensors could not be commanded inadvertently above 3% of maximum voltage.

Time-to-Digital-Converter (TDC) and Spectrum Analyzer Module (SAM). These two units are dedicated to converting raw IMS detector signals into TOF spectra and processing the data to identify ion species. They were discussed separately in Sections 6.5.2 and 6.5.3.

Grounding. CAPS is designed with single point secondary ground. The signal ground connection to chassis is located on the DPU mother-board close to the low voltage power supply. During testing after system integration it turned out that sensitive charge amplifiers near the ELS, IMS, and IBS detectors required additional grounding. As a consequence, the grounding concept was changed to allow additional connections between these signal grounds and chassis to optimize overall noise performance. The best performance was achieved by connecting signal-grounded internal cable shields to chassis at the sensor end of the interface cables. It should be noted that the isolation between primary and secondary power was not compromised by this arrangement.

Fault Protection. To prevent a failure in a subsystem from becoming catastrophic and damaging another subsystem or the entire instrument, all power lines exiting the DPU have been fused using dual redundant fuses conservatively rated for at least twice the nominal peak current. Custom-designed, current-limited solid-state switches protect power lines switched by the DPU, namely the ACT motor drive and latch, HVU1 and HVU2 converters, and supplemental heaters. Similar to the fusing philosophy, current limiting is set at twice the nominal peak current on a line.

In addition to fusing subsystem loads, other precautions were taken with the sensor high-voltage systems including SAFE/ARM controls implemented in software (in addition to hardware controls discussed above). Individual high- voltage converters must be armed and enabled by separate commands before high voltage will appear at the outputs. The more sensitive CAPS electrical circuits are protected from damage that may occur as the result of momentary ground loops associated with discharges by so-called "zap-traps." These provide a low inductance return path from the chassis structure to the origin of the high voltage, i.e., HV ground in the individual HV generators. Zap-traps consist of two parallel fast diodes mounted in opposite directions between HV return and structure.

The health and safety of the CAPS electrical system is monitored via ~90 housekeeping channels whose contents depend on instrument mode. During maintenance mode housekeeping data reflect primarily low-voltage power supply, temperature, and ACT status. During normal science, emphasis is placed on high voltage and system-status monitoring. The CPU1 contains a 12-bit ADC (Figure 39) that samples the HV output monitors, low-voltage supplies including total current drawn on the 28 V line from the spacecraft, the ACT position encoder, and the temperature at six locations within CAPS. To provide the required thermal environment, CAPS relies on two types of heating elements located in the DPU. Supplemental heaters are controlled by commands to CPU1 (Figure 39). They provide additional heat that may be needed when the instrument is operating in a reduced power state. Supplemental and replacement heaters are controlled by the spacecraft and are intended to maintain both the CAPS thermal environment and that of the FPP when CAPS is turned off.

7.3. Flight Software

7.3.1. *Software Description*

The primary functions of the CAPS flight software are to (1) maintain communication with the spacecraft telemetry and command systems, (2) control operation of the CAPS sensors and actuator, (3) generate data acquisition and telemetry products, and (4) carry out internal checks of system status. Figure 40 shows a detailed data flow diagram of the software control and data flow system.

The multi-processor hardware system (CPU1, CPU2, and SAM) distributes loosely coupled, fixed software tasks. The CPU1 is the "front end" of the data system and is the only processor that communicates directly with the spacecraft (Figure 40). The CPU1 collects data from all three sensors as well as housekeeping data, and formats it into "CCSDS" (Consultative Committee for Space Data Systems) standard transmission packets, which are then sent to the spacecraft command and data system (CDS) via the BIU. As can be imagined, CPU1 executes independently of CPU2 and SAM, which are slaved to the former. SAM in turn is slaved to CPU2. During operation the CPU2 extracts TOF data processed by SAM from a random access memory (RAM) shared between the two. CPU2 data are then communicated to CPU1 via another set of shared memory.

Spacecraft and instrument operational mode choices define the CAPS operating state, telemetry rate, data volume, data product type, and power allocation. As part of this scheme, the spacecraft operational mode and current physical data rate are communicated to CAPS once per second. CAPS telemetry is divided into two streams: science and housekeeping. The housekeeping stream has a nominal physical data rate of 22 bits/s dropping to 11 bits/s in maintenance mode. Science "logical" data rates can be varied by command from 16 kbps down to 0.25 kbps in factors of 2. However, the logical rates must always be smaller than or equal to one of two "physical" rates (16 and 2 kbps) set by the spacecraft CDS. Any data volume difference between physical and logical rates is filled with zero length

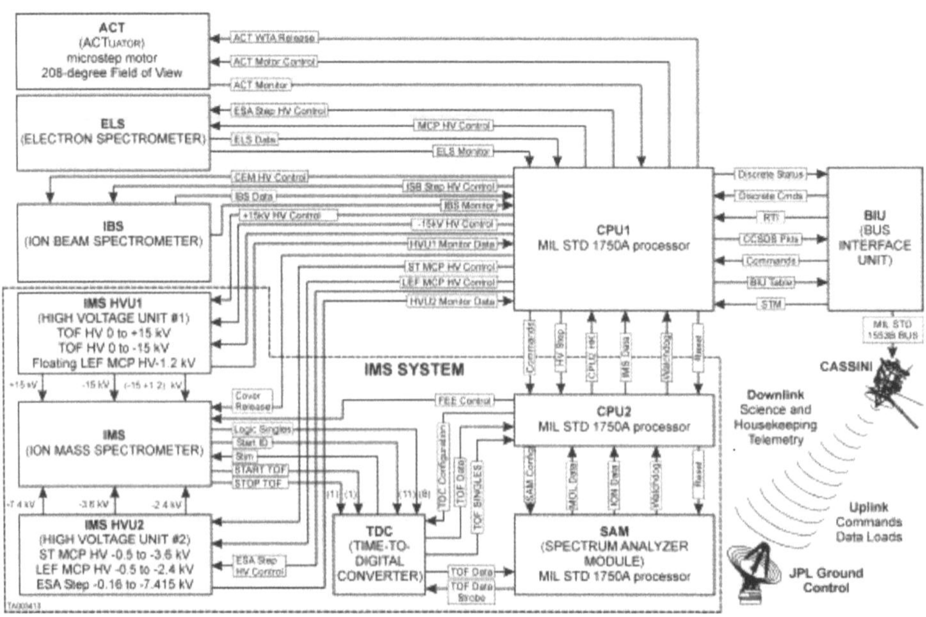

Figure 40. Detailed schematic of the CAPS data flow and command system.

packets (ZLP) so that the CDS retains the expected packet pick-up rate and is unaware of any differences in data volume. The ZLPs are recognized by the CDS and are not stored on the spacecraft solid-state recorder (SSR), thus allowing CAPS science data flow to be tailored to the allocated SSR volume without loss.

Commands or command sequences are uplinked to the spacecraft roughly every 28 days during the tour phase of the mission. Once uplinked they may be executed in real-time, stored in the spacecraft solid-state recorder (SSR), or stored in CAPS RAM memory as instrument-expanded (command) blocks (IEB′s). Command sequences containing time-ordered triggers for executing IEB's are stored in the SSR from which they are sent to CAPS. Real-time commanding is also available if necessary for time critical operations.

Data acquisition, high-voltage control, and spacecraft communications with CAPS (Figure 40) all take place at times determined by the spacecraft clock broadcast once per second; or with synchronization messages and real-time interrupts (RTI) sent once every 0.125 s. The CAPS hardware sample clocks are synchronized to the RTI signal during CPU1 initialization. Internally CAPS organizes data based on A- and B-cycles introduced earlier in Section 6.5.3.

7.3.2. *Software Operation*

CAPS flight software functions are split between PROM-resident and SSR-resident software depending on instrument mode. PROM resident code includes functions needed to implement maintenance, low power, and CPU2/SAM-Ready modes. Table XIII is a list of operational modes and their characteristics. The transitions and interrelationships between these modes are shown in Figure 41. When power is first applied, or when the DPU has been commanded to reset by the spacecraft, CAPS "boots up" from code stored in PROM and begins execution of memory checks, BIU initialization and RTI synchronization. It also determines the commanded operational mode and begins execution. In the initial "ON" state, code from the SSR is read to CPU1, which then initializes its own plus CPU2 and SAM operations, bringing all three processors to a "READY" state. From this the LOW POWER, NORMAL SCIENCE, or SLEEP modes can be reached. When CAPS is commanded to the "OFF" state it ramps down any high voltages that are on and then returns to LOW POWER mode before power is turned off.

Maintenance Mode. During parts of Cassini cruise phase the maintenance mode was run to exercise the ACT stepper motor and bearings once every 3 months during spacecraft periodic instrument maintenance (PIM) operations. No science data were taken. Only simple housekeeping packets containing ACT angular position as well as instrument current, voltage and temperature monitors were returned at 11 bits/s. During the first and third PIM periods following launch, wax thermal actuators released the ACT latch and IMS cover mechanisms, respectively.

Low-voltage Mode. The initial command from the spacecraft causes CAPS to boot up from PROM into a low-voltage state. Initially the instrument ignores any accidental commands to the ACT or high-voltage units as a safety precaution. Only

TABLE XIII
CAPS operational states.

	Power (W)		Status		Telemetry mode		Code source	
State	Average	Peak	ACT	HV	HK*	SCI*	PROM	SSR/Gnd
Maintenance	15.8	19.1	on	off	Maint.	ZLP	X	
Low voltage	12.5	12.5	off	off	Sci.	ZLP	X	X
CPU2/SAM ready	13.2	13.2	off	off	Sci.	ZLP		X
Sleep	13.2	13.2	off	on	Sci.	ZLP		X
Normal science w/o ACT	13.4	18.0	off	on	Sci.	Sci.		X
Normal science w/ ACT	16.4	21.0	on	on	Sci.	Sci.		X

*HK: housekeeping mode; SCI: science mode; Maint. data; sci. data; ZLP: zero length packet.

housekeeping is produced at 22 bits/s. From this mode CAPS can be commanded to enter the CPU2/SAM-ready mode (Figure 41), which is simply a transition state between PROM-based and RAM-based versions of the low-voltage mode. In the CPU2/SAM-ready mode, CPU1 releases the CPU2 processor, which then waits for program load commands from CPU1. Both CPU processors and the SAM are then ready to accept program loads in assisted load format (ALF) blocks from the SSR. After all three processors are loaded, CPU1 is commanded to begin execution while CPU2 is put back into reset. At this point CAPS is again in low-voltage mode, but code is now executing from instrument memory (instead from PROM) and CAPS can be commanded to enter normal science mode.

Normal Science Mode. Normal science is unique because it is the only one in which high voltages can be applied, the CPU2 and SAM are operational, and a full range of ACT motion is possible. Once CPU1 releases CPU2 from reset, the latter releases SAM and then initializes SAM and TDC based on a table controlled by ground command (SAM group tables were discussed in Section 6.5.3). If, during normal operations, the SAM or TDC configuration, or the SAM group table, are changed, then CPU2 puts SAM into reset.

In addition to controlling telemetry rates and other subsystems, CPU1 also controls sensor energy scans and scan rates. The scan table contents are set by ground command as are the rates. Normally ELS and IBS scans require 2 s whereas IMS requires 4 s. During normal science mode data are acquired in the following way: A-cycle data from IBS and ELS are accumulated and stored one-half of CPU1's double-buffered memory, while the other half of memory is being read out and processed. SAM data are gathered in the same way by CPU2 before being extracted by CPU1 during the next A-cycle. At the same time that A-cycle data are

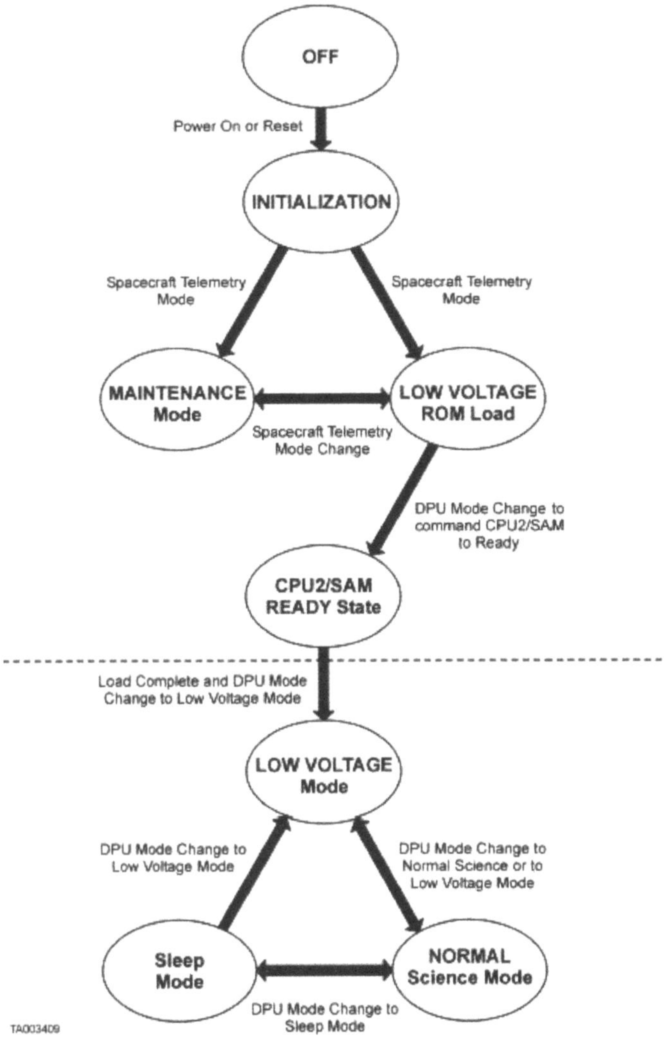

Figure 41. CAPS operational state diagram. Mode states above the dashed line are part of the CAPS initialization process; below the line are possible standard operational states.

acquired, CPU2 collects TOF data from SAM on the basis of a B-cycle (consisting of 8 A-cycles). Details of B-cycle data products were discussed in Section 6.5.3.

Sequential Event Mode. This mode is a diagnostic of the entire TOF system applicable only when CAPS acquires data at 16 kbits/s. In it SAM collects up to 16,000 TOF words sequentially at a fixed energy step over a period of two A-cycles. Each word corresponds to a single particle striking one of the detectors. During the next two A-cycles SAM computes ION data products corresponding to this data. The procedure is repeated for two more energy steps, acquiring 48,000 events in a single B-cycle.

Sleep Mode. Sleep is a reduced operations mode from which CAPS can quickly emerge (e.g., following the firing of the spacecraft's main engine during Saturn orbit insertion). In sleep mode the HVU1 voltages remain on at predetermined high, but safe, level (e.g., ±12 kV). All other supplies can be turned off. In addition CPU2 and SAM are held in reset and the ACT is parked and turned off. Only housekeeping data are produced.

7.3.3. *Fault Protection*

The CAPS flight software contains several features designed to detect and report faults and recover from them "gracefully." Among the most critical features, because it controls command and data transmission, is a watchdog timer that detects real-time interrupts (RTI) from the spacecraft. If one is missed, the timer generates an artificial RTI within CAPS. CPU2 also contains a watchdog circuit. If this expires, an interrupt is generated in CPU1 and an error reported in housekeeping. In the worst case, should the CPU1 watchdog expire, a reset of the CAPS system will occur automatically and a message will then be generated by the spacecraft telling the ground controllers that CAPS is in an OFF state. A second feature is on-board verification of all software and commands sent to CAPS. Commands are also checked for valid operations codes and appropriateness for the current instrument state. Detected command errors and their specific nature are reported in housekeeping. A third fault protection, mentioned earlier, is that CPU1 checks all high-voltage commands for validity before allowing execution to proceed.

7.4. Actuator

The primary function of the CAPS actuator (ACT) is to provide articulation of the three sensors and their fields-of-view over the lifetime of the Cassini mission (nominally 500,000 cycles). To this end the ACT rotates the instrument at a nominal rate of 1°/s around an axis parallel to the spacecraft *Z*-axis (Figures 3, 4, 37 and 38). Among the many requirements placed on the ACT was a lifetime of 5,00,000 cycles without failure. Although the maximum range of actuator motion possible is ±104 ° in azimuth centered on the spacecraft –*YZ* plane (Figure 4), the useful range is restricted to $-80^\circ \leq \psi \leq +104^\circ$, where at $\psi = 0^\circ$ the FOV is aligned with the spacecraft –*Y* axis. Most FOV encroachments that do occur are beyond −80° (Figure 5). At the $\psi = 0^\circ$ position all three sensors are aligned with the ORS instrument FOVs. At $\psi = +90^\circ$, they are aligned with the FOV of the INMS and the MIMI/CHEMS instruments. In spacecraft coordinates the rotation range corresponds to clock angles of 256° through 0° to 80°.

The actuator uses a 90 pole, two-phase, synchronous SAGEM motor that makes use of direct mechanical coupling. The motor is driven by a sine-cosine pair of phase voltages whose values are stored in the DPU and then sent to the actuator at a rate of 128 Hz via a parallel data interface. The phase voltage waveform table comprises one electrical cycle (4° of mechanical motion) with a resolution of 128 steps/degree. The same waveform pair is used for both directions of rotation as well as for acceleration and deceleration. Motor control electronics, DPU interface, and synchronization and timing logic are all located on a single circuit board housed inside the ACT (Figure 43). The DPU receives acceleration data that are used to determine the step size for the pointer of the phase voltage vector: If the received acceleration signal indicates that the actuator speed is decreasing due to changes in the detent torque of the motor, the nominal pointer increment of one step every 7.8125 ms is increased to two or three steps to increase motor torque and thus compensate for deceleration. Correspondingly, acceleration is compensated by not incrementing the pointer or even by taking a step "backwards" in the table.

Ideally the cumulative sum of step corrections should be zero. Due to possible offsets in the acceleration analog-to-digital converter this might not be the case. A cumulative error would cause the rate of rotation to deviate from the nominal 1°/s. To prevent this, a second control loop in the DPU integrates the step corrections and adds or deletes waveform pointer steps to keep the sum of corrections at zero. Because of Cassini's stringent pointing and stability requirements, ACT motion was designed to be mechanically very smooth. Scan motion is controlled by a direct-drive stepper motor and custom designed control electronics. Active position sensing, feedback and acceleration/deceleration ramps help further smooth its motion. A resistive encoder measures angular position which is fed as an analog signal to the DPU where it is differentiated twice, producing a signal proportional to angular acceleration, which is then used to control the motion.

The endpoints of the sweep are commandable by CPU1, which generates detailed micro step values sent to the motor control unit. Once the ACT is within 12° of the end point of a sweep deceleration begins. After reversing direction, the motor accelerates out to 12° and then sweeps back the other way in windshield wiper fashion. Pointing accuracy during this motion is $\pm$ 0.5°. By using the full ACT range combined with the ELS and IMS FOV, a 160° $\times$ 180° swath of the sky can be scanned approximately every 3 min, independent of spacecraft orientation.

The ACT also serves as the mechanical platform for CAPS and, during spacecraft ground testing and launch, holds the instrument "upside-down" on the spacecraft pallet (Figure 3). During launch the forces on the ACT reached 28 g's. These and other special requirements made the motor unit design very demanding. In addition to a minimum lifetime of 500,000 cycles, thermal gradients between CAPS and the spacecraft pallet pass through the bearings but must not cause significant increases in friction, or otherwise damage them. Finally, the ACT, which contains magnetic materials, had to meet a stray field requirement of 2.5 nT at 1.0 m, requiring the housing and bearing materials to be non-magnetic. One final important feature is a hollow shaft through the motor that allowed a flexible purge tube to be run from the spacecraft purge system up through the DPU to the sensors.

An important element of the ACT is bearings and bearing flanges, which are fitted together precisely to meet high concentricity requirements on their relative motion

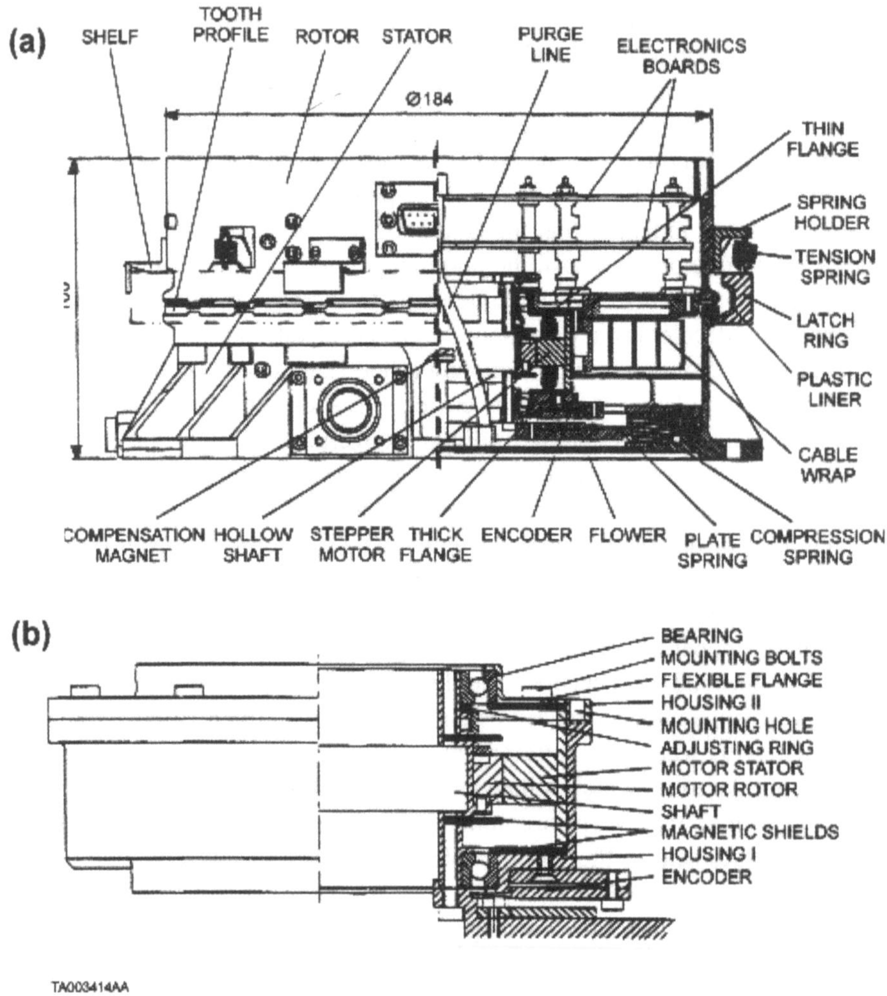

Figure 42. (a) Elevation view (left side) and cross-sectional view (right side) of the actuator. (b) Elevation and cross-sectional views of the ACT motor assembly.

(Figure 42). Because of residual magnetic field requirements there were only a limited number of materials that could be used. Aluminum alloys were avoided because of their high coefficient of thermal expansion. Moreover, common bearing steels are magnetic. We therefore chose instead AISI 316 steel as housing material and very precise non-separable angular contact ball bearings made of copper–beryllium alloy. Another important magnetic component of the ACT is the SAGEM hybrid stepper motor that provides 300 mNm of torque while requiring only 65 mA of current. Stray fields from the stepper motor were controlled by mounting μ-metal shields on mating housing parts and adding a small compensation magnet on the

Figure 43. Photograph of the ACT and associated launch latch mechanism (structure on right hand side of ACT) with the DPU mounting flange removed.

bottom of the housing to cancel the motor's first-order dipole field. The residual magnetic moment of the shielded ACT was measured to be 4 nT at 1 m slightly exceeding the requirement for 2.5 nT.

Bearing lubricant and rolling surfaces easily suffer severe damage if the balls do not roll but rather skid along the ball track. Therefore it was essential to ensure sufficient bearing preload and lubrication (ion plated lead coatings). Based on life testing the bearing preload value was set at 35 N while the motor friction torque was measured to be $\sim$5 mNm.

The ACT bearings could not carry CAPS launch loads, which are transferred through the ACT structure to the latch clamp ring. The latching device can be characterized as a Marman or V-band clamp (Figure 42) that works by pressing the ACT rotor and stator rigidly against each other. A plastic serrated ring made of PEEK (poly-ether-ether-ketone) was placed between the clamp and the ACT parts to prevent rotation under vibration loads when the ACT was latched.

The CAPS pointing direction is measured with a resistive encoder mounted between the motor housing and a non-rotating flange. The encoder voltage is converted to a digital signal used to adjust the pointer in the motor phase-voltage table. This arrangement results in a velocity controller that damps mechanical oscillations (caused by the proximity of the main torque ripple component of the motor at 1 Hz), to the resonance frequency of the mechanical system at 1.3 Hz. Another feature is that two limit switches, which give a signal at hard stops located at $\pm 110^{\circ}$ with respect to the sweep center, can be used for position calibration.

When being latched the rotor moves 2 mm downwards (see right cutaway view of Figure 42a) and the motor assembly moves with it. Compression springs keep the upper bearing preloaded to take up vibration loads on the shaft and flange during launch. The lower bearing carries no external loads. When unlatched, the rotor, shaft and flange all move back to their upper position and the flange rests against the ACT stator structure. In this way the latch compression springs do not load the bearings or motor assembly in the released state.

8. Planned Observations, Instrument Modes and Operations

8.1. Planned Observations

The Cassini orbital tour offers the opportunity to study essentially all regions of Saturn's magnetosphere and the interactions between the magnetosphere and Titan, icy satellites and ring particles. However, spacecraft orientation and limited data volume will prevent CAPS from obtaining the highest quality data at all times. In this section, we describe the main limitations placed on CAPS measurements and our plans to maximize scientific returns despite these limits.

One method of increasing the scientific return is through the use of magnetospheric and other models that can be updated as our knowledge base increases. Despite the fact that Saturn's magnetosphere is known only through three spacecraft encounters supplemented by Earth-based observations, enough information exists to create simple models that serve as an initial guide for planning observations. For now, a limited number of science models are available. A number of workers have addressed the problem of modeling the shape and plasma populations found in the saturnian magnetosphere. Among them are Richardson and Sittler (1990), Richardson (1992, 1995), Maurice *et al.* (1995, 1996a, b), Richardson *et al.* (1998), Slavin *et al.* (1983), and Behannon *et al.* (1983). Connerney *et al.* (1983), Davis and Smith (1990) and Maurice and Engle (1995) have constructed magnetic field configuration models. Finally, Maurice *et al.* (1996b) have constructed a usable model of the saturnian magnetosphere capable of giving the probable locations of saturnian plasma boundary features such as the magnetopause.

Reliable determinations of parameters such as ion density and flow velocity may not be possible when the peak of the velocity distribution is outside the instrument's roughly hemispheric field of view. Poor pointing of this sort will be a frequent occurrence during the mission because the spacecraft's remote sensing instruments are body-fixed, requiring spacecraft turns to observe and track various objects within the saturnian system. The orientations required for these remote sensing observations may or may not allow CAPS to observe the peak of the ion velocity distribution. More often than not, the pointing requirements are compatible, but only with an appropriate choice of the spacecraft's secondary axis (e.g., the rotation angle about the boresight of the remote sensing instruments). Spacecraft orientation is further restricted by the need to keep the Sun from illuminating the infrared instruments' radiators. Available data volume will also limit CAPS measurements. The spacecraft will, on average, return 1.75 Gbits of data per day throughout the tour. If CAPS operated at its full, 16 kbps data rate, it alone would generate 1.38 Gbits of data per day. Given that the Cassini spacecraft carries 12 instruments, CAPS clearly will have to operate on average at significantly lower rates for much of the mission.

To manage the restrictions on pointing and data volume, the Cassini MAPS (Magnetospheric And Plasma Science) instruments have adopted a strategy that mixes low-rate survey measurements with coordinated, focused observing campaigns. The survey provides a continuous, although limited, dataset covering the entire mission, whereas the campaigns provide full quality data in regions of interest. The MAPS instruments have planned various campaigns to study solar wind control of Saturn's aurora, magnetospheric boundaries, the magnetotail, Titan, the Titan torus, the inner magnetosphere, icy satellite and E ring interactions with the magnetosphere, and low-altitude crossings of auroral field lines. Table XIV shows the amount of time and the locations the spacecraft will be in various parts of the magnetosphere, as well as the anticipated times that will be part of MAPS campaigns. Figure 44 shows the arcs of the spacecraft trajectory corresponding to campaign periods. For orbits 0 through 22 (orbits with an apoapsis on the dawn side of the planet) the table and figures represent campaigns in the mission's integrated science operations plan. The science operations plan for later orbits is still under development at the time of this writing.

In survey mode, the CAPS instrument will alternate between 500 bps and 4 kbps data rates. Normally, there will be a repeating, $\sim$30-min pattern of six B-cycles at 500 bps followed by a single B-cycle at 4 kbps. This strategy is intended to provide a dataset without gaps, and which includes in the worst case at least one high-rate observation every $0.3R_S$ along the spacecraft's trajectory. During the various MAPS campaigns, CAPS will operate at data rates between 4 kbps and 16 kbps. The exceptions are solar wind/aurora campaigns, at times when the spacecraft is very far from Saturn and most probably in the solar wind. In these cases, CAPS will act as a solar wind monitor and use a specialized 2 kbps mode, which emphasizes the IBS rather than IMS ion data.

Difficulties resulting from unfavorable spacecraft orientations can be solved in several ways. When remote sensing instruments are taking data, the quality of the pointing for CAPS can be improved by selecting the secondary axis or by identifying which of several potential targets is more favorable to CAPS (e.g., observations by the remote sensing instruments of one ring ansa rather than another will frequently make the difference between good and bad pointing for CAPS.) A drawback of "riding along" in this manner is frequent turning and re-orienting of the spacecraft

TABLE XIV

Characteristics of the saturnian tour.

L shell crossings	Number in tour	Number in MAPS campaigns
Titan ($L = 20.25$) w/ |latitude| < 30°	124	37
Enceladus ($L = 3.94$) w/ |latitude| < 30°	69	22
G ring ($L = 2.72$) w/ |latitude| < 30°	14	8
"aurora" $L = 15$ w/ r < $5R_S$	22	13
Time in regions of the magnetosphere	Days in tour	Days in MAPS campaigns
Solar wind $r > 40\ R_S$, $0600 < LT < 1800$	277.64	107.56
Near magnetospheric boundaries $25 > r > 40\ R_S$, $0600 < LT < 1800$	313.19	195.77
Magnetotail $r > 25\ R_S$, $LT < 0600$ or $LT > 1800$	171.95	106.45
Titan Torus $10 < r < 25\ R_S$, |latitude|< 30°	374.34	62.53
Inner magnetosphere/Icy satellite torus $3 < r < 10\ R_S$, |latitude|< 30°	79.64	16.06
Inner magnetorphser/ring ionosphere $r < 3\ R_S$, |latitude|< 30°	1.67	0.95
Polar/auroral regions $r < 5R_S$, $L > 10R_S$	2.49	1.61

as remote sensing instruments observe different targets. This will complicate data analysis, but we do not expect it to affect data quality. When riding along with remote sensing observations is not possible, CAPS will occasionally act as the prime instrument and control spacecraft orientation. These periods will be spent either at a fixed and favorable orientation, or, around apoapsis periods, rolling the spacecraft. In addition, the spacecraft will spend an average of 9 h per day downlinking data to Earth. During these periods, the spacecraft will typically roll so that the CAPS sensors and other particle instruments can observe nearly 4π sr of velocity space. Since these rolls have a ~30-min period, they are not desirable when plasma conditions are likely to change rapidly. In the inner magnetosphere, such rolling downlinks will frequently be replaced with downlinks using a fixed orientation agreed upon by the particle instruments.

8.2. Operating Modes

Depending on mission phase and spacecraft power resources, CAPS will automatically go to one of two low-power states that operate out of PROM from which it can be commanded into one of three other states that operate out of software loaded from the ground or from the SSR. These states are discussed in Section 7.3.2. In only one of these (normal science) is CAPS able to acquire scientific data. In the sleep state, detector and ESA voltages are set to zero and the ±14.56 kV

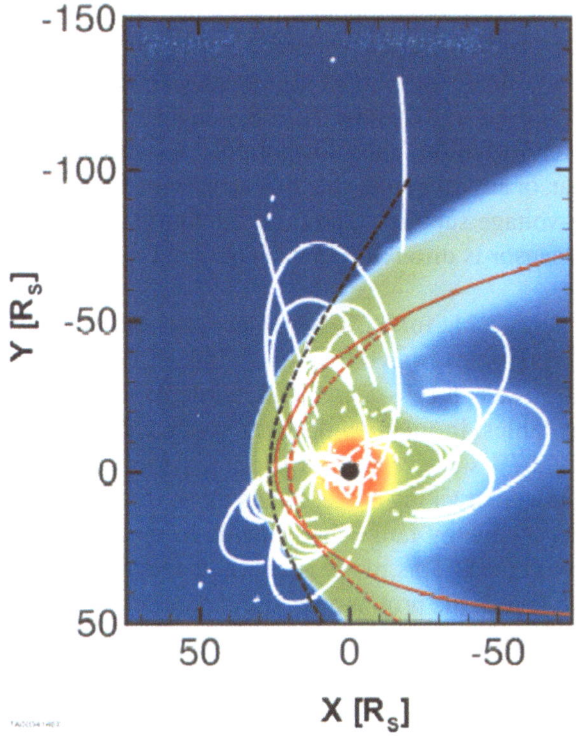

Figure 44. Cassini spacecraft trajectories corresponding to MAPS campaign periods during tour. The trajectory is overlaid on the results of a MHD simulation (Hansen, 2001) and model magnetospheric boundaries. Colors show density predicted by the MHD simulation, and the solid red line shows the magnetopause in this model. Dashed black and red lines show a Voyager-based model bow shock and magnetopause, respectively (Slaven *et al.*, 1983.)

voltages lowered to some safe value. The IBS and ELS sensors can be brought out of this state rather rapidly (~15 min). The IMS requires ~2 h. None of the other CAPS states incorporate high-voltage operations, nor can they generate science data (see Table XIII for operational modes). All spacecraft operating modes currently planned for tour allow CAPS to remain on and operating. Use of the CAPS sleep mode is expected to be limited to activities associated with Saturn orbit insertion and the Huygens probe mission.

Within the normal science state CAPS is capable of performing a wide range of table-driven programs devoted to science data gathering. As part of the science planning process, observational modes are specified that control the range of velocity space covered by the sensors, the rate at which velocity space is scanned, the conditions under which IMS data will be taken, and the data products to be returned from the three sensors. Voltage tables that control each sensor's ESA, as well as the range of ACT angles to be scanned, can be independently programmed thus controlling the ranges of velocity space covered by CAPS.

The three ELS voltage tables mentioned earlier are fixed. Those of IBS and IMS consist of 512 and 64 entries respectively that can be filled with step values corresponding to a desired voltage scan. As explained earlier, the step values for IBS and IMS may be chosen from among 12,288 micro-steps corresponding to digital to analog converter resolution. Thus it is possible to create tables corresponding to logarithmic, linear, or any other placement of steps within the allowed voltage range of the ESA high-voltage supplies. The time resolution and therefore the stepping interval for each sensor is different (Table I) and tied to the IMS sampling rate of 62.50 ms.

The default energy table for ELS, which controls the sweep generator in 16% decrements, covers 63 steps from 0.6 eV to 28.25 keV. For IMS, the default will be 63 steps from 1 to 50 keV in 16% decrements, while IBS will use a variety of mode-dependent energy tables. In solar wind search mode, IBS will sweep from 150 eV to 10 keV in 255, 1.67% steps. When the instrument identifies the proton peak, IBS will shift to a solar wind tracking mode. In this mode, IBS sweeps 127, 1.67% steps every second over an energy range that extends from one half to three times the energy of the proton peak. This range is adjusted every 256 s, on the basis of the previous 256 s of data. If the proton peak shifts outside the observed range, or if a change in spacecraft attitude obscures the solar wind, IBS will return to solar wind search mode. In magnetospheric scan mode, IBS will use two interleaved, 255-step sweeps, from 10 eV to 50 keV in 3.37% steps. The interleaved sweeps can be combined on the ground to give 1.67% energy resolution at a lower-time resolution. Custom, 255 step sweep tables will be used during close satellite encounters.

The range of actuator motion will be a compromise between time resolution and angular coverage. Sweeping through 180° (close to the maximum range) requires approximately 200 s, while sweeping through a 24° range takes only 48 s. The actuator may also be held at a constant position, producing a two-dimensional cut of velocity space at the 2- or 4-s energy sweep period of the sensors. This is the planned mode of operation during icy satellite encounters, which are very rapid (e.g., during an Enceladus encounter, the spacecraft crosses the satellite's geometric wake in under 50 s.)

Another important operational consideration is selecting the seven ion species reported by SAM. The SAM extracts spectra of up to 32 atomic species, of which up to seven are reported and returned in the CAPS data. Each ion group table contains 16 possible, 7-species combinations. Selection of ion species will be based on analysis of, and experience with, the IMS data, particularly the abundance of species indicated by B-cycle TOF data. The initial selections early in the mission will be based on theoretical models and earlier Voyager measurements. In addition, multiple group tables can be stored in the spacecraft SSR and loaded into SAM. This would allow group tables customized to different parts of Saturn's magnetosphere to be used without the need for uplinking new data tables.

The mix of data products placed in the telemetry stream by CAPS is determined by telemetry rate and is discussed in detail in the next section. Solid-state recorder memory on the spacecraft will be allocated on a day-by-day basis based on the detailed Cassini science plan. Depending on the CAPS allocation and particular measurement objectives during the day, one or more data rates and their duration are chosen to produce the required volume.

In addition to the normal science mode, CAPS has an engineering mode in which multiplier gains are checked, pulser stimulation signals are used to test and calibrate detector signal chains, and detector background is measured for an extended period. Before reaching Saturn the engineering mode will be operated as frequently as cruise data volume and pointing restrictions allow. After reaching Saturn this mode will be part of a periodic instrument calibration occurring approximately once every 50 days.

8.3. Data Products

At the maximum CAPS telemetry rate of 16 kbits/s all data products coming out of the science and calibration modes can be accommodated without the need for compression. The exception is the compressed extraction of ions by the SAM algorithm, and a semi-logarithmic collapse of all data words. The collapse replaces 16-bit data words stored in the DPU with 8-bit data words to be returned to Earth. For small data numbers, the 8-bit values are equal to the 16-bit values, but for higher values the scale is logarithmic. A similar 32- to 16-bit compression is used for TOF data. As a result of this compression, the uncertainty in the higher data numbers is roughly ± 0.015N rather than the statistical $\pm N^{1/2}$. No attempt was made to carry out more exotic on-board compression routines such as moment calculations or image-like compressions, with the exception of a run-length compression of the sparse, IBS data.

The contents of the CAPS data products at 16 kbits/s are distributed among the three sensors, the ACT and housekeeping channels as shown in Table XV. Data products are organized along A-cycle (32.0 s) boundaries. Acquisition and formation of B-cycle data products is more complex than the A-cycle process: The CPU2 extracts TOF data in the form of 512 channels each of ST and LEF data. In the default mode, adjacent energy steps are sampled to produce 2RES $\times$ 32E $\times$ 512TOF = 32,768 words. In the standard CAPS telemetry mode of 16 kbits/s, each word of B-cycle data is summed over 8 A-cycles, whereas for some lower rate modes it is summed over 16 A-cycles (i.e., the B-cycle is 8 A-cycles long for the 16 kbits/s mode and 16 A-cycles long for these other modes).

Most CAPS data products are generated at lower data rates by collapsing (summing) the 16 kbps data over adjacent energy, elevation and/or azimuth bins. In addition, snapshots (uncollapsed subsets of the 16 kbps data) may be included. The subset of data included in the snapshot can be determined on the spacecraft, so that the snapshot contains the peak of the velocity distribution. Leaving out certain products entirely produces the smallest possible datasets. The modes used by the CAPS

TABLE XV

Telemetry products in 16 kbits/s mode.

Product	EQ	EL	AZ	MQ	LOG	TOF	Total* bytes
	Data channels						
ELS	63	8	16	–	–	–	8068
IBS	255	3	16	–	–	–	12244
IMS ION	63	8	8	7	–	–	28288
IMS TDC LOG	63	–	8	–	4	–	2020
IMS TDC SNG	63	8	8	–	–	–	4036
IMS TOF LEF	32	–	–	–	–	512	16388
IMS TOF ST	32	–	–	–	–	512	16388
ACT	–	–	–	–	–	–	68
HOUSEKEEP	–	–	–	–	–	–	170

*Totals for each product include 4 bytes of fiducial information.

Acronyms: EQ: energy/charge, MQ: mass/charge, EL: elevation, LOG: logical, AZ: Azimuth, TOF: time-of-flight.

instrument were revised prior to the Cassini Jupiter encounter to include 0.25, 0.5, and 1 kbps rates, in addition to the original 2 and 16 kbps modes. These modes were further revised before reaching Saturn, to add 4 and 8 kbps modes and incorporate experience from analysis of the Jupiter data. Other data products that can be included as options (at the expense of sensor data) are memory readout of control tables for SAM and ESA stepping. Sequential event data that are used to verify IMS operations can also be included.

8.4. Ground Operations

In an attempt to reduce overall mission costs, the Cassini project has embraced the concept of distributed ground operations. For so-called principal investigator (PI) instruments this means that responsibility for instrument commanding and health and safety lies with the instrument teams and, in particular, with the PI institution. As a practical matter, the PI institution becomes an extension of the Cassini project ground system.

Ground system operations can be divided into two categories: Generating commands for uplink to the spacecraft and handling and processing data produced by the spacecraft during downlink.

Ground operations begin with a sequence-by-sequence list of measurement objectives developed by drawing on the Cassini science operations plan (SOP) and elaborations of the CAPS science objectives. The SOP describes spacecraft orbit, attitude, and allocation of resources such as power, pointing constraints, and telemetry. This plan was developed prior to reaching Saturn, and consists of conflict-free, flyable sequences. Later, if no significant changes are required, sequences will be taken from the SOP, updated, re-checked and sent to the spacecraft. For CAPS, the most important part of updating these sequences will be the revision or insertion of instrument-internal modes. This is the stage where experience from previous observations can be used to improve our choices of actuation angles, ion selection index, etc. In practice, significant changes to the SOP are likely. Unexpected events, such as satellite ephemeris changes or downlinks that cannot be scheduled at exactly the planned time, will force some changes and modification of the sequences. Changes in response to new discoveries will be highly desirable. The substantial effort required to develop the SOP prior to reaching Saturn is intended to allow for these changes. With flyable sequences in hand, the instrument teams can focus their efforts on science and on implementing desired changes.

Downlink processing begins when science and housekeeping data from CAPS, together with auxiliary data (spacecraft ephemeris, attitude, etc.), are extracted from the Cassini project data-base through the CAPS science operations and planning computer (SOPC) located at SwRI. Level 0 housekeeping data are immediately examined for any out-of-limit values that might affect CAPS health and safety. In the event that limits are exceeded, a contingency plan provides options for recovery from the fault. Non-emergency conditions that may require adjustments to system operating parameters will be incorporated in the normal command uplink cycle. Real-time operations can also be accommodated through the project's sequence virtual team (SVT) process. If necessary, emergency operations can be invoked in which commands will be sent to CAPS in real time. This same routine also checks the predicted against the actual instrument state vector to verify proper operation. Level 0 science data are also checked for trends that might indicate unanticipated changes in sensor operation such as shifts in detector gain or the appearance of unexpected sources of detector background.

In tandem with the ongoing limit checking process, level 0 data are decompressed, checked for quality, and formatted to create the CAPS level 1 products. Auxiliary data are collected into a parallel stream with the same file structure so that the two are compatible. Level 1a data are extracted and then converted to physical units, such as flux and velocity space density, which then constitute level 1b products. Finally, the derived level 2 data products, such as density, flow velocity and temperature are calculated. These data, together with the software tools needed to generate data displays, are then made available to the CAPS and Cassini teams. The interface with the data analysis system will be Web-based to maximize accessibility and opportunities for scientific study of the data. This interface allows wide access to a set of production survey tools, including various types of spectrograms and line plots, as well as orbit analysis tools. Survey plots (spectrograms and line plots) will be routinely produced as image files and stored on-line for rapid sequential and random access to facilitate browsing, surveys and comparisons with other Cassini observations.

9. Initial Operations and Performance

9.1. CAPS Initial Results During Earth Flyby

On August 18, 1999 the Cassini spacecraft flew past the Earth on its way to Saturn. While the purpose of the encounter with Earth was to gain velocity, it also opened the possibility for operation of the CAPS as calibration and test exercises. As Rymer *et al.* (2001) have described, the encounter pointing geometry had the spacecraft -*Z*-axis pointed toward the sun. In this orientation the center of the three sensors' elevation FOVs pointed orthogonal to the direction of motion. The fans (Figures 3 and 6) viewed forward (sunwards) as well as in the spacecraft direction of motion (i.e., the spacecraft ram direction). The latter proved favorable for viewing low-energy plasmas in the Earth's magnetosphere (a situation similar to what occurred during orbit insertion at Saturn). Although it is anticipated that the solar wind will be less intense at Saturn ($n_{SW} \sim R^{-2}$ where R is distance from the sun), other features inside the bow shock and magnetopause should be of broadly similar intensity. Thus Earth encounter presented a unique opportunity to test CAPS performance in a not too dissimilar environment from that of Saturn.

In Figure 45 we present energy–time spectrograms from all three CAPS sensors. The ordinate is particle energy, the abscissa is in units of hours UT, and color corresponds to detector counting rates. The upper panel corresponds to ELS data, the middle to IMS, and the lower to IBS. Cassini sliced through the near-equatorial magnetosphere near the sub-solar point. The encounter itself was unique in that the spacecraft speed of 16.1 km/s was about three times that of comparable orbiting spacecraft. With reference to Figure 45, at 0000 UT ELS and IMS measurements indicate that the spacecraft is clearly in the solar wind. At 0151 UT ELS shows most clearly and with the best time resolution that Cassini crosses the Earth's bowshock and passes into the magnetosheath (note the rapid increase in electron flux and the vertical barred pattern due to the motion of the CAPS actuator, indicating strong plasma flow in the magnetosheath). At 0225 UT Cassini crossed the magnetopause into the Earth's magnetosphere proper (sharp drops in electron and ion fluxes). Vertical yellow bars at ~0250 to 0300 are caused by penetrating particles from the Earth's radiation belts that enter the ELS and IBS detectors. IMS data require coincidences between the two detectors, which greatly reduced its background rates. From 0306 to 0351 the spacecraft was in the plasmasphere. What the figure shows is relatively cold (~few eV) plasma rammed into the ELS and IMS sensors when the spacecraft was inside the plasmasphere (red area below 10 eV from 0306 to 0351 UT) on the inbound and outbound trajectories. Because the IBS is set at a higher energy range, corresponding to solar wind proton energies >200 eV, it does not observe the plasmasphere. It is worth noting that the plasmasphere is similar in terms of density and temperature to the outer portion of Titan's ionosphere. Finally, the horizontal red bands in the IMS plot between approximately 0230 to 0310 UT and between 0345 to 0440 UT represent the Earth's ring current.

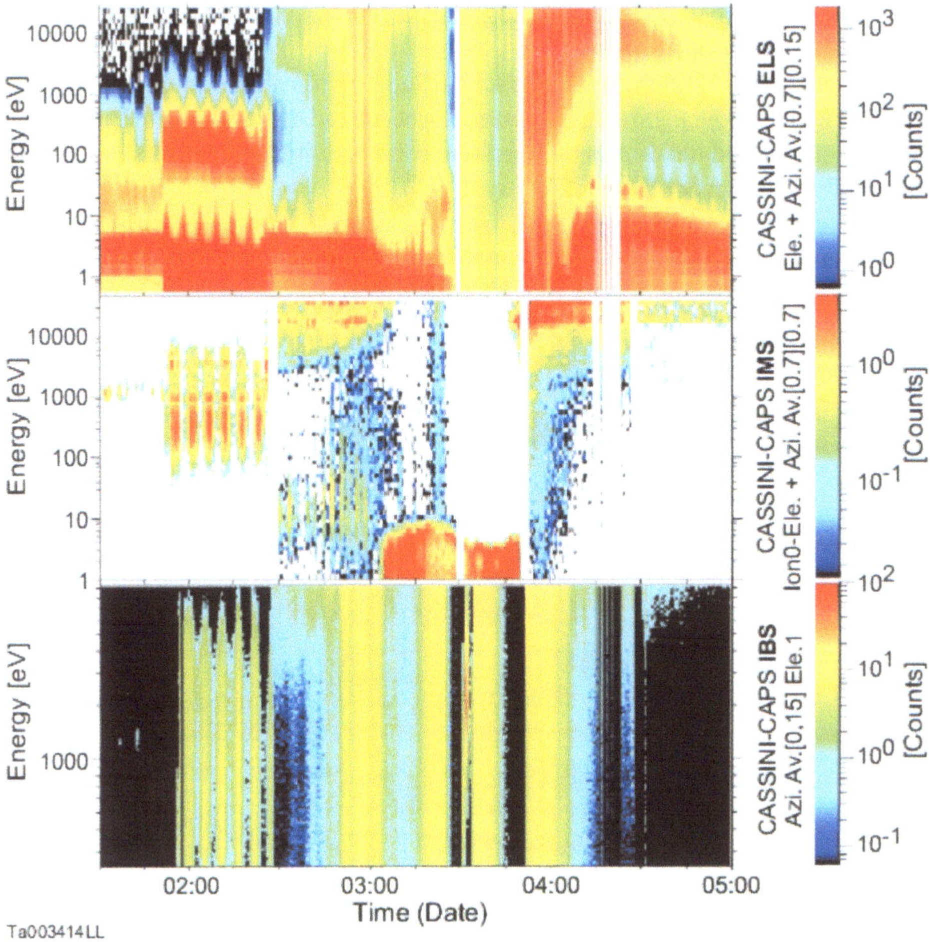

Figure 45. Composite energy-time spectrograms of ELS (upper panel), IMS (middle), and IBS (lower) taken during the Cassini Earth encounter. The color scale at the right is proportional to detector counting rates.

9.2. ELS Initial Results

Abel *et al.* (2001) and Rymer *et al.* (2001) give an overview and detailed discussion of ELS measurements at Earth. Figure 46 is an energy–time spectrogram of ELS data in the same format as Figure 45 except that the eight separate panels correspond to eight anodes and therefore to eight elevation directions (see Figure 10a, b). These data illustrate the high quality of ELS performance in all regions traversed by Cassini. More detailed analysis of bi-directional lobe electrons, including the first observation of a returning electron population in the magnetosheath, is described in Abel *et al.* (2001). Careful examination of Figure 46 shows many of the distributions to be anisotropic (e.g., in the magnetosheath between ~0200 and 0240 UT, or the plasmasphere near 0315 UT).

Initial analysis of measurements taken at Jupiter indicates high-quality data associated with the jovian bow shock, magnetosheath and magnetopause. In addition, ELS has monitored the effects of impulsive solar activity in the distant solar wind beyond 5 AU (from September 2000 onwards).

Data in Figure 46 also show that at low energies ($E < 30$ eV) ELS measurements are strongly affected by asymmetries in the spacecraft photoelectron sheath and, at Earth at least, by Cassini's high velocity (16.1 km/s) relative to the plasma (Rymer

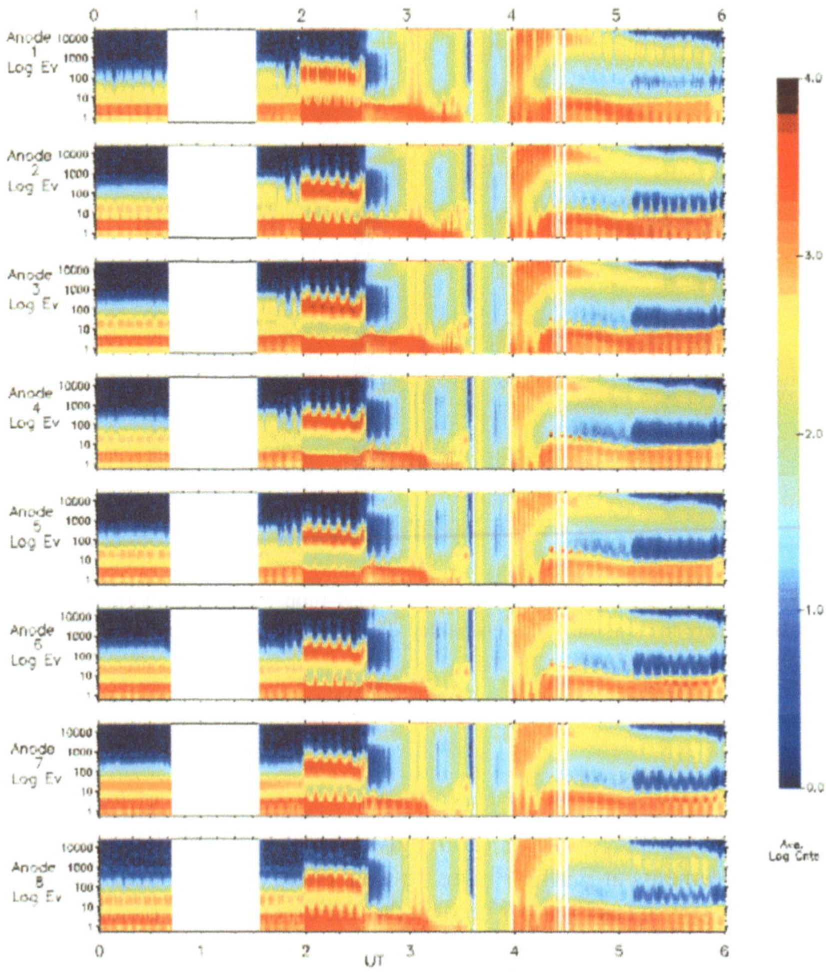

Figure 46. ELS energy-time spectrograms taken on August 18, 1999 during Cassini's flyby of the Earth. The eight separate plots represent the eight ELS elevation anodes, with anode eight looking closest to the direction of the spacecraft – *Z*-axis (i.e., toward the spacecraft antenna).

et al. (2001). At Titan although the spacecraft velocity is lower (~6 km/s) so are ambient electron thermal velocities. Thus we may expect an anisotropic sheath. Although photoelectron emission from the spacecraft is diminished by R^{-2} relative to that at Earth it is also expected to add to asymmetries. In other words, it is difficult to predict what the response of ELS will be during Titan encounters, although the one data point from Earth is a start for predicting that behavior.

9.3. IBS Initial Results

Because of its design, IBS data from the Earth encounter are less useful than those of the other two sensors. This can be understood with reference to the IBS section of this paper, but basically it is caused by the orientation of its FOV perpendicular to the ram flux coupled with the lack of resolution in the ram direction. On the other hand, during the Cassini encounter with Jupiter, when the sensors could be pointed toward the solar direction, its performance could be demonstrated.

Figure 47 is an example of one of ~40 bow shock crossings observed during encounters in late 2000/early 2001 (Szego *et al.*, 2003a, b) by IBS and the other CAPS sensors. In Figure 47 the jovian bow shock was crossed between 0430 ~ 0500 UT when Cassini was $576R_J$ (~4.1×10^7 km) from Jupiter. Protons in the solar wind upstream of the shock can be distinguished as a red band before 0430 UT. Above it is a thinner yellow band representing He^{++}. The shock crossing can be identified by the increase in ion temperature downstream. Comparison with magnetic field data indicates a parallel shock typical of turbulent shocks seen in the jovian magnetosheath. At various times after the shock the solar wind plasma temperature drops such that IBS is again able to resolve the He^{2+} peak. Data presented in Szego *et al.* (2003) dealing with the jovian bow shock and magnetosheath show convincingly that IBS is operating as anticipated.

9.4. IMS Initial Results

Figure 48 shows IMS data in energy versus TOF spectrogram format integrated over the month of March 2002 when Cassini was roughly 8 AU from the sun. The data represent the first observations of interstellar pickup ions made beyond 5 AU. Pickup ions originate from neutral gas in the interstellar medium (Gloeckler and Geiss, 1998). The pickup process occurs when gas is ionized by solar photons and accelerated by the solar wind electromagnetic field to energies as high as $2M\mathrm{v}^2\sin^2\alpha$ where v is the solar wind bulk velocity and α is the ion pitch angle. Interstellar ions have higher energies than solar wind ions as well as pitch angle distributions with particles arriving at angles other than the solar wind direction. Thus they effectively avoid the high-gain antenna of the spacecraft that prevents IMS from directly viewing the solar wind. In summary we would expect

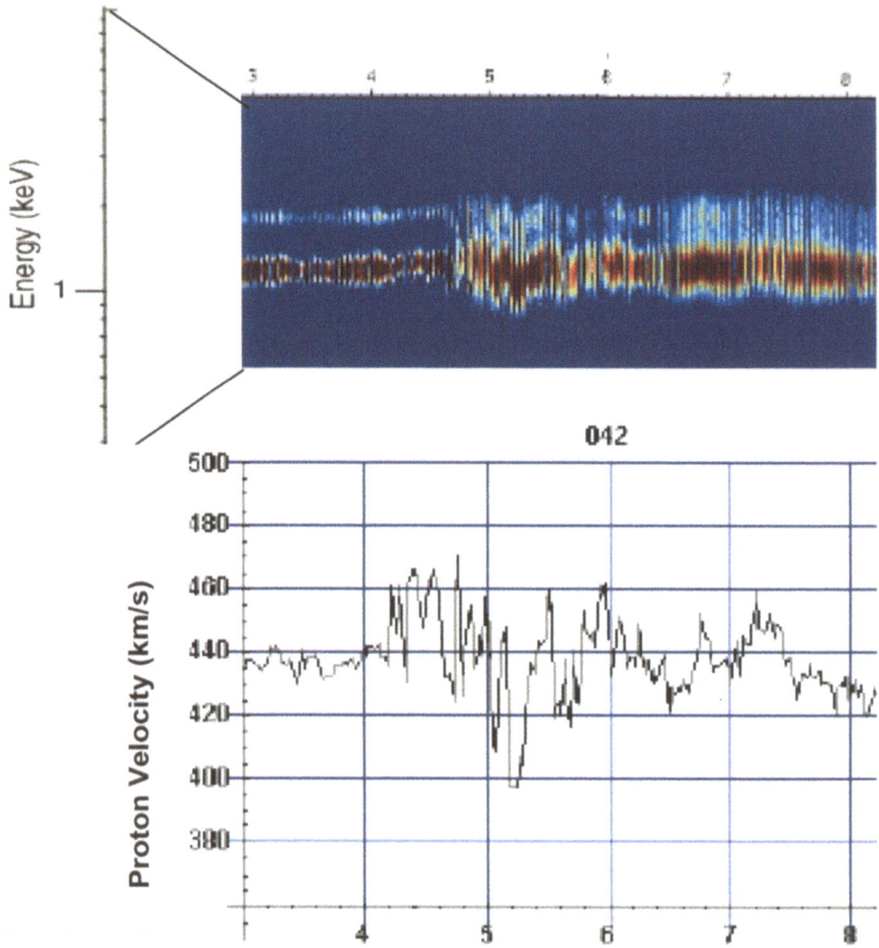

Figure 47. Upper panel shows energy–time spectrogram for the IBS taken between 0300 to 0900 UT on February 11, 2001. The energy scale covers 0.25 to 10.0 keV. The lower panel is a plot of solar wind proton velocity in km/s same period (from Szego *et al.*, 2003a).

to see ions arriving from directions unblocked by the HGA up to a cutoff energy of $2Mv^2\sin^2\alpha$. Complicating the picture is the fact that during March 2002 the spacecraft turned in such a way that there is some direct solar wind data mixed with data in Figure 48.

In the ST energy spectrum in Figure 48 H+ ranges up to 20 keV while the bulk of the distribution is at ~1 keV corresponding to ~440 km/s, roughly the average solar wind velocity. The highest H^+ energy (20 keV) suggests a maximum solar wind velocity during this period of 980 km/s, which is a high but still reasonable value. The second most intense peak in the ST spectrum, labeled LEF echo originates not

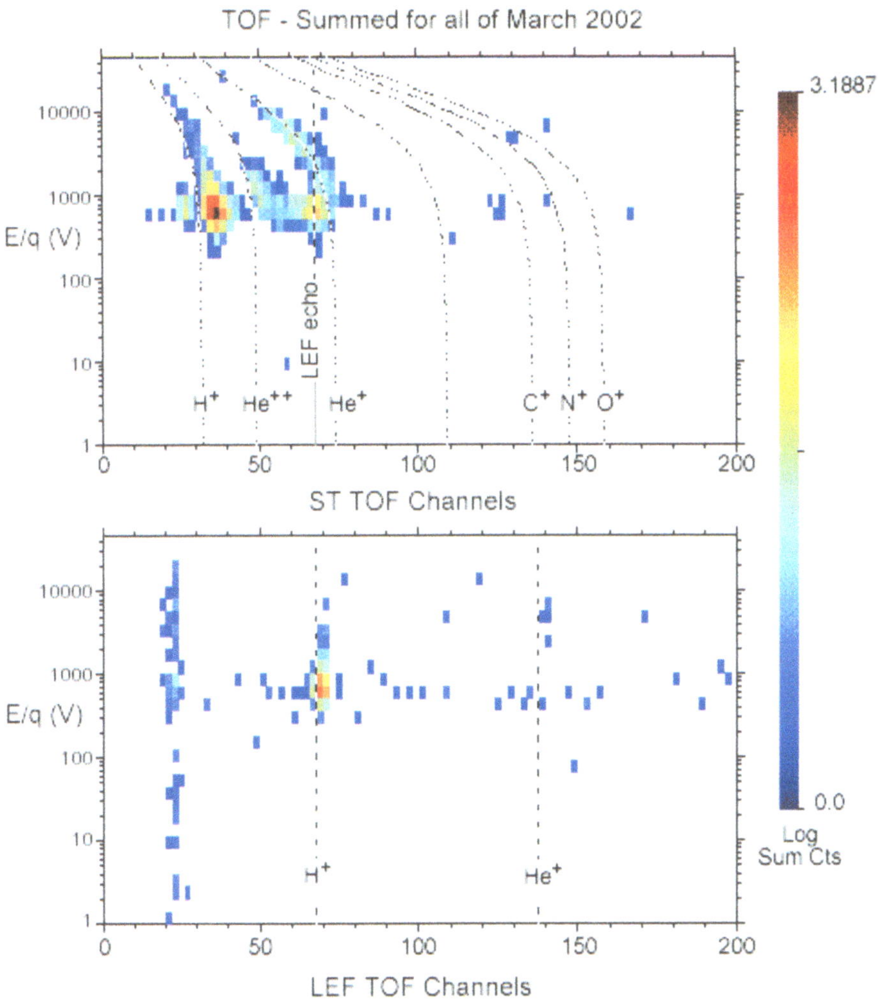

Figure 48. Color-coded logarithm of the sums of counts in the CAPS IMS ST (upper panel) and LEF (lower panel) detectors, respectively. The dashed lines are derived from a simple theoretical model of the instrument. Data are summed over all of March 2002. The summed counts are displayed as a function of energy-per-charge (vertical axes) and time-of-flight channel on the horizontal axis (1 channel = 3.125 ns).

from the nearby He^+ counts, but rather from an "echo" peak caused by H^+ ions (see discussion in Section 6.7 and Figure 38b).

The other two ions visible easily in the ST spectrum, in addition to H^+, originate from He^{++} and He^+ listed in order of TOF (see Equations 4–6). In order to understand the ST spectra it is necessary to consider charge exchange fractions leaving the foil (see Figure 21) as well as the center of mass velocity of particles leaving the foil. Thus, on passing through a foil, the two He charge states show up primarily as He^0. The LEF spectrum shows the H^+ peak, but, as a consequence of the low rates of He $\rightarrow$ He^+ and He $\rightarrow$ He^{++}, there are very few counts. We do expect, however, that by summing over longer periods of time and removing unwanted "background" that occurs when Cassini puts the IMS FOV in the solar wind direction, we will be able to distinguish and observe interstellar pickup ions.

Using data from the IMS McComas *et al.* (2004) have reported the first observations of interstellar pickup ions made beyond the orbit of Jupiter. The spatial distribution of the ion fluxes is consistent with gravitational focusing by the Sun. During the period reported' Cassini was located between 6.4 and 8.2 AU from the Sun in the downstream direction relative to incoming interstellar gas moving past the Sun. The IMS data also establish the existence of a hydrogen shadow in this same region where hydrogen is depleted relative to interstellar flow. A combination of radiation pressure on the hydrogen atoms together with trajectories passing close to the Sun that lead to photoionization and loss in the solar wind leads to the shadowing effect. We anticipate further observations of interstellar pickup ions as the spacecraft approaches Saturn and throughout the orbital mission when a large amount of time is spent in the solar wind (Table XIV).

10. Conclusions

After nearly 7 years in flight, Cassini went into orbit around Saturn on July 1, 2004. During the cruise to Saturn the CAPS instrument was operated successfully for periods of varying length during cruise and the Earth and Jupiter encounters. The long cruise period has given our team the time necessary to fully check out and test CAPS in the plasma environments of the solar wind and two magnetospheres. This gives us confidence that the instrument will perform successfully during the saturnian tour. In addition, the cruise period was used to develop a sophisticated ground system that enables and greatly assists the complex task of science planning for CAPS activities throughout the mission. The downlink portion of the ground system is equipped to reduce data rapidly once it is received on the ground and make it available as quickly as possible to the scientific community.

Acknowledgments

Initial development of the CAPS instrument began in 1989. During the intervening years design, fabrication, testing and calibration, as well as post-launch science planning and operations has been a collaborative effort involving the following institutions: Southwest Research Institute (overall responsibility for the investigation, system design, low voltage supplies, spacecraft and sensor interface boards, IBS and IMS high-voltage supplies, flight and ground system software, system calibration, flight operations, science planning lead), the Los Alamos National Laboratory (IMS sensor concept, design, integration, and calibration; IMS detectors, front end electronics, and time-to-digital converter; IBS concept, detectors, integration, and calibration, flight data processing), the Centre d'étude des Environnements Terrestre et Planetaires (IMS sensor mechanical design and fabrication, SAGEM motor for the actuator), the Goddard Space Flight Center (IMS high-voltage supplies, spectrum analyzer module hardware and software, and CPU2 software), the Jet Propulsion Laboratory (IMS calibration), the University College London (ELS concept, design, sensor fabrication, integration, and calibration, flight data processing), the University of Oulu (IBS optical design and analysis), the University of Virginia (IMS-related carbon foil characterization), the VTT Automation (IBS sensor mechanical design and fabrication, actuator subsystem design, fabrication, and test), the KFKI Research Institute for Particle and Nuclear Physics (electrical ground support equipment (EGSE) and EGSE software) the Norwegian Defense Research Establishment (ELS sensor management unit), and the Rutherford Appleton Laboratory (ELS high-voltage power supplies, electrical system, and detectors).

Many people at these and other institutions have given their best ideas and efforts to support CAPS development and we wish to thank them individually here:

At SwRI we particularly wish to thank M. Ahr, J. Alexander, S. Archey, N. Brinkmeyer, D. Burket, B. Gahagan, D. Guerrero, B. Gupta, D. LeBlanc, J. Ling, M. McGinnis, B. McLaren, G. Palacios, A. Ramirez, L. Roberson, S. Rogillio, M. Sattler, N. Sawka, T. Sawka, D. Strain, R. Thorpe, V. Torres, C. Urdiales, J. H. Waite, Jr. and R. Welch.

At LANL we particularly wish to thank D. Archuleta, J. Baldonado, S. Bame, P. Barker, R. Bramlett, D. Burr, P. Campbell, M. Dugan, R. Edwards, J. Elliot, D. Everett, K. Fuller, H. Funsten, I. Gonzales, V. Herrera, D. Laurence, I. Medina, E. Michel, K. Olivas, N. Olivas, P. Pazuchanics, D. Potter, R. Pressini, A. Roybal, P. Salazar, M. Shappirio, G. Smith, S. Storms, D. Suszcynsky, A. Ulibarri, M. Vigil, R. Vigil and V. Vigil.

At GSFC we particularly wish to thank C. Clatterbuck, D. Glenn, F. Hunsaker, W. Mish, K. Ogilvie, R. Vondrak, A. Williams, and K. Young.

At CETP we particularly wish to acknowledge the work of F. Leblanc, F. LeGoff, A. Ruatti and the ETEZ Company.

The ELS team leader, A. J. Coates, acknowledges the support of the Royal Society. We would like also to acknowledge the support of the rest of the ELS team including G. Abel, C. Alsop, T. Barsby, S. Burge, J. Coker, N. Flowers, L. Free, L. Gilbert, J. Godfrey, B. Hancock, J. Holmes, C. Jackman, A. Johnstone, D. Kelsh, P. Kendon, T. Murrell, J. Raymont, D. Reading, A. Rymer, J. Rose, P. Sheather, F.Thurlow, and L. Worth. Morgan-Matroc Limited, Rugby, U.K. supplied ELS anodes. ELS MCPs were supplied by Philips Components Ltd., France.

The IBS Collaborators would like to thank P. Kyrenius, T. Korpela, S. Kuitunen and J. Vidquist of VTT Automation.

The CAPS-EGSE was developed at KFKI-RMKI and we particularly wish to acknowledge the contributions of L. Foldy and I. Gladkih.

Work at the University of Virginia supported in part by NASA's Planetary Geology and Geophysics program.

The CAPS actuator (ACT) was developed and manufactured by the VTT under contract to SwRI. We particularly wish to thank members of the laboratory and workshops of VTT Automation: T. Antila, H. Berg, the personnel of the Fine-Mechanical Prototype Shop, and especially K. Siitari of Rejlers Inc. The ACT stepper motor was manufactured by SAGEM of France and provided by CETP. Starsys Research Inc., of Boulder, CO manufactured the wax thermal actuators and the IMS cover mechanism.

We wish to acknowledge the primary support for CAPS development and operations in the U.S. provided by NASA/Cal Tech/JPL through JPL contracts 958964, 959401, 959930, and 1243218 to the SwRI, and by JPL contract 1210586 to the University of Michigan. Initial R & D funding was made available by the Advisory Committee for Research at SwRI, and by the Department of Energy at Los Alamos National Laboratory. Work at CETP was supported by CNES grants 92/204 and 95/04; in Hungary by OMFB-07851 and a Hungarian Space Office grant MUI/TP/47; in Finland early development work on the ACT was supported by internal VTT funding, with further support provided by the Academy of Finland and the Ministry of Trade and Industry of Finland; in the U.K. by the Particle Physics and Astronomy Research Council; and in Norway by the Norwegian Defense Research Establishment and the Research Council of Norway.

Appendix: List of Acronyms

ACT	Actuator
ADC	Analog-to-Digital Converter
ALF	Assisted Load Format
amu/e	Atomic Mass Unit/Charge
AU	Astronomical Unit
AZ	Azimuth
BIU	Bus Interface Unit
CAPS	Cassini Plasma Spectrometer
CDS	Command and Data System
CEM	Channel Electron Multiplier
CH	TOF channel number (1 CH = 3.125 ns)
CPU	Central Processing Unit
DPU	Data Processing Unit
DAC	Digital-to-Analog Converter
E	Energy, energy step

(*Continued on next page*)

EL	Elevation
ELS	Electron Spectrometer
EPS	Energetic Particle Spectrometer (Voyager)
E/Q	Energy/Charge
ESA	Electrostatic Analyzer
FEE	Front End Electronics
FIFO	First-In-First-Out buffer
FOV	Field-of-View
FPP	Fields and Particles Pallet
FWHM	Full Width at Half Maximum
Gbits	Gigabits (10^9 bits)
HEEA	High-Energy Electron Analyzer (Cluster)
HVU	High-Voltage Unit
IBS	Ion Beam Spectrometer
ID	Identification bit
IEB	Instrument Expanded Block
IMS	Ion Mass Spectrometer
INMS	Ion Neutral Mass Spectrometer
kV	KiloVolt
keV	KiloelectronVolt
LEF	Linear Electric Field
LEMMS	Low Energy Magnetosphere Measurement System
LUT	Look-Up Table
MCP	Microchannel Plate
MIMI	Magnetospheric Imaging Investigation
MLI	Multi-layer Insulation
M/Q	Mass/charge
ms	Milliseconds
MSSL	Mullard Space Science Laboratory
ORS	Optical and remote sensing instruments
PEACE	Cluster Plasma Electron and Current Experiment
PI	Principal investigator
PLS	Plasma Science (Voyager)
PROM	Programmable Read-Only Memory
RAM	Random Access Memory
RES	Resolution (# of TOF channels: 512, 1024, 1536, 2048)
RTI	Real Time Interrupt
SAM	Spectrum Analyzer Module
SMU	Signal Management Unit
SOI	Saturn Orbit Insertion
SOP	Science Operations Plan

(*Continued on next page*)

SOPC	Science Operations and Planning Computer
SSR	Solid State Recorder
ST	Straight-Through (time-of-flight)
SwRI	Southwest Research Institute
SVT	Sequence Virtual Team
TDC	Time-to-Digital Converter
TOF	Time-of-Flight
1-D	1-Dimensional
2-D	2-Dimensional
3-D	3-Dimensional

References

Abel, G. A., Coates, A. J., Rymer, A. M., Linder, D. R., Thomsen, M. F., Young D. T., and Dougherty, M. K.: 2001, *J.Geophys.Res.* **106**, 30199.

Alsop C., Free, L., and Scott, S.: 1998, in R. F. Pfaff, J. E. Borovsky, and D. T. Young (eds.), *Measurement Techniques in Space Plasmas: Particles*, AGU Geophysics Monograph Series, Vol. 102, pp. 269–274.

Alton, G. D.: 1993 *Nucl. Instrum. Methods Phys. Res.* **B73**, 221–288.

Bame, S. J., Asbridge, J. R., Felthauser, H. E., *et al.*: 1978, *IEEE Trans. Geosci. Electron.* **GE-16**, 216.

Behannon, K. W., Lepping, R. P., and Ness, N. F.: 1983, *J. Geophys. Res.* **88**, 8791.

Bird, M. K., Dutta-Roy, R., Asmar, S., Rebold, T. A.: 1997, *Icarus* **130**, 426.

Blanc, M., *et al.*: 2004, *Space Sci. Rev.*, **114**, 1–112.

Brecht, H. S., Luhmann, J. G., and Larson, D. J.: 2000, *J. Geophys. Res.* **105**, 13119.

Bridge, H. S., *et al.*, *Space Sci. Rev.* **21**, 259.

Bridge, H. S., Belcher, J. W., Lazarus, A. J., Olbert, S., Sullivan, J. D. Bagenal, F., Gazis, P. R., Hartle, R. E., Ogilvie, K. W., Scudder, J. D., Sittler Jr., E. C., Eviatar, A., Siscoe, G. L., Goertz , C. K., and Vasyliunas, V. M.: 1981, *Science*, **212**, 217.

Broadfoot, A. L., Sandel, B. R., Shemansky, B. E., *et al.*: 1981, *Science* **212**, 206.

Burns, J. A., and Matthews, M. S.: 1986, *Satellites*, U. Arizona Press, Tucson.

Carlson C. W., Curtis, D. W., Paschmann, G., and Michael, W.: 1983, *Adv. Space Res.* **2**, 67.

Coates A. J., Alsop C., Coker A. J., Linder, D. R., Johnstone, A. D., Woodliffe, R. D., Grande, M., Preece, A., Burge, S., Hall, D. S., Narheim, B., Svenes, K., Young, D. T., Sharber, J. R., and Scherrer, J. R.: 1992, *J. British Interplanetary Soc.* **45**, 387.

Connerney, J. E. P., and Waite, J. H.: 1984, *Nature* **312**, 136.

Connerney, J. E. P., Acuna, M. H., and Ness, N. F.: 1983, *J. Geophys. Res.* **88**, 8779–8789.

Crary F. J., and Bagenal F.: 2000, *J. Geophys. Res.* **105**, 25379.

Cravens, T. E.; Lindgren, C. J., Ledvina, S. A.: 1998, *Planet. Space Sci.* **46**, 1193.

Davis, L., Jr., and Smith, E. J.: 1990, *J. Geophys. Res.* **95**, 15257–15261.

Delitsky, M. L., and A. L. Lane: 1997, *J. Geophys. Res.* **102**, 16385–16390.

Desch, M. D., and Rucker, J. H.: 1983, *J. Geophys. Res.* **88**, 8999.

Esposito, L. W. *et al.*: 2003, *Space Sci. Rev.* this volume.

Eviatar, A., and Richardson, J. D.: 1986, *J. Geophys. Res.* **91**, 3299.

Eviatar, A., Siscoe, G. L., Scudder, J. D., Sittler Jr., E. C., and Sullivan, J. D.: 1982, *J. Geophys. Res.* **87**, 8091.

Funsten, H. O.: 1995, *Phys. Rev. B* **52**, R8703.

Funsten, H. O., Barraclough, B. L., and McComas, D. J.: 1993b, *Nucl. Inst. Meth.* **B80/81**, 49–52.
Funsten, H. O., Barraclough, B. L., and McComas, D. J.: 1994, *Nucl. Inst. Meth. Phys. Res. B* **20**, 24–28.
Funsten, H. O., McComas, D. J., and Barraclough, B. L.: 1993a, *Optical Eng.* **32**, 3090–3095.
Funsten, H. O., Ritzau, S. M., and Harper, R. W.: 2001, *Phys. Rev. B* **63**, 155416.
Gloeckler, G., Balsiger, H., Bürgi, A., Bochsler, P., *et al.*: 1995, *Space Sci. Rev.* **71**, 79.
Gloeckler, G., Cain, J., Ipavich, F. M., Tums, E. O., *et al.*: 1998, *Space Sci. Rev.* **86**, 437–539.
Gloeckler, G., and J. Geiss: 1998, *Space Sci. Rev.* **86**, 127–159.
Gloeckler, G., and Hsieh, K. C.: 1976, *Nucl. Inst. Meth.* **165**, 537.
Goertz, C. K.: 1983, *Geophys. Res. Lett.* **10**, 455.
Gorney, D. J., *et al.*: 1981, *J. Geophys. Res.* **86**, 83.
Gosling, J. T., Asbridge, J. R., Bame, S. J., and Feldman, W. C.: 1978, *Rev. Sci. Instrum.* **49**, 1260.
Gurnett, D., *et al.*: 2003, *Space Sci. Rev.* **114**, 395–463.
Haff, P. K., Eviatar, A., and Siscoe, G. L.: 1983, *Icarus* **56**, 426.
Hamilton, D. C., Gloeckler, G., Ipavich, F. M., *et al.*: 1990, *Rev. Sci. Instrum.* **61**, 3104.
Hansen K. C.: 2001, *Ph.D. Dissertation*, Department of Atmospheric, Oceanic and Space Sciences, University of Michigan.
Hartle, R. E., Sittler, Jr., E. C., Ogilvie, K. W., Scudder, J. D., Lazarus, A. J., and Atreya, S. K.: 1982, *J. Geophys. Res.* **87**, 1383.
Hill, T. W.: 1979, *J. Geophys. Res.* **84**, 6554.
Hill, T. W., Dessler, A. J., and Maher, L. J.: 1981, *J. Geophys. Res.* **86**, 9020.
Hones, E. W., Jr.: 1979, S.-I. Akasofu (ed.), *Dynamics of the Magnetosphere*, D. Reidel, p. 545.
Hood, L. L.: 1985, *J. Geophys. Res.* **90**, 6295.
Horanyi, M., Burns, J. A., and Hamilton, D. P.: 1992, *Icarus* **97**, 248.
Hovestadt, D., *et al.*: 1995, *Solar Phys.* **162**, 441.
Huang, T. S., and Siscoe, G.: 1987, *Icarus* **70**, 366.
Ip, W.-H.: 1984, *J. Geophys. Res.* **89**, 8843.
Ip, W.-H.:1990, *Astrophys. J.* **362**, 354.
Ip, W.-H.: 1992, *Proceedings of the Symposium on Titan*, **ESA SP-338**, 243.
Ip, W.-H.: 1995, *Icarus* **115**, 295.
Johnson, R. E.: 1990, *Energetic Charged-Particle Interactions with Atmospheres and Surfaces*, Springer-Verlag, NY.
Johnson, R. E.: 1994, *Space Sci. Rev.* **69**, 215–233.
Johnson, R. E.: 1998, *Sputtering of Ices in the Solar System II*, Kluwer, Amsterdam, 1998.
Johnson, R. E., and Jesser, W. A.: 1996, *Astrophys. J.* **480**, L79–L82.
Johnson, R. E., Pospieszalska, M. K., Sittler Jr., E. C., Cheng, A. F., Lanzerotti, L. J., and Sieveka, E. M.: 1989, *Icarus* **77**, 311.
Johnson, R. E., and Quickenden, T. I.: 1997, *J. Geophys. Res.* **102**, 10985–10996.
Johnson, R. E., and Sittler Jr., E. C.: 1990, *Geophys. Res. Lett.***17**, 1629.
Johnstone, A. D., Alsop, C., Burge, S., Carter, P. J., Coates, A. J., Coker, A. J. Fazakerley, A. N., Grande, M., Gowen, R. A., Gurgiolo, C., Hancock, B. K., Narheim, B., Preece, A., Sheather, P. H. Winningham, J. D., and Woodliffe, R. D.: 1997, *Space Sci. Rev.* **79**, 351.
Johnstone, A. D., Coates, A. J.,Wilken, B., *et al.*: 1987, *J. Phys. E: Sci. Instrum.* 795–805.
Jurac, S., Johnson, R. E., Richardson, J. D.: 2001, *Icarus* **149**, 384.
Jurac, S., McGrath, M. A., Johnson, R. E., Richarson, J. D., Vasyliunas, V. M., and Eviatar, A.: 2002, *Geophys. Res. Letts.* **29**, 2172.
Kaiser, M. L., Desch, M. D., and Connereny, J. E. P.: 1984b, *J. Geophys. Res.* **89**, 2371.
Kaiser, M. L., Desch, D., Kurth, W. S., Lecacheux, A., Genova, F., Pederson, B. M., and Evans, D. R.: 1984a, *Saturn*, University of Arizona Press, p. 378.
Keller, C. N., and Cravens, T. E.: 1994, *J. Geophys. Res.* **99**, 6527.

Keller, C. N., Cravens, T. E., and Gans, L.: 1994, *J. Geophys. Res.* **99**, 6511.
Klimas, A. J.: 1985, in B. T. Tsurutani, and E. Stone (eds.), *Collisionless Shocks in the Heliosphere: Reviews of Current Research*, AGU Geophysics Monograph Series, Vol. 35, p. 237.
Knight, S.: 1973, *Plan. Space Sci.* **21**, 741.
Kopp, A., and Ip, W.-H.: 2001, *J. Geophys. Res.* **106**, 8323.
Krimigis, S. M., *et al.*: 2004, *Space Sci. Rev.* **21**, 329.
Ledvina, S., and Cravens, T. E.: 1998, *Planet. Space Sci.* **46**, 1175.
Ledvina, S., *et al.*: 2000, *Adv. Space Res.* **26**, 1691.
Linder, D. R., Coates, A. J., Woodliffe, R. D., Alsop, C., Johnstone, A. D., Grande, M., Preece, A., Narheim, B., Svenes, K., and Young, D. T.: 1998, in F. Pfaff, J. E. Borovsky, and D. T. Young (eds.), *Measurement Techniques in Space Plasmas: Particles*, AGU Geophysics Monograph Series, Vol. 102, pp. 257–262.
Lindgren, C. J., Cravens, T. E., and Ledvina, S. A.: 1997, *J. Geophys. Res.* **102**, 17395–17406.
Luhmann, J. G., Russell, C. T., Schwingenschuh, K., and Yeroshenko, Y. E.: 1991, *J. Geophys. Res.* **96**, 11199.
Luhmann, J. G.: 1996, *J. Geophys. Res.* **101**, 29,387.
Luna, H., Michael, M., Shah, M. B., Johnson, R. E., Latimer, C. J., and McConkey, J. W.: 2003, *J. Geophys. Res.* **108**, 10.1029/2002JE001950.
Maurice, S., and Engle, I. M.: 1995, *J. Geophys. Res.* **100**, 17143.
Maurice, S., Engle, I. M., Blanc, M., and Skubis, M.: 1996a, *J. Geophys. Res.* **101**, 27053.
Maurice, S., Sittler, E. C., Cooper, J. F., Mauk, B. H., Blanc, M., and Selesnick, R. S.: 1996b, *J. Geophys. Res.*, **101**, 15211–15232.
McComas, David J., and Jane E. Nordholt: 1990, *Rev. Sci. Inst.* **61**, 3095–3097.
McComas, D. J., Nordholt, J. E., Samuel, J., Bruce, B., Barraclough, L., and Gosling, J. T.: 1990, *Proc. Natl. Acad. Sci., U S A*, **87**, 5925–5929.
McComas, D. J., Nordholt, J. E., Young, D. T., and Berthelier, J. J.: 1998, in R. F. Pfaff, J. E. Borovsky, and D. T. Young (eds.), *Measurement Techniques in Space Plasmas: Particles*, AGU Geophysics Monograph Series, Vol. 102, pp. 187–193.
McComas, D. J., Schwardron, N. A., Crary, F. J., Elliott, H. A., Young, D. T., Gosling, J. T., Thomsen, M. F., Sittler, E., Berthelier, J. J., Szego, K., and Coates, A. J.: 2004, *J. Geophys. Res.* (*Space Phys.*), **109**, A02104.
Mizera, P. F., *et al.*: 1981, *J. Geophys. Res.* **86**, 2329.
Mizera, P. F., Fennel, J. F., Croley, D. R., *et al.*: 1981, *J. Geophys. Res.* **86**, 2329.
Morfill, G., Havnes, O., and Goertz, C. K.: 1993, *J. Geophys. Res.* **98**, 11285.
Nagy, A. F., and Cravens, T. E.: 1998, A Review, *Planet Space Sci.* **46**, 1149.
Nagy, A. F., Liu, Y., Hansen, K. C., Kabin, K., *et al.*: 2001, *J. Geosphys. Res.* **106**, 6151.
Ness, N. F., Acuna, M. H., and Behannon, K. W.: 1982a, *J. Geophys. Res.* **87**, 1369.
Ness, N. F., Acuña, M. H., Behannon, K. W., and Neubauer, G. M.: 1982b, *J. Geosphys. Res*, **87**, 1369.
Noll, K. S., Roush, T. L., Cruikshank, D. P., Johnson, R. E., and Pendleton, Y. L.: 1997, *Nature* **388**, 45–47.
Nordholt, J. E., Berthelier, J.-J., Burr, D. M., Funsten, H. O., Goldstein, R., Illiano, J. M., McCabe, K. P., McComas, D. J., Potter, D. M., and Young, D. T.: 1998, in R. Pfaff, J. Borovsky, and D. Young (eds.), *Measurement Techniques in Space Plasmas: Particles*, AGU Geophysics Monograph Series Vol. 102, pp. 209–214.
Pontius, D. H., and Hill, T. W.: 1982, *Geophys. Res. Lett.* **12**, 1321.
Reiff, P. H., H. L. Colin, J. D. Craven, *et al.*: 1988, *J. Geophys. Res.* **93**, 7441.
Richardson, J. D.: 1986, *J. Geophys. Res.* **91**, 1381.
Richardson, J. D.: 1992, *J. Geophys. Res.* **97**, 13705.
Richardson, J. D.: 1995, *Geophys. Res. Lett.* **22**, 1177.

Richardson, J. D.: 1998, *Rev. Geophys.* **36**, 501.

Richardson, J. D., and Eviatar, A.: 1987, *Geophys. Res. Lett.* **14**, 999.

Richardson, J. D., Eviatar, A., McGrath M. A., and Vasyliunas, V. M.: 1998, *J. Geophys. Res.*, **103**, 20245.

Richardson, J. D., Eviatar, A., and Siscoe, G. L.: 1986, *J. Geophys. Res.* **91**, 8749.

Richardson, J. D., and Sittler, E. C.: 1990, *J. Geophys. Res.*, **95**, 12019.

Ritzau, S. M., and Baragiola, R. A.: 1998, *Phys. Rev. B* **58**, 2529.

Rymer, A. M., Coates, A. J., Abel, G. A., Linder, D. R., Svenes, K., Narheim, B., Thomsen, M., and Young, D. T.: 2001, *J. Geophys. Res.* **16**, 30177.

Shemansky, D. E., and Hall, D. T.: 1992, *J. Geophys. Res.* **97**, 4143.

Shemansky, D. E., Matheson, P., Hall, D. T., Hu, H.-Y., and Tripp, T. M.: 1993, *Nature* **363**, 329.

Shemansky, D. E., Smith, G. R., and Hall, D. T.: 1985, *EOS Trans. AGU* **66**, 1108.

Shi, M., Baragiola, R. A., Grosjean, D. E., Johnson, R. E., Jurac, S., and Schou, J.: 1995a, *J. Geophys. Res.* **100**, 26387.

Shi, M., Grosjean, D. E., Schou J., and Baragiola, R. A.: 1995b, *Nucl. Instrum. Methods* **B96**, 524.

Scarf, F. L., Frank, L. A., Gurnett, D. A., Lanzerotti, L. J., Lazarus, A., and Sittler Jr., E. C.: 1984, in T. Gehrels, and M. S. Matthews (eds.), *Saturn*, University of Arizona Press, Arizona, pp. 318–353.

Shematovich, V. I., Johnson, R. E., Micheal, M., and Luhmann, J. G.: 2003, *J. Geophys. Res.*, **108**(E8), 5087.

Sittler, E. C., Jr.: 1993, *Rev. Sci. Instrum.* **64**, 2771, 1993.

Sittler, E. C., Jr., and Hartle, R. E.: 1996, *J. Geophys. Res.* **101**, 10863.

Sittler, E. C., Jr., Scudder, J. D., and Bridge, H. S.: 1981, *Nature* **292**, 711.

Slavin, J. A., Smith, E. J., Gazis, P. R., and Mihalov, J. D.: 1983, *Geophys. Res. Lett.* **10**, 9.

Stevenson, D. J.: 1982, *Nature* **298**, 142–144

Szego, K., Young, D. T., Barraclough, B., Berthelier, J.-J., Coates, A. J., *et al.*: 2003, *J. Geophys. Res.* **108**, 11–1.

Thomsen, M. F.: 1985, in B. T., Tsurutani and E. Stone (eds.), *Collisionless Shocks in the Heliosphere: Reviews of Current Research*, AGU Geophysics Monograph Series, Vol. 35, 253.

Vasyliunas, V. M.: 1983, in A. J. Dessler (ed.), *Physics of the Jovian Magnetosphere*, Cambridge University Press, Cambridge, Chap. 11.

Vilppola, J. H., Keisala, J. T., Tanskanen P. J., and Huomo, H.: 1993, *Rev. Sci. Instrum.* **64**, 2190.

Vilppola, J. H., Tanskanen, P. J., Barraclough, B. L., and McComas, D. J.: 2001, *Rev. Sci. Instrum.* **72**, 3662.

Vilppola, J. H., Tanskanen, P. J., Huomo, H., and Barraclough, B. L.: 1996, *Rev. Sci. Instrum.* **67**, 1494.

Waite, J. H., *et al.*: 2004, *Space Sci. Rev.*, **114**, 113–231.

Warwick, J. W., Pearce, J. B., Evan, D. R., *et al.*: 1981, *Science* **212**, 239.

Wollnik, H.: 1971, *Nucl. Instrum. Method* **95**, 453.

Woodliffe R. D.: 1991, Design of Space-Borne Plasma Analysers by Computer Simulation, *Ph.D. Thesis*, University of London.

Woodliffe R. D., and A. D. Johnstone: 1998, in R. Pfaff, J. Borovsky, D. Young (eds.), *Measurement Techniques in Space Plasmas: Particles*, AGU Monograph Series **102**, 263–268.

Young, D. T., Bame, S. J., Thomsen, M. F., Martin, R. M., Burch, J. L., Marshall, J. A., and Reinhard, B.: 1988, *Rev. Sci. Instrum.* **59**, 743.

Young, D. T., Barraclough, B. L., Berthelier, J. J., Blanc, M., Burch, J. L., Coates, A. J., Goldstein, R., Grande, M., Hill, T. W., Illiano, J. M., Johnson, M. A., Johnson, R. E., Baragiola, R. A., Kelha, V., Linder, D., McComas, D. J., Narheim, B. T., Nordholt, J. E., Preece, A.,

Sittler, E. C., Svenes, K. R., Szalai, S., Szegö, K., and Tanskanen, P.: 1998, in R. F. Pfaff, J. E. Borovsky, and D. T. Young (eds.), *Measurement Techniques in Space Plasmas: Particles*, AGU Geophysics Monograph Series **102**, 237–242.

Young, D. T., Barraclough, B. L., McComas, D. J., Thomsen, M. F., McAbe, K., and Vigil, R.: 1991, *J. Spacecraft Rockets* **29**, 596–598.

THE CASSINI ION AND NEUTRAL MASS SPECTROMETER (INMS) INVESTIGATION

J. H. WAITE[1,*], JR., W. S. LEWIS[2], W. T. KASPRZAK[3], V. G. ANICICH[4], B. P. BLOCK[1], T. E. CRAVENS[5], G. G. FLETCHER[1], W.-H. IP[6], J. G. LUHMANN[7], R. L. MCNUTT[8], H. B. NIEMANN[3], J. K. PAREJKO[1], J. E. RICHARDS[3], R. L. THORPE[2], E. M. WALTER[1] and R. V. YELLE[9]

[1] *University of Michigan, Ann Arbor, MI, U.S.A.*
[2] *Southwest Research Institute, San Antonio, TX, U.S.A.*
[3] *NASA Goddard Space Flight Center, Greenbelt, MD, U.S.A.*
[4] *NASA Jet Propulsion Laboratory, Pasadena, CA, U.S.A.*
[5] *University of Kansas, Lawrence, KS, U.S.A.*
[6] *National Central University, Chung-Li, Taiwan*
[7] *University of California, Berkeley, CA, U.S.A.*
[8] *Johns Hopkins University Applied Physics Laboratory, Laurel, MD, U.S.A.*
[9] *University of Arizona, Flagstaff, AZ, U.S.A.*
*(*Author for correspondence, E-mail: hunterw@umich.edu)*

(Received 13 August 1998; Accepted in final form 17 February 2004)

Abstract. The Cassini Ion and Neutral Mass Spectrometer (INMS) investigation will determine the mass composition and number densities of neutral species and low-energy ions in key regions of the Saturn system. The primary focus of the INMS investigation is on the composition and structure of Titan's upper atmosphere and its interaction with Saturn's magnetospheric plasma. Of particular interest is the high-altitude region, between 900 and 1000 km, where the methane and nitrogen photochemistry is initiated that leads to the creation of complex hydrocarbons and nitriles that may eventually precipitate onto the moon's surface to form hydrocarbon–nitrile lakes or oceans. The investigation is also focused on the neutral and plasma environments of Saturn's ring system and icy moons and on the identification of positive ions and neutral species in Saturn's inner magnetosphere. Measurement of material sputtered from the satellites and the rings by magnetospheric charged particle and micrometeorite bombardment is expected to provide information about the formation of the giant neutral cloud of water molecules and water products that surrounds Saturn out to a distance of $\sim$12 planetary radii and about the genesis and evolution of the rings.

The INMS instrument consists of a closed ion source and an open ion source, various focusing lenses, an electrostatic quadrupole switching lens, a radio frequency quadrupole mass analyzer, two secondary electron multiplier detectors, and the associated supporting electronics and power supply systems. The INMS will be operated in three different modes: a closed source neutral mode, for the measurement of non-reactive neutrals such as N_2 and CH_4; an open source neutral mode, for reactive neutrals such as atomic nitrogen; and an open source ion mode, for positive ions with energies less than 100 eV. Instrument sensitivity is greatest in the first mode, because the ram pressure of the inflowing gas can be used to enhance the density of the sampled non-reactive neutrals in the closed source antechamber. In this mode, neutral species with concentrations on the order of $\geq 10^4$ cm^{-3} will be detected (compared with $\geq 10^5$ cm^{-3} in the open source neutral mode). For ions the detection threshold is on the order of 10^{-2} cm^{-3} at Titan relative velocity (6 km sec^{-1}). The INMS instrument has a mass range of 1–99 Daltons and a mass resolution $M/\Delta M$ of 100 at 10% of the mass peak height, which will allow detection of heavier hydrocarbon species and of possible cyclic hydrocarbons such as C_6H_6.

Space Science Reviews **114:** 113–231, 2004.

The INMS instrument was built by a team of engineers and scientists working at NASA's Goddard Space Flight Center (Planetary Atmospheres Laboratory) and the University of Michigan (Space Physics Research Laboratory). INMS development and fabrication were directed by Dr. Hasso B. Niemann (Goddard Space Flight Center). The instrument is operated by a Science Team, which is also responsible for data analysis and distribution. The INMS Science Team is led by Dr. J. Hunter Waite, Jr. (University of Michigan).

Keywords: Cassini, Titan, Saturn, Huygens, Mass Spectrometry

1. Introduction and Scientific Background

The Cassini spacecraft will enter Saturn orbit in July 2004. During Cassini's 4-year exploration of the Saturn system, the INMS instrument will acquire data needed to address two major scientific objectives: (1) characterize the composition, structure, and chemical behavior of Titan's upper atmosphere and its interaction with the plasma of Saturn's magnetosphere and (2) investigate the neutral and plasma environments of the rings and icy satellites and their interactions with the magnetosphere. Key mission events for the INMS investigation include Saturn Orbit Insertion (SOI), during which the spacecraft will pass at low altitude over the ring system and cross the ring plane, providing a unique opportunity for in-situ measurement of the ring environment; over 40 planned close encounters with Titan; and close flybys of the icy satellites Enceladus, Dione, and Rhea. The primary measurement objective during the icy satellite encounters will be the detection of neutral species that have been sputtered from the satellite surfaces by charged particle bombardment. Measurement of this material will provide information about the surface composition of the satellites and about the formation of the neutral water/water product cloud surrounding Saturn. Of the Titan flybys, the sixth, nearly 9 months after SOI, will be particularly important, as it will occur just 3 months after the descent of the Huygens Probe through Titan's nitrogen–methane atmosphere. INMS data acquired near 950-km altitude during this time (Flyby T5: April 16, 2005) will provide detailed composition data on the upper atmosphere, which can be compared with Probe data on the lower atmospheric composition.

1.1. The INMS Titan Investigation

During the Orbiter's Titan encounters, the INMS instrument will gather data on the composition, density, and temperature of Titan's upper neutral atmosphere and on the composition and density of the ionosphere. These data, together with complementary neutral atmosphere data from the Orbiter Radio Science Subsystem (RSS), UV Imaging Spectrograph (UVIS), Visible and Infrared Mapping Spectrometer (VIMS), and Composite Infrared Spectrometer (CIRS) experiments and the Probe atmospheric experiments and ionospheric data from the Cassini Plasma

Spectrometer (CAPS), the Radio Plasma Wave Spectrometer (RPWS), and the RSS investigations, will be used (1) to characterize Titan's neutral atmosphere and ionosphere, with particular emphasis on the chemistry that initiates the formation of complex hydrocarbon and nitrile compounds present in the aerosol haze layers and perhaps in liquid and solid form on the surface; and (2) to study the interaction of Titan's upper atmosphere/ionosphere with Saturn's magnetosphere and, occasionally, the solar wind. In the following, we will review the science background for the INMS Titan investigation and outline the measurements that the INMS instrument will make to address the investigation's science objectives. Before turning to this discussion, however, we would like to note that the INMS measurements have operational as well as scientific importance. INMS data on neutral atmosphere densities acquired during the Orbiter's first pass (at a conservative altitude of 1200 km) will be used to assess possible atmospheric drag effects on the spacecraft and will be factored into planning Orbiter trajectories for subsequent Titan encounters. It is hoped that the analysis of the density measurements made during the first Titan encounter will indicate that the sixth Titan flyby can take place at a lower altitude, $\sim$950 km, allowing for measurement deeper in the atmosphere at a time when the atmosphere will not have undergone significant changes since the Probe data were taken. This would provide a valuable basis for correlative studies, based on INMS and Probe data, of the upper and lower regions of Titan's atmosphere and of the processes coupling them.

1.1.1. *Titan's Neutral Upper Atmosphere*

Our present knowledge of Titan's neutral atmosphere rests wholly on remote-sensing (infrared imaging and radio and ultraviolet occultation) data acquired during the Voyagers 1 and 2 Saturn flybys and on observations with various ground-based telescopes, and, since 1997, with the European Space Agency's Infrared Space Observatory (ISO). The INMS instrument will make the first in-situ measurements of Titan's upper atmosphere and, with over 40 planned flybys of the moon, will achieve the extensive seasonal, spatial, and temporal coverage needed to understand the atmosphere's varying response to solar and magnetospheric energy inputs. The INMS data will provide altitude-dependent composition profiles both for the major atmospheric gases (N_2 and CH_4) and for the hydrocarbons and nitriles (e.g., C_2H_2, C_2H_6, HCN, etc.) created through methane and nitrogen photochemistry. The non-methane hydrocarbons significantly influence the energetics and thermal structure of the atmosphere and are sensitive probes of upper atmospheric chemistry. Such profiles will thus be of crucial importance in elucidating the coupled chemistry and dynamics of Titan's upper atmosphere. The INMS will also measure isotope ratios (H/D and $^{14}N/^{15}N$) and argon (if present). These data, together with information about atmospheric escape and loss resulting from the upper atmosphere's interaction with Saturn's magnetosphere and other processes, will provide evidence about the formation and evolution of Titan's atmosphere over geologic time. Finally, as presently planned, the INMS will not sample Titan's atmosphere below 950 km;

however, the information that the INMS can provide—both through direct measurement of neutral densities and through measurement of ion densities from which neutral densities can in some cases be deduced—on Titan's thermosphere will be of direct relevance to questions of aerosol formation, neutral composition and chemistry, atmospheric dynamics, and radiative transfer in the lower atmosphere. INMS measurements in the upper atmosphere will thus provide a valuable and necessary complement to data on the lower atmosphere from the Probe Aerosol Collector and Pyrolysor (ACP), Huygens Atmospheric Structure Instrument (HASI), and Gas Chromatograph/Mass Spectrometer (GC/MS) and the Orbiter UV Imaging Spectrograph (UVIS), Visibile and Infrared Mapping Spectrometer (VIMS), and Composite Infrared Spectrometer (CIRS) investigations.

1.1.1.1. Composition and Photochemistry. Titan has a massive neutral atmosphere, with a surface pressure of ~1.5 bar. The major neutral constituents are nitrogen, methane, and hydrogen. Nitrogen, with a mixing ratio in the troposphere of ≥90%, is the dominant species. Its presence in Titan's atmosphere was proposed by Lewis (1971) and Hunten (1972) on theoretical grounds and was established with the Voyager UVS instrument (Broadfoot *et al.*, 1981), which detected extreme ultraviolet emissions from molecular and atomic nitrogen.

Methane was the first constituent to be identified in Titan's atmosphere; its detection—by Kuiper (1944)—confirmed the suggestion made over three decades earlier by Comas Solá (1908) that Titan had an atmosphere. Based on their analysis of Voyager 1 IRIS data, Samuelson *et al.* (1997) posit a methane-supersaturated troposphere with a methane mixing near the surface of 0.057 at the equator and of 0.02 at higher latitudes. Courtin *et al.* (1995) estimate the stratospheric methane mole fraction to lie between 0.026 and 0.045. In the thermosphere, where the INMS measurements will be made, methane mixing ratios of 0.06 ± 0.01 and 0.20 ± 0.02 at altitudes of 1000 and 1400 km, respectively, have been inferred from solar EUV occultation measurements made with the Voyager 1 UVS (Strobel *et al.*, 1992). However, recent re-analysis of the Voyager UVS occultation results indicates lower CH_4 and higher N_2 densities than previously reported (Vervack, 1997; Vervack *et al.*, 1999). The implications of these new and not yet widely disseminated findings for the chemistry and thermal structure of Titan's atmosphere are still being investigated.

The presence of molecular hydrogen in Titan's troposphere, tentatively identified by Trafton (1972) on the basis of ground-based observations, was confirmed by IRIS spectra (Samuelson *et al.*, 1981). The most recent estimate of the tropospheric H_2 mole fraction, based on a re-analysis of the IRIS data by Samuelson *et al.* (1997), is ~0.0011.

Although it has not been detected, argon is considered to be a possible major constituent of Titan's atmosphere, i.e., present at the few percent level. A wide range of abundances has been proposed. To account for the atmosphere's mean molecular mass of 28.6 Daltons estimated from the Voyager radio occultation and

IRIS data, Samuelson *et al.* (1981) invoked the presence of argon and estimated its mole fraction at the surface to be 11.6%. Based on Voyager UVS data, Broadfoot *et al.* (1981) estimated an upper limit of 6% for the argon mixing ratio in the thermosphere; extrapolation from this value yields a mixing ratio at Titan's surface of 30% for ^{36}Ar (Hunten *et al.*, 1984). Lellouch *et al.* (1989) consider a mole fraction of 21% to be the upper limit on argon abundance at the surface. Calculations by Strobel *et al.* (1993), based on the Voyager UVS solar occultation data, indicate that a value of $\leq$10% represents a reasonable upper limit on the argon mixing ratio at the tropopause; and recent re-analyses of IRIS data by Courtin *et al.* (1995) and Samuelson *et al.* (1997) yield argon mole fractions in the troposphere of <0.06 and 0.026 ± 0.045, respectively. The measurement of argon is important because of its relevance to the question of the origin and evolution of Titan's atmosphere. A large argon-to-nitrogen ratio (0.01–0.1) would support the hypothesis (Owen, 1982; Strobel and Shemansky, 1982) that the N_2 component of Titan's atmosphere was derived from clathrates, and was thus present from the beginning, rather than the product—accumulated over geologic time—of NH_3 photodissociation (Courtin *et al.*, 1995, and references therein).

The Voyager IRIS data confirmed the presence of several non-methane hydrocarbons (C_2H_6, C_2H_2, and C_2H_4) that had been detected earlier with ground-based observations and revealed the presence of other hydrocarbons not previously detected (C_3H_4 and C_3H_8), along with presence of the nitriles HCN, HC_3N, and C_2N_2 (Hanel *et al.*, 1981; Maguire *et al.*, 1981; Kunde *et al.*, 1981). (Recent estimates of the abundances of these and other species are given in Table I.) Limited information about non-methane hydrocarbons in Titan's upper atmosphere was also provided by the Voyager UVS solar occultation data, which revealed a sharp increase in the C_2H_2/N_2 mixing ratio in the altitude range 725–825 km above Titan's surface from 0.1–0.3% below 725 km to 1–2% above 825 km (Smith *et al.*, 1982); however, as in the case of CH_4, the re-analysis of these data by Vervack (1997) and Vervack *et al.* (1999) has yielded lower C_2H_2 densities than those retrieved by Smith *et al.* The UVS occultation data also indicated the possible presence of a polymer layer between 675 and 1025 km at both the evening and morning terminators.

The oxygen-bearing compounds CO_2 (Samuelson *et al.*, 1983), CO (Lutz *et al.*, 1983), and—most recently—H_2O (Coustenis *et al.*, 1998) have also been detected in Titan's atmosphere. The principal source of oxygen in Titan's atmosphere is believed to be water vapor released by sublimation from infalling icy meteoroids (Samuelson *et al.*, 1983; Yung *et al.*, 1984). Photolysis of the water vapor thus introduced yields H, which is lost from the atmosphere, and the hydroxyl radical. OH reacts with the methyl radical CH_3 (from the photodissocation of methane) to produce CO and molecular hydrogen. Some of the CO then reacts with OH to form CO_2 and H. Based on the data acquired with the ISO Short Wavelength Spectrometer, Coustenis *et al.* (1998) estimate the mole fraction of water vapor at 400 km to be $\sim 8 \times 10^{-9}$ and the meteoritic water influx at an altitude of 700 km to be $(0.8–2.8) \times 10^6$ mol cm^{-2} sec^{-1}.

TABLE I

Abundances of atmospheric constitutents in Titan's atmosphere.

Species	Mole fraction	Reference
	Troposphere ($z \leq 40$ km)	
N_2	0.90–0.99	Gautier and Raulin (1997)
CH_4	~0.057	Samuelson *et al.* (1997)
Ar	0.026 ± 0.045	Samuelson *et al.* (1997)
H_2	~0.0011	Samuelson *et al.* (1997)
	Stratosphere ($z \sim 40$–325 km)	
CH_4	0.026–0.045	Courtin *et al.* (1995)
CO	2.9×10^{-5} at 60 km; 2.4×10^{-5} at 175 km	Hidayat *et al.* (1998)
C_2H_6	9.4×10^{-6} at $z \sim$ 120–350 km	Kostiuk *et al.* (1997)
C_2H_2	2.2×10^{-6} at 130 km	Coustenis *et al.* (1989)
C_3H_8	7.0×10^{-7}	Coustenis *et al.* (1989)
HCN	3.5×10^{-7} at $z \geq 200$ km, decreasing below 200 km to $\sim 10^{-8}$ at the condensation level (~80 km)	Hidayat *et al.* (1997)
C_2H_4	9.0×10^{-8} at 125 km	Coustenis *et al.* (1989)
CO_2	1.4×10^{-8} at 110 km	Coustenis *et al.* (1989)
C_3H_4	4.4×10^{-9} at 105 km	Coustenis *et al.* (1989)
C_4H_2	1.4×10^{-9} at 110 km	Coustenis *et al.* (1989)
HC_3N	$\leq 1.5 \times 10^{-9}$ at 105 km	Coustenis *et al.* (1989)
C_2N_2	$\leq 1.5 \times 10^{-9}$ at 105 km	Coustenis *et al.* (1989)
	Mesosphere ($z \sim 325$–600 km)	
CO	4.8×10^{-6} at 350 km	Hidayat *et al.* (1998)
H_2O	8×10^{-9} at 400 km	Coustenis *et al.* (1998)
	Thermosphere ($z \sim 600$ km)	
CH_4	0.06 ± 0.01 at 1000 km; 0.20 ± 0.02 at 1400 km	Strobel *et al.* (1992)
C_2H_2	0.001–003 at $\geq$725 km; 0.01–0.02 at $\geq$825 km	Smith *et al.* (1982)

Note: Except for argon, which has not been detected but is postulated to be present, the values listed above are observation-derived estimates of the abundances of species that have been identified in Titan's atmosphere. For details of the models used to derive specific abundances from the data, and for uncertainties and differences from values reported in earlier studies, the reader should consult the references cited. The altitude ranges for the diffferent atmospheric layers are based on the thermal structure model of Yelle *et al.* (1997). The mixing ratios given by Smith *et al.* (1982) for acetylene in the thermosphere are based on the N_2 density rather than the total atmospheric density. This table is a modified and updated version of Table I of Toublanc *et al.* (1995).

The interpretation of Voyager IRIS data has yielded some limited information about variations with latitude and altitude in the distribution of certain atmospheric constituents. Coustenis *et al.* (1991) find that the abundances of the hydrocarbons C_2H_2, C_4H_2, and C_3H_4 and of the nitriles HC_3N and C_2N_2 (and to a lesser degree of

HCN) are greater in Titan's north polar stratosphere than at the equator. (Coustenis *et al.*, speculate that the cold temperature of the polar stratosphere, compared to that of the equatorial stratosphere, may be attributable to increased IR cooling resulting from the enhanced hydrocarbon abundance.) No significant latitudinal differences were seen in the mixing ratios of the three most abundant hydrocarbons (ethane, acetylene, and propane), while the abundance of CO_2 was a factor of two lower at the pole than at the equator. The reason for these variations is not known but may have to do with seasonal and latitudinal variability in the solar ultraviolet flux at Titan and with differences in the photodissociation rates of the various species. Analysis of the north polar data (vertical distribution information cannot be derived from the equatorial data) also reveals an increase in the mixing ratio with altitude for the nitriles and some of the hydrocarbons. As Coustenis *et al.*, point out, the positive gradient in mole fraction with increasing altitude is explicable in terms of the photochemical production of these species in Titan's thermosphere.

The non-methane hydrocarbons and nitriles identified in the stratosphere by Voyager are created by complex photochemical reactions occurring among the products of CH_4 and N_2 dissociation (Strobel and Shemansky, 1982; Yung *et al.*, 1984; Yung, 1987; Toublanc *et al.*, 1995; Lara *et al.*, 1996; Wilson, 2002). In Titan's thermosphere and mesosphere, CH_4 is photolyzed directly by solar extreme ultraviolet photons. The principal products of direct methane photolysis at high altitudes are C_2H_2 (the most abundant non-methane hydrocarbon in Titan's atmosphere) and C_2H_4. In the stratosphere, where the bulk of the CH_4 destruction takes place, CH_4 dissociation occurs catalytically, in conjunction with the photodissociation of acetylene (transported downward from its production region in the upper atmosphere); the main products of methane photochemistry in the stratosphere are ethane and propane. The hydrogen produced by methane dissociation and in other reactions escapes from Titan's atmosphere to space and contributes to a cloud of neutral molecular and atomic hydrogen that extends from $\sim$25 R_S inward to the top of Saturn's atmosphere (Shemansky and Hall, 1992). (Hydrogen escape makes possible the irreversible destruction of methane and its conversion to heavier hydrocarbons.) The principal loss process for the hydrocarbons produced by methane photolysis is condensation at the tropopause and precipitation onto Titan's surface, possibly resulting in the formation, over the age of the solar system, of a hydrocarbon ocean (Lunine *et al.*, 1983; Lunine, 1993, 1994).

N_2 is dissociated in Titan's upper atmosphere both photolytically, by EUV photons ($\lambda < 80$ nm), and by the impact of energetic electrons from Saturn's magnetosphere. (Prior to the re-analysis by Strobel *et al.* (1991, 1992) of Titan's EUV airglow emissions, electron impact was held to be the dominant mechanism for nitrogen dissociation. Recent calculations (Toublanc *et al.*, 1995, Table II and associated discussion) indicate that the dissociation rate for EUV photolysis is about five times that for electron impact). Nitrogen is also dissociated, in the stratosphere, by cosmic ray absorption (Capone *et al.*, 1983; Lara *et al.*, 1996). Nitrogen atoms produced by the dissociation of N_2 react with methyl and methylene radicals produced by methane photolysis to form the most abundant nitrile, hydrogen cyanide (Yung *et al.*, 1984; Yung, 1987). HCN then participates in photochemical reactions that, together with reactions between dissociated nitrogen atoms and acetylene, lead to the formation of the nitriles HC_3N, C_2N_2, C_4N_2, and CH_3CN. Like the non-methane hydrocarbons, the nitriles are lost through condensation at the tropopause.

In addition to the creation of the lighter hydrocarbons, CH_4 and N_2 photochemistry is expected to lead to the synthesis, at altitudes above 500 km, of heavier, more complex molecules that may play an important role in aerosol formation. For example, polyacetylenes ($C_{2n}H_2$), created through insertion reactions involving ethynyl (C_2H) and butadiynyl (C_4H) radicals (from the photolysis of C_2H_2 and C_4H_2, respectively), have been proposed as one source of the aerosols that constitute Titan's detached haze layer, which lies above the main layer at altitudes between 350 and 400 km (Chassefière and Cabane, 1995). Chassefière and Cabane also suggest that C–N oligomers synthesized from dicyanoacetylene (C_4N_2) may be an additional source of aerosols for the detached haze layer comparable in importance to the polyacetylenes. Production of both $C_{2n}H_2$ and C–N oligomers occurs at altitudes of ~500–800 km.

A further contribution to the detached haze layer may come from C–H–N polymers created at higher altitudes (900–1000 km) by energetic electron bombardment of Titan's upper atmosphere. According to Chassefière and Cabane's two-layer haze model, however, C–H–N polymers are primarily important for the formation of the main haze layer. This layer is distinct from the detached layer in terms of its composition, formation region, and formation mechanism and consists predominantly of aerosols created from C–H–N polymers produced by energetic-particle-initiated chemistry at altitudes between 350 and 400 km. (Such C–H–N polymers are conventionally termed "tholins" following the usage of the researchers at Cornell who have synthesized them in the laboratory under conditions approximating those at Titan and consider them likely to be a significant source for Titan's haze (e.g., Thompson *et al.*, 1994, and references therein.)

Photochemical models of Titan's neutral atmosphere have been developed by Yung *et al.* (1984), Toublanc *et al.* (1995), Lara *et al.* (1996), Wilson (2002), and, as a background atmosphere for their ionosphere model, by Fox and Yelle (1997) (cf. Figure 1). Coupled neutral-charged particle models have also been recently developed by Banaszkiewicz *et al.* (2000), as an extension of the Lara *et al.* model, and Wilson (2002). The model by Yung *et al.* (with updated nitrile chemistry; Yung, 1987) has until recently been the standard model for Titan photochemistry. Neutral abundances calculated with this model were in reasonable agreement with those determined from the early analyses of the data acquired during the Voyager encounters in 1980 and 1981. However, more extensive analysis of the Voyager data by Coustenis *et al.* (1989) indicated that, with the exception of CO_2 and C_2H_4, the model predicted concentrations of minor species that were higher than the values observed in Titan's equatorial stratosphere. Coustenis *et al.* suggested that the discrepancy between model-generated and observed values could be reduced by adjustment of the eddy diffusion coefficient or through adoption of a revised rate constant for the reaction $H + C_4H_2$ that would increase the concentration of atomic hydrogen available to suppress the formation of heavier hydrocarbons. Analysis of IRIS data for Titan's north polar stratosphere likewise showed that the model of Yung *et al.* overestimated the abundances of most minor species in that region (Coustenis *et al.*, 1991).

The recently published models by Toublanc *et al.* (1995) and Lara *et al.* (1996) attempt to achieve better agreement with the Voyager data than is possible with the Yung *et al.* model. Toublanc *et al.* incorporate a new scheme in their model for methane photolysis, improved photolysis rates and radiative transfer calculations, and a revised eddy diffusion coefficient constrained to fit an HCN profile derived from recent ground-based observations. The Wilson (2002) model uses updated rate coefficients, cross sections, and quantum yields, with a radiative transfer scheme used in the study of Jovian atmospheres (Edgington *et al.*, 1998), to obtain chemical abundances in a coupled ion–neutral environment. With this model, a best-fit eddy diffusion profile yields chemical abundances, with the exception of C_2H_4, that are underpredicted by no more than a factor of three compared to observational constraints, which may be misinterpreted on the high end at the given altitudes as a result of uniform mole fraction assumptions inherent in the observations, as in the case of CO_2 (Coustenis *et al.*, 1989) and C_6H_6 (Cousteinis *et al.*, 2002).

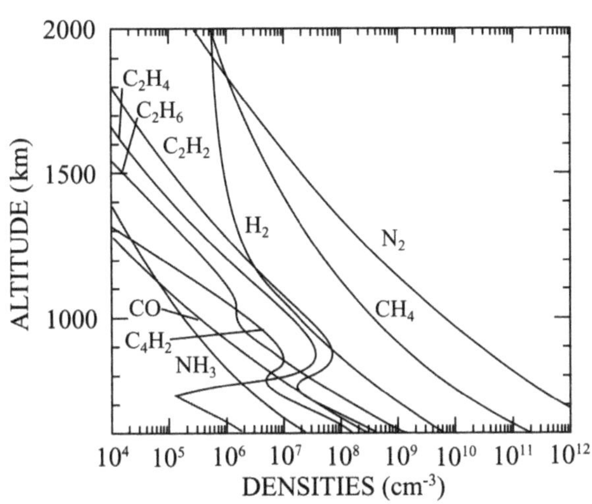

Figure 1. Neutral densities as a function of altitude in Titan's upper atmosphere (from Fox and Yelle, 1997). Hydrocarbon profiles are based on the neutral atmosphere model of Yung *et al.* (1984); N_2 and CH_4 profiles are from Strobel *et al.* (1992). The minimum altitude at which INMS measurements will be made is 950 km. In closed source mode, the INMS will be able to measure neutrals with concentrations $\geq 10^4$ cm^{-3}; in the open source, densities $\geq 10^5$ cm^{-3} can be measured.

These models obtained better fits to the observations for some species than were achieved by Yung *et al.* Numerous differences remain, however, both between observed values and those calculated by each model and between the models themselves. These differences are attributable to insufficient observational data, the lack of accurate rate coefficients for some key reactions, and uncertainty about the eddy diffusion coefficient. Reconciling these differences and obtaining improved agreement with observational constraints is important both for our understanding of Titan's neutral atmosphere and also for the development of model ionospheres (cf. Keller *et al.*, 1998). Thus, evaluation of existing neutral atmosphere models and development of improved models is one of the key tasks undertaken by the INMS Science Team as part of its planning for the neutral atmosphere and ion composition measurements to be made during the Orbiter's Titan flybys. (A preliminary list of target constituents, based on the present state of model development and currently available data, is shown in Table II.)

TABLE II

Expected neutral species in Titan's upper atmosphere.

Molecular weight	Closed source	Open source neutral
2	H_2	
3	HD	
4	He	
14		N
15		NH
16	CH_4	CH_4, O
17	$^{13}CH_4$	OH
18	H_2O	H_2O
26	C_2H_2	
27	HCN	HCN
28	N_2, C_2H_4, CO	N_2, C_2H_4, CO
29	$^{15}N^{14}N$, $^{13}C_2H_4$	
30	C_2H_6	
36	(^{36}Ar)	
39		CHCN
44	CO_2, C_3H_8	
50		C_3N
51	CH_3CN, HC_3N	
52	C_2N_2	
74		C_6H_2
76	C_4N_2	
78	C_6H_6	

1.1.1.2. Thermal Structure. The temperature of Titan's upper atmosphere between 1265 and 1525 km above the surface has been inferred from Voyager UVS measurements of N_2 density variations to be 186 $\pm$ 20 K (Smith *et al.*, 1982). Below an altitude of 200 km, the thermal profile can be derived from RSS and IRIS data, while constraints on temperatures in the region between 200 and 450 km are provided by IRIS measurements and stellar occultation data (cf. Lellouch *et al.*, 1990, and references therein). Analysis of these data and modeling studies reveals an atmospheric structure characterized by negative temperature gradients in the troposphere and mesosphere and positive gradients in the stratosphere and thermosphere. The principal heat source for the stratosphere is the absorption of solar radiation by organic haze particles, while the thermosphere and exosphere are heated by the absorption of solar UV radiation by CH_4 and, above $\sim$1000 km, by nitrogen (Lellouch *et al.*, 1990; Yelle, 1991). Magnetospheric electron precipitation is expected to contribute to thermospheric heating as well, but to a lesser and still undetermined degree (cf. Strobel *et al.*, 1991, 1992 on thermospheric energy sources). (The relative contributions of the solar EUV and magnetospheric energy inputs to Titan aeronomy will vary with Titan's orbital position and degree of exposure to the magnetospheric electron flux, which is maximized on the wake side). The thermally active hydrocarbons CH_4, C_2H_2, and C_2H_6 serve both as heating agents and coolants, their particular role varying with altitude. HCN has been shown to be an extremely important coolant.

Four models of the vertical structure of Titan's upper atmosphere have been developed to date. Friedson and Yung (1984) made the first attempt to interpret the Voyager UVS measurements of Titan's exospheric temperature by constructing an aeronomical model that included solar EUV and FUV radiation, thermal conduction, and non-LTE cooling by hydrocarbons (principally C_2H_2) in the mid-IR. Hydrocarbon cooling rates were estimated by assuming that the emissions were optically thin. Friedson and Yung were able to match the Voyager measurements quite well; however, the solar heating rates used in their model were later discovered to be in error (Lellouch *et al.*, 1990). The correct solar heating rates were found to yield a temperature near the exobase of $\sim$300 K, which is in serious disagreement with the observations. Using an aeronomical model similar to Friedson and Yung's, Lellouch *et al.* succeeded in producing a thermospheric temperature profile with an exospheric temperature consistent with that derived from Voyager UVS data, but only by adopting unreasonably low heating efficiencies (0.015–0.20) and an unreasonably large C_2H_2 cooling rate.

The third model of the thermal structure of Titan's upper atmosphere is a sophisticated aeronomical model by Yelle (1991) that includes non-LTE radiative transfer in the mid-IR vibration–rotation bands of CH_4, C_2H_2, and C_2H_6, a detailed treatment of vibrational chemistry, and, most importantly, cooling by rotational emissions from HCN (Figure 2). Neither HCN nor ethane was included in the models of Friedson and Yung and Lellouch *et al.* Yelle's model predicts that HCN rotational cooling is sufficient to balance solar UV heating of the thermosphere and

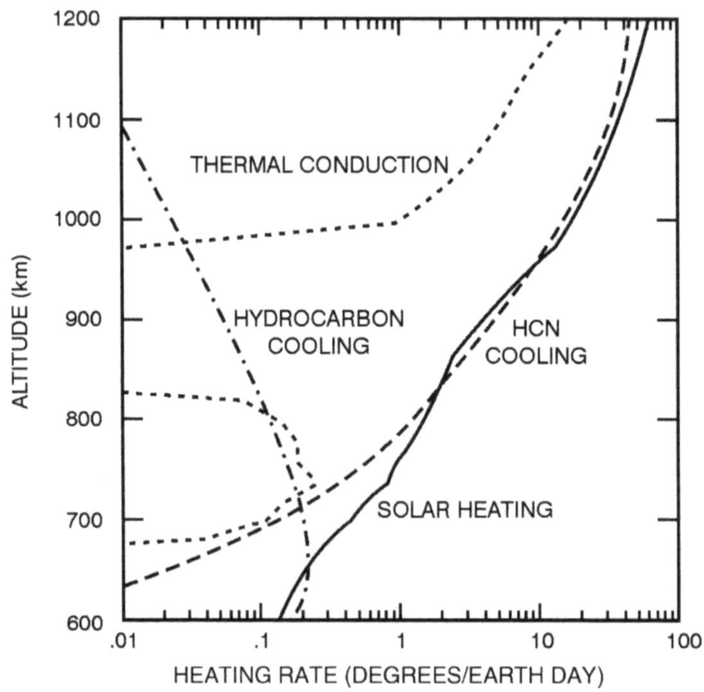

Figure 2. Plot illustrating the contributions of various heating and cooling processes to the thermal structure of Titan's atmosphere (Yelle, 1991).

to maintain the thermospheric temperature near the value derived from the Voyager UVS data (186 ± 20 K) and that neither low heating efficiencies nor enhanced C_2H_2 cooling rates are required. This result suggests that Titan's thermal structure is radiatively controlled at all levels and distinguishes Titan's atmosphere from most other atmospheres, whose thermospheres are dominated by thermal conduction. The important influence of HCN cooling on Titan's thermospheric structure predicted by the Yelle model can be confirmed and characterized with INMS measurements of HCN abundance and its variability with time and latitude.

The fourth model is the engineering model developed by Yelle *et al.* (1997). It is an empirical model based on the Yelle (1991) physical model (Figure 3).

All four of the above models predict a well-developed mesopause. Friedson and Yung (1984) calculated a very cold mesopause temperature of ~110 K at an altitude of 736 km. As Lellouch *et al.* (1990) point out, however, if hydrostatic equilibrium is to be maintained, such a low mesopause temperature requires higher stratospheric temperatures than are consistent with the IRIS data. In the model of Lellouch *et al.*, the mesopause occurs at an altitude of ~800 km and has a temperature of 135 K. Yelle's model predicts a comparable mesopause temperature (135–140 K) but locates the mesopause at an altitude of ~600 rather than 800 km. The Yelle model indicates that CH_4—not acetylene, as assumed by Friedson and Yung and Lellouch *et al.*—is the primary coolant near the mesopause and, surprisingly, that

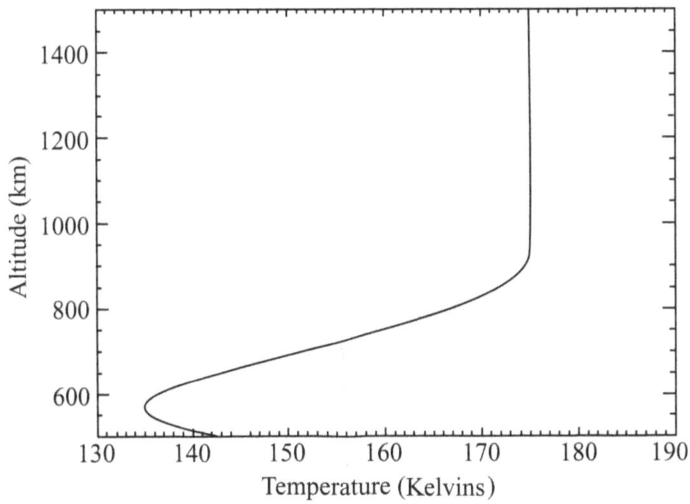

Figure 3. Temperature profile with altitude for Titan's upper atmosphere ($z > 500$ km), according to the engineering model of Yelle *et al.* (1997). The mesopause, with a temperature of ~135 K, is located at around 570 km. The model atmosphere above 900 km is isothermal; in actuality, thermospheric temperature will show both temporal and spatial variations. The engineering model uses a nominal altitude of 1040 km for the homopause. It should be noted, however, that there is some disagreement among the models, with some models indicating that Titan does not have a homopause. The exobase is located at an altitude of 1400 km.

ethane—not considered by either Friedson and Yung or Lellouch *et al.*—is the dominant heating agent in this region.

The 1D models described above are useful for identifying important physical processes and representing a global average atmosphere, but Titan's atmosphere is three dimensional and 3D Thermospheric General Circulation Models (TGCMs) are needed to study the full complexity of the upper atmosphere. Such models have been presented by Müller-Wodarg *et al.* (2000) and Müller-Wodarg and Yelle (2002) while initial estimates on the importance of dynamics have been presented by Rishbeth *et al.* (2000). The models incorporate heating and cooling processes, dynamical processes, and simplified chemistry. The comprehensive nature of the models makes them ideal for interpretation of INMS measurements of temperature and densities across Titan's globe. They predict variations in latitude and local time that are well within the INMS measurement capability.

Müller-Wodarg *et al.* (2000) presented the first Titan TGCM and calculated a circulation pattern with upwelling slightly after local noon, horizontal flow from the day to night side, and subsidence at night (see Figure 4). These authors calculated day/night temperature difference of 20 K and a maximum wind speed of 60 m sec^{-1}, of the same order as winds in the Earth's thermosphere. Dynamics strongly affect the temperature field, greatly reducing gradients over radiative solutions; however, the predicted temperature differences of 20 K are easily measured by INMS. On

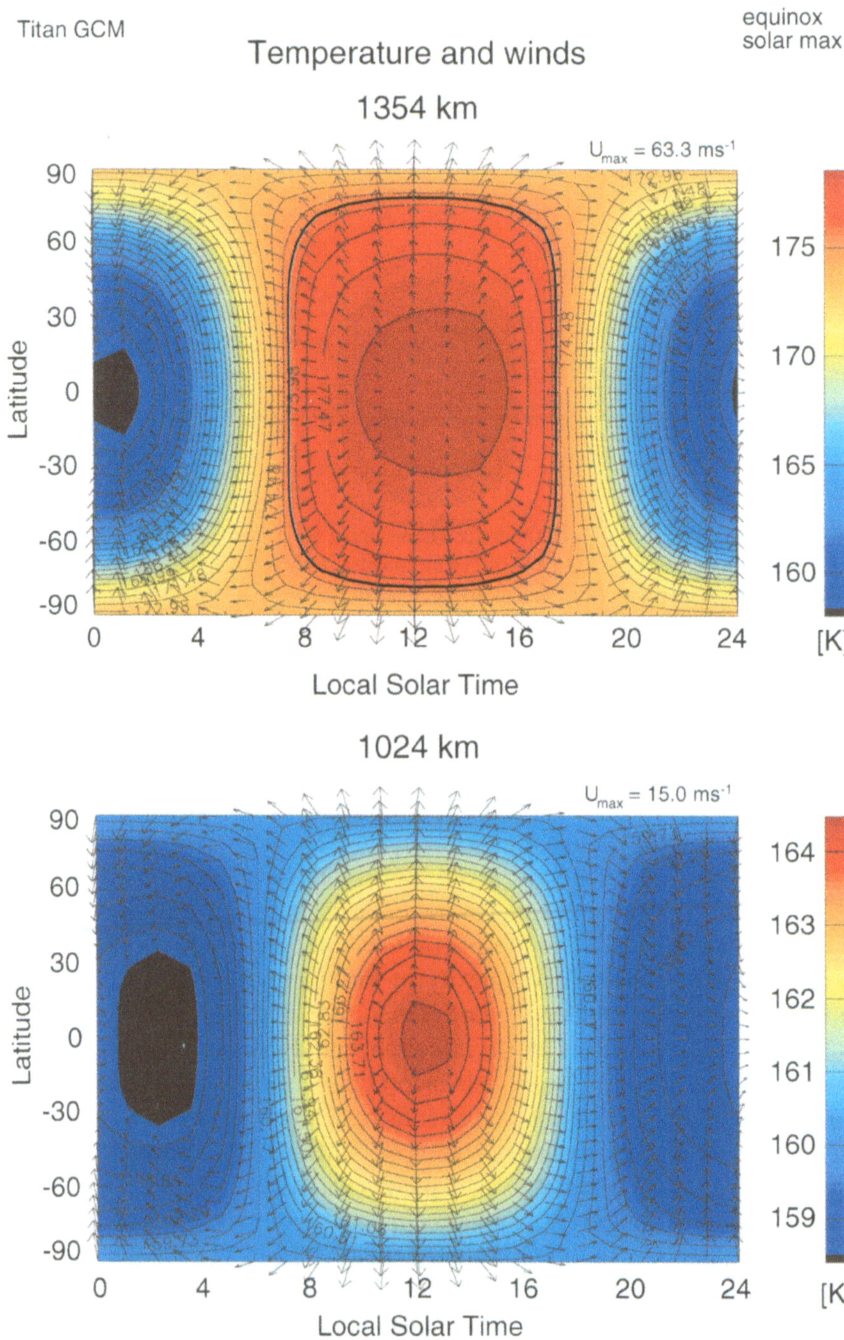

Figure 4. Simulations of temperature and wind fields at two altitudes in Titan's thermosphere. The temperatures and winds are calculated with a 3D thermospheric general circulation model that solves the continuity, momentum, and energy equations, including only solar input.

Earth, pressure gradient forces are balanced by molecular viscosity and ion drag. On Titan, molecular viscosity is weak because of the low temperature and densities while ion drag is weak because of the weak or absent magnetic field (Rishbeth *et al.*, 2000). Thus, despite the low solar heating rate, Titan has strong winds which will strongly affect the distribution of molecular constituents in its atmosphere.

Extension of the Titan TGCM to include composition shows that the dynamics in the upper atmosphere cause large changes in the distribution of minor constituents. Müller-Wodarg *et al.* (2000) show that solar driven dynamics can cause a factor of several change in the CH_4 mole fraction with local time. This is caused primarily by the interaction of the wind field with diffusive flows in the upper atmosphere. Similar phenomena have been seen in the upper atmospheres of the Earth and Venus. The calculated CH_4 distribution exhibits a maximum in late afternoon over the region of maximum subsidence and a minimum in early morning over the region of maximum upwelling. These variations also alter the heating rates in the upper atmosphere. Horizontal variations are likely for other species as well with species whose mean molecular weight differs substantially from N_2 showing the greatest variation. Analysis of horizontal variations of long-lived constituents will provide the best means for diagnosing winds in the upper atmosphere.

Fortunately, because N_2 and CH_4 are nonreactive gases, the more accurate closed source INMS mode can be used to make the measurements from which the thermal structure will be derived. Closed source measurements have the further advantage of being less dependent than the open source measurements on instrument pointing direction. An exospheric temperature of 175 K is used for present planning purposes. This value gives an atmospheric scale height of 81 km near 1000 km. The tangent path of the geometry increases the path length through the atmosphere by a factor of approximately 10, which, with a spacecraft velocity of 6.1 ± 0.3 km sec^{-1}, gives a time of $\sim$150 sec for traversing one scale height near closest approach. Determination of the temperature requires a minimum of five samples over this time interval or, in other words, a sample of major constituent densities every 30 sec. This temporal/spatial resolution can be increased 1000 fold if required.

1.1.2. *Titan's Ionosphere*

The characterization of Titan's ionosphere, and of the interaction of the ionosphere and neutral upper atmosphere with Titan's plasma environment, is a major objective of the INMS investigation and the Cassini mission. To achieve this objective, the INMS instrument will measure ionospheric composition as a function of altitude at various locations on Titan (day side, night side, ram side, wake side). Most of these data will be acquired within Saturn's magnetosphere; however, measurements may also be made when Titan, at times of high solar wind dynamic pressure, lies outside the subsolar magnetopause, in Saturn's magnetosheath or the solar wind. The INMS measurements, along with CAPS/Ion Beam Spectrometer (IBS) ion velocity and temperature measurements and CAPS/Ion Mass Spectrometer composition measurements, will be the first in situ measurements of the ionosphere of an outer planetary body. INMS ion composition and density data, together with complementary data from other Orbiter investigations (e.g., CAPS, UVIS, MAG, RSS, RPWS, etc.), will be used to address questions relating to the composition, chemistry, structure, and dynamics of Titan's ionosphere; the interactions between the ionosphere and neutral atmosphere; the relative roles of solar EUV and magnetospheric energy inputs in ion production and in ionospheric energetics and dynamics; and the loss of atmospheric material through hydrodynamic outflow, scavenging by the magnetospheric plasma flow, and the production of fast neutrals through ionospheric photochemistry. Finally, it is possible that the densities of important but difficult-to-measure neutrals—such as CH_3—can be deduced from ion densities measured with the INMS and CAPS, allowing important chemical pathways to be distinguished by which, for example, HCN and polyacetylenes are formed in the upper atmosphere.

1.1.2.1. Structure and Composition. Observational data on Titan's ionosphere are limited to Voyager 1 radio occultation measurements. No definitive detection of an ionosphere was reported; however, initial analysis of these data placed an upper limit on the peak ionospheric electron density of 3×10^3 cm^{-3} near the evening terminator and of 5×10^3 cm^{-3} near the morning terminator (Lindal *et al.*, 1983; McNutt and Richardson, 1988). Re-analysis of the Voyager ingress data by Bird *et al.* (1997) indicates that the Voyager measurements yield a "marginal detection" of an ionosphere with an electron density peak of 2400 ± 1100 cm^{-3} at an altitude of 1180 ± 150 km. These values (for the evening terminator) are consistent with those computed by Keller *et al.* (1992) for the terminator ionosphere (3030 cm^{-3} at an altitude of 1175 km). On the dayside, with an average solar zenith angle of 60°, the electron density maximum is predicted to be larger and the altitude of the density peak lower. For the dayside case, Keller *et al.* calculated a density maximum of 6150 cm^{-3} at 1055 km, which compares well with the 7.5×10^3 cm^{-3} at 1040 km from the recent model by Fox and Yelle (1997).

The composition of Titan's ionosphere is determined by a complex chemistry involving neutrals, ions created by photoionization and electron impact (e.g., N_2^+, N^+, CH_4^+, CH_3^+, etc.), and ions formed in ion–neutral reactions. Recent modeling studies (by Fox and Yelle, 1997; Keller *et al.*, 1998) indicate that the major ionospheric constituents are N_2^+, hydrocarbon ions (including higher-mass ions with three or more carbon atoms), and ions containing H, C, and N. Of the nitrile ions, $HCNH^+$ is the most common and is predicted by several ionospheric models to be the dominant ion species near the ionospheric peak (~1100 km). The dominance of $HCNH^+$ is questioned by Fox and Yelle, whose model provides for relatively rapid loss of $HCNH^+$ through ion–neutral reactions not included in earlier models and predicts an $HCNH^+$ density in the ionospheric peak region several times lower than that previously computed (e.g., Ip, 1990; Keller *et al.*, 1992). Keller *et al.* (1998) incorporate similar chemistry in an updated version of their 1992 model and calculate a somewhat lower density for $HCNH^+$ as well (compared with their earlier calculation). However, the inclusion of the new loss channels notwithstanding, $HCNH^+$ remains the major ion near the ionospheric peak in their model, with a density three times that estimated by Fox and Yelle. As Keller *et al.* point out, this disagreement results from differences between the neutral background atmospheres used in the two models.

The discrepancy between the Fox/Yelle and Keller *et al.* models regarding the density of the important ionospheric species $HCNH^+$ illustrates the sensitivity of Titan ionosphere models to assumptions about the neutral atmosphere. Our limited knowledge of the latter is a major source of uncertainty in the former. A further source of uncertainty is the lack of laboratory data on important ion–neutral reactions (see Anicich and McEwan, 1997 for a summary of the relevant laboratory data currently available). Thus new observational data on the structure and composition of Titan's neutral atmosphere, together with new laboratory data on the relevant ion–neutral and neutral–neutral chemistry, are essential for developing more accurate models of the ionosphere. Conversely, new observational data acquired with the INMS instrument on ionospheric composition (e.g., on $HCNH^+$ density at the peak) will help constrain models of the neutral atmosphere. As in the case of the neutral atmosphere models discussed in the preceding section, assessment of existing model ionospheres and development of improved models (e.g., through the incorporation of new laboratory data on key ion–neutral reaction rates and pathways) will be an integral part of the planning of INMS operations during the Titan flybys. For initial planning purposes, both the Fox and Yelle model and the updated Keller *et al.* model will be used. Representative ionospheric density profiles from both models are shown in Figure 5, and a list of the most abundant species expected to be present in Titan's ionosphere is given in Table III.

1.1.2.2. Ionization Mechanisms. Titan's ionosphere is produced by both solar EUV photoionization and electron impact ionization. Magnetospheric electron impact was originally considered to be the dominant ionization mechanism (e.g., Ip, 1990). However, later modeling studies (Keller *et al.*, 1992, 1994; Keller and Cravens, 1994) have demonstrated that solar EUV photons and photoelectrons are in many cases more important than magnetospheric electrons in ion production and that the relative roles of photoionization and magnetospheric electron impact ionization vary according to the degree of insolation and the extent to which magnetospheric electrons have access to Titan's lower ionosphere. The former depends on Titan's orbital position (Figure 6). The latter is thought to be a function of the geometry of the Saturnian field lines that drape around Titan and will be greater on the wake side than on the ram side (Keller and Cravens, 1994), as the radial orientation of the field lines in the wake region allows magnetospheric electrons to reach the lower ionosphere without significant attenuation from extended passage through the neutral atmosphere. Model calculations indicate that the rate of ion production by magnetospheric electron impact can be as much as 100 times greater on the wake side than on the ram side and is comparable to the rate of ion

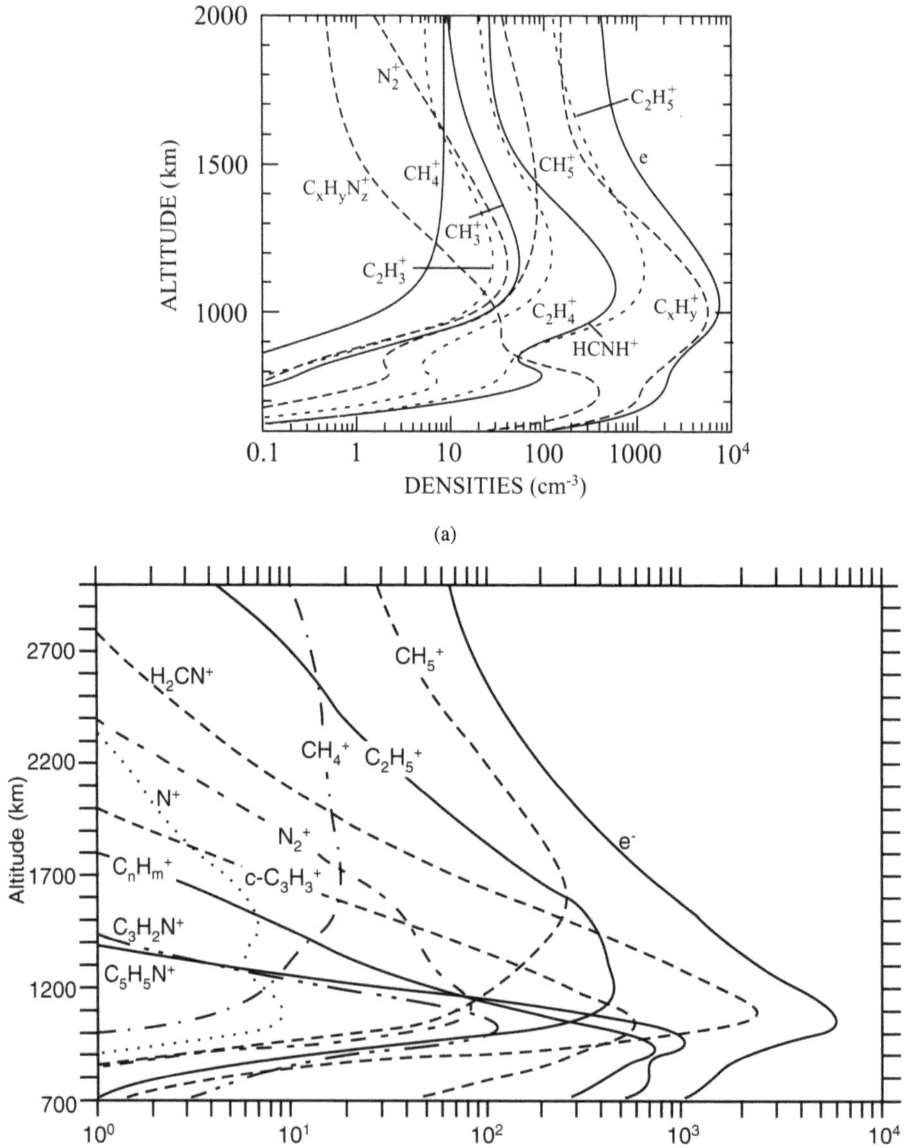

Figure 5. Density vs. altitude profiles for the major ion species in Titan's ionosphere as calculated with the models of (a) Fox and Yelle (1997) and (b) Keller *et al.* (1998). Both models use neutral atmospheres based on the model of Yung *et al.* (1984). (Keller *et al.* also calculate ion profiles—not shown here—using a more recent neutral atmosphere model developed by Toublanc *et al.*, 1995). Both models predict that heavy hydrocarbons (with three or more carbon atoms) are important ionospheric species. Fox and Yelle obtain a lower value for the $HCNH^+$ density near the ionospheric peak than do Keller *et al.* because the background atmosphere used in their model contains three times less HCN than is contained in the neutral atmosphere used by Keller *et al.*

TABLE III

Most abundant ion species in Titan's ionosphere.

Mass Group	Mass Number	Species
Light	1	H^+
	2	H_2^+
	3	H_3^+
Medium	14	N^+, CH_2^+
	15	CH_3^+, NH^+
	16	CH_4^+
	17	**CH_5^+**
Heavy	27	$C_2H_3^+$
	28	N_2^+, $C_2H_4^+$, **$HCNH^+$**
	29	N_2H^+, **$C_2H_5^+$**
Very heavy	39	*cis*-$C_3H_3^+$
	41	$C_3H_5^+$
	51	$C_4H_4^+$
	52	$C_3H_2N^+$
	53	**$C_4H_5^+$**
	65	**$C_5H_5^+$**
	67	**$C_5H_7^+$**
	69	$C_5H_9^+$
	77	$C_6H_5^+$
	79	$C_6H_7^+$, **$C_5H_5N^+$**
	91	$C_7H_7^+$

Note: Bold type identifies the major species. Recent modeling studies by Fox and Yelle (1997) and Keller *et al.* (1998) indicate that higher mass hydrocarbons are important constituents, accounting for approximately 50% of the predicted ion density.

production by EUV photoionization on the day side (Gan *et al.*, 1992; Keller *et al.*, 1994; Keller and Cravens, 1994; Cravens *et al.*, 1992). Auger electron spikes in the measured (CAPS) superthermal electron fluxes can help distinguish magnetospheric electrons from solar radiation-produced photoelectrons. Auger electrons will be produced by K-shell ionization from carbon (electron energies near 280 eV) and nitrogen (electron energies near 400 eV) (Craven *et al.*, 2004). C and N are found in the CH_4, and N_2 molecules, respectively.

Maximum ionization occurs when the wakeside ionosphere is in sunlight, i.e., when both magnetospheric electrons and solar EUV photons contribute comparably to ion production (Roboz and Nagy, 1994). In the case of the sunlit ramside ionosphere, on the other hand, 80% of the ionization at the ionospheric peak is due to photoionization. The remaining 20% results predominantly from photoelectron

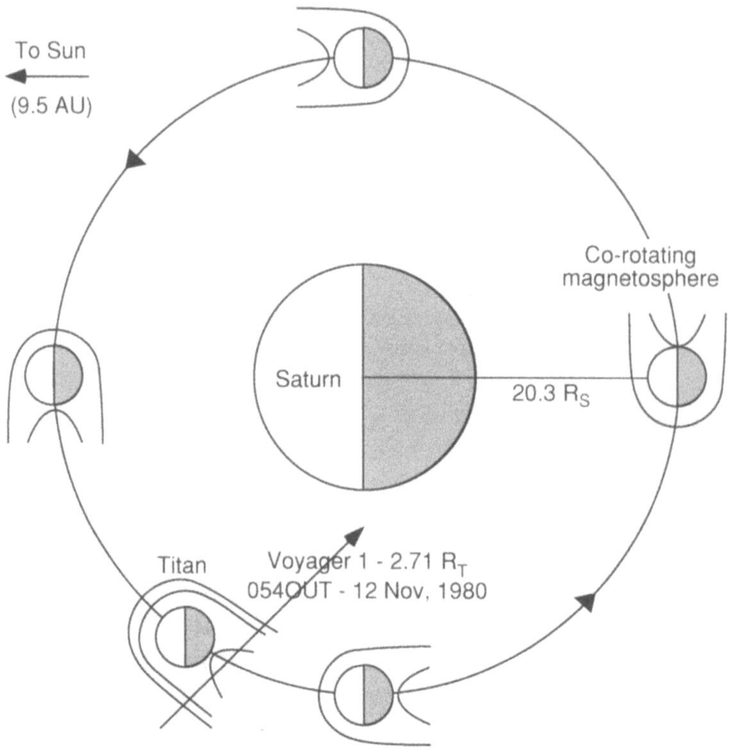

Figure 6. Solar EUV radiation and magnetospheric electron precipitation provide the energy that drives the chemistry and dynamics and determines the structure of Titan's atmosphere. Although solar EUV radiation is thought to be the dominant energy source for heating, ion production, and photochemistry, magnetospheric electrons also play an important role. The relative contributions of the two energy sources to Titan aeronomy will vary according to the intensity of the solar radiation, which is a function of latitude and Titan's orbital phase, and the degree of exposure to magnetospheric electron flux, which studies have shown to be maximized on the wake side (Keller and Cravens, 1994). The sketch above (based on Keller and Cravens, 1994) illustrates the changing geometry of solar zenith angle, ram direction, and orientation of the draped field lines (shown here rotated 90° out of their actual plane for ease of illustration). Maximum ionization will occur when both solar EUV and magnetospheric electrons contribute to ion production. Magnetospheric electrons will be the sole energy source for Titan's atmosphere when Titan is in Saturn's shadow.

impact, with a small contribution (1%) from magnetospheric electrons (Keller *et al.*, 1992). When Titan is outside Saturn's magnetosphere (see below), the solar EUV flux is the principal ionization source; the contribution from solar wind electrons is minor. (This case has recently been modeled by Galand *et al.* (1999), who also consider the case where Titan is in Saturn's shadow and magnetospheric electron impact is the sole ionization source.) Changes in the intensity of the solar EUV flux (with orbital phase) and the magnetospheric electron flux (from ram side to wake side) and in their relative contribution to ion production will lead to variations in ionospheric densities, temperatures, and dynamics (Keller *et al.*, 1994; Roboz and Nagy, 1994; Nagy and Cravens, 1998). To adequately characterize this variability it will be necessary to acquire INMS data at a variety of local times, latitudes, and plasma ram angles. Thus INMS measurements during a number of low-altitude Titan passes are required (cf. Section 3.3.4 and Table VII).

1.1.2.3. Titan's Interaction with Its Plasma Environment. Titan's interaction with the surrounding plasma is highly complex, and its details depend on the state of the ionosphere and the plasma environment. As discussed above, the former varies according to changing fluxes of solar EUV photons and magnetospheric electrons incident on Titan's upper atmosphere. The latter, as discussed by Wolf and Neubauer (1982), varies with Saturn local time and solar wind conditions. At its orbital distance of $\sim$20 R_S, Titan usually remains within Saturn's outer magnetosphere and thus interacts with the partially co-rotating magnetospheric plasma. However, whenever increased solar wind ram pressure compresses Saturn's magnetosphere, Titan may be located for a portion of its orbit outside the dayside magnetopause. In this case, the interaction is with the magnetosheath or the upstream solar wind. (Under relatively undisturbed interplanetary conditions, the stagnation point is located at a nominal standoff distance of $\sim$23 R_S. When conditions are disturbed, the magnetopause can move inward as far as $\sim$17 R_S, as observed during the Pioneer 11 encounter; Schardt *et al.*, 1984.) Even within the magnetosphere, Titan's plasma environment can be expected to vary according to the satellite's location in the magnetosphere. Estimated parameters for four representative plasma environments (solar wind, magnetosheath, outer magnetosphere, magnetotail) are given by Wolf and Neubauer (1982); however, observational data are available only for the case when Titan was located in the dayside outer magnetosphere.

Our picture of Titan's interaction with its surrounding plasma environment (Figure 7) is based on data acquired by Voyager 1 during a single pass downstream of Titan and on theoretical models of the interaction of a flowing magnetized plasma with an unmagnetized body possessing a significant atmosphere/ionosphere. During its traversal of the Saturn magnetosphere in November 1980, Voyager flew within <3 R_T of Titan. The plasma science (PLS) instrument detected a wake of relatively cool, dense plasma that may have been dominated by mass 28 ions of N_2^+/H_2CN^+ (Hartle *et al.*, 1982). The magnetometer revealed perturbations in the Saturnian magnetic field near Titan that have been interpreted as the signature of draped field lines that form a Venus-like bipolar induced magnetic tail; there was no clear evidence for the presence of an intrinsic magnetic field (Ness *et al.*, 1982; Kivelson and Russell, 1983; Luhmann *et al.*, 1991). The plasma wave (PWS) instrument observed plasma wave emissions apparently associated with the acceleration of pickup ions and the escape down the tail of plasma from Titan's ionosphere (Gurnett *et al.*, 1981, 1982). PLS data also indicated a significant slowing of the magnetospheric plasma flow near the tail as a result of mass loading by the

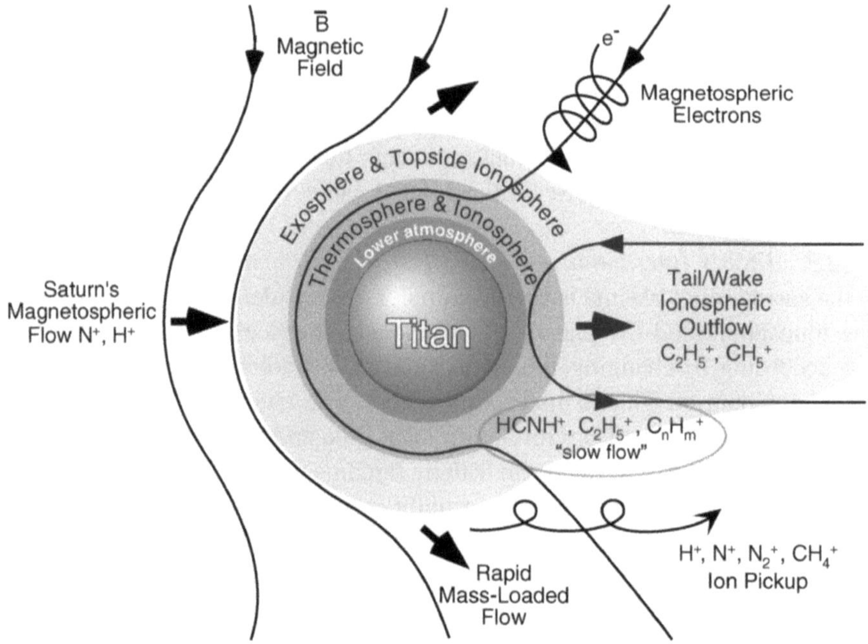

Figure 7. Sketch of Titan's interaction with Saturn's subsonic co-rotating magnetospheric plasma. Horizontal transport moves cold ionospheric plasma (principally N^+, N_2^+, and CH_4^+) from altitudes above 1700 km toward the downstream side of Titan, where it can become entrained in the flowing plasma, while a polar wind-type hydrodynamic outflow is responsible for the escape of ionized material down the wake. N^+ and H^+ are produced by the photoionization of neutral atomic nitrogen and hydrogen in Titan's extended exosphere and picked up by the external plasma flow. At altitudes of 2500 km and less, ion–neutral friction and mass loading slow the external plasma flow, resulting in strong field line draping and the production of a magnetic barrier on Titan's ram side (Cravens *et al.*, 1998).

addition of heavy ions such as N_2^+, $HCNH^+$, and/or N^+. The velocity differential between the mass-loaded flow and the faster flow of the background plasma is thought to lead to the draping of Saturn's magnetic field lines, which are frozen into the background plasma, around Titan to form the induced tail. Asymmetries observed in the orientation and structure of the tail are likely due to the day–night differences in Titan's ionosphere on the ram side or to the influence of finite pickup ion gyroradii (Ness *et al.*, 1982; Gurnett *et al.*, 1981; Luhmann, 1996; Brecht *et al.*, 2000). As revealed by the Voyager observations, Titan's interaction with Saturn's magnetospheric plasma is similar to the interaction of Venus with the solar wind. There are important differences, however. No bow shock forms at Titan because the co-rotating magnetospheric plasma is submagnetosonic (Ness *et al.*, 1982), and there is little draping of the field observed in the "magnetosheath" region compared to that in the induced magnetotail.

Models developed to examine the effects of Titan's interaction with the magnetospheric plasma flow on the dynamics and structure of the ionosphere predict that the piling up of field lines forms a magnetic barrier around Titan's ramside and flanks from which magnetic flux convects downward into the ionosphere (Keller *et al.*, 1994; Cravens *et al.*, 1998). As a result, the ionosphere down to relatively low altitudes is expected to be magnetized. Under some conditions, the entire ionosphere may be magnetized, in which case magnetic flux may "leak" into the lower atmosphere and the field strength may be non-zero at Titan's surface. (It should be noted that the models of the Titan interaction with Saturn's magnetospheric plasma assume that the ionospheric magnetic fields and tail are induced by the interaction. The presence of an intrinsic magnetic field at Titan must still be considered a possibility, however, in light of the Galileo discovery of an internally generated field at Ganymede and possibly at Io; Kivelson *et al.*, 1996; Khurana *et al.*, 1997.)

Based on the in-situ Voyager data, the magnetospheric plasma incident on Titan's neutral atmosphere and ionosphere is thought to consist of H^+ and N^+ (in the number ratio ~1:2) flowing at a velocity of ~80–150 km sec^{-1} (Hartle *et al.*, 1982). According to a recent two-dimensional MHD modeling study of the plasma interaction (Cravens *et al.*, 1998), at radial distances from Titan's center of <2.5 R_T this flow is slowed by ion–neutral friction and mass loading to

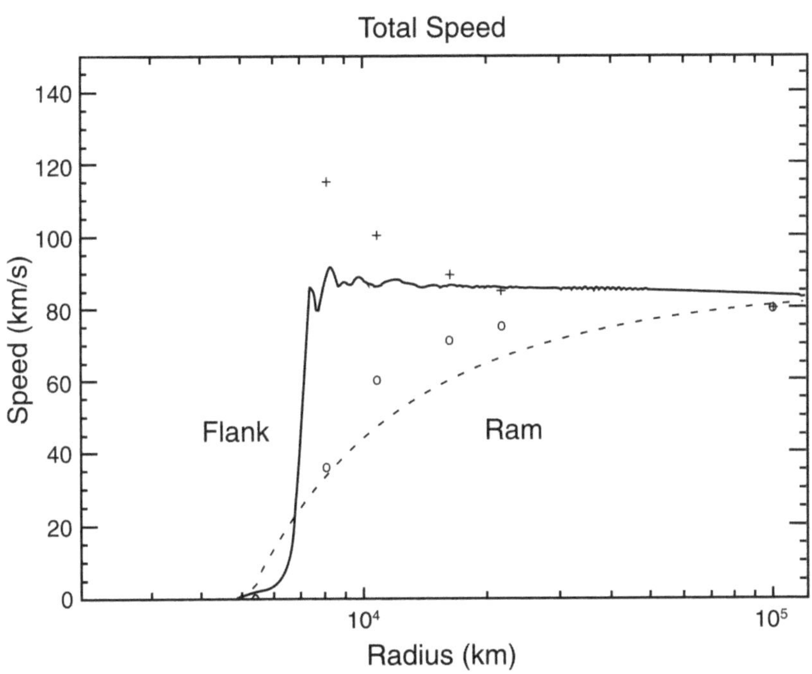

Figure 8. Predicted plasma flow speed vs. radial distance from Titan for the magnetospheric plasma ram direction (dashed curve) and the flank direction (solid line), calculated from the 2D plasma interaction model of Cravens *et al.* (1998). According to this model, the flow speed inside 2 R_T is <1 km sec^{-1}, so that ions should enter the INMS instrument essentially in the Orbiter ram direction (2 R_T is about 5000 km). At greater distances, alteration of the incoming plasma direction from the ram direction may complicate the alignment of the INMS open source look direction with the ion flow direction.

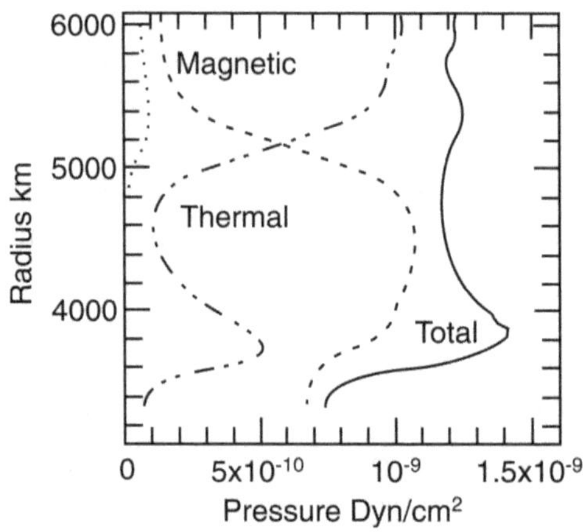

Figure 9. Comparison of the contributions of magnetic pressure and ionospheric thermal pressure to the momentum balance at Titan (from Cravens *et al.*, 1998). At radial distances of $<2\ R_T$, the magnetic field on Titan's ram side and flanks is enhanced as a result of the draping and pile up of the Saturnian magnetic field lines. As this plot indicates, within $\sim 2\ R_T$ it is the enhanced magnetic field strength rather than the thermal pressure of the ionosphere that makes the greater contribution to the obstacle that Titan presents to the external plasma flow.

velocities of under 1 km sec^{-1} (Figure 8). The slowing of the external plasma flow results in the piling up of field lines on the ram side, producing the magnetic barrier, which forms the principal obstacle to the flow (Figure 9). The induced magnetic fields are found to have little effect beyond 2 R_T, in the fast flow region upstream of Titan, but according to three-dimensional MHD models of the plasma interaction (Ledvina and Cravens, 1998; Kabin *et al.*, 1999), they produce effects many radii downstream. The actual length of Titan's wake is not yet known.

Within the ionosphere, below an altitude of around 2000 ± 500 km, plasma flow speeds are on the order of a few meters per second and the flow direction is principally vertical (Keller *et al.*, 1994; Cravens *et al.*, 1998). At these altitudes, horizontal transport plays a relatively unimportant role, and the chemical lifetimes of the ionospheric species are much shorter than the time scales for vertical transport. In other words, photochemistry is more important than dynamics in determining ionospheric density distributions. In the topside ionosphere, however, horizontal flows (of more than a kilometer per second) are expected to dominate over photochemistry. Ionospheric material—consisting of the primary ionization products (N^+, N_2^+, CH_4^+) rather than ions produced at lower altitudes by photochemical processing—will be removed as a result of pick up and entrainment in the external plasma flow (Keller *et al.*, 1994). The extent to which this is a fluid-like, as opposed to a kinetic pick-up process, remains to be determined (e.g., Luhmann *et al.*, 1996), but new "hybrid" models of the Titan–plasma interaction that preserve the particle aspects of the ion motion are beginning to address this question (Brecht *et al.*, 2000).

1.1.3. *Atmospheric Loss: Titan as a Source of Material for Saturn's Magnetosphere*

Titan supplies both neutrals and plasma to Saturn's outer magnetosphere (Figure 10). Neutral hydrogen, produced by the photolytic destruction of methane, is rapidly lost from Titan's atmosphere, principally by Jeans escape. Photochemical modeling by Yung *et al.* (1984) predicts global escape rates for H and H_2 of 4.6×10^{27} sec^{-1} and 1.2×10^{28} sec^{-1}, respectively. Nonthermal removal of nitrogen and other neutrals occurs as well, through the sputtering of Titan's neutral atmosphere by magnetospheric or solar wind particles and/or pickup ions and through production of fast neutrals by ion and ion–neutral chemistry. The rate of neutral loss due to sputtering has been estimated to be 3×10^{25} N atoms sec^{-1} (for sputtering by solar wind protons) and 7×10^{26} N atoms sec^{-1} (for sputtering by magnetospheric N^+ and protons) (Lammer and Bauer, 1993). The escape of fast neutrals produced by electron impact dissociation, dissociative recombination, and certain ion–neutral reactions within the ionosphere is predicted to lead to a global loss of 2.5×10^{25} N atoms (in the form of N, N_2, HCN, NH, and CN) and 1.7×10^{25} carbon atoms

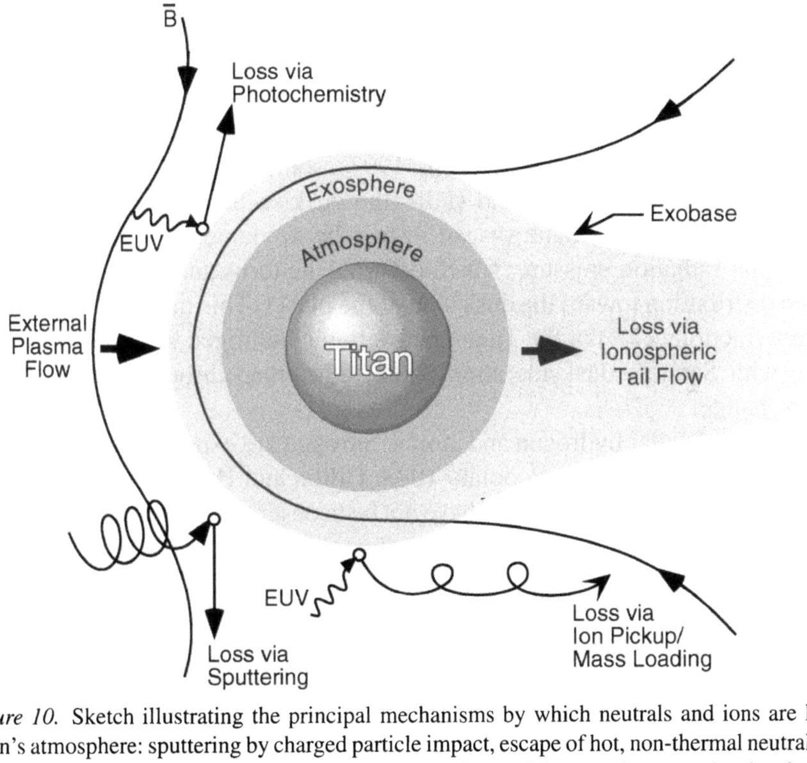

Figure 10. Sketch illustrating the principal mechanisms by which neutrals and ions are lost from Titan's atmosphere: sputtering by charged particle impact, escape of hot, non-thermal neutrals created by dissociative recombination and ion–neutral chemistry, pick-up and scavenging by the external plasma flow, and hydrodynamic escape down the tail. Determining the relative efficiency of each loss process and assessing the strength of Titan as a source of neutrals and plasma for Saturn's magnetosphere are two objectives of the INMS Titan investigation.

(principally in the form of CH_4) per second (Cravens *et al.*, 1997). Additional loss occurs as a result of the photoionization of hot exospheric neutrals and their subsequent pickup and removal by the external plasma flow. Lammer and Bauer (1991) calculate a loss of 10^{24} heavy ions (N_2^+ or H_2CN^+) s^{-1} as a result of pick up by Saturn's co-rotating magnetospheric plasma and a loss of 2.5×10^{22} N_2^+ s^{-1} as result of solar wind scavenging when Titan is outside Saturn's magnetosphere. (These values are upper limits.) More is said about loss due to ion pick up below.

Atomic hydrogen that has escaped from Titan's exosphere contributes to a cloud of neutral hydrogen atoms that occupies a sizeable region within Saturn's magnetosphere. Initial analysis of Voyagers 1 and 2 UVS observations of resonantly scattered H Lyman α appeared to indicate that the escaping hydrogen atoms form a symmetrical torus with well-defined inner and outer boundaries at 8 and 25 R_S, respectively, a vertical thickness of 14–16 R_S, and a density of $\sim$20 cm^{-3} (Broadfoot *et al.*, 1981; Sandel *et al.*, 1982). However, a more recent analysis using a larger UVS data set indicates that the neutral cloud is neither symmetrically toroidal nor predominantly Titanogenic. The distribution of atomic hydrogen in the neutral cloud appears instead to be highly nonuniform, with strong enhancements in the dusk and dawn sectors, and to extend inward, with increasing density, to the top of Saturn's atmosphere (Shemansky and Hall, 1992). Shemansky and Hall conclude that Saturn rather than Titan is the major source of the neutral atomic hydrogen cloud, although Titan is the dominant source at 18 R_S and beyond. According to a numerical simulation of the orbital behavior of hydrogen atoms originating in the Titan exosphere (Smyth and Marconi, 1993), some of the features of the neutral cloud noted by Shemansky and Hall—the dusk sector enhancement and the increasing density toward Saturn—can in part be accounted for in terms of the effect of solar radiation pressure, which deforms the torus and pushes the atomic hydrogen distribution toward the dusk side of the planet. This model predicts that a significant fraction (27%) of the Titanogenic atomic hydrogen will be lost through collision with Saturn's duskside atmosphere rather than through ionization and charge exchange.

Escaping molecular hydrogen and atomic nitrogen are expected to form neutral tori as well (e.g., Eviatar and Podolak, 1983; Hilton and Hunten, 1988; Ip, 1992; Smyth and Marconi, 1993). Such tori have not been observed, however. It is believed that solar radiation pressure will have little effect on the spatial distribution of these species in the magnetosphere so that the corresponding tori will be approximately symmetrical, in contrast to the atomic hydrogen torus (Smyth and Marconi, 1993).

Titan supplies plasma to Saturn's magnetosphere both directly, through the removal of ions from the topside ionosphere as the result of a polar wind type hydrodynamic outflow from Titan's wakeside ionosphere (Keller and Cravens (1994) calculate a total ion loss by this process of (6–20) $\times$ 10^{24} sec^{-1}. This rate is consistent with earlier Voyager-based estimates of Titan's plasma source strength by Bridge *et al.* (1981), Gurnett *et al.* (1982), and Neubauer *et al.* (1984)), and indirectly, through the ionization of material contributed to the neutral atomic hydrogen cloud and in the H_2 and N tori. As noted above, Lammer and Bauer (1991) estimate that ions created from "hot" exospheric neutrals (mass 28) are lost due to the pick-up process at a rate of 10^{24} sec^{-1}, which is comparable to the estimated rate of loss by hydrodynamic outflow. Direct removal of plasma occurs through pick up by Saturn's co-rotating magnetospheric plasma flow (or by the solar wind, when Titan is outside the magnetosphere).

The contribution of ionized Titanogenic neutral cloud/torus material to Saturn's magnetospheric plasma has received relatively limited attention in the literature and has been addressed principally in two papers by Barbosa (1987, 1990). Assuming that ions are produced at the same rates at which the corresponding neutrals escape from Titan's atmosphere, Barbosa (1990) uses production rates of 10^{27} and 10^{26} sec^{-1} for H^+ and N^+, respectively, in his study of diffusive ion transport in Saturn's magnetosphere. In the earlier study, Barbosa (1987) proposes that the ions created by ionization of Titan's postulated atomic nitrogen torus "supply the bulk of the mass and energy input to the magnetosphere at a power level of 2×10^{11} W".

Finally, we note that assessing the efficiency of the various loss processes is important for our understanding not just of the present state of Titan's upper atmosphere but also of its evolution. Lammer and Bauer (1993) estimate, for example, that sputtering by magnetospheric charged particles may have resulted in the loss of 20% of the mass of Titan's atmosphere over the age of the solar system.

1.1.4. *Titan Science Questions and Measurement Objectives*

The reference tour selected by the Cassini Project Science Group (tour T18-5, reference trajectory 041001) provides for 44 targeted flybys of Titan. Forty-three of the Titan encounters take place at altitudes below 8000 km; as currently planned, the altitude of closest approach for over half the passes will be 950 km. (The actual minimum altitude will be established on the basis of the atmospheric densities determined from measurements made during the initial flyby.) It is possible that the closest approach altitude will be decreased to 850 km for a few of the Titan passes. Orbiter velocities during Titan flybys will range from 5.8 to 6.2 km sec^{-1}.

Three orbital sequences (cf. Sections 3.3.2–3.3.4) have been designed for INMS measurements during the Titan encounters: a Titan Exploratory Sequence for the initial two passes, a High-Altitude Sequence for flybys with closest approach altitudes above the exobase ($\sim$1465 km), and a Low-Altitude Aeronomy Sequence for passes with a closest approach at 950 km, below the homopause and ionospheric peak (Strobel *et al.*, 1992; Fox and Yelle, 1997). The Titan orbital sequences provide for a combination of mass survey scans and scans of selected neutral and ion species. Neutral species with densities $\geq 10^4$ cm^{-3} can be measured in the INMS closed source mode, and with densities $\geq 10^5$ cm^{-3} in the open source neutral mode; positive ions with densities above 10^{-2} cm^{-3} can be measured in the open source ion mode (cf. Figures 1 and 5, which indicate predicted neutral and ion densities over the lower part of the altitude range that the Orbiter will traverse during the Titan encounters).

When the INMS is operating in the open source ion mode, optimal sampling requires that the inflowing ions enter the instrument within 8.6° of the ram direction. According to the recent modeling study of Titan's interaction with the external plasma flow cited above (Cravens *et al.*, 1998), the plasma is expected to flow rather slowly (<1 km sec^{-1}) inside a radial distance of 2.5 R_T (altitude = 3800 km), and ions should enter the INMS instrument essentially in the Orbiter ram direction. Outside this distance, however, INMS thermal plasma measurements will be complicated by plasma flow directions that lie outside the open source acceptance angle and by ion flow speeds that exceed the instrument's upper energy limit (100 eV). CAPS will be the more appropriate instrument for measurements in this region. Understanding Titan's interaction with the external plasma flow and the effects of this interaction on the dynamics, structure, and energetics of Titan's upper atmosphere will require a combination of INMS data from the region inside 2.5 R_T, CAPS data on external ion flow velocities and on ion and electron temperatures, and magnetic field data from the MAG experiment. For example, both INMS and CAPS data will be needed, together with models, to calculate escape rates from both dissociative recombination of ionospheric ions and ion pickup.

Key specific science questions to be addressed by the INMS Titan investigation and measurement requirements are summarized in the following.

- What is the thermal structure of Titan's upper atmosphere? (*Needed measurements: high-time-resolution INMS measurements of densities as a function of altitude for the principal neutral and ion species known or expected to be present in Titan's atmosphere; complementary UVIS and RSS data.*)
- What is the bulk composition of Titan's neutral upper atmosphere and ionosphere? What are the key chemical processes determining the composition of the ionosphere and neutral upper atmosphere? How does the ionosphere affect the composition of the neutral atmosphere and vice-versa? For example, how is the ion composition determined by the densities of neutral species such as C_2H_2, HCN, and nitriles? What neutrals, including complex hydrocarbon species, are produced by the ion chemistry? How does the chemistry occurring in Titan's upper atmosphere contribute to the production of aerosol hazes in the stratosphere? More specifically, how are complex hydrocarbons and nitriles generated in Titan's upper atmosphere and ionosphere, and do magnetospheric inputs affect this? (*Needed measurements: detailed INMS ion and neutral composition measurements over the complete mass range (1–99 Daltons); complementary UVIS occultation measurements and VIMS and CIRS data on the composition of the lower atmosphere.*)
- How do Saturn's magnetosphere (when Titan resides in it) and the solar wind (when Titan is located outside Saturn's magnetopause) interact with Titan?

How are the structure and dynamics of Titan's ionosphere affected by this interaction? How are the energetics of the ionosphere, and upper atmosphere as a whole, affected by this interaction? Is the magnetospheric interaction or solar EUV radiation a more important energy source for Titan's ionosphere and upper atmosphere? How does this depend on location on Titan (that is, versus solar zenith angle, on the ram or wake side, etc.)? How does the plasma interaction depend on the relative angles of the subsolar and subflow points on Titan? (*Needed measurements: measurements of total ion density versus altitude for a wide range of solar zenith angles and ram angles; energetic particle measurements from CAPS and MIMI, complementary MAG and PWS data, and RSS occultation measurements of the ionosphere; INMS, CAPS, MAG, PWS measurements at Titan whenever Titan is near noon Saturn time in the event that high solar wind dynamic pressure compresses magnetopause, placing Titan in the magnetosheath or solar wind.*)

- Is Titan's ionosphere magnetized? And if so, by a small intrinsic field or by fields induced by the interaction with Saturn's magnetosphere? How does this field interact with the thermal plasma in the ionosphere? (The ionospheric electrical conductivity will determine whether or not an induced magnetic field can penetrate into the lower atmosphere.) (*Needed measurements: INMS neutral and ion density measurements to determine ionospheric electrical conductivity as a function of altitude, local time, ram angle, etc.; complementary MAG and RSS data.*)
- How is the upper boundary of the ionosphere affected by the plasma interaction? Is the scale height altered by plasma erosion on the ram side? Are there ionospheric "tail rays" in the wake as seen at comets, Venus, and Mars? (*Needed measurements: global coverage of the upper ionosphere region to examine 3D shape of the ionosphere boundary; CAPS data; PWS Langmuir Probe measurements and MAG measurements to understand geometry of plasma interaction-related features.*)
- What are the relative contributions of various loss processes—ion scavenging, sputtering by co-rotating particles, the escape of hot, non-thermal neutrals produced by ion–neutral and dissociative recombination reactions, and the bulk flow of ionospheric plasma out the tail/wake—to the loss of material from Titan's neutral upper atmosphere and ionosphere? What are the evolutionary consequences for Titan? (*Needed measurements: INMS measurements of neutral and ion densities vs. altitude and location, e.g., ram side versus wake side; CAPS measurements of flux and composition of the incident particles.*)
- How many and what neutrals and ions does Titan supply to Saturn's magnetosphere and how does this Titan source affect the outer magnetosphere of Saturn? How does Titan's ionosphere contribute to the extended exosphere of that satellite as well as to the neutral torus around Saturn near Titan's orbit? (*Needed measurements: same as in preceding paragraph.*)

1.2. Saturn's Inner Magnetosphere: Rings and Icy Satellites

A secondary focus of the INMS investigation is on the neutral and thermal plasma environment of Saturn's inner magnetosphere ($L < 10$) and its association with the rings and icy satellites. Sputtering of material from the rings and the surfaces of the moons by charged particle and micrometeorite bombardment is believed to create an extensive neutral cloud of water molecules and water dissociation products in the inner magnetosphere. Subsequent ionization of this neutral material through electron impact, EUV radiation, and charge exchange is the primary source for the thermal plasma (protons and heavy ions) observed in the inner magnetosphere by Pioneer 11 (Frank *et al.*, 1980) and Voyagers 1 and 2 (Bridge *et al.*, 1981, 1982; Lazarus and McNutt, 1983). The relative contributions of the neutral sources, their adequacy to account for neutral densities determined from recent HST observations, and the contributions and efficiencies of different production mechanisms (e.g., plasma sputtering, photosputtering, micrometeoroid bombardment) have been the subject of several modeling studies. Such studies, together with the HST observations and new analysis of Voyager plasma data, have added significantly to our understanding of the neutral and plasma environment of Saturn's inner magnetosphere. There remain a number of important questions, however, whose resolution must await the arrival of Cassini in the Saturn system. (For a review of the relevant observations and model results, see Richardson (1998). See also the recent studies by Jurac *et al.* (2001a,b).)

INMS measurements in the inner magnetosphere will be challenging because of the low neutral densities, intense radiation background, and the co-rotational energy acquired by the thermal ions. However, long accumulation periods and the co-adding of mass scans will be used to increase counting statistics, and thermal ion energies will be low enough in certain regions of interest (e.g., above and at the ring plane during Saturn Orbit Insertion, icy satellite wakes) to permit useful data to be acquired.

1.2.1. *The Neutral and Plasma Environment of the Main Ring System*

Prior to the deployment of the Hubble Space Telescope, the only direct information on the neutral gas population in the inner magnetosphere was provided by remote-sensing observations of Lyman α radiation resonantly scattered by neutral atomic hydrogen. These observations were interpreted as evidence for the presence of a tenuous ring atmosphere with an atomic hydrogen number density of 400–600 cm^{-3} (Weiser *et al.*, 1977; Judge *et al.*, 1980; Carlson, 1980; Broadfoot *et al.*, 1981). Such a neutral atmosphere is thought to be created and maintained by the ejection of water vapor and water products from the icy ring particles as a result principally of micrometeorite bombardment and is expected to consist of H_2O, OH, and O in addition to atomic hydrogen (Ip, 1984, 1995; Pospieszalska and Johnson, 1991). In addition to the atomic hydrogen associated with the rings, atomic hydrogen apparently originating in Saturn's exosphere also exists within the inner magnetosphere in significant amounts and with densities as high as 100 cm^{-3}

(Shemansky and Hall, 1992). As Hall *et al.* (1996) point out, the presence of this exospheric component and of contributions from Titan and other sources makes it difficult to distinguish an atomic hydrogen population uniquely associated with the ring system. Thus estimates of the density and properties of the ring atmosphere based on the Lyman α data (e.g., Broadfoot *et al.*, 1981) should be used with caution.

The signature of an OH component of the ring atmosphere was detected with the HST Faint Object Spectrograph (FOS) during edge-on observations made at the time of the Earth's ring plane crossing in August 1995 (Hall *et al.*, 1996). Five observations of UV fluorescence emissions from the OH A–X band were made close to the ring plane, at altitudes ranging from 0.28 to 0.60 R_S above the plane and at equatorial distances between 1.9 and 2.3 R_S. Emission brightness was found to decrease with altitude above the ring plane and to increase with increasing equatorial distance, with the brightest emissions occurring at an altitude of 0.28 R_S and an equatorial distance of 2.3 R_S. Based on the brightness of these emissions, Hall *et al.* estimate a scale height of 0.45 R_S and a number density of 150–700 cm^{-3} for the OH component of the ring atmosphere. (The larger value for the OH number density represents an upper limit because it may include an OH contribution from non-ring sources in the inner magnetosphere.) A much lower OH density ($\sim$30 cm^{-3}) near the rings has been calculated by Richardson *et al.* (1998), whose model is constrained by HST OH observations from three epochs (1992 and 1994 as well as 1995) and Voyager atomic hydrogen and plasma data (see their Figures 2 and 4). The higher value obtained by Hall *et al.* (1996) may reflect the preliminary nature of their analysis.

According to the Richardson *et al.* model (their Figure 8), OH accounts for $\sim$50% of the neutral density in the vicinity of the rings (at a radial distance of 2.3 R_S), which implies a total neutral density of $\sim$60 cm^{-3}. This value is consistent with that given by Ip (1995) for the density near 2.5 R_S but is lower than that predicted by the ring neutral cloud model of Pospieszalska and Johnson (1991) ($\sim$300 cm^{-3} at 2.3 R_S and at an altitude of 0.28 R_S above the ring plane; see their Figure 1a). The differences among the model predictions are attributable in part to uncertainties in our knowledge of the micrometeorite flux at Saturn and of the rate of gas production by hypervelocity impact.

Indirect evidence for the existence of a neutral ring atmosphere was provided by the detection of a population of cold (10 eV), dense (100 cm^{-3}) heavy ions during the Voyager 2 ring plane crossing at an equatorial distance of $\sim$2.7 R_S, just outside the outer edge of the main rings ($\sim$2.3 R_S) (Bridge *et al.*, 1982; Lazarus and McNutt, 1983; Gan-Baruch *et al.*, 1994). It is probable that this cold dense plasma consists of water group ions (O^+, OH^+, H_2O^+, and H_3O^+) created by the ionization of the water vapor and water dissociation products that make up the neutral atmosphere of the ring system. The ring atmosphere—which one model has shown can extend well beyond the edge of the main rings—would thus constitute the "missing" source of the thermal plasma that exists inside the orbit of Enceladus and is inadequately accounted for by other sources (Pospieszalska and Johnson, 1989,

1991). In addition to the creation of plasma through the ionization of the sputtered neutrals, some plasma will also be produced directly by the micrometeorite impacts upon the ring material (Morfill *et al.*, 1983).

Eviatar and Richardson (1990, 1992) suggest that the cold dense plasma associated with the rings and lying between the outer edge of the main rings and the orbit of Mimas ($\sim$3 R_S) should be considered the outer "ionosphere" of the ring system, a population distinct from the warmer plasma associated with the icy satellites. Voyager 2 plasma measurements in the ring plane at 2.7 R_S indicate that the ring ionosphere in this region is rather dense ($\sim$100 cm^{-3}). No observational data exist on the density of the plasma in the immediate vicinity of the main rings (i.e., between 1.2 and 2.3 R_S); however, extrapolation from the Voyager data suggests a density of $\sim$30 heavy (e.g., O^+, OH^+, H_2O^+, and H_3O^+) ions cm^{-3} (25–50 Daltons) at altitudes between 0.2 and 0.4 R_S above the ring plane (Richardson and Sittler, 1990, Figure 9).

Because the rings are magnetically connected with Saturn's ionosphere, it is expected that plasma transport between the ionosphere and the ring system (and between the ring system and the ionosphere) will occur and that the plasma in the vicinity of the rings will thus consist of both ring-generated plasma (mainly water group ions) and ionospheric plasma (principally H^+) (Wilson and Waite, 1989). Model calculations by Wilson and Waite indicate that the relative proportions of the two populations will vary according to radial distance and source strength; for example, for one set of model parameters, ionospheric H^+ (densities $\geq 1.0\ cm^{-3}$) is found to be the dominant species near the C ring, while the plasma environment of the B and A rings is dominated by heavy ions with densities as high as 50 cm^{-3}. The Wilson and Waite model also demonstrates that the flux of ring-generated water group ions from the B ring into the ionosphere can be significant, $>2 \times 10^7\ cm^{-2}\ sec^{-1}$, and could contribute to the low electron densities observed in the ionosphere at latitudes magnetically connected to the B ring (cf. Connerney and Waite, 1984). Since absorption of ionospheric material by the rings can also produce such features (Luhmann and Walker, 1981), in-situ measurements are needed to determine what exchanges between the rings and ionosphere are taking place.

1.2.2. *The E Ring and Icy Satellite Environments*

Neutrals produced by micrometeorite bombardment of the main rings are likely to contribute significantly to the neutral population inside the orbit of Enceladus (Pospieszalska and Johnson, 1991). However, the principal source of neutral gas and thermal plasma in Saturn's inner magnetosphere is the sputtering of water products from the surfaces of the icy satellites and from the icy particles of the diffuse E ring (3–8 R_S) (Shi *et al.*, 1995; Jurac *et al.*, 2001a). Neither Pioneer 11 nor Voyagers 1 and 2 were instrumented to make in-situ neutral measurements during their passes through the inner magnetosphere. However, the measurement of heavy ions with the Pioneer and Voyager plasma instruments near the orbits of the icy satellites implied the presence of an extended neutral cloud of water molecules and water dissociation products that could serve as the source for the observed ions. Using Voyager LECP particle flux data and laboratory data on the sputtering of water ice, Lanzerotti *et al.* (1983) demonstrated that the sputtering of the inner moons by charged particle bombardment could produce such an extended neutral source. Calculations of neutral and plasma densities surrounding the inner moons were reported by Richardson *et al.* (1986), and a model of neutral cloud morphology and of the inner heavy ion torus was developed by Johnson *et al.* (1989). In both models, charged particle bombardment was assumed to be the primary means by which water and water products are sputtered from the satellite surfaces to form the neutral cloud. Both models predicted relatively low neutral densities in the inner magnetosphere: $<7\ cm^{-3}$ in the Tethys–Dione torus (Richardson *et al.*, 1986, Table Vb) and 44–60 cm^{-3} at the orbit of Tethys (4.88 R_S) (Johnson *et al.*, 1989, Figure 5). In contrast, Shemansky and Hall (1992, Table V) estimated the densities of water group neutrals (H_2O, OH, O) to be as high as $\sim$470 cm^{-3} at a distance of 4.5 R_S.

The first direct observation of neutral gas associated with the inner moons was the detection with the HST FOS of a surprisingly large abundance of OH (mean density $= 160 \pm 50\ cm^{-3}$) at 4.5 R_S—near the orbit of Tethys—by Shemansky *et al.* (1993). Model calculations constrained by these and subsequent HST observations (from 1994 and 1995) indicate even higher OH densities, $>$700 cm^{-3} at 4.5 R_S and on the order of 500 cm^{-3} at 3 R_S (Richardson *et al.*, 1998; S. Jurac, private communication, 2001). These findings confirm the prediction by Shemansky and Hall (1992) of large amounts of heavy neutrals in Saturn's inner magnetosphere and call into question the adequacy of the principal neutral source mechanism—charged particle sputtering—assumed by the earlier models of Richardson *et al.* (1986) and Johnson *et al.* (1989). By using revised yields for sputtering by keV O^+ and including sputtering of the icy E-ring grains as an important neutral source in the region between Enceladus and Tethys, Shi *et al.* (1995) were able to increase the total source rate for charged particle sputtering by a factor of two compared with that used by Johnson *et al.* (1989). Even so, the new production rate ($1.7 \times 10^{26}\ sec^{-1}$) is roughly 10 times smaller than the source rate required by Shemansky *et al.* (1993) ($2 \times 10^{27}\ sec^{-1}$) and more recently by Richardson *et al.* (1998) ($1.4 \times 10^{27}\ sec^{-1}$) to account for the HST OH observations.

The discrepancy between the neutral densities implied by the OH observations and those predicted by models that invoke charged particle sputtering as the major neutral source may be attributable in part to underestimates of the actual sputter yields from the icy satellites and/or of the magnitude of energetic particle fluxes between 4 and 8 R_S (cf. Shi *et al.*, 1995). This discrepancy may also be partly accounted for in terms of the contribution of processes other than charged particle sputtering to the formation of the neutral cloud. For example, vapor production by micrometeorite impact is usually considered a relatively minor source of neutrals at the inner moons (as opposed to at the rings) (Johnson *et al.*, 1989). There is great uncertainty about micrometeorite fluxes at Saturn, however, and thus Shemansky *et al.* (1993) suggest that, with sufficiently large fluxes, micrometeorite bombardment of the satellites could be a plausible additional source of neutrals for the inner magnetosphere.

Following up on a suggestion by Hamilton and Burns (1993), Ip (1997) considers as another possible source the impact of E-ring particles, in highly eccentric orbits, on Enceladus and suggests that this mechanism could contribute substantially to the neutral cloud in the vicinity of this satellite. According to Ip's model, a combination of charged particle sputtering and sputtering of Enceladus by E-ring particles (estimated source strength = 6×10^{25} sec^{-1}) gives an OH density at 4.5 R_S of 300 cm^{-3} and a total heavy neutral density of 4×10^3 cm^{-3}. In addition, Ip notes that E-ring particles in eccentric orbits could also collide with the A ring and contribute to the production of a dense ring system atmosphere (cf. the OH densities at 2.3 R_S reported by Hall *et al.*, 1996).

Another possible solution to the problem of the discrepancy between the modeled sources and the observed OH densities is offered by Jurac *et al.* (2001a,b), who posit the existence of a population of submicron-sized ice grains distributed among the larger E ring grains near the orbit of Enceladus. According to their model, this population represents a stronger source of sputtered neutrals than either the satellites or the regular E ring grains, and the three sources together are sufficient to account for the measured OH densities.

Models show that neutral densities peak close to the icy moons (Johnson *et al.*, 1989; Ip, 1997). For example, estimated densities of sputtered H_2O and O_2 at an altitude of 560 km above the surface of Dione are $\sim 7 \times 10^4$ cm^{-3} and 10^4 cm^{-3}, respectively (Johnson, 1998, Figure 9). This is approximately the altitude of closest approach during the Dione encounter as specified in the Cassini reference tour T18-5. Thus some direct sampling with the INMS of satellite neutral atmospheres may be possible during targeted (i.e., low-altitude) icy satellite encounters. Data acquired during these encounters can be used to assess the neutral source strength of a particular moon and to obtain some information about satellite surface composition. Determination of satellite surface composition from sputter products will be significantly improved by the high-sensitivity CAPS measurements of sputter-produced plasma (cf. Johnson and Sittler, 1990).

1.2.3. *INMS Inner Magnetosphere Science Objectives and Measurement Requirements*

The Cassini Orbiter is the first outer planet probe instrumented to make in-situ neutral composition measurements. It thus has the potential to provide unique data on the neutral gases that recent HST observations have shown to be abundant in Saturn's inner magnetosphere. Such data would be of extraordinary value in our efforts to characterize the neutral environments of the rings and icy satellites, to understand the interactions of these bodies with the magnetospheric plasma or, in the case of the rings, with Saturn's ionosphere, and to assess the relative contributions of the ring system and inner moons to the neutral and thermal plasma populations of the inner magnetosphere. INMS neutral (and thermal plasma) data would contribute as well to efforts to address planetological questions, e.g., relating to the age and evolution of the rings, the composition of the surfaces of the icy satellites, etc.

INMS measurements of neutral gases in the inner magnetosphere are expected to be challenging, however, because, as is evident from the discussion in the preceding sections, available data and modeling results indicate that neutral densities in the inner magnetosphere do not exceed a few times 10^3 cm^{-3} at the most. Neutral densities are thus generally below the detection level of the INMS in both the closed source (10^4 cm^{-3}) and open source neutral (10^5 cm^{-3}) modes. These measurements are further complicated by the intense radiation background in the inner magnetosphere. To mitigate this latter problem, the INMS was modified to include a 0.23-cm-thick tantalum shield around the detectors (cf. Section 2.1). To address the problem of densities below the neutral detection threshold, long accumulation periods of up to 50 min will be employed to allow for the co-adding of mass scans. Five mass spectra consisting of 15–255 co-added mass scans each will be acquired during each of the four measurement cycles (at least one in each INMS operating mode) that constitute the inner magnetosphere orbital sequence (cf. Sections 3 and 3.3.6).

The INMS can also provide, under certain circumstances, valuable data on the thermal ions in the inner magnetosphere. Although thermal plasma densities in the equatorial plane near the rings and satellites (e.g., heavy ion densities of 20–25 cm^{-3} and proton densities of 2 cm^{-3} near the orbit of Dione; Richardson, 1986) lie above the detection threshold (10^{-2} cm^{-3}), the co-rotational energy acquired by the ions places them at times outside the instrument's energy range (1–100 eV) at radial distances greater than $\sim$3.5 R_S. (Co-rotational energy is approximately given by $E \sim AR^2/2$ eV, where A is the atomic mass unit and R is radial location in Saturn radii. However, this general rule must be supplemented by information on the spacecraft's orbital velocity to set the actual limit on detection.) Following Saturn Orbit Insertion (SOI), the Orbiter will cross the ring plane through the gap between the F and G rings, at $\sim$2.5 R_S. Prior to this, the Orbiter will fly over the ring plane at altitudes ranging from 0.5 to 0.13 R_S for $\sim$1.5 hr after the main engine burn associated with SOI. The ring plane crossing and, depending on the plasma scale height, the ring overflight thus offer a key opportunity for INMS open source ion measurements of the ring plasma environment. Additional opportunities for thermal ion measurements are expected to occur during targeted flybys of the icy satellites, whenever the Orbiter is located in plasma stagnation regions in the satellite wakes.

Representative science questions to be addressed with INMS data acquired in the inner magnetosphere and measurement requirements include the following:

- What is the composition and density of the neutral atmosphere and ionosphere of the ring system? How are the rings coupled to the Saturnian ionosphere? Is there active water injection from the ring systems into Saturn's atmosphere? (*Needed measurements: ion composition, density, and temperature and open source neutral measurements during ring overflight and at ring plane crossing; complementary ion measurements with CAPS.*)

- What are the interactions between icy satellites and the magnetospheric plasma? Do Enceladus, Dione, and/or Rhea have exospheres and ionospheres? If so, how are they generated, by charged particle sputtering or micrometeorite bombardment? What can be inferred from the neutral/plasma environment of the satellites about their surface composition? Is there any current outgassing activity (cyro-volcanism) on these icy satellites (i.e., Enceladus and/or Dione)? Which satellite is most important in supplying neutral material to the magnetosphere? What is the contribution of the E ring to the neutral cloud? What are the density distribution and composition of the neutral gas cloud in the vicinity of the satellites? (*Needed measurements: ram-pointed neutral density and composition measurements at closest approaches to icy satellites for possible detection of exospheres and to set limits on exosphere sizes; ion density and composition measurements at wake crossings; complementary measurements with CAPS, MAG, and plasma wave experiments; MIMI imaging of neutral tori.*)

1.3. Plasma Outflow from Saturn's High-Latitude Ionosphere

Hamilton *et al.* (1983) reported the detection with the Voyager LECP instrument of H_2^+ and H_3^+ in Saturn's magnetosphere and postulated that such ionized molecular hydrogen most likely originates in Saturn's high-latitude ionosphere. Outflows of thermal ions from the Earth's high-latitude ionosphere are known to be a major source of plasma for the terrestrial magnetosphere (cf. Chappell *et al.*, 1987), and calculations by Nagy *et al.* (1986) demonstrated that polar ion outflows could in principle supply substantial amounts of plasma to the Jovian magnetosphere as well. Evidence for an ionospheric plasma source at Jupiter was provided by the observation of Doppler-shifted Lyman α emissions from the Jovian auroral zone, which Clarke *et al.* (1989) attributed to fast atomic H produced by interactions with upflowing 10–20 eV protons and possibly H_2^+ and H_3^+ as well. If similar outflows occur at Saturn, ion flow energies would likely be <10 and <1 eV in the case of a polar-wind-type outflow. Such energies fall within the energy range of the INMS for open source ion measurements. Thus an additional objective of the INMS investigation is to attempt to detect the postulated H_2^+ and H_3^+ outflows from Saturn's ionosphere during the Orbiter's passes through the polar magnetosphere and, if such outflows are detected, to assess the strength of the ionospheric source. INMS measurements will be coordinated with CAPS observations, which will provide information on plasma flow directions and energy distributions.

2. Instrument Description

2.1. Overview

The INMS instrument (Kasprzak *et al.*, 1996) is a modification of the Neutral Gas and Ion Mass Spectrometer (NGIMS) instrument designed for the Comet Rendezvous Asteroid Flyby Mission (CRAF). Its heritage includes similar sensors designed by GSFC for such missions as Atmospheric Explorer, Dynamics Explorer, Pioneer Venus, and the Galileo Probe. The Cassini instrument consists of two separate ion sources for sampling ambient neutrals and ions, an ion deflector/trap, four hot-filament electron guns, an electrostatic quadrupole switching lens that selects between the sources, various focusing lenses, a quadrupole mass analyzer, and two secondary electron multiplier (SEM) detectors. Instrument control is provided by the Flight Computer, according to the values entered in various software tables (Section 3.1). A sketch of the key INMS components is shown in Figure 11, and the primary instrument parameters are listed in Table IV.

The gas densities at Titan and other INMS targets are nearly optimal for direct sampling without ambient pressure reduction. Two separate ion sources—a closed source and an open source—rather than a single combined quasi-open ion source are used in the INMS instrument in order to optimize interpretation of the neutral species (Figure 12). In the closed source mode, the ram pressure of the inflowing gas creates a density enhancement in the source antechamber, allowing the sampled species to be measured with relatively high precision and sensitivity. This mode will be used to measure species, such as N_2 and CH_4, which do not react with the antechamber surfaces. The open source has the advantage that it can measure reactive neutral radicals, such as atomic nitrogen, and ions. In this mode, the ambient neutral gas density is sampled directly with no stagnation enhancement and no collisions with the surfaces of the instrument. For open source ion measurements, the INMS angular response can be increased beyond the geometric view cone (8.6° cone half angle) by adjusting the voltages on the plates in the ion deflector/trap and the exit aperture lens (top plate lens). For neutral sampling in the open source mode, the ion trap removes incoming ions and electrons, which could cause spurious ionization of neutral species, and allows only neutrals to pass into the ionization region. In both the closed and open source modes, impacting electrons emitted from the hot-filament electron guns ionize the sampled neutrals.

Electrostatic lenses are used to focus the ambient ions and those created from ambient neutrals by electron impact into the quadrupole switching lens (Mahaffy and Lai, 1990), an electrostatic device that steers ions from either the closed or open source through a system of focusing lenses into a dual radio frequency (RF) quadrupole mass analyzer. The mass analyzer selectively filters the ions according to their mass-to-charge ratio. Two secondary electron multipliers operating in pulse-counting mode cover the dynamic range required. The INMS mass range was increased from its initial value of 1–66 to 1–99 Daltons (atomic mass units)

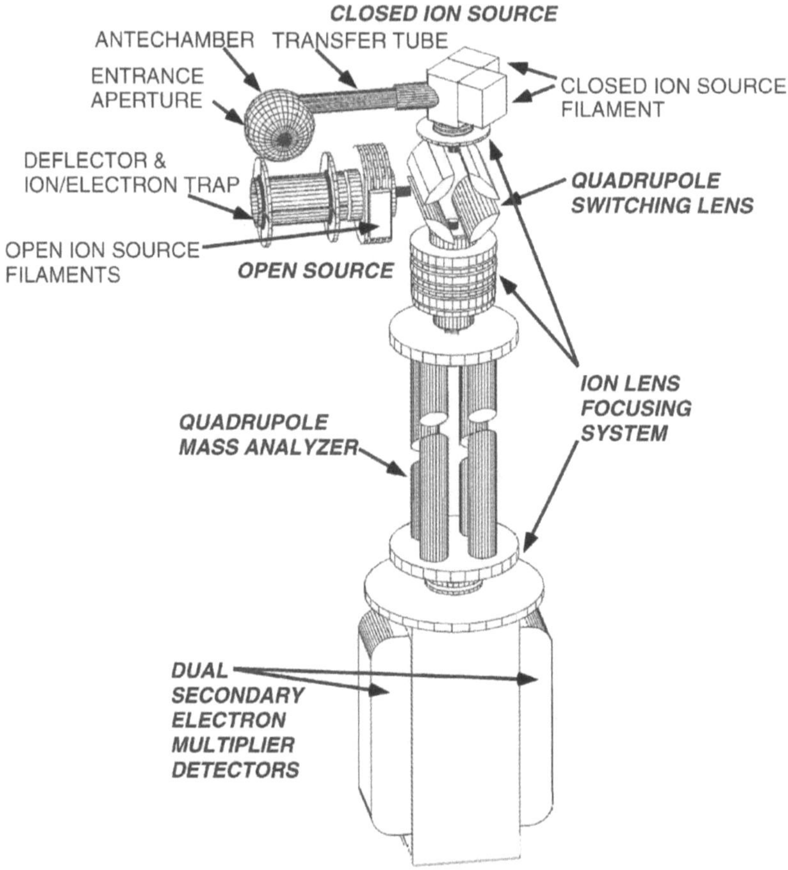

Figure 11. Schematic illustrating the principal components of the Cassini Ion and Neutral Mass Spectrometer (INMS).

to allow detection of heavier hydrocarbon species and possible pre-biotic cyclic hydrocarbons such as C_6H_6. Using two different radio frequencies and scanning the mass-to-charge ratios from 1 to 8 and then from 12 to 99 Daltons accomplish this.

The INMS instrument is mounted on the Cassini's Fields and Particles Pallet (FPP). The outward normal to both the open and closed source INMS apertures lies in the spacecraft $-X$ direction. The open source geometric field of view is about 8.6° cone half angle. This limits the angular response for neutrals and ions measured in the open source mode, although, as noted above and discussed in greater detail below, the angular response for the measurement of ambient ions can be improved by adjusting the voltage applied to the open source ion deflectors. In contrast, the closed source has a much wider geometric field of view of approximately 2π steradians. The open source is vented to lower the ion source and analyzer pressures (increasing the ion mean free path) during a Titan pass when the spacecraft ram is

TABLE IV

Cassini Ion and Neutral Mass Spectrometer (INMS) parameter summary.

Neutral gas sampling systems	(1) Open source (molecular beaming) with energy discrimination; (2) closed source
Ion sampling system	Thermal and suprathermal positive ions
Sample system switching	Electrostatic quadrupole deflector
Viewing angle (angle of response)	(1) Open source 8.6° cone half angle; (2) closed source $\sim 2\pi$ steradians; (3) exhaust vent $\sim 2\pi$ steradians
Neutral mode ionization sources	Electron impact ionization; for the primary filament Open source: 25 and 70 eV; Closed source: 27 and 71 eV
Mass analyzer	Quadrupole mass filter, 0.5 cm field radius, 10 cm rod length Radio frequencies: 1.64 and 3.57 MHz
Mass range	1–8, 12–99 Daltons nominal; high pass filter mode
Scan modes	(1) Survey: scan mass range in 1/8 or 1 Dalton steps; (2) adaptive mode: select mass values
Resolution/crosstalk	10^{-6} for adjacent masses
Detector system	Two secondary electron multiplier detectors operating in pulse counting mode (detector noise < 1 count per minute in laboratory) Dynamic range of two detector system for 1 integration period sample $\sim 10^8$
Sensor sensitivity	(1) Ion flux sensitivity for nominal 3 eV energy, entrance aperture diameter 0.864 cm He^+ $\sim 9 \times 10^{-4}$ (counts/sec)/(ions/cm^2/sec) Ar^+ $\sim 7 \times 10^{-4}$ (counts/sec)/(ions/cm^2/sec) Kr^+ $\sim 5 \times 10^{-4}$ (counts/sec)/(ions/cm^2/sec) (2) Neutral mode sensitivity for mass 28, N_2, primary filament Closed source, detector 1 $\sim 6 \times 10^{-4}$ (counts/sec)/(particle/cm^3) Closed source, detector 2 $\sim 3.6 \times 10^{-7}$ (counts/sec)/(particle/cm^3) Open source, detector 1 $\sim 1.9 \times 10^{-4}$ (counts/sec)/(particle/cm^3) Open source, detector 2 $\sim 1.3 \times 10^{-7}$ (counts/sec)/(particle/cm^3) (3) Maximum energy for ion beaming mode: 54 eV; for neutral beaming mode: 103 eV.
Density/flux for 1 count per integration period, no background	Minimum neutral ion source density Closed source $\sim 5 \times 10^4$ particles/cm^3 Open source $\sim 2 \times 10^5$ particles/cm^3 Maximum neutral ion source density $\sim 10^{12}$ particles/cm^3 Maximum closed source ram enhancement factor for N_2 at 6 km/sec = 50 Minimum ion flux and density for 3 eV ions $\sim$(7–10) $\times 10^3$ ions/cm^2/sec for nom. 3 eV ions $\sim 10^{-3}$ particles/cm^3 (6 km/sec speed)

(*Continued on next page.*)

TABLE IV
(*Continued*).

Data rate	Sample integration period = 31.1 ms; total sample period = 34.0 ms
Spatial resolution	~200 meters along spacecraft track per sample period
Instrument control	Microprocessor: MA31750, RAM: 64 Kbytes, IORAM: 32 Kbytes, PROM: 64 kbytes
Telemetry	Science data rate: 1498 bps; housekeeping data rate: 12 bps
	Reduced science packet production mode implemented
Deployment mechanism	Metal ceramic breakoff cap, pyrotechnically activated
Power (current best estimate)	Neutral mode: average 23.3 W
	Ion mode: average 20.9 W
	Sleep: average 13.1 W
	Off: 4 W replacement heater
Size	Maximum envelope (cruise): height 20.3 cm (8.0 in), length 42.2 cm (16.6 in), width 36.5 cm (14.4 in)
Weight (current best estimate)	10.29 kg including ~1.4 kg for tantalum radiation shield 0.23 cm (0.090 in) thick

approximately along the −X direction. Venting occurs at right angles to the −X axis.

The INMS sensor and supporting electronics are packaged as shown in Figures 13 and 14. The aluminum electronics box is mechanically strong and lightweight and provides electrostatic shielding for the instrument as well as protection against micrometeoroids and high-energy particles. A 0.23-cm (0.09-in) thick tantalum shield shields the secondary electron multipliers (SEM) against magnetospheric background radiation, which could interfere with the measurement of low-density gases in the vicinity of Saturn's icy satellites. The tantalum shielding reduces the modeled background radiation flux to an acceptable level of less than 10^{-11} particles cm^{-2} sec^{-1}. Multilayer insulation (MLI) is used for thermal shielding and micrometeoroid protection of the sensor and electronics package, and is attached to the brackets as shown in Figure 13. A thermal radiator panel is used to dissipate internally generated heat and is not covered by the MLI. The package design is such that the entrance apertures of the sensor protrude beyond the edge of the FPP platform and MLI insulation. This configuration prevents contamination from spacecraft out gassing and provides the maximum field of view.

The INMS sensor is made of titanium and is free of organic materials. It was baked to about 280°C for vacuum clean up. During the launch and cruise phases, a getter pump is used to maintain the sensor interior at a low pressure; internal pressure is monitored with a miniature ion gauge and a thermistor. In this sealed configuration the INMS retains a residual gas atmosphere (composed primarily of helium, argon, methane, carbon monoxide, carbon dioxide, hydrogen, and water).

Figure 12. Photo showing the INMS closed and open ion sources.

This residual atmosphere can be used for sensor testing. The INMS will remain sealed and at near vacuum interior pressures to preserve instrument cleanliness until ambient atmosphere measurements can be made.

The INMS entrance apertures are masked by a single plate and are covered by a metal–ceramic break-off hat that can be pyrotechnically separated from the main sensor. Ejection of the break-off hat and opening of the sensor to the external

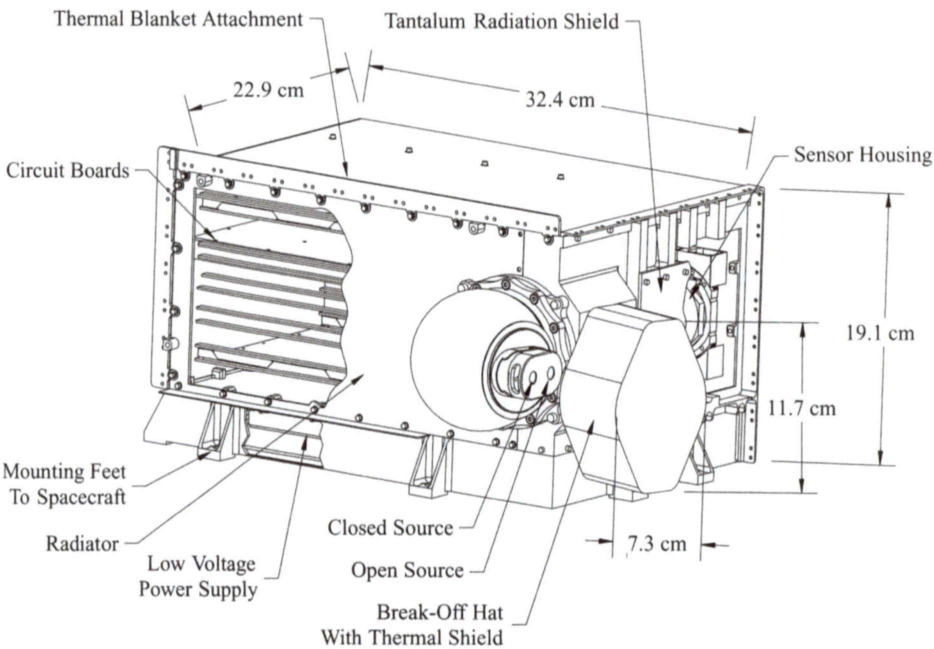

Figure 13. Schematic showing the INMS with the electronic and mechanical subsystems.

Figure 14. The INMS sensor and electronics.

environment are planned to occur after SOI main engine burn but prior to the descending ring plane crossing. The break-off device has been used successfully on Earth-orbiting satellites as well as on the Pioneer Venus Orbiter.

2.2. INMS Components and Operating Principles

As discussed above, the INMS instrument uses two ion sources, a closed source for the measurement of non-reactive neutrals and an open source for the measurement of reactive neutrals and ions. The INMS is thus operated in three basic modes: neutral closed source, open source neutral beaming, and open source ion. In this section, we describe in detail the major INMS components and the principles of operation for each of the three modes. (A fourth mode, open source thermal neutral, is possible but would only be used as a backup in the event that the closed source fails. It will not be discussed in this paper. In the following, whenever reference is made to the "open source neutral mode," the "neutral beaming mode" is meant unless otherwise specified.)

2.2.1. *Closed and Open Ion Sources*

The closed source for ionization of neutrals consists of a spherical antechamber with an entrance orifice for the ambient gas flux and a cylindrical tube that connects this antechamber to the entrance of the ionization region (Figure 15). The incoming gas makes many collisions with the antechamber surfaces and thermally accommodates to the surface temperature. A ram enhancement of the antechamber pressure above that of the ambient gas is achieved by limiting gas conductance from the antechamber into the ion source while maintaining a much higher gas conductance through the entrance aperture. Once in the ion source, an electrostatically collimated electron beam emitted from one of two redundant electron gun assemblies ionize the neutral gas. The ions thus formed are focused into the quadrupole mass analyzer by a series of cylindrical electrostatic lenses and by the quadrupole switching lens.

The entrance aperture of the open source ion (Figure 15) leads into the ion deflector/trap, a cylindrical antechamber that contains four electrodes in equally spaced segments. The four deflectors can be used to trap incoming ions and electrons during neutral measurements or to focus incoming ions into an aperture in the top plate lens (TPL). In the ion mode, the deflectors and TPL are set at ground potential as the default setting; the deflector electrode and TPL lens potentials can be programmed to vary as a function of species mass and angle of attack to improve the instrument's angular response (see Section 2.3). The open source ionization region contains two redundant opposing hot-filament electron guns that are used to ionize the neutral beam; in the open source ion mode, the electron gun filaments are switched off. Four open source lenses (designated OL1–4) focus the ion beam into the quadrupole-switching lens. OL4 serves as the entrance lens to the switching lens; its potential is set at −5 V in the open source neutral mode and at −30 V

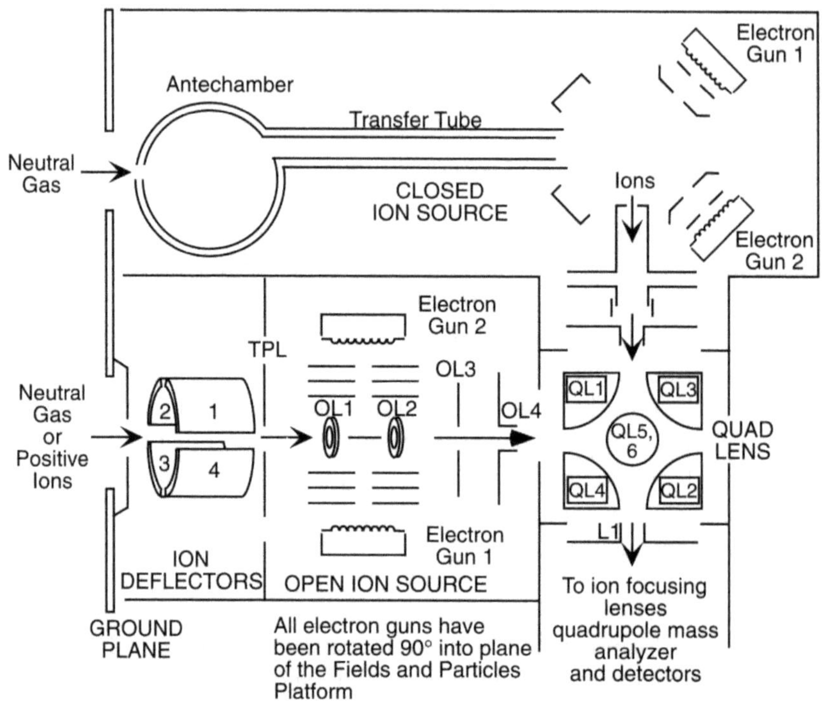

Figure 15. Schematic showing both INMS ion sources, the deflector plates in the ion trap, the open source focusing lens (OL1–4), and the switching lens electrodes (QL1–4).

in the ion mode. These settings reflect the different requirements for measuring neutrals or ions and determine the mean energy of the ions being transmitted into the switching lens.

In both the closed source and open source neutral modes, impacting electrons emitted from hot-filament electron guns ionizes sampled neutrals. Electron impact ionization offers significant efficiency for almost all neutral species and produces unique fractionation patterns that aid in the identification of molecules. In some cases, fragment ions created by the dissociative ionization of the more abundant primary species can mask the presence of trace constituents. However, this difficulty can, to a large degree, be overcome by using two ionization energies to produce different fractionation patterns that can be exploited to minimize interference. Nominal electron energies of 70 and 25 eV have been selected for INMS operation. The electron guns have filaments made of a tungsten–rhenium alloy (0.0076 cm diameter; 97% tungsten, 3% rhenium) in a coiled configuration, which are heated to emit an electron beam. The beam is collimated and focused by electrostatic lenses.

The electron guns are controlled according to information stored in the Cycle Tables (Section 3.1). These tables specify whether the filaments are on or off and indicate the ionization energy to be used. Normally the higher electron energy is used; however, sometimes the lower energy may be used to obtain simpler fractionation patterns. Changing ionization energies requires that the filaments be switched off while the energy is being changed and then turned back on again. This prevents the filaments from being driven to their maximum current during the transition. The process takes about 10 sec.

2.2.2. *Quadrupole Switching Lens*

The quadrupole switching lens (Mahaffy and Lai, 1990) consists of four circular rod sectors (QL1-4) mounted in a cube assembly and two end plates (OL 5/6). As commanded by the Flight Computer, the switching lens selects ions from either the closed source or the open source and directs them into the common entrance lens system of the quadrupole mass analyzer. When the INMS is operating in the closed source mode, the quad lens simply serves as part of a straight-line focusing lens system. In the open source mode, the switching lens provides a nearly hyperbolic electrostatic field that deflects by 90° ions entering the mass analyzer from the open source. The switching lens rod potentials can be set to voltages ranging from −103 to +101 V as commanded by the Flight Computer according to the relevant parameters in the flight software control tables (cf. Section 3.1). Because the deflection angle of the incoming ions is influenced by ion mass, spacecraft velocity, and charge, the voltages applied to the switching lens rods must be adjusted to compensate for these factors. In the open source ion mode, the switching lens can be used to provide estimates of ion energy distributions (over the energy range 1–100 eV) by varying the rod voltages while dwelling on a particular mass to selectively direct particles of varying kinetic energies into the analyzer. The energy resolution ($\Delta E/E$) is albeit a rather coarse 0.3 (see Figure 37).

In the open source mode, the switching lens rods 1–4 are slaved such that QL1 = QL2 and QL3 = QL4. Voltages on the rods are adjusted to allow transmission of an ion beam of a given energy according to the equation QL1 = QL2 = −(QL3 + 2K), where K is the absolute value of the potential applied to the entrance lens OL4 (and also to the exit lens L1), i.e., K equals approximately 5 V in the neutral beaming mode and 30 V in the ion mode. The potential (in volts) applied to QL3 varies with the ion energy. In the closed source mode, switching lens voltages are set as follows: QL1 and 3 = −88 V; QL2 and 4 = 0 V. End plate voltages are set at 0 V in the open source mode and at +2.59 V in the closed source mode.

Laboratory studies using an ion beam (see Section 4.2) with energies ranging from several eV to about 20 eV were performed to characterize switching lens operation in the open source neutral and open source ion modes (see Section 4.3), and the results were incorporated in the Switching Tables (Section 3.1).

2.2.3. *Quadrupole Mass Analyzer*

The quadrupole mass analyzer (or mass filter) allows ions transmitted through the switching lens and ion focusing lens to a limited range of mass-to-charge ratios, typically a fraction of a mass unit. It consists of four precision-ground hyperbolic rods mounted in a mechanical assembly. The rod spacing parameter, r_0, is 0.58 cm and the rod length is 10 cm. The quadrupole rods are excited by radio frequency (RF) and direct current (DC) potentials that together create a dynamic electrostatic field within the quadrupole region that controls the transmitted mass (mass/charge ratio), the resolution, and the transmission efficiency. A mass scan is effected by varying the RF potential amplitude, V_{ac}, to satisfy the relationship $M = 0.55\ V_{ac}/f^2$; V_{ac} is in Volts, f is the RF frequency in MHz, and M is mass in atomic mass units (Daltons). The nominal mass range for the INMS is 1–99 Daltons. Two separate RF frequencies are used; a high frequency (3.57 MHz) for the mass range 1–8 Daltons and a low frequency (1.64 MHz) for the mass range 12–99 Daltons. The control voltages (V_{ac} and V_{dc}) applied to the quadrupole filter rods are calculated by the flight software as a function of Daltons and the RF frequency selected. Because frequency drift can cause mass peaks to be slightly offset, RF frequency is monitored, allowing the control voltages to the rods to be corrected for drift. Resolution is controlled over each mass range by programming the V_{dc}/V_{ac} ratio to maintain the resolving power as defined by a cross-talk criterion appropriate for that mass range. The resulting flat-topped peaks and narrow widths allow a mass scan mode in which each mass is monitored by a single step to achieve the lowest detection limit in a specified period. In addition to performing mass scans, the quadrupole analyzer also operates in "total transmission" or high-pass filter mode in which V_{dc} is reduced to zero and all masses above a specified Dalton/charge ratio are transmitted to the detectors.

When the INMS is operating in the open source mode, a quadrupole bias voltage, as specified in the Focus Table (Section 3.1), is added to the DC voltage applied to the RF mass analyzer rods to slow down incoming ions and increase their residence time in the analyzer's RF field. In the case of ions transmitted from the closed source, the ion energy is similar for all species because the gas has been thermalized in the antechamber. In the case of the open source, the ion energy is a function of the species mass and the relative velocity vector, and is higher than that of the closed source ions. This higher energy into the analyzer results in a wider mass peak width (Kasprzak *et al.*, 1987). Slowing the ions down by applying a quadrupole retarding bias potential increases the mass resolution by reducing the peak width (Figure 16). In the open source neutral and ion modes, the quad bias voltage = KE − 1 V (KE is the ion energy in eV); in the closed source mode, a quad-bias voltage of −1 V is applied.

2.2.4. *Secondary Electron Multipliers*

Ions exiting the quadrupole mass filter are detected by one of two secondary electron multipliers (continuous channel electron multipliers). They are electronically biased such that most of the ions are deflected into the primary detector with a much smaller fraction (on the order of several per thousand) being scattered into the secondary detector. This dual system is identical to that used in the GC/MS instrument on the Huygen's probe (Niemann *et al.*, 1997) and is used to increase the overall dynamic range of signal detection by about a factor of ~1500 for INMS (see Table IV). Charge pulses at the anode of the multiplier are amplified and counted.

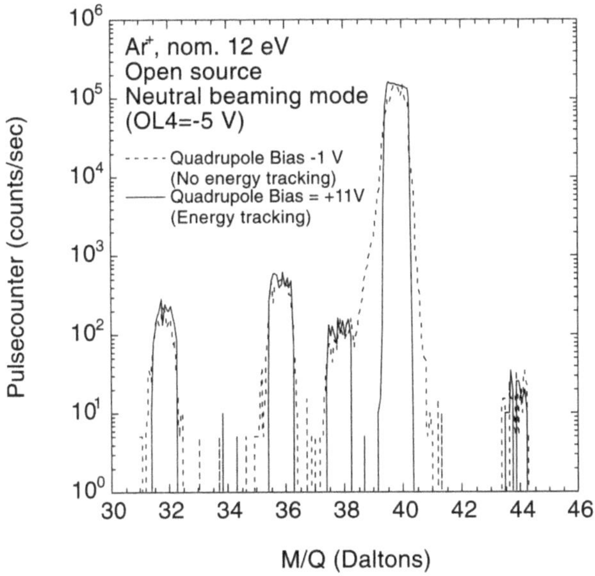

Figure 16. Spectrum showing the reduction in the width of the mass peak for Ar^+ that results from the application of a +11 V quad bias to the mass analyzer. The quad bias compensates for the energy of the ion in the open source and equals KE (the kinetic energy of the ion) −1 V.

The detection threshold is determined by the background noise in the multiplier (approximately 1 count per minute in the laboratory). The upper count rate of each detector system is about 10 MHz, limited by the product of the multiplier pulse width and gain bandwidth of the pulse amplifier counter system. (There is a non-linear response that occurs in the range of 1–10 MHz). Ion counts above this value can be measured directly as an analog current. Multiple sample periods can be combined to lower the detection threshold to the background noise level (signal/noise ratio = 1). Assuming a maximum 1 MHz counting rate, the dynamic range of the two detector system is $\sim 10^8$.

2.3. Sensor Sensitivity and Detectability Thresholds

The sensitivity to neutral species varies with species because of the different ionization efficiencies for neutrals in the ion sources, the differential transmission of the quadrupole switching lens and mass filter, and the conversion efficiency at the secondary electron multiplier. In general, when INMS is operated in the neutral closed source mode, neutrals with densities of $\sim 5 \times 10^4$ cm^{-3} can be detected; in the open source neutral mode, the detection threshold is higher, $\sim 2 \times 10^5$ cm^{-3}. In the case of ions, the instrument's sensitivity varies with ion species because of different spacecraft equivalent energies for the different masses, the differential transmission of the quadrupole switching lens and filter, and conversion efficiency at the secondary electron multiplier detector. Spacecraft potential will also influence INMS sensitivity to ions by modifying the incoming flux and ion trajectory directions relative to the spacecraft. At a spacecraft velocity of 6 km sec^{-1}, the minimum ion density is $\sim 10^{-2}$ cm^{-3}.

In the neutral closed source mode, the background count rate due to gases adsorbed onto the surface of the source may be comparable to the minimum detectable density in certain regions of the spectrum, particularly below about 50 Daltons. For both the closed and open sources, interference at some mass numbers by other ambient gases present in high concentrations also presents a problem (cf. Section 4.2 and Figure 32). In the closed source, the calculated ambient density is lower than the ion source density because of the velocity ram enhancement. The maximum ion source density is limited by mean free path conditions in the ion source and analyzer regions.

There are several potential sources of a detector background. Section 2.1 discusses the use of an external tantalum shield around the secondary electron multiplier section of the sensor housing to reduce the effect of magnetospheric particle radiation. For high-energy electrons and ions, the physical geometry of the path from the open or closed source entrance apertures to the detectors involves a right angle bend, conducting surfaces with many constricting apertures and positive voltages that are a fraction of a keV. The probability that a photon or a high-energy particle will make it to the detectors is extremely difficult to estimate or simulate in the laboratory, but is quite low. The net effect of either photons or highly energetic ions/electrons is that they will not be mass/charge separated in the quadrupole section of the sensor because of their high energy or uncharged condition. This will create a mass independent background in the detectors. In performing a full range mass scan (1–99 Daltons), those mass slots not occupied by ambient ions or neutrals can be used to estimate the effect of this mass-independent background, which can be subtracted off from the ambient mass peaks.

2.4. Instrument Response

In order for the INMS to make valid ion and neutral density measurements, the spacecraft velocity vector plus the drift velocity of any neutral or plasma flows (crosswinds) must be within the field of response of the appropriate source.

2.4.1. *Closed Source Mode*

The closed source response for several gas species and different molecular masses as a function of the angle of attack (angle between the aperture normal and the spacecraft velocity vector) is given in Figure 17. Particles entering through the aperture collide many times with the surfaces of the ion source electrodes and enclosure, and thermally accommodate to the surface temperature before leaving through the transfer tube. For angles of attack $<90°$ this results in an enhancement of the number density in the antechamber over that in the ambient atmosphere.

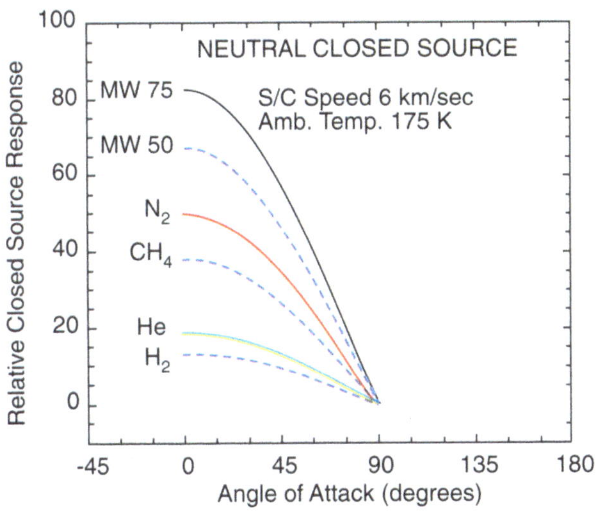

Figure 17. Plot showing the response of the INMS in the closed source mode as a function of the angle of attack. MW: molecular mass (in Dalton).

The maximum ram enhancement factor for N_2 and a spacecraft velocity of ~6 km sec^{-1} is 50. The signal varies approximately with the cosine of the angle of attack, and is proportional to the product of the spacecraft speed and the square root of the ratio of the molecular weight divided by the gas temperature. A balance of the incoming number flux at the spacecraft speed and the outgoing gas flux at the antechamber wall temperature determines the number density in the antechamber. The relationship between the ion source density, incoming flux, and ambient density is predictable from kinetic theory. Non-reactive species (e.g., N_2 and CH_4) will be measured in this mode. In the closed source mode, ambient density measurements can be made at angles of attack ranging from about 0 to 90°, depending on the gas background.

2.4.2. *Open Source Neutral Mode*

In the open source neutral mode, the ambient particle density is measured directly. The quasi-open source geometry allows ambient gas to enter the ionization region directly and permits measurement of chemically active species. The angular response is limited by the geometric field of view. The theoretical response of the open source neutral mode as a function of angle of attack is given in Figure 2.4.2 for a view cone half angle of 8°. Optimum open source neutral mode measurements will occur at Titan closest approach, where the density is sufficiently large and the angle of attack is approximately within the geometric view cone. During many Titan flybys the spacecraft will provide tracking of local nadir for the Radar instrument. In this case the INMS angle of attack needed for the open source operation ranges

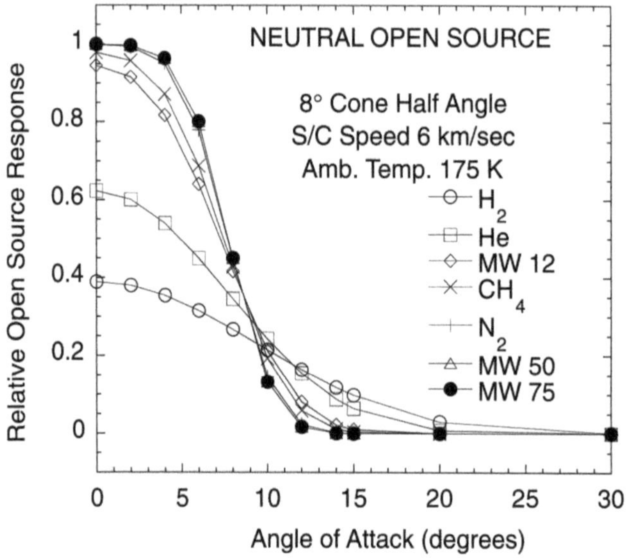

Figure 18. Plot showing the response of the INMS in the open source neutral beaming mode as a function of the angle of attack.

from 0 to 8° within about 1.4 min of closest approach at a spacecraft speed of 6 km sec^{-1} (see Figure 2.4.2).

For measurements in open source neutral mode, neutrals that have been thermally accommodated to the ion source walls must be distinguished from the direct beaming component. This is achieved by (1) maximizing the gas conductance from the ionizing region into the vent region at right angles to the open source axis, which reduces the velocity ram stagnation density enhancement; (2) minimizing the ion acceleration into the switching lens; and (3) using the quadrupole switching lens transmission, which is a function of energy, to filter the ion energy.

Numerical studies (Swaminathan *et al.*, 1996) indicate that <2% of mass 28 ions from thermally accommodated N_2 gas are transmitted through the mass analyzer. This is with the switching lens potential set to transmit a 28-Daltons ionized neutral beam at spacecraft energies (the spacecraft equivalent energy for a 6 km sec^{-1} spacecraft velocity is 5.22 eV compared to 0.015 eV for a thermally accommodated gas at 300 K temperature). A laboratory estimate of this ratio was made using thermal gas ions and operating the switching lens in neutral beaming mode with the entrance lens potential (OL4) set to −5 V. The mass was set at 28 in the total mode ($V_{dc} = 0$ V and V_{ac} set for mass 28). Figure 20 shows the energy scans for a thermalized neutral gas, with the open source lenses OL1 and OL2 set at 0 V, and for a simulated neutral beam (OL1 = OL2 = 2 V). The contribution of the thermal ions to the more energetic 2 V ions at signal maximum is about 10^{-4} for equal signal amplitudes. Comparison of the overlap area of the two curves with the total area under the neutral beaming curve yields a similar value, 7×10^{-3}. As in

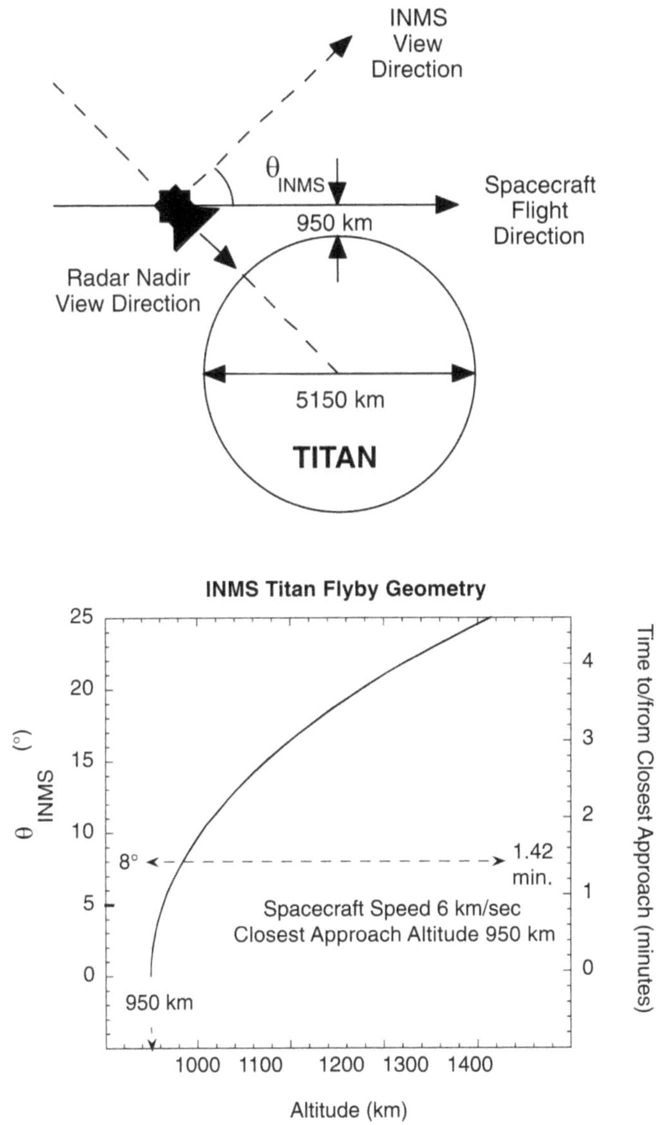

Figure 19. Titan flyby geometry and the change in INMS angle of attack as function of altitude with the Orbiter in Radar Mode.

the closed source, the thermalized gas component is formed in the open source by scattering off the walls of the ion source; little ram enhancement occurs because the gas is vented. If we assume, as an extreme case, that the density of the open source thermalized component is equivalent to the maximum ram enhancement of about 50 (for N_2 at an angle of attack of 0°), the thermal contribution would be less than or equal to $50 \times 10^{-4} = 5{\times}10^{-3}$, which is negligible for mass 28. Neutral

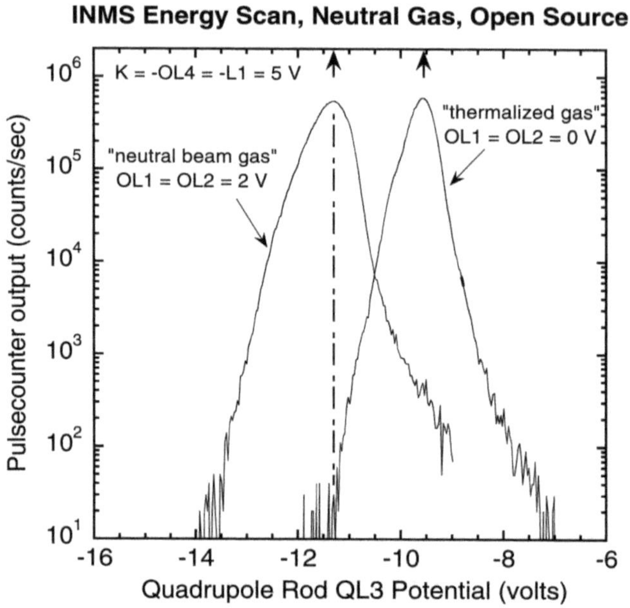

Figure 20. Results of an energy scans performed in the neutral thermal (OL1 = OL2 = 0 V) and simulated neutral beaming (OL1 = OL2 = 2 V) modes. The contribution of the thermalized population to the beaming population, at signal maximum (dashed line), is about 10^{-4}.

beam tests on the engineering unit will provide more complete data for this and other species.

2.4.3. *Open Source Ion Mode*

To determine the response of the INMS in the open source ion mode, a laboratory study was performed of the angular response with the deflector electrodes 1–4, the top plate lens, OL1 and OL2 set at ground potential. Figure 21 shows the relative transmission as a function of angle for two ion masses (Ar^+ and Kr^+) and two energies (8 and 13.9 eV) that closely correspond to a 6-km sec^{-1} spacecraft speed for these masses. The open source ion mode response has also been studied by numerical simulation using the Sarnoff BEAM 3D (Swaminathan *et al.*, 1996) software to model the lens system from the entrance aperture to the entrance of the quadrupole mass analyzer. Results of that simulation are also shown in Figure 21. Although the conditions of the simulation and the laboratory data are not exactly identical, both sets of data show a similar decrease of transmission with increasing angle except for the Ar^+ *Z*-axis rotation (X's), which most likely reflects experimental difficulties. The INMS scale is arbitrary, while for the numerical simulation data it is the total lens system transmission.

A more advanced numerical study of the angular response (Swaminathan *et al.*, 1996) is shown in Figure 22 (left-hand scale) for a spacecraft speed of 6 km^{-1} and

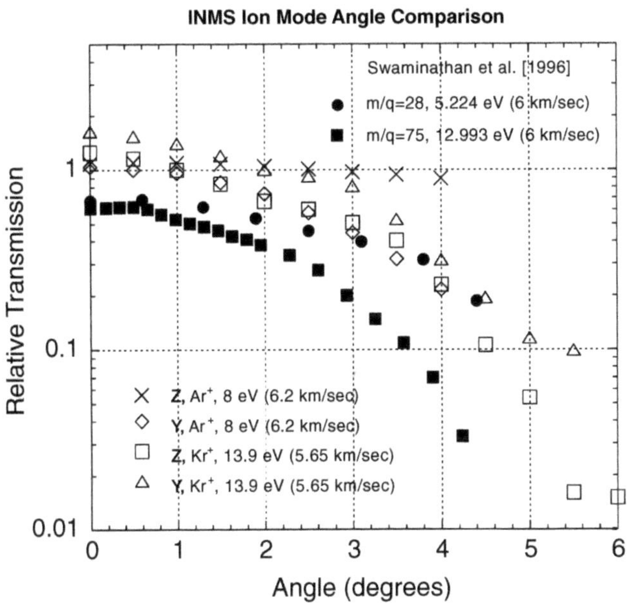

Figure 21. Laboratory data showing the relative transmission in the open source ion mode for Kr^+ and Ar^+ corresponding to a spacecraft velocity of 6 km sec^{-1}. The laboratory data are compared with similar data from a numerical study by Swaminathan *et al.* (1996). The INMS *Y* and *Z* correspond to rotations about the space *Y* and *Z* axes (outward INMS source normal is $-X$).

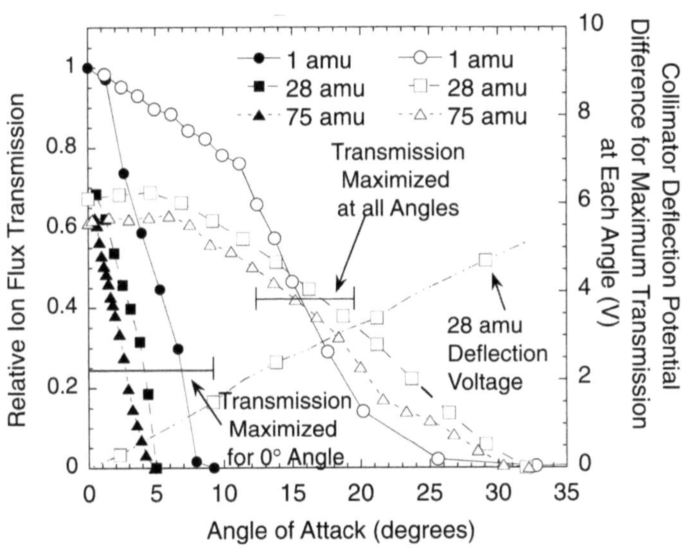

Figure 22. Results of a numerical study (Swaminathan *et al.*, 1996) of the angular response of the open source in the ion mode. The deflector plate voltages were adjusted to optimize the transmission for various angles of attack. Adjusting the voltages on the deflector plates makes it possible to increase open source acceptance beyond the nominal 8.3° half angle. Ion energies for 1, 28, and 75 Dalton were 0.187, 5.224, and 12.993 eV, respectively.

no thermal ion energy spread. If the transmission is optimized only at 0° angle of attack, then the half-width half-maximum (HWHM) point is about 3° for 28 Daltons. Optimizing the transmission by adjusting the potential on paired deflectors at each angle of attack results in a HWHM of about 15° for 28 Daltons. The difference in potential between opposite pairs of deflector electrodes needed to optimize the transmission is approximately linear with increasing angle of attack (Figure 22, right-hand scale). The advantage of being able to program the deflector voltages to increase the angular response beyond the geometric view cone is important for increasing the coverage in altitude of thermal ions during a Titan pass. For example, optimization of the angle coverage for N_2^+ for a 950 km closest approach altitude to Titan is sufficient to permit measurements over the ionospheric peak (Keller *et al.*, 1998) up to about 1050 km. In evaluating the use of the four-deflector electrodes to steer low-energy ions, it was found that the digital-to-analog converters (DAC) for the plates did not have adequate voltage resolution. A post-characterization change was made to introduce a non-linearity into the response (Figure 23) such that there is higher voltage resolution around 0 V as might be required by the Figure 22 tracking.

The INMS can also measure suprathermal ions with energies up to about 100 eV. Ion energy and direction can only be inferred by scans of the deflector electrodes and switching lens that require a substantial number of data samples. The suprathermal

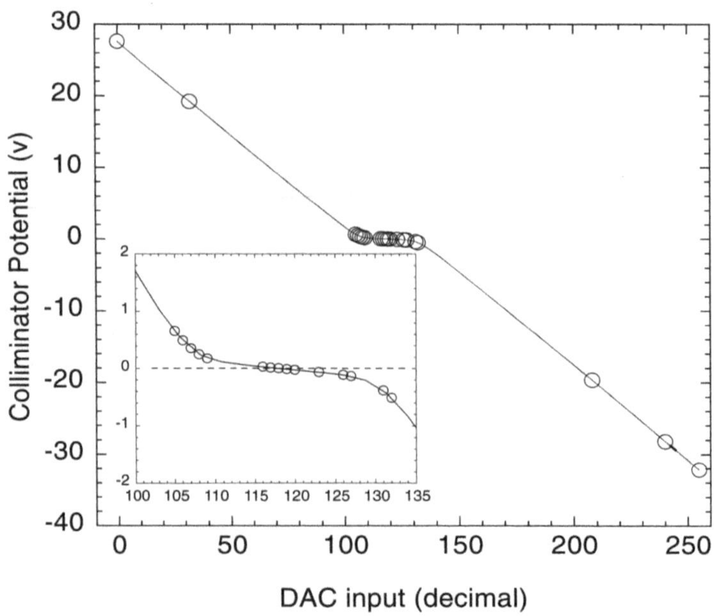

Figure 23. Deflector plate potentials as a function of the decimal values in the software table that controls digital-to-analog converters (DAC) that supply the voltage to the deflectors. (See Section 3 for a discussion of software control tables.) A nonlinearity was introduced into the response to improve the voltage resolution around 0 V.

ion mass-to-charge ratio can, however, be uniquely determined on a per sample basis by the quadrupole analyzer fields. Even if the particles are energetic and cause mass peak spreading, the peak center will be at the appropriate mass-per-charge value for each 34-ms sample period. This unique mass determination will complement measurements with the CAPS/IBS, which, for trans-sonic plasma regimes, will have difficulty resolving closely spaced mass groups.

2.5. INMS Electronics

The INMS electronics system (Figure 24) is based on designs used for the Huygen's Probe GC/MS instrument. A low-voltage (LV) power supply converts spacecraft power to well-regulated DC voltages that are supplied to the instrument electronics. A pulse-width-modulated converter allows efficient generation of multiple secondary voltages while providing secondary-to-primary isolation. A large number of voltages are required to bias the various focus electrodes as well as to supply DC voltages for the secondary electron multipliers. Analog modules are used for regulating the emission of the electron guns, for providing fixed and programmable voltages to set lens potentials, for supplying RF and DC for the quadrupole mass

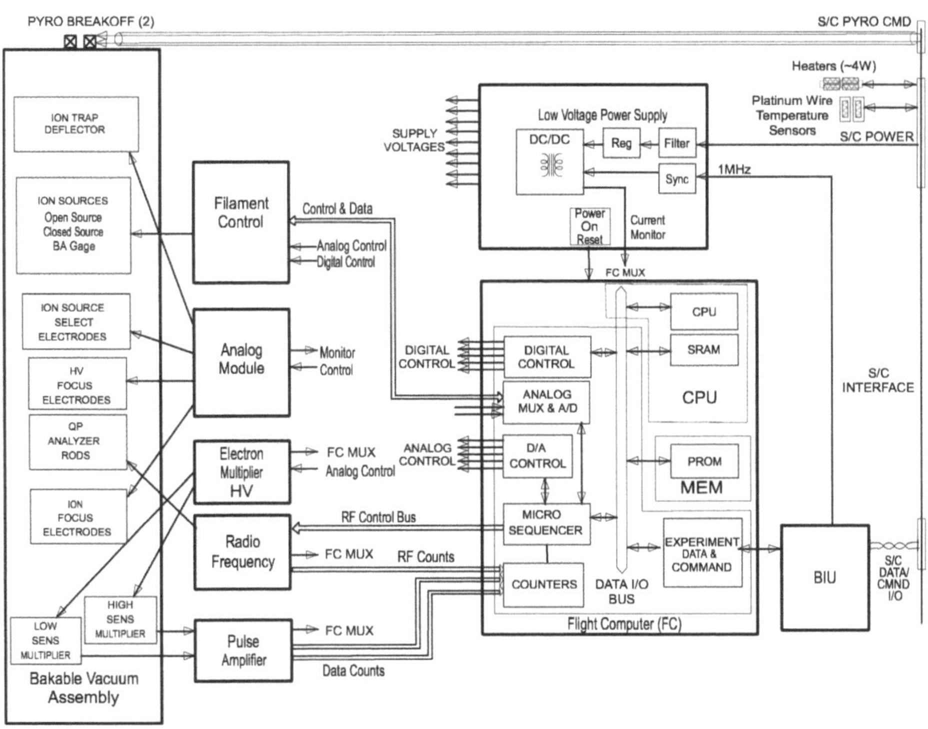

Figure 24. Schematic of the INMS electronic subsystems.

analyzer, for supplying high voltages for the detectors, and for the pulse-counting circuits. The digital electronics includes a single microprocessor, a spacecraft bus interface circuit, and an interface between the CPU and the analog modules. Major portions of the electronics are packaged in hybrid circuits to save weight and space.

A radio frequency generator drives the quadrupole at two resonant frequencies in order to reduce the need for large amplitude potential for the required mass range (1–99 Daltons). A solid-state switched bandpass filter performs frequency selection. The DC voltage is created by high-voltage operational amplifiers and is superimposed on the RF amplitude. Digital-to-analog converters program both the RF and DC amplitudes.

Charge pulses at the anode of the electron multiplier are converted by a pulse amplifier into voltage pulses that are counted if they are above a pre-set threshold. Analog measurement of the multiplier current is used to determine the in-flight multiplier gain.

The Flight Computer uses a 16-bit Marconi MA31750 microprocessor running at 10 MHz, with 64 K primary RAM, 64 K ROM, and 32 K "extra" RAM (used only for data storage, not for execution of flight software). The computer controls the INMS measurement sequence, counts the detector pulses, provides analog-to-digital conversion of the detector current, and monitors instrument housekeeping parameters. The computer is programmed in Ada as the target language with some use of assembly language to handle time-critical functions, input/output, and interrupts. The instrument ROM/RAM contains the default measurement and test sequences without requiring memory upload.

The INMS has a number of different states (Figure 2.5) which determine what information is returned from the instrument. After startup, the INMS typically waits in the Sleep_0 state for further commands. The Test and ALF states are used for configuration, memory loads, and testing. The Pressure Check is a short test to determine the pressure in the chamber while the cover is still on. Default Science is equivalent to running Science Sequence number 1 (described in Section 3.3.1). All modes except Reset return housekeeping packets, but not all modes return science packets.

Telemetry data transmitted from the INMS to the Orbiter consist of (1) housekeeping packets that contain normal analog-to-digital converter channel data or memory dump data if science packets are not being collected and (2) science packets, which contain normal science data, memory dump data, or special test data. Standard CCSDS packetization formats are used for both housekeeping and science data.

3. Science Operations: Measurement Strategies and Sampling Methods

INMS measurement strategies and sampling methods are determined by the investigation's science objectives and must take into account the region and species

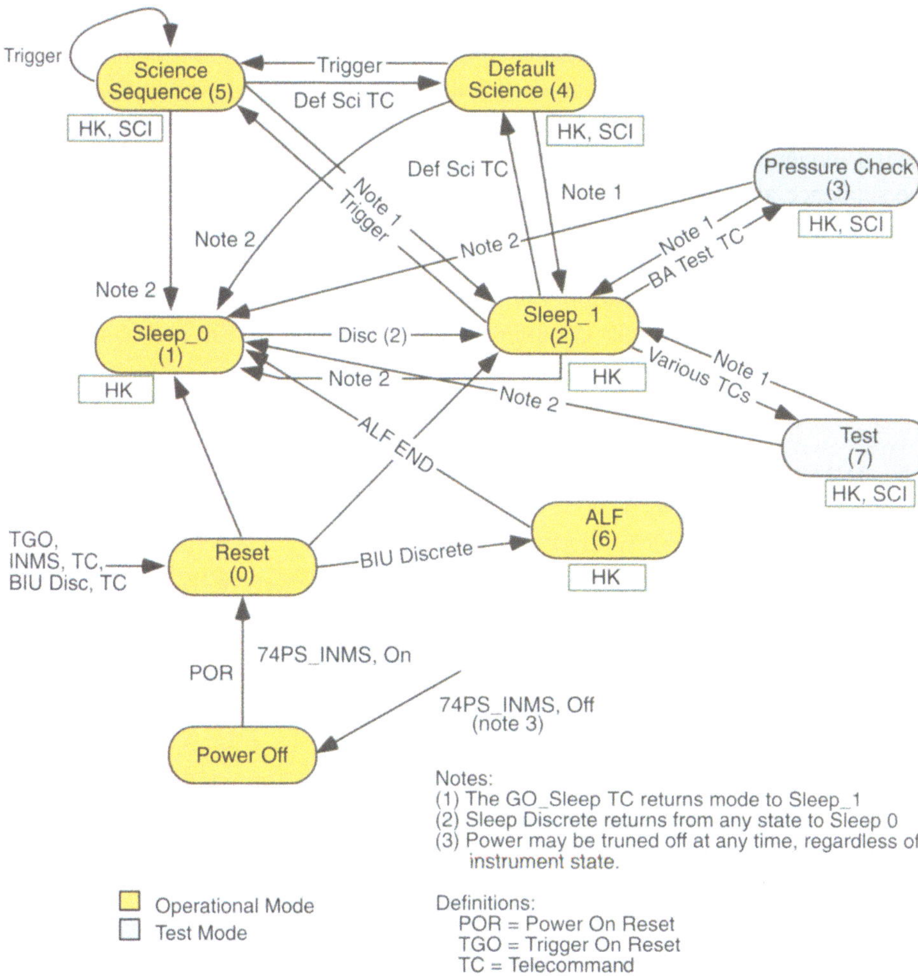

Figure 25. Diagram of the INMS instrument states. The values in parentheses correspond to the internal state number. Arrows denote allowed state changes and the commands which produce those changes, and boxes denote the types of packets that are produced.

being sampled. The basic sampling sequence is the "scan," which is a series of 68 mass/charge measurements; the mass numbers to be sampled in each scan are specified by a particular "Mass Table" (see the following section for a discussion of the INMS software tables). Each measurement period or "integration period" (IP) lasts 34 ms (a 31-ms counting period + ~3 ms for set up and read out). Each scan therefore requires 2.3 sec (= 34 ms per sample × 68 IPs or samples). The different types of scans and the rationale for each will be discussed below. A scan or series of scans to be repeated constitutes a "cycle." The operation of the instrument for each scan in a cycle is defined by a "Cycle Table" which indicates the Mass Table and other control tables to be used for a particular scan. One or more cycles make up a "science sequence." A science sequence is initiated by receipt of a time-tagged "trigger" command from the Orbiter. Trigger commands will be sent from the ground to the Orbiter and stored in the Solid State Recorder (SSR) for later execution; under some circumstances, it may be possible to command an orbital sequence from the ground in real time. The cycles to be performed during the sequence are identified in a "Sequence Table," which also specifies a velocity constant used to modify the quad lens voltages for velocity compensation in the open source mode. Several science sequences were defined prior to SOI. These are discussed in Section 3.3. Additional sequences are expected to be designed and uploaded to the INMS flight computer once exploration of the Saturn system is under way.

3.1. INMS Software Tables

The information needed by the Flight Computer to command INMS operations is contained in eight different types of software tables stored in the computer's memory. In addition to the Mass, Cycle, and Sequence tables mentioned above, the software tables include Ion Trap tables, Switching tables, Focus tables, D/A tables, and a single Velocity table. The Velocity table contains velocity constants (256 constants of which 64 are distinct velocity values) used in calculations performed to determine voltage adjustments required for velocity (energy) compensation in the open source mode. The Trap, Switching, Focus, and D/A tables are "control tables" (Table V) which contain control parameters. These values are read by the Flight Computer, which writes them to the digital-to-analog converters (DAC) that control the voltages applied to the INMS optics (or, in the case of the D/A tables, to the detectors).

Mass tables, too, are control tables. Instead of voltages, however, they specify the masses to be selected by the quadrupole mass analyzer (see Table VI) and the science mode to operate in, when velocity compensation is used. Mass Table values are used by the software to calculate the AC and DC voltages and RF oscillator frequency required to "tune" the quadrupole analyzer rods to select a particular mass. Up to 96 Mass Tables, each containing 68 records, can be stored in the Flight Computer's memory. Each one-word record contains three fields to indicate the science operating mode—open source ion (OSI), open source neutral thermal (OSNT), open source neutral beaming (OSNB) or closed source neutral (CSN)—energy scan mode, and mass number, quantized to 1/8 Daltons.

There are 64 Cycle and 64 Sequence tables. As indicated above, Sequence tables define the cycles that constitute a particular measurement sequence, and Cycle Tables define the scans to be performed during a particular cycle. A single Sequence table record contains 1–31 cycles, and each cycle contains the number of scans to perform per cycle, the velocity value to use for velocity compensation calculations, and the Cycle Table number to execute in that cycle. Each Cycle table

TABLE V

INMS control tables.

Table Type	No. of Table	Records/ Table	Words/ Record	No. of Words	Function
Sequence	64	31/1[a]	2/1	4032	Controls execution of all tables through the Cycle Tables (see Figure 3.1)
Cycle	64	31/1[a]	3/1	6016	Controls execution of trap, switching, mass, focus and D/A tables (see Figure 3.1)
Mass	64	68	1	4352	Controls mass selection
Trap	32	68	3	6528	Controls collimator/deflector, TPL, and OL4 voltages
Switching	32	68	3	6528	Controls quad lens rod voltages
Focus	32	68	3	6528	Controls OL 1–3, end plate, ion lenses and quad bias voltages
D/A	16	1	3	48	Controls electron multiplier, detector threshold settings
Velocity	1	256	1	256	Used in voltage calculations for velocity compensation in open source mode

[a]Sequence and Cycle Tables have one header record per table.

contains 1–31 scans, and each scan is a record with the numbers of the Mass, Switching, Trap, Focus, and D/A tables to be used during the scan. Additional fields in the record indicate whether the open and/or closed source electron gun filaments are to be on during the scan and specify the electron energy to be used. The relationships among the various software tables are illustrated schematically in Figure 26.

3.2. Mass Scans

The basic INMS sampling sequence is a scan consisting of 68 mass/charge measurements as specified in a Mass Table. Masses to be measured during a scan are entered into the Mass Table for that scan either as integers from 1 to 8 and from 12 to 99 Daltons or, in 12 tables, as fractional values, in increments of 1/8 Dalton, from 0.5 to 8.5 and 11.5 to 99.5. The Mass Tables include, in addition to the mass numbers for the primary species, mass numbers for isotopic species as well as those for possible cracking products. (A summary of currently defined Mass Tables is presented in Table VI. References to a particular Mass Table in this and the following section and in Table VII refer to this table.) During one 68-step scan, the INMS can either (i) sequentially sweep or survey, in increments of 1 or 1/8 Dalton, the entire mass range covered by the instrument or a portion of that range, or (ii) sample selected masses. The first type of scan is known as a "survey" scan and can

TABLE VI

Updated mass tables.

	0	1	2	3	4	5	6	7	8	9	10	11	12	13	14	15	16	17	18	19	20	21
IP	SOM0	SOM1	SOM1	SOM1	SOM1	SOM1	SOM1	SOM1	SOM1	SOM1	SOM1	SOM1	SOM1	SOM1	SOM1	SOM1	SOM1	SOM1	SOM1	SOM1	SOM1	SOM1
1	0	1	0.500	11.500	19.500	27.500	35.500	43.500	51.500	59.500	67.500	75.500	83.500	91.500	12	28	2	2	2	2	2	2
2	0	2	0.625	11.625	19.625	27.625	35.625	43.625	51.625	59.625	67.625	75.625	83.625	91.625	13	28	16	16	16	16	4	4
3	0	3	0.750	11.750	19.750	27.750	35.750	43.750	51.750	59.750	67.750	75.750	83.750	91.750	14	28	17	17	17	17	7	7
4	0	4	0.875	11.875	19.875	27.875	35.875	43.875	51.875	59.875	67.875	75.875	83.875	91.875	15	28	28	28	28	28	12	12
5	0	5	1.000	12.000	20.000	28.000	36.000	44.000	52.000	60.000	68.000	76.000	84.000	92.000	16	28	29	29	29	29	13	13
6	0	6	1.125	12.125	20.125	28.125	36.125	44.125	52.125	60.125	68.125	76.125	84.125	92.125	17	28	1	52	12.625	25.625	14	14
7	0	7	1.250	12.250	20.250	28.250	36.250	44.250	52.250	60.250	68.250	76.250	84.250	92.250	18	28	2	53	12.750	25.750	15	15
8	0	8	1.375	12.375	20.375	28.375	36.375	44.375	52.375	60.375	68.375	76.375	84.375	92.375	19	28	3	54	12.875	25.875	16	16
9	0	12	1.500	12.500	20.500	28.500	36.500	44.500	52.500	60.500	68.500	76.500	84.500	92.500	1	28	4	55	13.000	26.000	24	24
10	0	13	1.625	12.625	20.625	28.625	36.625	44.625	52.625	60.625	68.625	76.625	84.625	92.625	2	28	5	56	13.125	26.125	25	25
11	0	14	1.750	12.750	20.750	28.750	36.750	44.750	52.750	60.750	68.750	76.750	84.750	92.750	3	28	6	57	13.250	26.250	26	26
12	0	15	1.875	12.875	20.875	28.875	36.875	44.875	52.875	60.875	68.875	76.875	84.875	92.875	4	28	7	58	13.375	26.375	27	27
13	0	16	2.000	13.000	21.000	29.000	37.000	45.000	53.000	61.000	69.000	77.000	85.000	93.000	5	28	8	59	13.500	26.500	28	28
14	0	17	2.125	13.125	21.125	29.125	37.125	45.125	53.125	61.125	69.125	77.125	85.125	93.125	6	28	12	60	13.625	26.625	29	29
15	0	18	2.250	13.250	21.250	29.250	37.250	45.250	53.250	61.250	69.250	77.250	85.250	93.250	7	28	13	61	13.750	26.750	30	30
16	0	19	2.375	13.375	21.375	29.375	37.375	45.375	53.375	61.375	69.375	77.375	85.375	93.375	8	28	14	62	13.875	26.875	36	36
17	0	20	2.500	13.500	21.500	29.500	37.500	45.500	53.500	61.500	69.500	77.500	85.500	93.500	12	28	15	63	14.000	27.000	37	37
18	0	21	2.625	13.625	21.625	29.625	37.625	45.625	53.625	61.625	69.625	77.625	85.625	93.625	13	28	2	2	2	2	38	38
19	0	22	2.750	13.750	21.750	29.750	37.750	45.750	53.750	61.750	69.750	77.750	85.750	93.750	14	28	16	16	16	16	39	39
20	0	23	2.875	13.875	21.875	29.875	37.875	45.875	53.875	61.875	69.875	77.875	85.875	93.875	15	28	17	17	17	17	40	40
21	0	24	3.000	14.000	22.000	30.000	38.000	46.000	54.000	62.000	70.000	78.000	86.000	94.000	16	28	28	28	28	28	1	52
22	0	25	3.125	14.125	22.125	30.125	38.125	46.125	54.125	62.125	70.125	78.125	86.125	94.125	17	28	29	29	29	29	2	53
23	0	26	3.250	14.250	22.250	30.250	38.250	46.250	54.250	62.250	70.250	78.250	86.250	94.250	18	28	16	64	14.125	27.125	3	54
24	0	27	3.375	14.375	22.375	30.375	38.375	46.375	54.375	62.375	70.375	78.375	86.375	94.375	19	28	17	65	14.250	27.250	4	55
25	0	28	3.500	14.500	22.500	30.500	38.500	46.500	54.500	62.500	70.500	78.500	86.500	94.500	20	28	18	66	14.375	27.375	5	56
26	0	29	3.625	14.625	22.625	30.625	38.625	46.625	54.625	62.625	70.625	78.625	86.625	94.625	21	28	19	67	14.500	27.500	6	57
27	0	30	3.750	14.750	22.750	30.750	38.750	46.750	54.750	62.750	70.750	78.750	86.750	94.750	22	28	20	68	14.625	27.625	7	58

(Continued).

	0	1	2	3	4	5	6	7	8	9	10	11	12	13	14	15	16	17	18	19	20	21
IP	SOM0	SOM1	SOM1	SOM1	SOM1	SOM1	SOM1	SOM1	SOM1	SOM1	SOM1	SOM1	SOM1	SOM1	SOM1	SOM1	SOM1	SOM1	SOM1	SOM1	SOM1	SOM1
28	0	31	3.875	14.875	22.875	30.875	38.875	46.875	54.875	62.875	70.875	78.875	86.875	94.875	23	28	21	69	14.750	27.750	8	59
29	0	32	4.000	15.000	23.000	31.000	39.000	47.000	55.000	63.000	71.000	79.000	87.000	95.000	24	28	22	70	14.875	27.875	12	60
30	0	33	4.125	15.125	23.125	31.125	39.125	47.125	55.125	63.125	71.125	79.125	87.125	95.125	25	28	23	71	15.000	28.000	13	61
31	0	34	4.250	15.250	23.250	31.250	39.250	47.250	55.250	63.250	71.250	79.250	87.250	95.250	26	28	24	72	15.125	28.125	14	62
32	0	35	4.375	15.375	23.375	31.375	39.375	47.375	55.375	63.375	71.375	79.375	87.375	95.375	27	28	25	73	15.250	28.250	15	63
33	0	36	4.500	15.500	23.500	31.500	39.500	47.500	55.500	63.500	71.500	79.500	87.500	95.500	12	28	26	74	15.375	28.375	16	64
34	0	37	4.625	15.625	23.625	31.625	39.625	47.625	55.625	63.625	71.625	79.625	87.625	95.625	13	28	27	75	15.500	28.500	17	65
35	0	38	4.750	15.750	23.750	31.750	39.750	47.750	55.750	63.750	71.750	79.750	87.750	95.750	14	28	2	2	2	2	18	66
36	0	39	4.875	15.875	23.875	31.875	39.875	47.875	55.875	63.875	71.875	79.875	87.875	95.875	15	28	16	16	16	16	19	67
37	0	40	5.000	16.000	24.000	32.000	40.000	48.000	56.000	64.000	72.000	80.000	88.000	96.000	16	28	17	17	17	17	20	68
38	0	41	5.125	16.125	24.125	32.125	40.125	48.125	56.125	64.125	72.125	80.125	88.125	96.125	17	28	28	28	28	28	21	69
39	0	42	5.250	16.250	24.250	32.250	40.250	48.250	56.250	64.250	72.250	80.250	88.250	96.250	18	28	29	29	29	29	22	70
40	0	43	5.375	16.375	24.375	32.375	40.375	48.375	56.375	64.375	72.375	80.375	88.375	96.375	19	28	28	76	15.625	28.625	23	71
41	0	44	5.500	16.500	24.500	32.500	40.500	48.500	56.500	64.500	72.500	80.500	88.500	96.500	28	28	29	77	15.750	28.750	24	72
42	0	45	5.625	16.625	24.625	32.625	40.625	48.625	56.625	64.625	72.625	80.625	88.625	96.625	29	28	30	78	15.875	28.875	25	73
43	0	46	5.750	16.750	24.750	32.750	40.750	48.750	56.750	64.750	72.750	80.750	88.750	96.750	30	28	31	79	16.000	29.000	26	74
44	0	47	5.875	16.875	24.875	32.875	40.875	48.875	56.875	64.875	72.875	80.875	88.875	96.875	31	28	32	80	16.125	29.125	27	75
45	0	48	6.000	17.000	25.000	33.000	41.000	49.000	57.000	65.000	73.000	81.000	89.000	97.000	32	28	33	81	16.250	29.250	28	76
46	0	49	6.125	17.125	25.125	33.125	41.125	49.125	57.125	65.125	73.125	81.125	89.125	97.125	33	28	34	82	16.375	29.375	29	77
47	0	50	6.250	17.250	25.250	33.250	41.250	49.250	57.250	65.250	73.250	81.250	89.250	97.250	34	28	35	83	16.500	29.500	30	78
48	0	51	6.375	17.375	25.375	33.375	41.375	49.375	57.375	65.375	73.375	81.375	89.375	97.375	35	28	36	84	16.625	29.625	31	79
49	0	52	6.500	17.500	25.500	33.500	41.500	49.500	57.500	65.500	73.500	81.500	89.500	97.500	12	28	37	85	16.750	29.750	32	80
50	0	53	6.625	17.625	25.625	33.625	41.625	49.625	57.625	65.625	73.625	81.625	89.625	97.625	13	28	38	86	16.875	29.875	33	81
51	0	54	6.750	17.750	25.750	33.750	41.750	49.750	57.750	65.750	73.750	81.750	89.750	97.750	14	28	39	87	17.000	30.000	34	82
52	0	55	6.875	17.875	25.875	33.875	41.875	49.875	57.875	65.875	73.875	81.875	89.875	97.875	15	28	2	2	2	2	35	83
53	0	56	7.000	18.000	26.000	34.000	42.000	50.000	58.000	66.000	74.000	82.000	90.000	98.000	16	28	16	16	16	16	36	84
54	0	57	7.125	18.125	26.125	34.125	42.125	50.125	58.125	66.125	74.125	82.125	90.125	98.125	17	28	17	17	17	17	37	85
55	0	58	7.250	18.250	26.250	34.250	42.250	50.250	58.250	66.250	74.250	82.250	90.250	98.250	18	28	28	28	28	28	38	86
56	0	59	7.375	18.375	26.375	34.375	42.375	50.375	58.375	66.375	74.375	82.375	90.375	98.375	19	28	29	29	29	29	39	87
57	0	60	7.500	18.500	26.500	34.500	42.500	50.500	58.500	66.500	74.500	82.500	90.500	98.500	36	28	40	88	17.125	30.125	40	88
58	0	61	7.625	18.625	26.625	34.625	42.625	50.625	58.625	66.625	74.625	82.625	90.625	98.625	37	28	41	89	17.250	30.250	41	89

(Continued on next page.)

TABLE VI
(*Continued*).

	0	1	2	3	4	5	6	7	8	9	10	11	12	13	14	15	16	17	18	19	20	21
IP	SOM0	SOM1	SOM1	SOM1	SOM1	SOM1	SOM1	SOM1	SOM1	SOM1	SOM1	SOM1	SOM1	SOM1	SOM1	SOM1	SOM1	SOM1	SOM1	SOM1	SOM1	SOM1
59	0	62	7.750	18.750	26.750	34.750	42.750	50.750	58.750	66.750	74.750	82.750	90.750	98.750	38	28	42	90	17.375	30.375	42	90
60	0	63	7.875	18.875	26.875	34.875	42.875	50.875	58.875	66.875	74.875	82.875	90.875	98.875	39	28	43	91	17.500	30.500	43	91
61	0	64	8.000	19.000	27.000	35.000	43.000	51.000	59.000	67.000	75.000	83.000	91.000	99.000	40	28	44	92	17.625	30.625	44	92
62	0	65	8.125	19.125	27.125	35.125	43.125	51.125	59.125	67.125	75.125	83.125	91.125	99.125	41	28	45	93	17.750	30.750	45	93
63	0	66	8.250	19.250	27.250	35.250	43.250	51.250	59.250	67.250	75.250	83.250	91.250	99.250	42	28	46	94	17.875	30.875	46	94
64	0	67	8.375	19.375	27.375	35.375	43.375	51.375	59.375	67.375	75.375	83.375	91.375	99.375	43	28	47	95	18.000	31.000	47	95
65	0	68	8.500	19.500	27.500	35.500	43.500	51.500	59.500	67.500	75.500	83.500	91.500	99.500	44	28	48	96	18.125	31.125	48	96
66	0	69	28.000	28.000	28.000	28.000	28.000	28.000	28.000	28.000	28.000	28.000	28.000	28.000	45	28	49	97	18.250	31.250	49	97
67	0	70	16.000	16.000	16.000	16.000	16.000	16.000	16.000	16.000	16.000	16.000	16.000	16.000	46	28	50	98	18.375	31.375	50	98
68	0	71	14.000	14.000	14.000	14.000	14.000	14.000	14.000	14.000	14.000	14.000	14.000	14.000	47	28	51	99	18.500	31.500	51	99

(*Continued on next page.*)

TABLE VI
(*Continued*).

	22	23	24	25	26	27	28	29	30	31	32	33	34	35	36	37	38	39	40	41	42
IP	SOM1	SOM1	SOM1	SOM1	SOM3	SOM3	SOM3	SOM3	SOM3	SOM3	SOM3	SOM3	SOM3	SOM3	SOM3	SOM3	SOM3	SOM3	SOM3	SOM3	SOM3
1	1	2	2	2	1	0.500	11.500	19.500	27.500	35.500	43.500	51.500	59.500	67.500	75.500	83.500	91.500	12	2	2	2
2	2	12	12	12	2	0.625	11.625	19.625	27.625	35.625	43.625	51.625	59.625	67.625	75.625	83.625	91.625	13	16	16	12
3	4	14	14	13	3	0.750	11.750	19.750	27.750	35.750	43.750	51.750	59.750	67.750	75.750	83.750	91.750	14	17	17	14
4	7	16	16	14	4	0.875	11.875	19.875	27.875	35.875	43.875	51.875	59.875	67.875	75.875	83.875	91.875	15	28	28	16
5	12	27	27	15	5	1.000	12.000	20.000	28.000	36.000	44.000	52.000	60.000	68.000	76.000	84.000	92.000	16	29	29	27
6	13	28	28	16	6	1.125	12.125	20.125	28.125	36.125	44.125	52.125	60.125	68.125	76.125	84.125	92.125	17	1	52	28
7	14	2	2	17	7	1.250	12.250	20.250	28.250	36.250	44.250	52.250	60.250	68.250	76.250	84.250	92.250	18	2	53	2
8	15	12	12	18	8	1.375	12.375	20.375	28.375	36.375	44.375	52.375	60.375	68.375	76.375	84.375	92.375	19	3	54	12
9	16	14	14	26	12	1.500	12.500	20.500	28.500	36.500	44.500	52.500	60.500	68.500	76.500	84.500	92.500	1	4	55	14
10	24	16	16	27	13	1.625	12.625	20.625	28.625	36.625	44.625	52.625	60.625	68.625	76.625	84.625	92.625	2	5	56	16
11	25	27	27	28	14	1.750	12.750	20.750	28.750	36.750	44.750	52.750	60.750	68.750	76.750	84.750	92.750	3	6	57	27
12	26	28	28	32	15	1.875	12.875	20.875	28.875	36.875	44.875	52.875	60.875	68.875	76.875	84.875	92.875	4	7	58	28
13	27	2	2	44	16	2.000	13.000	21.000	29.000	37.000	45.000	53.000	61.000	69.000	77.000	85.000	93.000	5	8	59	2
14	28	12	12	2	17	2.125	13.125	21.125	29.125	37.125	45.125	53.125	61.125	69.125	77.125	85.125	93.125	6	12	60	12
15	29	14	14	12	18	2.250	13.250	21.250	29.250	37.250	45.250	53.250	61.250	69.250	77.250	85.250	93.250	7	13	61	14
16	30	16	16	13	19	2.375	13.375	21.375	29.375	37.375	45.375	53.375	61.375	69.375	77.375	85.375	93.375	8	14	62	16
17	36	27	27	14	20	2.500	13.500	21.500	29.500	37.500	45.500	53.500	61.500	69.500	77.500	85.500	93.500	12	15	63	27
18	37	28	28	15	21	2.625	13.625	21.625	29.625	37.625	45.625	53.625	61.625	69.625	77.625	85.625	93.625	13	2	2	28
19	38	2	2	16	22	2.750	13.750	21.750	29.750	37.750	45.750	53.750	61.750	69.750	77.750	85.750	93.750	14	16	16	2
20	39	12	12	17	23	2.875	13.875	21.875	29.875	37.875	45.875	53.875	61.875	69.875	77.875	85.875	93.875	15	17	17	12
21	40	14	2	18	24	3.000	14.000	22.000	30.000	38.000	46.000	54.000	62.000	70.000	78.000	86.000	94.000	16	28	28	14
22	76	16	14	26	25	3.125	14.125	22.125	30.125	38.125	46.125	54.125	62.125	70.125	78.125	86.125	94.125	17	29	29	16
23	1	27	17	27	26	3.250	14.250	22.250	30.250	38.250	46.250	54.250	62.250	70.250	78.250	86.250	94.250	18	16	64	27
24	2	28	28	28	27	3.375	14.375	22.375	30.375	38.375	46.375	54.375	62.375	70.375	78.375	86.375	94.375	19	17	65	28
25	4	2	29	32	28	3.500	14.500	22.500	30.500	38.500	46.500	54.500	62.500	70.500	78.500	86.500	94.500	20	18	66	2
26	7	12	2	44	29	3.625	14.625	22.625	30.625	38.625	46.625	54.625	62.625	70.625	78.625	86.625	94.625	21	19	67	12

(*Continued on next page.*)

TABLE VI
(*Continued*).

	22	23	24	25	26	27	28	29	30	31	32	33	34	35	36	37	38	39	40	41	42
IP	SOM1	SOM1	SOM1	SOM1	SOM3	SOM3	SOM3	SOM3	SOM3	SOM3	SOM3	SOM3	SOM3	SOM3	SOM3	SOM3	SOM3	SOM3	SOM3	SOM3	SOM3
27	12	14	14	2	30	3.750	14.750	22.750	30.750	38.750	46.750	54.750	62.750	70.750	78.750	86.750	94.750	22	20	68	14
28	13	16	17	12	31	3.875	14.875	22.875	30.875	38.875	46.875	54.875	62.875	70.875	78.875	86.875	94.875	23	21	69	16
29	14	27	28	13	32	4.000	15.000	23.000	31.000	39.000	47.000	55.000	63.000	71.000	79.000	87.000	95.000	24	22	70	27
30	15	28	29	14	33	4.125	15.125	23.125	31.125	39.125	47.125	55.125	63.125	71.125	79.125	87.125	95.125	25	23	71	28
31	16	2	2	15	34	4.250	15.250	23.250	31.250	39.250	47.250	55.250	63.250	71.250	79.250	87.250	95.250	26	24	72	2
32	24	12	14	16	35	4.375	15.375	23.375	31.375	39.375	47.375	55.375	63.375	71.375	79.375	87.375	95.375	27	25	73	12
33	25	14	17	17	36	4.500	15.500	23.500	31.500	39.500	47.500	55.500	63.500	71.500	79.500	87.500	95.500	12	26	74	14
34	26	15	28	18	37	4.625	15.625	23.625	31.625	39.625	47.625	55.625	63.625	71.625	79.625	87.625	95.625	13	27	75	15
35	27	16	29	26	38	4.750	15.750	23.750	31.750	39.750	47.750	55.750	63.750	71.750	79.750	87.750	95.750	14	2	2	16
36	28	27	2	27	39	4.875	15.875	23.875	31.875	39.875	47.875	55.875	63.875	71.875	79.875	87.875	95.875	15	16	16	27
37	29	28	14	28	40	5.000	16.000	24.000	32.000	40.000	48.000	56.000	64.000	72.000	80.000	88.000	96.000	16	17	17	28
38	30	2	17	32	41	5.125	16.125	24.125	32.125	40.125	48.125	56.125	64.125	72.125	80.125	88.125	96.125	17	28	28	2
39	36	12	28	44	42	5.250	16.250	24.250	32.250	40.250	48.250	56.250	64.250	72.250	80.250	88.250	96.250	18	29	29	12
40	37	14	29	2	43	5.375	16.375	24.375	32.375	40.375	48.375	56.375	64.375	72.375	80.375	88.375	96.375	19	28	76	14
41	38	16	2	12	44	5.500	16.500	24.500	32.500	40.500	48.500	56.500	64.500	72.500	80.500	88.500	96.500	28	29	77	16
42	39	27	14	13	45	5.625	16.625	24.625	32.625	40.625	48.625	56.625	64.625	72.625	80.625	88.625	96.625	29	30	78	27
43	40	28	17	14	46	5.750	16.750	24.750	32.750	40.750	48.750	56.750	64.750	72.750	80.750	88.750	96.750	30	31	79	28
44	76	2	28	15	47	5.875	16.875	24.875	32.875	40.875	48.875	56.875	64.875	72.875	80.875	88.875	96.875	31	32	80	2
45	1	12	29	16	48	6.000	17.000	25.000	33.000	41.000	49.000	57.000	65.000	73.000	81.000	89.000	97.000	32	33	81	12
46	2	14	2	17	49	6.125	17.125	25.125	33.125	41.125	49.125	57.125	65.125	73.125	81.125	89.125	97.125	33	34	82	14
47	4	16	14	18	50	6.250	17.250	25.250	33.250	41.250	49.250	57.250	65.250	73.250	81.250	89.250	97.250	34	35	83	16
48	7	27	17	26	51	6.375	17.375	25.375	33.375	41.375	49.375	57.375	65.375	73.375	81.375	89.375	97.375	35	36	84	27
49	12	28	28	27	52	6.500	17.500	25.500	33.500	41.500	49.500	57.500	65.500	73.500	81.500	89.500	97.500	12	37	85	28
50	13	2	29	28	53	6.625	17.625	25.625	33.625	41.625	49.625	57.625	65.625	73.625	81.625	89.625	97.625	13	38	86	2
51	14	12	2	32	54	6.750	17.750	25.750	33.750	41.750	49.750	57.750	65.750	73.750	81.750	89.750	97.750	14	39	87	12
52	15	14	14	44	55	6.875	17.875	25.875	33.875	41.875	49.875	57.875	65.875	73.875	81.875	89.875	97.875	15	2	2	14
53	16	16	17	2	56	7.000	18.000	26.000	34.000	42.000	50.000	58.000	66.000	74.000	82.000	90.000	98.000	16	16	16	16

TABLE VI

(*Continued*).

	22	23	24	25	26	27	28	29	30	31	32	33	34	35	36	37	38	39	40	41	42
IP	SOM1	SOM1	SOM1	SOM1	SOM3	SOM3	SOM3	SOM3	SOM3	SOM3	SOM3	SOM3	SOM3	SOM3	SOM3	SOM3	SOM3	SOM3	SOM3	SOM3	SOM3
54	24	27	28	12	57	7.125	18.125	26.125	34.125	42.125	50.125	58.125	66.125	74.125	82.125	90.125	98.125	17	17	17	27
55	25	28	29	13	58	7.250	18.250	26.250	34.250	42.250	50.250	58.250	66.250	74.250	82.250	90.250	98.250	18	28	28	28
56	26	2	2	14	59	7.375	18.375	26.375	34.375	42.375	50.375	58.375	66.375	74.375	82.375	90.375	98.375	19	29	29	2
57	27	12	14	15	60	7.500	18.500	26.500	34.500	42.500	50.500	58.500	66.500	74.500	82.500	90.500	98.500	36	40	88	12
58	28	14	17	16	61	7.625	18.625	26.625	34.625	42.625	50.625	58.625	66.625	74.625	82.625	90.625	98.625	37	41	89	14
59	29	16	28	17	62	7.750	18.750	26.750	34.750	42.750	50.750	58.750	66.750	74.750	82.750	90.750	98.750	38	42	90	16
60	30	27	29	18	63	7.875	18.875	26.875	34.875	42.875	50.875	58.875	66.875	74.875	82.875	90.875	98.875	39	43	91	27
61	36	28	2	26	64	8.000	19.000	27.000	35.000	43.000	51.000	59.000	67.000	75.000	83.000	91.000	99.000	40	44	92	28
62	37	2	14	27	65	8.125	19.125	27.125	35.125	43.125	51.125	59.125	67.125	75.125	83.125	91.125	99.125	41	45	93	2
63	38	12	17	28	66	8.250	19.250	27.250	35.250	43.250	51.250	59.250	67.250	75.250	83.250	91.250	99.250	42	46	94	12
64	39	14	28	32	67	8.375	19.375	27.375	35.375	43.375	51.375	59.375	67.375	75.375	83.375	91.375	99.375	43	47	95	14
65	40	16	29	44	68	8.500	19.500	27.500	35.500	43.500	51.500	59.500	67.500	75.500	83.500	91.500	99.500	44	48	96	16
66	76	27	28	18	69	28.000	28.000	28.000	28.000	28.000	28.000	28.000	28.000	28.000	28.000	28.000	28.000	45	49	97	27
67	28	28	29	18	70	16.000	16.000	16.000	16.000	16.000	16.000	16.000	16.000	16.000	16.000	16.000	16.000	46	50	98	28
68	28	28	28	16	71	14.000	14.000	14.000	14.000	14.000	14.000	14.000	14.000	14.000	14.000	14.000	14.000	47	51	99	28

(*Continued on next page*.)

TABLE VI
(*Continued*).

	43	44	45	46	47	48	49	50	51	52	53	54	55	56	57	58	59	60	61	62	63
IP	SOM3	SOM3	SOM0	SOM0	SOM0	SOM0	SOM0	SOM2	SOM2	SOM2	SOM2	SOM2	SOM2	SOM2	SOM2	SOM2	SOM2	SOM2	SOM2	SOM2	SOM2
1	2	2	0	0	0	0	0	1	0.500	27.500	12	2	2	2	2	2	2	1	2	2	2
2	12	12	0	0	0	0	0	2	0.625	27.625	13	16	16	16	16	4	4	2	12	12	12
3	14	13	0	0	0	0	0	3	0.750	27.750	14	17	17	17	17	7	7	4	14	14	13
4	16	14	0	0	0	0	0	4	0.875	27.875	15	28	28	28	28	12	12	7	16	16	14
5	27	15	0	0	0	0	0	5	1.000	28.000	16	29	29	29	29	13	13	12	27	27	15
6	28	16	0	0	0	0	0	6	1.125	28.125	17	1	52	12.625	25.625	14	14	13	28	28	16
7	2	17	0	0	0	0	0	7	1.250	28.250	18	2	53	12.750	25.750	15	15	14	2	2	17
8	12	18	0	0	0	0	0	8	1.375	28.375	19	3	54	12.875	25.875	16	16	15	12	12	18
9	14	26	0	0	0	0	0	12	1.500	28.500	1	4	55	13.000	26.000	24	24	16	14	14	26
10	16	27	0	0	0	0	0	13	1.625	28.625	2	5	56	13.125	26.125	25	25	24	16	16	27
11	27	28	0	0	0	0	0	14	1.750	28.750	3	6	57	13.250	26.250	26	26	25	27	27	28
12	28	32	0	0	0	0	0	15	1.875	28.875	4	7	58	13.375	26.375	27	27	26	28	28	32
13	2	44	0	0	0	0	0	16	2.000	29.000	5	8	59	13.500	26.500	28	28	27	2	2	44
14	12	2	0	0	0	0	0	17	2.125	29.125	6	12	60	13.625	26.625	29	29	28	12	12	2
15	14	12	0	0	0	0	0	18	2.250	29.250	7	13	61	13.750	26.750	30	30	29	14	14	12
16	16	13	0	0	0	0	0	19	2.375	29.375	8	14	62	13.875	26.875	36	36	30	16	16	13
17	27	14	0	0	0	0	0	20	2.500	29.500	12	15	63	14.000	27.000	37	37	36	27	27	14
18	28	15	0	0	0	0	0	21	2.625	29.625	13	2	2	2	2	38	38	37	28	28	15
19	2	16	0	0	0	0	0	22	2.750	29.750	14	16	16	16	16	39	39	38	2	2	16
20	12	17	0	0	0	0	0	23	2.875	29.875	15	17	17	17	17	40	40	39	12	12	17
21	2	18	0	0	0	0	0	24	3.000	30.000	16	28	28	28	28	1	52	40	14	2	18
22	14	26	0	0	0	0	0	25	3.125	30.125	17	29	29	29	29	2	53	76	16	14	26
23	17	27	0	0	0	0	0	26	3.250	30.250	18	16	64	14.125	27.125	3	54	1	27	17	27
24	28	28	0	0	0	0	0	27	3.375	30.375	19	17	65	14.250	27.250	4	55	2	28	28	28
25	29	32	0	0	0	0	0	28	3.500	30.500	20	18	66	14.375	27.375	5	56	4	2	29	32
26	2	44	0	0	0	0	0	29	3.625	30.625	21	19	67	14.500	27.500	6	57	7	12	2	44
27	14	2	0	0	0	0	0	30	3.750	30.750	22	20	68	14.625	27.625	7	58	12	14	14	2

TABLE VI

	43	44	45	46	47	48	49	50	51	52	53	54	55	56	57	58	59	60	61	62	63
IP	SOM3	SOM3	SOM0	SOM0	SOM0	SOM0	SOM0	SOM2	SOM2	SOM2	SOM2	SOM2	SOM2	SOM2	SOM2	SOM2	SOM2	SOM2	SOM2	SOM2	SOM2
28	17	12	0	0	0	0	0	31	3.875	30.875	23	21	69	14.750	27.750	8	59	13	16	17	12
29	28	13	0	0	0	0	0	32	4.000	31.000	24	22	70	14.875	27.875	12	60	14	27	28	13
30	29	14	0	0	0	0	0	33	4.125	31.125	25	23	71	15.000	28.000	13	61	15	28	29	14
31	2	15	0	0	0	0	0	34	4.250	31.250	26	24	72	15.125	28.125	14	62	16	2	2	15
32	14	16	0	0	0	0	0	35	4.375	31.375	27	25	73	15.250	28.250	15	63	24	12	14	16
33	17	17	0	0	0	0	0	36	4.500	31.500	12	26	74	15.375	28.375	16	64	25	14	17	17
34	28	18	0	0	0	0	0	37	4.625	31.625	13	27	75	15.500	28.500	17	65	26	15	28	18
35	29	26	0	0	0	0	0	38	4.750	31.750	14	2	2	2	2	18	66	27	16	29	26
36	2	27	0	0	0	0	0	39	4.875	31.875	15	16	16	16	16	19	67	28	27	2	27
37	14	28	0	0	0	0	0	40	5.000	32.000	16	17	17	17	17	20	68	29	28	14	28
38	17	32	0	0	0	0	0	41	5.125	32.125	17	28	28	28	28	21	69	30	2	17	32
39	28	44	0	0	0	0	0	42	5.250	32.250	18	29	29	29	29	22	70	36	12	28	44
40	29	2	0	0	0	0	0	43	5.375	32.375	19	28	76	15.625	28.625	23	71	37	14	29	2
41	2	12	0	0	0	0	0	44	5.500	32.500	28	29	77	15.750	28.750	24	72	38	16	2	12
42	14	13	0	0	0	0	0	45	5.625	32.625	29	30	78	15.875	28.875	25	73	39	27	14	13
43	17	14	0	0	0	0	0	46	5.750	32.750	30	31	79	16.000	29.000	26	74	40	28	17	14
44	28	15	0	0	0	0	0	47	5.875	32.875	31	32	80	16.125	29.125	27	75	76	2	28	15
45	29	16	0	0	0	0	0	48	6.000	33.000	32	33	81	16.250	29.250	28	76	1	12	29	16
46	2	17	0	0	0	0	0	49	6.125	33.125	33	34	82	16.375	29.375	29	77	2	14	2	17
47	14	18	0	0	0	0	0	50	6.250	33.250	34	35	83	16.500	29.500	30	78	4	16	14	18
48	17	26	0	0	0	0	0	51	6.375	33.375	35	36	84	16.625	29.625	31	79	7	27	17	26
49	28	27	0	0	0	0	0	52	6.500	33.500	12	37	85	16.750	29.750	32	80	12	28	28	27
50	29	28	0	0	0	0	0	53	6.625	33.625	13	38	86	16.875	29.875	33	81	13	2	29	28
51	2	32	0	0	0	0	0	54	6.750	33.750	14	39	87	17.000	30.000	34	82	14	12	2	32
52	14	44	0	0	0	0	0	55	6.875	33.875	15	2	2	2	2	35	83	15	14	14	44
53	17	2	0	0	0	0	0	56	7.000	34.000	16	16	16	16	16	36	84	16	16	17	2
54	28	12	0	0	0	0	0	57	7.125	34.125	17	17	17	17	17	37	85	24	27	28	12
55	29	13	0	0	0	0	0	58	7.250	34.250	18	28	28	28	28	38	86	25	28	29	13
56	2	14	0	0	0	0	0	59	7.375	34.375	19	29	29	29	29	39	87	26	2	2	14
57	14	15	0	0	0	0	0	60	7.500	34.500	36	40	88	17.125	30.125	40	88	27	12	14	15
58	17	16	0	0	0	0	0	61	7.625	34.625	37	41	89	17.250	30.250	41	89	28	14	17	16

(*Continued on next page.*)

TABLE VI

(*Continued*).

	43	44	45	46	47	48	49	50	51	52	53	54	55	56	57	58	59	60	61	62	63
IP	SOM3	SOM3	SOM0	SOM0	SOM0	SOM0	SOM0	SOM2	SOM2	SOM2	SOM2	SOM2	SOM2	SOM2	SOM2	SOM2	SOM2	SOM2	SOM2	SOM2	SOM2
59	28	17	0	0	0	0	0	62	7.750	34.750	38	42	90	17.375	30.375	42	90	29	16	28	17
60	29	18	0	0	0	0	0	63	7.875	34.875	39	43	91	17.500	30.500	43	91	30	27	29	18
61	2	26	0	0	0	0	0	64	8.000	35.000	40	44	92	17.625	30.625	44	92	36	28	2	26
62	14	27	0	0	0	0	0	65	8.125	35.125	41	45	93	17.750	30.750	45	93	37	2	14	27
63	17	28	0	0	0	0	0	66	8.250	35.250	42	46	94	17.875	30.875	46	94	38	12	17	28
64	28	32	0	0	0	0	0	67	8.375	35.375	43	47	95	18.000	31.000	47	95	39	14	28	32
65	29	44	0	0	0	0	0	68	8.500	35.500	44	48	96	18.125	31.125	48	96	40	16	29	44
66	28	18	0	0	0	0	0	69	28.000	28.000	45	49	97	18.250	31.250	49	97	76	27	28	18
67	29	18	0	0	0	0	0	70	16.000	16.000	46	50	98	18.375	31.375	50	98	28	28	29	18
68	28	16	0	0	0	0	0	71	14.000	14.000	47	51	99	18.500	31.500	51	99	28	28	28	16

SOM: science operating mode; 0: OSNT, 1: CSN, 2: OSNB, 3: OSI.

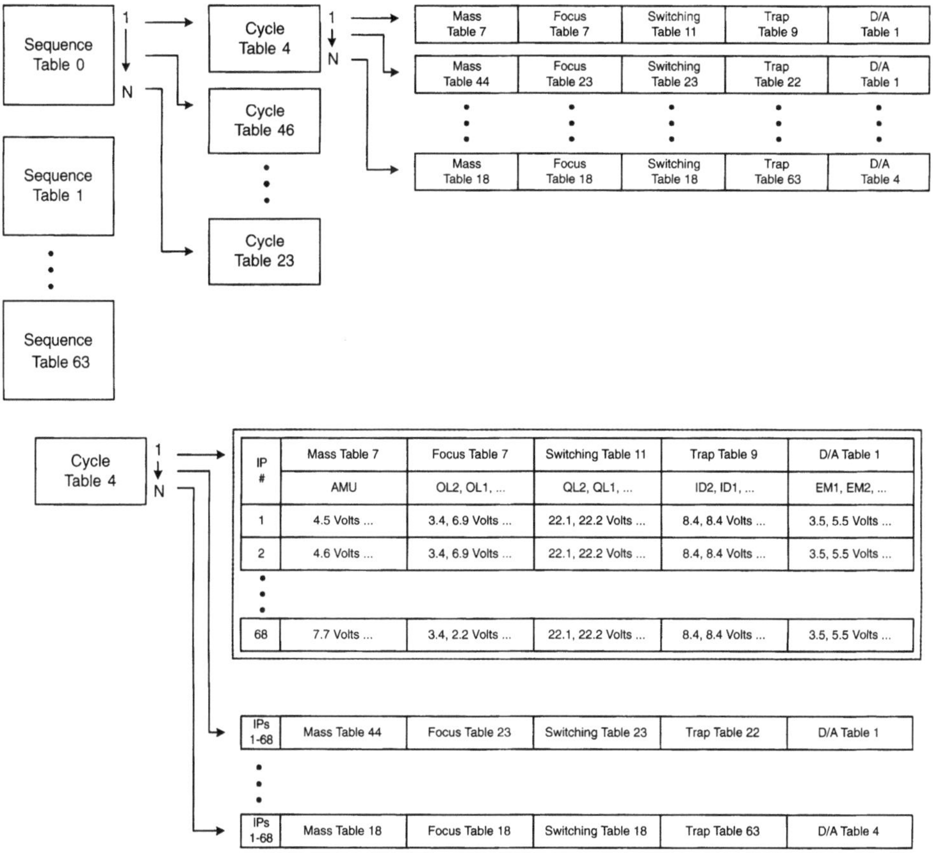

Figure 26. Schematic illustrating the relationships of various INMS software tables.

take the form of either a unit or 1/8-Dalton scan. Survey scans will allow detection of species not predicted by models to be present (or expected to be present only at subdetectable levels) in Titan's upper atmosphere or Saturn's magnetosphere. One-eighth-Dalton surveys are principally engineering scans performed to check instrument tuning and resolution; however, they also help provide ground-truth verification of mass peaks identified in the INMS data. The second type of scan is known as an "adaptive" scan and will be used for high-time-resolution sampling of species known to be present in a particular region of interest. Adaptive scan data will allow scale heights to be established with a spatial resolution of ~3 km or less and will thus be especially important for studies of Titan's atmospheric structure. In an adaptive scan, either a set of selected masses (e.g., Mass Table 25: "Outer Magnetosphere CSN") or just one mass (e.g., Mass Table 15: "Only Mass 28 CSN") will be repeatedly sampled throughout the entire 2.3-sec scan. The INMS may also perform adaptive scans in which the measurement of a given set of selected masses alternates with sweeps in 1-Dalton increments during the same

TABLE VII
INMS science sequences.

Cycle	Cycle Table: Description	Mode	Velocity (km/sec)	Total Scans	Mass Table	Comments
1: Default Science—1498 bps						
1	1: Unitary survey—CS and OSI	CS	0.000	778	1	Default Science, scaled to 6 bps. Note that this sequence takes ~5 hr to run once through. The co-added result will be 5 Unitary scans followed by one of each 1/8-Dalton survey scan in CS then in OSI.
		OSI			26	
2	2: 1/8-Dalton survey—OSI and CS	OSI	0.000	24	27	This should be used with a co-add rate of 240 scans = 554 sec.
		OSI			28	
		OSI			29	
		OSI			30	
		OSI			31	
		OSI			32	
		OSI			33	
		OSI			34	
		OSI			35	
		OSI			36	
		OSI			37	
		OSI			38	
		CS			2	
		CS			3	
		CS			4	
		CS			5	
		CS			6	
		CS			7	
		CS			8	
		CS			9	
		CS			10	
		CS			11	
		CS			12	
		CS			13	
4: Default Science—100 bps						
1	21: Unitary survey—CS	CS	0.000	300	1	Default Science, scaled to 100 bps. Note that this sequence takes ~35 min to run once through.

(*Continued on next page.*)

TABLE VII
(*Continued*).

Cycle	Cycle Table: Description	Mode	Velocity (km/sec)	Total Scans	Mass Table	Comments
						The co-added result will be 20 Unitary scans followed by one of each 1/8-Dalton survey scan in CS then in OSI.
2	35: 1/8-Dalton survey CS (1 of 12)	CS	0.000	15	2	
3	36: 1/8-Dalton survey CS (2 of 12)	CS	0.000	15	3	This should be used with a co-add rate of 15 scans = 34 sec.
4	37: 1/8-Dalton survey CS (3 of 12)	CS	0.000	15	4	
5	38: 1/8-Dalton survey CS (4 of 12)	CS	0.000	15	5	
6	39: 1/8-Dalton survey CS (5 of 12)	CS	0.000	15	6	
7	40: 1/8-Dalton survey CS (6 of 12)	CS	0.000	15	7	
8	41: 1/8-Dalton survey CS (7 of 12)	CS	0.000	15	8	
9	42: 1/8-Dalton survey CS (8 of 12)	CS	0.000	15	9	
10	43: 1/8-Dalton survey CS (9 of 12)	CS	0.000	15	10	
11	44: 1/8-Dalton survey CS (10 of 12)	CS	0.000	15	11	
12	45: 1/8-Dalton survey CS (11 of 12)	CS	0.000	15	12	
13	46: 1/8-Dalton survey CS (12 of 12)	CS	0.000	15	13	
14	22: Unitary survey—OSI	OSI	0.000	300	26	
15	23: 1/8-Dalton survey OSI (1 of 12)	OSI	0.000	15	27	
16	24: 1/8-Dalton survey OSI (2 of 12)	OSI	0.000	15	28	
17	25: 1/8-Dalton survey OSI (3 of 12)	OSI	0.000	15	29	
18	26: 1/8-Dalton survey OSI (4 of 12)	OSI	0.000	15	30	
19	27: 1/8-Dalton survey OSI (5 of 12)	OSI	0.000	15	31	

(*Continued on next page.*)

TABLE VII

(*Continued*).

Cycle	Cycle Table: Description	Mode	Velocity (km/sec)	Total Scans	Mass Table	Comments
20	28: 1/8-Dalton survey OSI (6 of 12)	OSI	0.000	15	32	
21	29: 1/8-Dalton survey OSI (7 of 12)	OSI	0.000	15	33	
22	30: 1/8-Dalton survey OSI (8 of 12)	OSI	0.000	15	34	
23	31: 1/8-Dalton survey OSI (9 of 12)	OSI	0.000	15	35	
24	32: 1/8-Dalton survey OSI (10 of 12)	OSI	0.000	15	36	
25	33: 1/8-Dalton survey OSI (11 of 12)	OSI	0.000	15	37	
26	34: 1/8-Dalton survey OSI (12 of 12)	OSI	0.000	15	38	
6: Default Science—50 bps						
1	21: Unitary survey—CS	CS	0.000	600	1	Default Science, scaled to 50 bps. Note that this sequence takes ~70 min to run once through. The co-added result will be 20 Unitary scans followed by one of each 1/8-Dalton Survey scan in CS then in OSI.
2	35: 1/8-Dalton survey CS (1 of 12)	CS	0.000	30	2	
3	36: 1/8-Dalton survey CS (2 of 12)	CS	0.000	30	3	This should be used with a co-add rate of 30 scans = 69 sec.
4	37: 1/8-Dalton survey CS (3 of 12)	CS	0.000	30	4	
5	38: 1/8-Dalton survey CS (4 of 12)	CS	0.000	30	5	
6	39: 1/8-Dalton survey CS (5 of 12)	CS	0.000	30	6	
7	40: 1/8-Dalton survey CS (6 of 12)	CS	0.000	30	7	
8	41: 1/8-Dalton survey CS (7 of 12)	CS	0.000	30	8	
9	42: 1/8-Dalton survey CS (8 of 12)	CS	0.000	30	9	
10	43: 1/8-Dalton survey CS (9 of 12)	CS	0.000	30	10	

(*Continued on next page.*)

TABLE VII
(*Continued*).

Cycle	Cycle Table: Description	Mode	Velocity (km/sec)	Total Scans	Mass Table	Comments
11	44: 1/8-Dalton survey CS (10 of 12)	CS	0.000	30	11	
12	45: 1/8-Dalton survey CS (11 of 12)	CS	0.000	30	12	
13	46: 1/8-Dalton survey CS (12 of 12)	CS	0.000	30	13	
14	22: Unitary survey—OSI	OSI	0.000	600	26	
15	23: 1/8-Dalton survey OSI (1 of 12)	OSI	0.000	30	27	
16	24: 1/8-Dalton survey OSI (2 of 12)	OSI	0.000	30	28	
17	25: 1/8-Dalton survey OSI (3 of 12)	OSI	0.000	30	29	
18	26: 1/8-Dalton survey OSI (4 of 12)	OSI	0.000	30	30	
19	27: 1/8-Dalton survey OSI (5 of 12)	OSI	0.000	30	31	
20	28: 1/8-Dalton survey OSI (6 of 12)	OSI	0.000	30	32	
21	29: 1/8-Dalton survey OSI (7 of 12)	OSI	0.000	30	33	
22	30: 1/8-Dalton survey OSI (8 of 12)	OSI	0.000	30	34	
23	31: 1/8-Dalton survey OSI (9 of 12)	OSI	0.000	30	35	
24	32: 1/8-Dalton survey OSI (10 of 12)	OSI	0.000	30	36	
25	33: 1/8-Dalton survey OSI (11 of 12)	OSI	0.000	30	37	
26	34: 1/8-Dalton survey OSI (12 of 12)	OSI	0.000	30	38	
9: Default Science—6.2 bps						
1	21: Unitary survey—CS	CS	0.000	1200	1	Default Science, scaled to 6 bps. Note that this sequence takes ~5 hr to run once through. The co-added result will be 5 Unitary scans followed by one of each 1/8-Dalton Survey scan in CS then in OSI.

(*Continued on next page.*)

TABLE VII
(*Continued*).

Cycle	Cycle Table: Description	Mode	Velocity (km/sec)	Total Scans	Mass Table	Comments
2	35: 1/8-Dalton survey CS (1 of 12)	CS	0.000	240	2	
3	36: 1/8-Dalton survey CS (2 of 12)	CS	0.000	240	3	This should be used with a co-add rate of 240 scans = 554 sec.
4	37: 1/8-Dalton survey CS (3 of 12)	CS	0.000	240	4	
5	38: 1/8-Dalton survey CS (4 of 12)	CS	0.000	240	5	
6	39: 1/8-Dalton survey CS (5 of 12)	CS	0.000	240	6	
7	40: 1/8-Dalton survey CS (6 of 12)	CS	0.000	240	7	
8	41: 1/8-Dalton survey CS (7 of 12)	CS	0.000	240	8	
9	42: 1/8-Dalton survey CS (8 of 12)	CS	0.000	240	9	
10	43: 1/8-Dalton survey CS (9 of 12)	CS	0.000	240	10	
11	44: 1/8-Dalton survey CS (10 of 12)	CS	0.000	240	11	
12	45: 1/8-Dalton survey CS (11 of 12)	CS	0.000	240	12	
13	46: 1/8-Dalton survey CS (12 of 12)	CS	0.000	240	13	
14	22: Unitary survey—OSI	OSI	0.000	1200	26	
15	23: 1/8-Dalton survey OSI (1 of 12)	OSI	0.000	240	27	
16	24: 1/8-Dalton survey OSI (2 of 12)	OSI	0.000	240	28	
17	25: 1/8-Dalton survey OSI (3 of 12)	OSI	0.000	240	29	
18	26: 1/8-Dalton survey OSI (4 of 12)	OSI	0.000	240	30	
19	27: 1/8-Dalton survey OSI (5 of 12)	OSI	0.000	240	31	
20	28: 1/8-Dalton survey OSI (6 of 12)	OSI	0.000	240	32	
21	29: 1/8-Dalton survey OSI (7 of 12)	OSI	0.000	240	33	

(*Continued on next page*.)

TABLE VII
(*Continued*).

Cycle	Cycle Table: Description	Mode	Velocity (km/sec)	Total Scans	Mass Table	Comments
22	30: 1/8-Dalton survey OSI (8 of 12)	OSI	0.000	240	34	
23	31: 1/8-Dalton survey OSI (9 of 12)	OSI	0.000	240	35	
24	32: 1/8-Dalton survey OSI (10 of 12)	OSI	0.000	240	36	
25	33: 1/8-Dalton survey OSI (11 of 12)	OSI	0.000	240	37	
26	34: 1/8-Dalton survey OSI (12 of 12)	OSI	0.000	240	38	
10: Titan Exploratory—TA Sequence—1498 bps						
1	47: Adaptive survey—CS and OSNB for TA	CS	5.625	620	16	This is the sequence that will be used during TA. It should be started at 1611 sec before CA (10,000 km): start 2004-300T15:03:51 and end 2004-300T15:56:55 based on current trajectory information. 1498 bps data rate required.
		CS			17	
		OSNB			54	It performs a combined adaptive and 1–99 scan between CS and OSNB until ~18 sec before closest approach, then follows a more complicated sequence (see Cycle Table 4) for ~30 sec, then switches back to the first cycle.
		OSNB			55	
2	47: Adaptive survey—CS and OSNB for TA	CS	5.875	68	16	The velocities are selected to attempt to match the radial velocity over different parts of the period, while keeping velocity compensation constant through the expected location of the magnetosphere/thermal crossing.
		CS			17	
		OSNB			54	
		OSNB			55	

(*Continued on next page.*)

TABLE VII
(*Continued*).

Cycle	Cycle Table: Description	Mode	Velocity (km/sec)	Total Scans	Mass Table	Comments
3	4: 1/8 Adaptive—CS and OSNB—TA	CS	6.000	16	18	
		CS			16	
		OSNB			56	
		CS			17	
		CS			19	
		CS			16	
		OSNB			57	
		CS			17	
4	47: Adaptive survey—CS and OSNB for TA	CS	5.875	68	16	
		CS			17	
		OSNB			54	
		OSNB			55	
5	47: Adaptive survey—CS and OSNB for TA	CS	5.625	620	16	
		CS			17	
		OSNB			54	
		OSNB			55	
11: Titan High-Altitude Ionosphere Flyby—1498 bps						
1	3: Adaptive survey—CS and OSI—Titan high altitude	CS	5.875	4	16	This is a very simple sequence, to be repeated over the course of all high-altitude flybys (>1500 km). It contains a combined adaptive and 1–99 scan. 1498 bps data rate required.
		CS			17	
		OSI			40	The currently selected velocity is 5.875 km/sec, which is ideal for T4.
		OSI			41	
12: Titan Low-Altitude 006TI_T5—1498 bps						
1	5: OSI for outer magnetosphere and low altitude Titan	OSI	5.500	778	44	This sequence has been worked to match the timing of the 006TI_T5 Titan pass. It should be started 1 hr before closest approach: 2005-106T18:05:57 by the current ephemeris information. 1498 bps data rate required.

(*Continued on next page.*)

TABLE VII
(*Continued*).

Cycle	Cycle Table: Description	Mode	Velocity (km/sec)	Total Scans	Mass Table	Comments
2	48: Unitary survey—CS and OSNB for T5	CS	5.750	520	1	
		OSNB			50	(Times are before closest approach, altitudes given in height from surface)
		OSNB			62	Cycle 1: 1 hr (18449 km) → 30 min (8439 km)
		CS			24	Cycle 2: 30 min (8439 km) → 10 min (2400 km)
3	7: Unitary and adaptive—CS and OSNB for T5	CS	5.875	144	1	Cycle 3: 10 min (2400 km) → 263 sec (1260 km)
		OSNB			50	Cycle 4: 263 sec (1260 km) → 52.5 sec (962 km)
		CS			23	Cycle 5: 52.5 sec (962 km) → 34 sec (950.5)
		OSNB			62	Cycle 6: 34 sec (950.5) → CA
4	8: Adaptive and 1–99—CS and OSNB for T5	CS	6.000	90	20	
		CS			21	Going out is exactly the reverse, except that Cycle Table 6 (switching to CSN/OSI) starts at CA + 492 sec (~2000 km) and the later timings are shifted accordingly.
		OSNB			54	
		OSNB			55	The selected velocities are for T5 specifically, to match the changing radial velocity.
		CS			22	
		OSNB			60	
5	9: Adaptive—CS and OSNB for T5 pre-CA	CS	6.000	8	20	
		OSNB			58	
		CS			21	
		OSNB			59	
6	10: Adaptive large and 1/8 Dalton—CS and OSNB	CS	6.000	32	2	
		CS			22	
		OSNB			51	
		OSNB			60	
		CS			5	
		CS			22	
		OSNB			52	

(*Continued on next page.*)

TABLE VII
(*Continued*).

Cycle	Cycle Table: Description	Mode	Velocity (km/sec)	Total Scans	Mass Table	Comments
		OSNB			60	
7	9: Adaptive—CS and OSNB for T5 pre-CA	CS	6.000	8	20	
		OSNB			58	
		CS			21	
		OSNB			59	
8	8: Adaptive and 1–99—CS and OSNB for T5	CS	6.000	90	20	
		CS			21	
		OSNB			54	
		OSNB			55	
		CS			22	
		OSNB			60	
9	7: Unitary and adaptive—CS and OSNB for T5	CS	5.875	100	1	
		OSNB			50	
		CS			23	
		OSNB			62	
10	6: Unitary survey—CS and OSI for T5	CS	5.750	261	1	
		OSI			26	
		OSI			43	
11	5: OSI for outer magnetosphere and low altitude Titan	OSI	5.500	1082	44	
20: Outer magnetosphere—ion 1498 bps						
1	20: CS outer magnetosphere OS fil off	CS	2.750	1	25	Outer magnetosphere sequence, alternating CS and OSI, not scaled.
2	5: OSI for outer magnetosphere and low altitude Titan	OSI	10.000	1	44	This should be used with no co-adding.
23: Outer magnetosphere—ion 100 bps						
1	20: CS outer magnetosphere OS fil off	CS	2.750	15	25	Outer magnetosphere sequence, alternating CS and OSI, scaled to 100 bps.
2	5: OSI for outer magnetosphere and low altitude Titan	OSI	10.000	15	44	This should be used with a co-add rate of 15 scans = 34 sec.
25: Outer magnetosphere—ion 50 bps						
1	20: CS outer magnetosphere OS fil off	CS	2.750	30	25	Outer magnetosphere sequence, alternating CS and OSI, scaled to 50 bps.

(*Continued on next page*.)

TABLE VII
(*Continued*).

Cycle	Cycle Table: Description	Mode	Velocity (km/sec)	Total Scans	Mass Table	Comments
2	5: OSI for outer magnetosphere and low altitudeTitan	OSI	10.000	30	44	This should be used with a co-add rate of 30 scans = 69 secs.
28: Outer magnetosphere—ion 6.2 bps						
1	20: CS outer magnetosphere OS fil off	CS	2.750	240	25	Outer magnetosphere sequence, alternating CS and OSI, scaled to 6 bps.
2	5: OSI for outer magnetosphere and low altitude Titan	OSI	10.000	240	44	This should be used with a co-add rate of 240 scans = 554 sec.
30: Outer magnetosphere—neutral 1498 bps						
1	11: CS outer magnetosphere OS fil on	CS	2.750	1	25	Outer magnetosphere sequence, alternating CS and OSNB, not scaled.
2	12: OSNB outer magnetosphere	OSNB	2.750	1	63	This should be used with no co-adding.
33: Outer magnetosphere—neutral 100 bps						
1	11: CS outer magnetosphere OS fil on	CS	2.750	15	25	Outer magnetosphere sequence, alternating CS and OSNB, scaled to 100 bps.
2	12: OSNB outer magnetosphere	OSNB	2.750	15	63	This should be used with a co-add rate of 15 scans = 34 sec.
35: Outer magnetosphere—neutral 50 bps						
1	11: CS outer magnetosphere OS fil on	CS	2.750	30	25	Outer magnetosphere sequence, alternating CS and OSNB, scaled to 50 bps.
2	12: OSNB outer magnetosphere	OSNB	2.750	30	63	This should be used with a co-add rate of 30 scans = 69 sec.
38: Outer magnetosphere—neutral 6.2 bps						
1	11: CS outer magnetosphere OS fil on	CS	2.750	240	25	Outer magnetosphere sequence, alternating CS and OSNB, scaled to 6 bps.
2	12: OSNB outer magnetosphere	OSNB	2.750	240	63	This should be used with a co-add rate of 240 scans = 554 sec.
40: Inner magnetosphere—ion 1498 bps						
1	18: CS inner magnetosphere OS fil off	CS	4.000	1	14	Inner magnetosphere sequence, alternating CS and OSI, not scaled.

(*Continued on next page.*)

TABLE VII
(Continued).

Cycle	Cycle Table: Description	Mode	Velocity (km/sec)	Total Scans	Mass Table	Comments
2	19: OSI inner magnetosphere	OSI	10.000	1	39	This should be used with no co-adding.
43: Inner magnetosphere—ion 100 bps						
1	18: CS inner magnetosphere OS fil off	CS	4.000	15	14	Inner magnetosphere sequence, alternating CS and OSI scaled to 100 bps.
2	19: OSI inner magnetosphere	OSI	10.000	15	39	This should be used with a co-add rate of 15 scans = 34 sec.
45: Inner magnetosphere—ion 50 bps						
1	18: CS inner magnetosphere OS fil off	CS	4.000	30	14	Inner magnetosphere sequence, alternating CS and OSI, scaled to 50 bps.
2	19: OSI inner magnetosphere	OSI	10.000	30	39	This should be used with a co-add rate of 30 scans = 69 sec.
48: Inner magnetosphere—ion 6.2 bps						
1	18: CS inner magnetosphere OS fil off	CS	4.000	240	14	Inner magnetosphere sequence, alternating CS and OSI, scaled to 6 bps.
2	19: OSI inner magnetosphere	OSI	10.000	240	39	This should be used with a co-add rate of 240 scans = 554 sec.
50: Inner magnetosphere—neutral 1498 bps						
1	14: CS inner magnetosphere OS fil on	CS	4.000	1	14	Inner magnetosphere sequence, alternating CS and OSNB, not scaled.
2	13: OSNB inner magnetosphere	OSNB	4.000	1	53	This should be used with no co-adding.
53: Inner magnetosphere—neutral 100 bps						
1	14: CS inner magnetosphere OS fil on	CS	4.000	15	14	Inner magnetosphere sequence, alternating CS and OSNB, scaled to 100 bps.
2	13: OSNB inner magnetosphere	OSNB	4.000	15	53	This should be used with a co-add rate of 15 scans = 34 sec.
55: Inner magnetosphere—neutral 50 bps						
1	14: CS inner magnetosphere OS fil on	CS	4.000	30	14	Inner magnetosphere sequence, alternating CS and OSNB, scaled to 50 bps.
2	13: OSNB inner magnetosphere	OSNB	4.000	30	53	This should be used with a co-add rate of 30 scans = 69 sec.

(Continued on next page.)

TABLE VII
(*Continued*).

Cycle	Cycle Table: Description	Mode	Velocity (km/sec)	Total Scans	Mass Table	Comments
58: Inner magnetosphere—neutral 6.2 bps						
1	14: CS inner magnetosphere OS fil on	CS	4.000	240	14	Inner magnetosphere sequence, alternating CS and OSNB, scaled to 6 bps.
2	13: OSNB inner magnetosphere	OSNB	4.000	240	53	This should be used with a co-add rate of 240 scans = 554 sec.
60: Ring Overflight for SOI—1498 bps						
1	17: SOI: magnetic field particles—CS and OSI	CS	20.000	156	14	To be started at 2004-183T03:54:30 and ended at 04:14:15.
		OSI			39	This sequence is designed to have 705 sec of B-field pointing (neutrals and ions) followed by 480 sec of neutral ram pointing (only neutrals). 1498-bps data rate required.
2	17: SOI: magnetic field particles—CS and OSI	CS	15.000	150	14	
		OSI			39	This mode may not be correct for the entire SOI period. It is correct for the SOI INMS prime period pointing. For time outside that region, careful use of the inner magnetosphere scans should suffice. Also note that the OS filament should be on (warmed up) prior to this observation, because it is on for the second cycle, which is after >12 min of being off.
3	16: SOI: corotation/ RAM—OSNB and CS	CS	5.000	108	14	
		OSNB			53	The velocities selected are designed to attempt to catch a range of particles, moving relative to the spacecraft.
4	16: SOI: corotation/ RAM—OSNB and CS	CS	6.000	100	14	
		OSNB			53	

CS: closed source (neutral); OSI: open source ion; OSN: open source neutral; CA: closest approach. A velocity of 0 represents the use of lens values precomputed at 6.0 km/sec.

scan (e.g., Mass Tables 16: "Titan low/high altitude part 1—CSN" and 17: "Titan low/high altitude part 2—CSN"). Finally, a third type of scan is the "high-pass" scan, in which the total signal above a given mass number (mass-to-charge ratio) is measured using the total transmission operation of the quadrupole analyzer. As noted in Section 2.2.3 above, for high-pass scans the DC mass filter voltage is set to zero. High-pass scans are used to extend the mass range of the INMS instrument.

In the case of regions where densities are expected to be very low, long accumulation periods will be required and mass scans will be co-added. The INMS software allows co-adding of up to 255 mass scans. The actual number of scans that can be co-added is determined by the count rate. That is, if the count rate is high, then fewer scans will be co-added. Otherwise, the capacity of the counter (32 bits) will be exceeded and data will be lost. However, count rates in the regions where co-adding of mass scans will be employed (e.g., the inner magnetosphere) will likely be sufficiently low that the maximum allowable number of scans can be co-added without risk of exceeding counter capacity.

3.3. Science Sequences and Cycles

Twenty-four science sequences have been defined. These sequences and their constituent cycles are discussed in the following and summarized in Table VII. Because of the exploratory nature of the Cassini mission, it is expected that new science sequences will be defined in response to discoveries made during the orbital tour phase. In this case, the appropriate Sequence and Cycle tables will be uploaded.

3.3.1. *Default Science Sequence*

The "Default Science—1498 bps" Sequence is the basic sequence executed by the INMS unless another orbital sequence has been commanded. This sequence comprises two cycles. In Cycle 1, the instrument performs two unitary survey scans from 1 to 8 and 12 to 70 Daltons (Mass Table I for CSN and 26 for OSI), the first in the closed source mode and the second in the open source ion mode. Cycle 1 is performed in 4.6 sec and repeated for ~30 min (389 scans in each mode). Cycle 2 consists of alternating survey scans in the open source ion mode and in the closed source mode. Mass Tables 2–13 for CSN and 27–38 for OSI, covering the mass ranges 0.5–8.5 and 11.5–99.5, are used for both surveys in Cycle 2. Cycle 2 takes 55.2 sec to execute. The sequence is looped until a different sequence is commanded.

There are three other Default Science Sequences, each tailored to a specific data rate: 100, 50 and 6.2 bits per second (bps). These rates are designed to make use of the co-adding function of the INMS, while keeping a very similar measurement order and timing to the full rate Default Science mode. The 100 and 50-bps modes are co-added 15 and 30 times, respectively, and each will return 20 packets each containing one unitary scan, followed by a complete 1/8-Dalton scan before changing to the other scan mode. The data rates most commonly assigned to the INMS during the mission are 100 and 50 bps. The 6.2-bps mode is co-added 240 times and will return five packets, each containing one unitary scan, before performing a complete 1/8-Dalton scan. This data rate was selected to allow bursting to full rate when the data volume allocated to the INMS is small.

3.3.2. *Titan Exploratory—TA Sequence*

The "Titan Exploratory—TA" Sequence will be executed during the Orbiter's first two flybys of Titan (Ta: October 2004, Tb: December 2004) and will occur at an altitude of approximately 1250 km. The initial flybys will take place prior to the descent of the Hugyens Probe (Tc: January 2005). Execution of this sequence will initiate the INMS investigation of Titan's thermosphere and ionosphere, which is the primary science objective of the INMS experiment. As noted earlier (Section 1.1), INMS measurements of atmospheric density made during the initial flybys will be operationally as well as scientifically important because they will allow assessment of atmospheric drag effects on the Orbiter during subsequent flybys at lower altitudes. Initial INMS measurements at relatively high altitudes (1250 km) will determine densities and temperatures at and above this altitude. This information will be used to extrapolate densities to lower altitudes, which will be verified on subsequent flybys. If the extrapolated densities are found to be too high or too low, the orbital tour will need to be re-designed to protect the spacecraft while at the same time ensuring adequate science return. (Based on current atmospheric models, a minimum operational altitude of 950 km has been assumed for planning purposes.) A preliminary run of this extrapolation process has been completed using simulated INMS data from an atmospheric model. The results were satisfactory: density and temperature values extrapolated using two different methods agreed with the simulated density to within 15%. A more detailed run is planned for March.

The Titan Exploratory Sequence is composed of five cycles and is designed to characterize the major neutral species in Titan's upper atmosphere. The INMS will execute Cycle 1 from an altitude of $\sim$10,000 km until $\sim$180 sec before Titan closest approach. Two scans will be performed in sequence, first in the closed source mode (using Mass Tables 16 and 17) and then in the open source neutral mode (using Mass Tables 54 and 55). As specified by these two tables, the INMS will alternate sampling of masses 2, 16, 17, 28, and 29 with mass surveys in 1-Dalton increments until the entire mass range of 1–99 Daltons (excluding 9–11 Daltons) has been covered. Repeated measurement of masses 2, 16, 17, 28, and 29 during the two scans will provide high-temporal-resolution data on the density profiles of the principal neutral and ion species known or expected to be present in Titan's atmosphere: H_2 (2), CH_4 (16), N_2 (28), H_2CN^+ (28), CH_5^+ (17), and $C_2H_5^+$ (29). With these data, scale heights can be calculated with a resolution of $\leq$3 km, thus allowing the detailed structure of Titan's upper atmosphere to be determined. After $\sim$1435 sec, Cycle 2 will start, performing the same mass scans as Cycle 1 but with a slightly different velocity compensation value to reflect the changing Titan-relative radial velocity of Cassini. Cycle 3 will start $\sim$18 sec before closest approach. In

Cycle 3 the INMS will perform an alternating sequence of adaptive/unitary scans (Mass Tables 16 and 17 for CSN) and adaptive/fractional scans (Mass Tables 18/19 for CSN and 56/57 for OSNB). Throughout all of these scans, masses 2, 16, 17, 28, and 29 will be sampled at the same rate, to keep a consistent measurement of the primary constituents of Titan's atmosphere. At ~18 sec after closest approach the INMS will perform Cycle 2 followed by Cycle 1 ~160 sec later in an exact mirror of the beginning of the sequence.

3.3.3. *Titan High-Altitude Ionosphere Flyby*

The "Titan High-Altitude Ionosphere Flyby" Sequence will be used during Titan flybys at altitudes above 1500 km, i.e., above the exobase (~1425 km). This sequence consists of a single repeated cycle identical to Cycle 1 in the "Titan Exploratory—TA" Sequence with OSI replacing OSNB mode. It will thus provide both the survey data needed to characterize the composition of Titan's exosphere and ionosphere and the high-temporal-resolution data on the expected major constituents (masses 2, 16, 17, 28, and 29) needed to establish the structure of the upper atmosphere. The neutral composition measurements will complement Probe in situ data and Orbiter remote-sensing data on atmospheric composition at lower altitudes. The ion measurements will yield insights into the photochemical and energetic-particle-induced processes occurring in Titan's upper atmosphere. Because Titan's upper atmosphere, and especially the ionosphere, are expected to exhibit day/night, plasma ram/wake, and latitude/longitude asymmetries, high-altitude flybys of Titan at several points in its orbit about Saturn will be required to obtain the coverage needed to adequately characterize Titan's upper atmosphere and understand its variability.

The region in which the INMS will invoke this orbital sequence is the site of the interaction between Titan's upper atmosphere (cf. Section 1.1.2) and the external plasma environment, as well as the region from which atmospheric escape occurs as the result of several different loss processes, including sputtering, scavenging by the external plasma flow, and photochemical loss by ion–neutral reactions (cf. Section 1.1.3). INMS data acquired during the Orbiter's high-altitude passes will thus be particularly important for characterizing Titan's interaction with its external plasma environment, establishing the relative importance of different atmospheric loss mechanisms, and assessing Titan's role as a source of neutrals and plasma for the Saturnian magnetosphere. Moreover, knowledge of atmospheric escape processes and rates acquired through the analysis of high-altitude INMS data, as well as from other Cassini investigations, will be of significant value in our efforts to understand the origin and evolution of Titan's atmosphere.

3.3.4. *Titan Low-Altitude 006TI_T5*

The "Titan Low-Altitude 006TI_T5" Sequence is designed for composition measurements at altitudes as low as is consistent with Orbiter safety (this version is specifically tailored to the 5th Titan Pass). As explained above, minimum safe altitudes will be determined on the basis of INMS density and temperature data obtained during the initial Titan flybys. Several such low-altitude passes, with spacecraft orientation optimized to point the open source aperture into the spacecraft ram direction, are required for successful completion of the INMS science investigation. A minimum flyby altitude of 950 km has been selected for these passes, based on densities predicted by theoretical models. Flybys at this altitude will allow for data acquisition well below the ionospheric peak and the homopause—both of which are predicted to occur at ~1050 km (Strobel *et al.*, 1992; Fox and Yelle, 1997; Keller *et al.*, 1998)—and well into the region where the photochemical production of complex hydrocarbons and nitriles is initiated. At this altitude, the INMS will be able to measure with maximum sensitivity minor species, including short-lived chemically active neutral and ion species that play an important role in Titan's photochemistry and ion–neutral chemistry.

A minimum of 18 low-altitude aeronomy passes are needed to fully achieve INMS Titan science objectives. Low-altitude flyby scenarios proposed by the INMS Science Team provide for sampling of the atmosphere and ionosphere at noon, midnight, dusk, and dawn and at varying latitudes on both the ram side and the wake side (see Table VIII). Detailed ion and neutral composition measurements under these different conditions will yield a comprehensive picture of variability

TABLE VIIIA

INMS combined latitude coverage—open source ion measurements.

	Day	Dusk	Dawn	Night
Ram		T26out		T5out
		T32out		T21out
Flank	T18out			
	T36out			
	T39out			
	T40out			
	T42out			
Wake	T17out		T40in	
	T37out			

TABLE VIIIB

INMS combined ram/wake coverage—open source neutral measurements.

	Day	Dusk	Dawn	Night
North		T18in	T5in	T32in
South	T39in		T36in	
	T42in			
Equatorial		T17in	T37in	T26in

in the composition and structure of Titan's upper atmosphere, including possible seasonal effects, and will help elucidate the relative roles of solar EUV radiation and magnetospheric electron precipitation in neutral chemistry and ion production.

The INMS science team has determined that, as a minimum set, INMS must have control of primary spacecraft axis pointing on nine low-altitude passes (nine half passes each for neutral and ions). This plan relies on INMS having secondary axis pointing on many of the other low-altitude flybys to complete its science objectives. The nine passes are a revised minimum established by the INMS science team to cover season and latitude, as well as extremes of solar and magnetospheric production of ions and radicals that lead to the complex chemistry seen in the lower atmosphere. The flybys currently allocated to INMS comprise four separate phases of the mission: investigation validation (flyby 5), northern latitude survey (flybys 17, 18, 21), terminator study (flybys 26, 32) and southern latitude survey (flybys 36, 37, 39, 40, 42) (see Figure 27).

One of the major issues identified to date through the Titan Orbiter Science Team (TOST) process is the pointing conflict between INMS and RADAR. The INMS closed source field of view is such that in some cases it may be possible to take measurements during the same encounter, most likely within ± 8 min of closest approach. In this case the angle of attack of the closed source sensor would be changing throughout the measurement period. The INMS team is performing a study to determine the angular dependency of the closed source. The results of this study will determine in part the extent of INMS Closed Source/RADAR compatibility.

The INMS open source geometrical field of view cone half-angle is 8.6°. An assessment of the geometry of an INMS/RADAR pass suggests that INMS is in a preferred pointing direction only for ± 1 min near closest approach (out of a required ± 12 min) if RADAR is optimally pointed. Furthermore, mutually and maximally compromised pointing of both INMS and RADAR increases that window to only ± 2 min. Therefore, geometrically there is no way to satisfy the open source pointing requirements of INMS and RADAR at the same time. Two remaining possibilities are "dithering" from an INMS-optimum to a RADAR-optimum pointing continuously through a pass or using the electrostatic deflection built into the open source to obtain the ion portion of the data set. However, both of these options will require extensive calibration and simulation by both teams to demonstrate feasibility.

There has been extensive discussion within the MAPS, atmospheres, and INMS working groups to examine the INMS ionospheric and neutral atmospheric requirements. The current list of flybys includes: 5 (INMS inbound, ride with RADAR outbound), 17 outbound, 18 (full), 21 (outbound), 26 (full), 32 (outbound), 36 (full, although INMS may lose this flyby to atmospheres), 37 (starting -7 min through outbound), 39 (outbound), 40 (full) and 42 (still under discussion as to which leg will be INMS). Below is an attempt to demonstrate the importance of each of

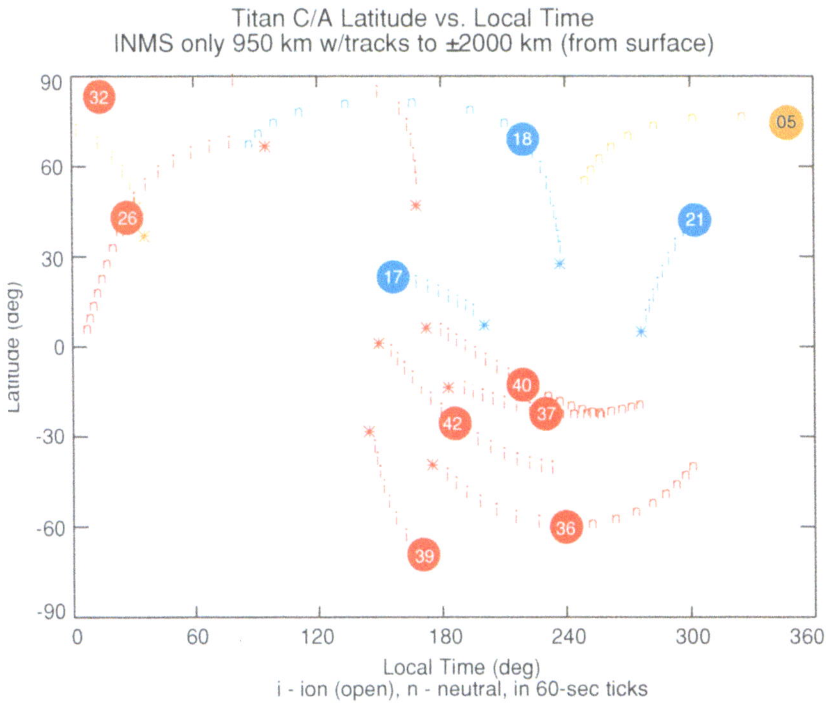

Figure 27. This figure shows latitude versus local time (Titan sub-spacecraft longitude relative to the Titan sub-sun longitude) of the Cassini spacecraft for all Titan encounters. The local time (*X*-axis of the plot), is expressed in degrees. The circular points indicate where closest approach occurs. The number within the circle indicates which Titan encounter it represents. The Cassini orbit track is shown for periods when the spacecraft is 2000 km or less from Titan's surface. The '*' indicates which is the outbound leg.

these orbits on a per flyby basis, throughout the discussion reference is made to the color-coded figure of Titan's orbital phases.

1. *Investigation validation phase—Pass 5.* Pass 5 is the first low-altitude pass in the tour that is available for INMS operations. Therefore, it is extremely important for determining the minor neutral and ion densities. Pass 5 is also important as a first step in sampling the global composition of the thermosphere and ionosphere and the thermal structure as a function of local time and latitude under varying conditions of magnetospheric input. It occurs during the "yellow" (a) segment of Titan's orbit (see Figure 28), which is of primary importance for studying magnetic induction in subsurface layers. From a magnetospheric point of view, it is a good flux tube crossing (see Table VIII and Figure 29). The team thought that it was imperative from both a technical and a scientific point of view to

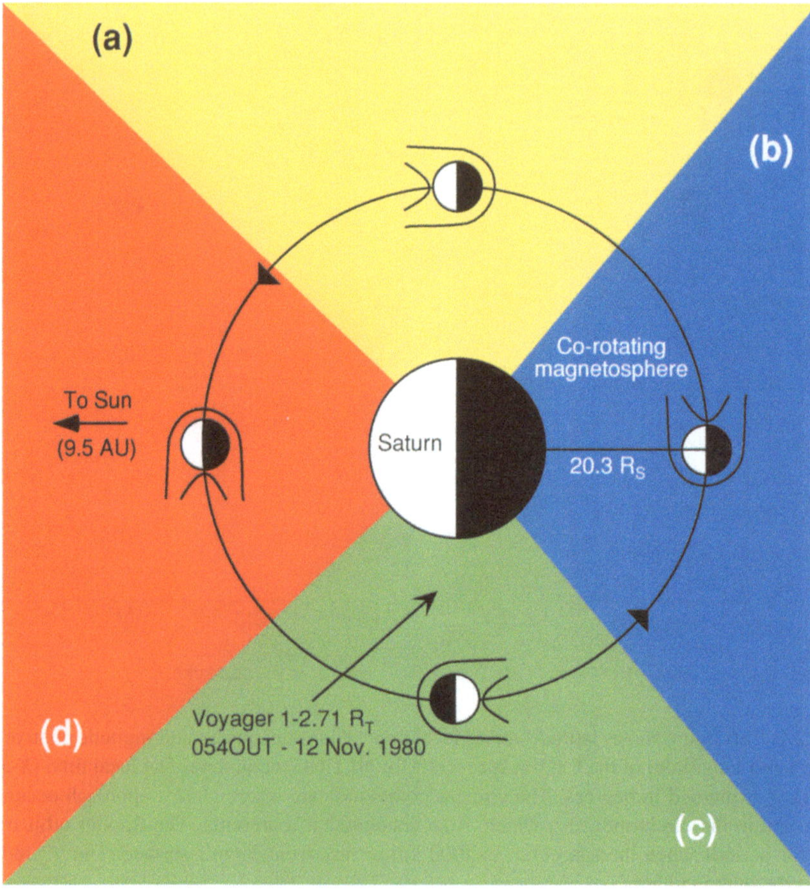

Figure 28. Shows four quadrants of Saturn's magnetosphere in sun-fixed coordinates.

have a full pass early in the mission to perform both open ion and open neutral measurements in a similar environment. This approach will best determine the performance of the instrument and will also make possible the study of the complex chemical coupling between the atmosphere and ionosphere needed to optimize the instrument performance for the rest of the mission.

2. *Northern latitude survey—Passes 17, 18, and 1/2 of 21*. Passes 17 and 18 are both passes within several hours of noon local time that provide coverage of the northern latitude under the same conditions of magnetospheric interaction (see Figures 27 and 29). They are both representative of the "blue" (b) orbital phase (see Figure 28). They occur near local noon and provide an excellent sampling of the flank of the magnetospheric interaction region where local ion pickup is thought to maximize, which will provide significant loss to the ionosphere and modification of the ion neutral chemistry (see Figure 29 and Table VIII). Pass 21

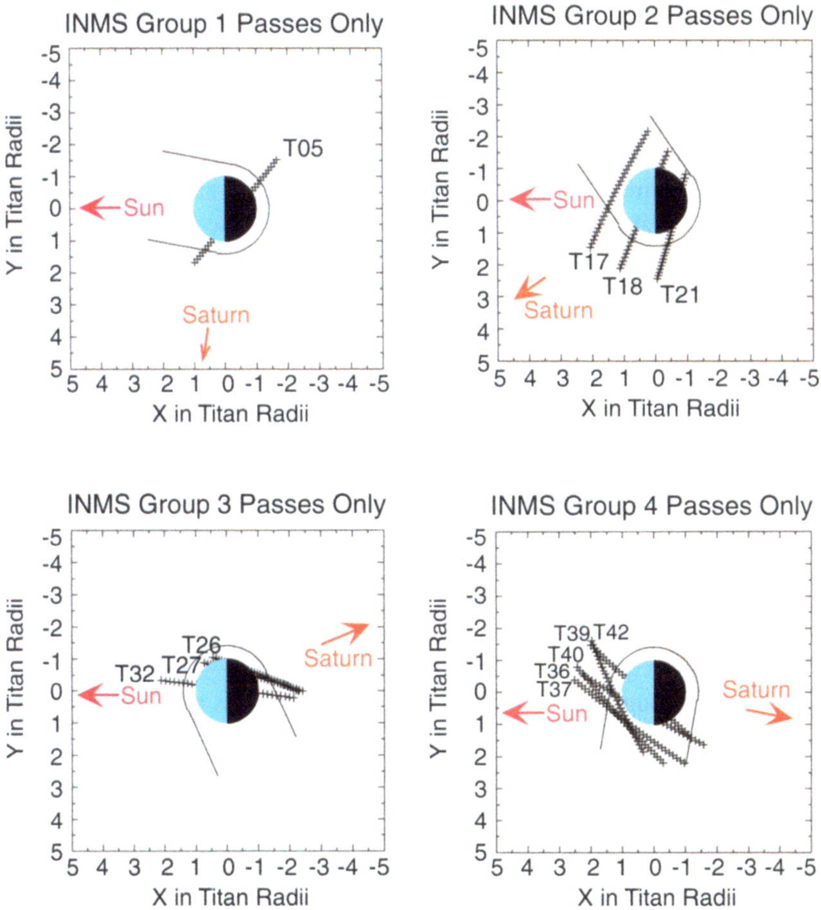

Figure 29. The position of Cassini and where it crosses the magnetospheric boundaries relative to Titan (in Titan-centered coordinates) broken up into four groups based on Titan's local time relative to Saturn.

fills in the latitude sampling, but more importantly, samples the ramside and flux tube of the magnetospheric interaction region at a local time near dusk and thus provides important measurements of the production and loss of ions and radicals that drive the ion neutral chemistry (see Table VIII and Figure 29). This subset will provide a fairly complete assessment of the factors affecting the northern latitude thermosphere and ionosphere.

3. *Terminator study—Passes 26, and 1/2 of Pass 32*. These passes are a two-orbit survey of the mid- to high-latitude terminator. Pass 26 takes place near 40° latitude and pass 32 occurs near 75° latitude (see Figure 27). They are both from the "red" (d) orbital phase (see Figure 28) and sample the flux tube where electron precipitation may have strong effects on the

heating and chemistry of the thermosphere and ionosphere (see Table VIII and Figure 29).

4. *Southern latitude survey—Passes 36, 37, 1/2 of 39, 40, and 42*. The set of passes 36, 37, 39, and 42 provides an adequate sampling of the southern latitudes for the "red" (d) orbital segment (see Figure 28) in both the open ion and open neutral modes (see Figure 27). They occur again all within an hour or two of local noon and thus sample the flank of the magnetospheric interaction region (see Figure 29), where energetic ion precipitation will have maximum effect on the thermosphere and ionosphere. Here pickup and energetic ion precipitation will be very different owing to the asymmetry produced by finite gyro radius effects of the magnetospheric ion interaction. Pass 40 is also important because it provides sampling of the low-latitude southern hemisphere (see Figure 27), which will provide continuity between the northern and southern latitude surveys. Pass 36 is unique in that it is the only pass selected for INMS that provides sampling of the important magnetospheric wake region where ion loss and electron influx will strongly affect the topside ionosphere and thermosphere (see Figure 29).

The Aeronomy Sequence used in these low-altitude flybys is the most complex of the INMS science sequences. It consists of seven cycles (1–11), each for a different altitude region through which the Orbiter will pass on its approach to Titan. The cycles are presented in Table VII in the order in which they will be executed during the entire flyby. Note that the inbound and outbound legs are not symmetric. The cycles prescribe a combination of unitary survey scans (for bulk composition measurement) and adaptive scans (for high-temporal-resolution measurement of key neutral and ion species expected to be present within a given altitude range). The target species for the adaptive scans to be performed during this sequence are summarized in Table IX.

3.3.5. *Outer Magnetosphere Sequence*

The INMS instrument will provide compositional information on neutral and ionized species in Saturn's outer magnetosphere. Of particular interest is the region near the orbit of Titan, as data acquired here will be used to assess Titan's contribution of neutrals and plasma to the outer magnetosphere (cf. Section 1.1.3). Because densities are expected to be low, long accumulation periods will be used and the mass scans co-added to improve counting statistics. Mass Tables 25 (CSN) and 44 (OSNB) are used for exclusively neutral measurements and 25 (CSN) and 63 (OSI) are used to sample ions; masses of particular interest are 14 (N, N^+), 16 (O, O^+), 17 (OH, OH^+), 18 (H_2O, H_2O^+), and 28 (N_2/H_2CN, N_2^+/H_2CN^+).

There are two different versions of the Outer Magnetosphere Sequence, one for purely neutral measurements (used during the inbound leg of the orbit) and one for ion and neutral measurements (used during the outbound leg of the orbit). Each of these measures the same mass values in the same order, but uses OSNB or OSI mode,

TABLE IX

This table lists the Dalton range, and in some case the species sampled, for scans executed during the various INMS orbital sequences detailed in Table VII.

Mass Table(s)	Scan Type	Description (source listed where applicable)
1, 26, 50	Unitary	1–8 and 12–70 Daltons
2–13, 27–38	Survey	Closed source/open source ion: 1/8-Dalton surveys covering the mass range 0.5 to 99.5
14, 39, 53 (inner magnetosphere)	Adaptive	Four water fragmentation product scans from 12 to 19 Daltons embedded in a unitary scan from 1 to 48 Daltons
16/17, 40/41, 54/55 (Titan low/high altitude parts 1 and 2)	Adaptive	Eight scans of N_2, CH_4 and H_2 embedded in unitary scans from 1 to 8 and 12 to 99 Daltons
18/19, 56/57 (TA adaptive and 1/8)	Adaptive/Survey	Closed source/open source neutral: eight scans of N_2, CH_4 and H_2 embedded in 1/8 Dalton scans from 12.625 to 18.500 and 25.625 to 30.500. Used in TA to check the mass peak selection accuracy
20/21, 58/59 (Titan low altitude parts 1 and 2)	Adaptive/Unitary	Closed source: H_2, He, CH_4, C_2H_2, HCN, $N_2/C_2H_4/H_2CN/(CO)$, $^{15}N^{14}N/^{13}C_2H_4$, C_2H_6, ^{36}Ar, C_4N_2, cracking products (12–15, 24–25, 37–40 Daltons); sweep of 1–51 and 52–99 Daltons
		Open source neutral: same species as are measured in closed source mode plus N, NH, O, CHCN; sweep of 1–51and 52–99 Daltons
25, 44, 63 (outer magnetosphere)	Adaptive	Open source ion: H_2^+, C^+, CH^+, N^+/CH_2^+, CH_3^+, CH_4^+/O^+, CH_5^+/OH^+, H_2O^+, $C_2H_2^+/CN^+$, $HCN^+/C_2H_3^+$, $N_2^+/H_2CN^+/C_2H_4^+/(CO^+)$, $C_3H_8^+/(CO_2^+)$
		Closed source/open source neutral: same species as above, but neutrals
22, 60 (Titan low altitude adaptive large)	Adaptive	Closed source: H_2, He, CH_4, C_2H_2, HCN, $N_2/C_2H_4/H_2CN/(CO)$, $^{15}N^{14}N/^{13}C_2H_4$, C_2H_6, ^{36}Ar, C_4N_2, cracking products (12–15, 24–25, 37–40 Daltons)
		Open source neutral: same species as are measured in closed source mode plus N, NH, O, CHCN
23, 42, 61 (Titan low altitude adaptive 1)	Adaptive	Closed source/open source neutral: H_2, CH_4, HCN, $N_2/C_2H_4/H_2CN/(CO)$, cracking products (12 and 14 Daltons)
		Open source ion: same species as above, but neutrals.

(*Continued on next page.*)

TABLE IX

(*Continued*).

Mass Table(s)	Scan Type	Description (source listed where applicable)
24, 43, 62 (Titan low altitude adaptive 2)	Adaptive	Open source ion: H_2^+, C^+, N^+/CH_2^+, CH_4^+/O^+, CH_5^+/OH^+, $HCN^+/C_2H_3^+$, $N_2^+/H_2CN^+/C_2H_4^+/(CO^+)$, $C_2H_5^+/N_2H^+/(HCO^+)$ Closed source/open source neutral: H_2, N, CH_4/O, OH/$^{13}CH_4$, HCN, N_2/C_2H_4/(CO), $^{15}N^{14}N$, $^{13}C_2H_4$

Parentheses indicate species whose presence is possible but unlikely.

respectively, for the second set of measurements. Velocity compensation values were selected to account for expected spacecraft-relative velocities of particles in Keplerian, corotating or magnetic-field-aligned orbits. There are also four different data rate modes currently available, just as there are for Default Science: 1498, 100, 50, and 6.2 bps, with 1, 15, 30, and 240 co-added scans, respectively.

3.3.6. *Inner Magnetosphere*

While Cassini will spend most of its time in the outer magnetosphere of Saturn, the higher densities in the inner magnetosphere will be much more conducive to INMS measurements. As discussed in Section 1.2, the inner magnetosphere is characterized by the presence of water group neutrals and ions associated with the rings and icy satellites and believed to be produced from these bodies by charged particle sputtering and micrometeorite bombardment. With predicted densities as high as 4×10^3 cm^{-3} near the orbit of Enceladus (Ip, 1997), this material, through the complex plasma–gas–dust interactions in which it participates, is thought to play an extremely important role in the chemistry, dynamics, and structure of the inner magnetosphere. The Mass Tables used—14 for CSN, 39 for OSI and 53 for OSNB—involve repeated measurement of masses 12–19 interleaved with measurements of the mass ranges 1–8, 20–27, 28–35, and 26–47 Daltons. This provides for the repeated sampling during each scan of the water group neutrals O (16), OH (17), and H_2O (18), and ions O^+ (16), OH^+ (17), H_2O^+ (18), and H_3O^+. Although the densities of these species are expected to be at a maximum near the predicted source regions, they will still be at the lower end of the INMS sensitivity.

The organization of these sequences is very similar to that of the "Outer Magnetosphere" sequences: a single mass range sampled alternately in CSN and OSNB or CSN and OSI modes. The choice of neutral or ion and neutral is the same as described above—neutral for inbound and ion and neutral for outbound—and the four data rates are organized in the same way. Neutral particles in Keplerian orbits closer to Saturn will move faster, and the velocity compensation values were increased accordingly.

3.3.7. *Ring Overflight for SOI*

The "Ring Overflight for SOI" Sequence is designed to sample the neutral and plasma environments of the rings and icy satellites in Saturn's inner magnetosphere and will be executed during the overflight of the rings following SOI. A modified version of this sequence could also be used during the planned flybys of Iapatus, Enceladus, Dione, and Rhea. Because the INMS team has primary spacecraft axis control during a period after SOI, a specific sequence was designed to cover this period. The Mass Tables used are the same as those used in the "Inner Magnetosphere" sequences, but the timing is different. The first ~700 sec of the measurement period centers on magnetic-field-aligned and corotating ions, while the next ~480 sec will measure neutrals corotating and in Keplerian orbits. Velocity compensation values were chosen to match the Cassini-relative velocities of particles in each type of orbit at that time period, based on estimated particle masses and energies.

4. Flight Unit Characterization

4.1. General

The characterization of the INMS flight unit was performed at Goddard Space Flight Center using a high-vacuum test station with both thermal neutral and ion sources (Figure 30). A neutral beam system was not available at the time of the INMS

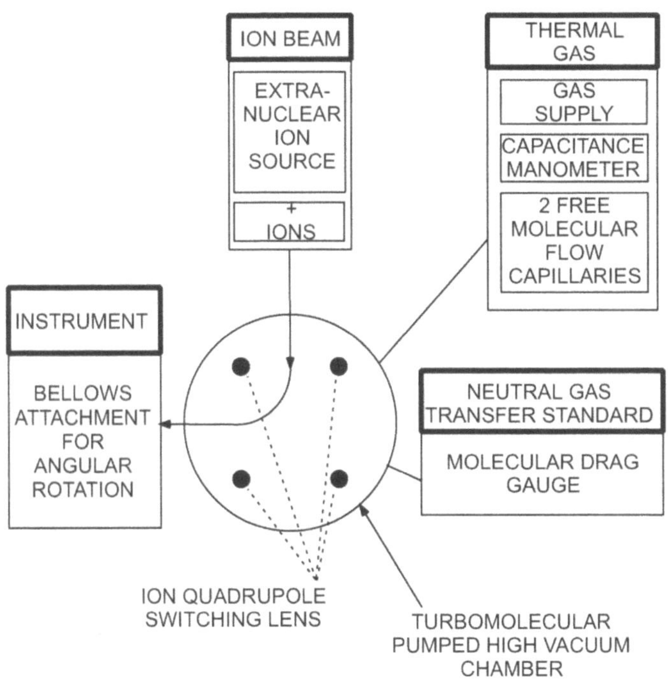

Figure 30. Schematic of the high-vacuum test station used for INMS characterization.

testing. Thus the ion beam was also used to characterize instrument performance in the open source neutral mode; for these tests, however, the INMS entrance lens (OL4) potential was set at −5 V, as required for the neutral beaming mode (cf. Section 2.2.1). The test station was designed so that all the INMS operational modes could be characterized without breaking the vacuum and thus necessitating re-baking the sensor. Neutral gases and ions used for characterization testing were introduced into the main vacuum chamber, to which the INMS was attached by a flexible bellows with two degrees of rotational freedom for angles up to about 5°. The instrument could be translated to allow appropriate positioning of the source being tested (i.e., of the open source with respect to the ion beam). Pressures inside the main vacuum chamber were kept below $\sim 10^{-6}$ hPa in order to prevent possible damage to the secondary electron multipliers. Thus the operation of the instrument at higher pressures, i.e., up to mid-10^{-5} hPa, the estimated ram pressure at Titan closest approach, was not tested. Laboratory support electronics were used for early testing; flight electronics were used for the final characterization. The characterization of INMS performance will continue during the post-launch period with testing of the engineering unit.

4.2. Neutral Mode Characterization

INMS performance in the closed source and open source neutral thermal modes was characterized using the principal nonreactive species expected to be encountered at Titan: H_2, He, N_2, CH_4, C_2H_2, C_2H_4, and Ar. Tests were performed with both individual target gases and gas mixtures. Pure Kr was used as a high-mass reference gas, and a noble gas mixture was used to check instrument tuning. Because the geometries and efficiencies of the primary and secondary filaments in the ionization region are slightly different, test runs were conducted for both sets of filaments. Tests were also performed at both high and low electron energies. In each test run, the first measurements—both unit and 1/8-Dalton survey scans—were made at main chamber gas background pressure with no inflow of gas from the thermal gas source. Following acquisition of these background data, both survey (unit and 1/8-Dalton) scans and programmed mass scans were performed for various gas samples at different pressure levels up to 10^{-6} hPa. Representative results from thermal gas characterization tests are presented in Figures 31–36 and are discussed in the following paragraphs.

The INMS mass range is from 1 to 8 and 12 to 99 Daltons. The capability of the INMS to measure high-mass species is important because of the expected presence of long-chain hydrocarbons (e.g., polyacetylenes) in Titan's upper atmosphere. To investigate the instrument's performance up to mass numbers near 100, data were acquired using the closed source for a sample consisting of Ne (10%), Ar (10%), Kr (10%), Xe (10%), and He (60%). The resulting spectrum, with clearly defined mass peaks for He^+ and singly and doubly charged Ar and Kr, is shown in Figure 31. Xe is not measured at the parent peak near mass 132, but is observed as a doubly

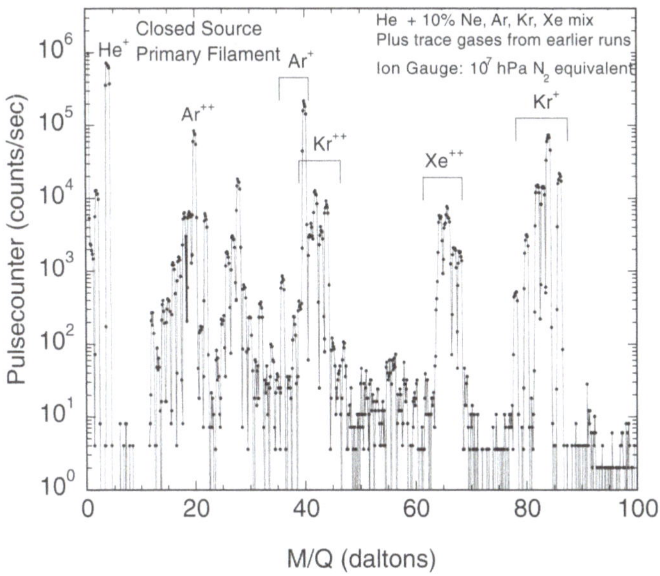

Figure 31. Mass spectrum for a mixture of noble gases illustrating the INMS performance at high mass numbers. Data were acquired in the closed source mode.

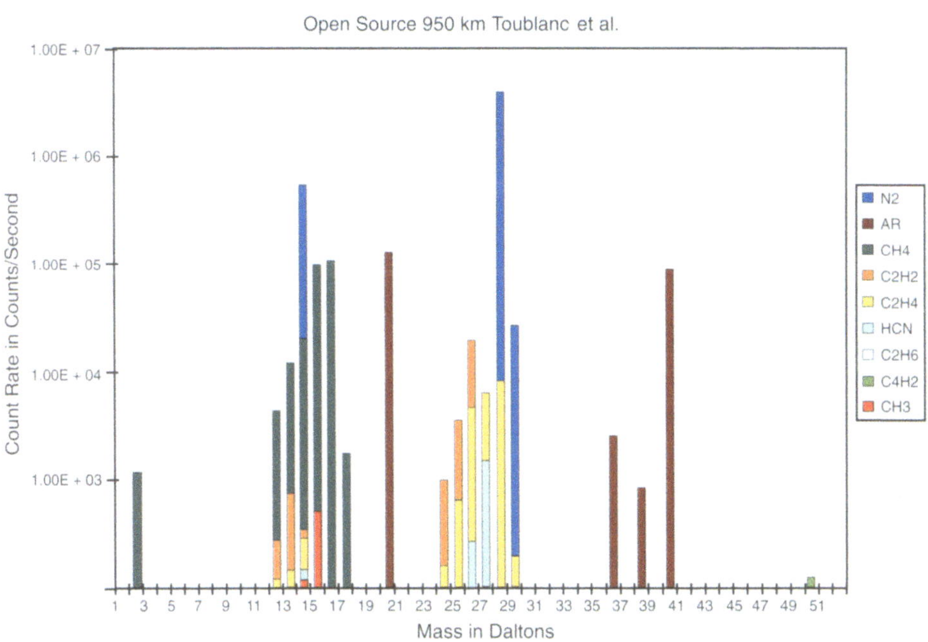

Figure 32. Simulated open source spectrum illustrating the masking of parent species and cracking products by other more abundant species with the same mass numbers. Concentrations are based on densities at Titan closest approach according to the neutral atmosphere model of Toublanc *et al.* (1995).

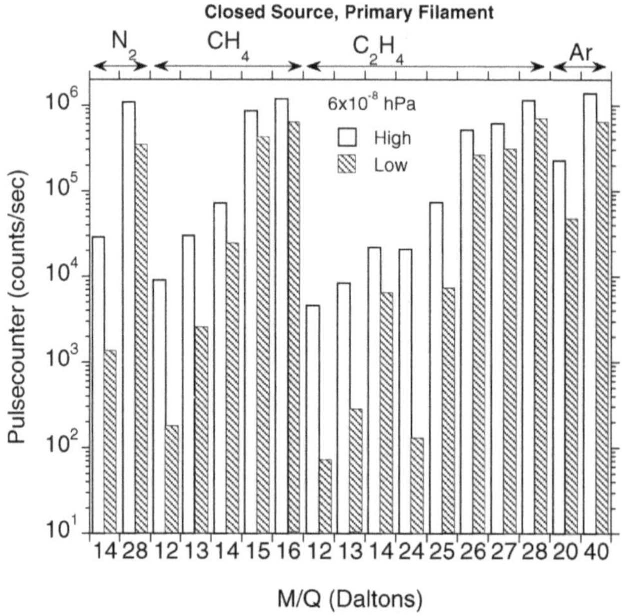

Figure 33. Counts obtained with the INMS closed source using both high and low electron energies for N_2, CH_4, C_2H_4, and Ar at a nominal pressure of 6×10^{-8} hPa. Low electron energy reduces the signal amplitude and changes the relative pattern of the various mass peaks. The ionization cross sections at low electron energy change more rapidly than at higher electron energy. The ionizing electron beam energy has an energy spread of several eV, so using low electron energy does not create a sharp cutoff of some mass peaks that might be expected otherwise.

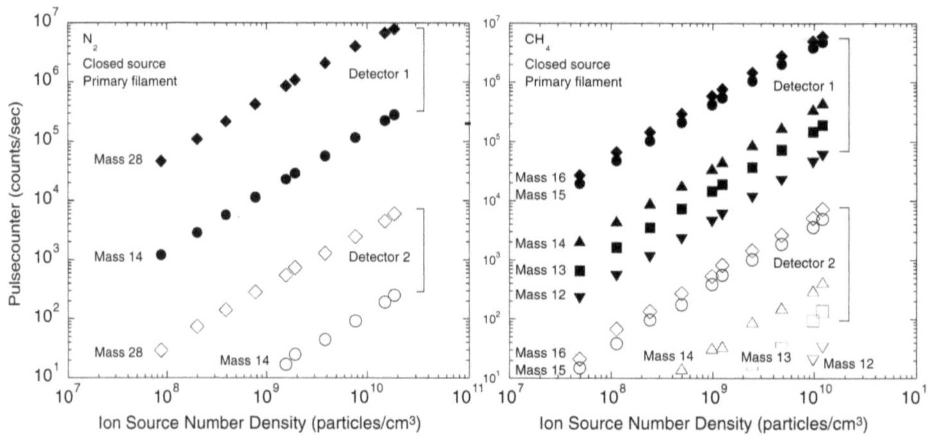

Figure 34. Pulse counter output for Detectors 1 and 2 as a function of ion source density (=main chamber density). Mass peaks for N_2 and CH_4 (both parent species and fragments) are shown.

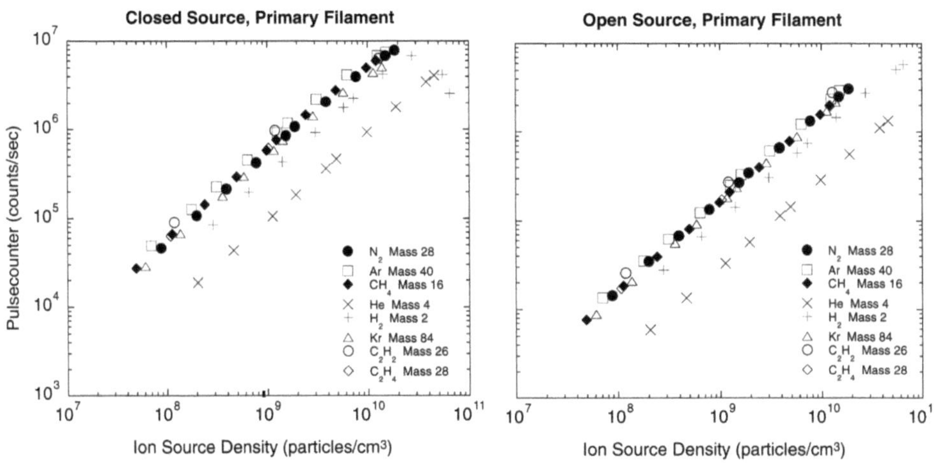

Figure 35. Summary of the pulse counter output as a function of ion source density (=main chamber density) for the parent species of all gases used in the characterization of the INMS closed source and open source neutral thermal mode. The data points for CH_4 and C_2H_4, which have about the same sensitivity, overlap. Only three data points were obtained for C_2H_2 and C_2H_4 because of time limitations.

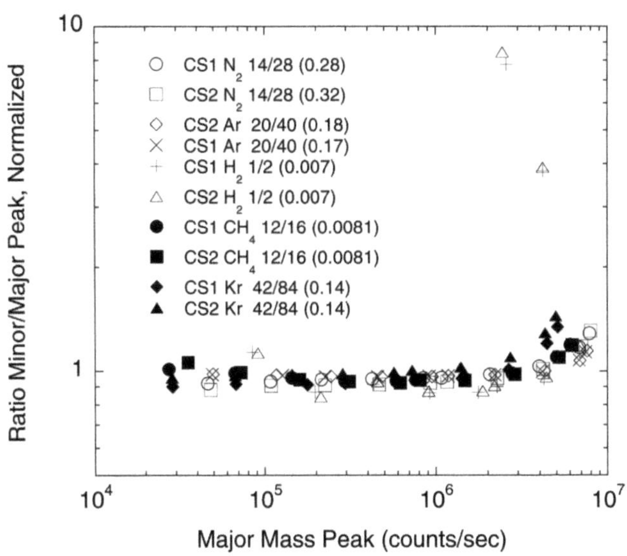

Figure 36. Minor peak to major peak ratios for several species plotted as a function of the parent peak count rate. Data were taken in the closed source (CS) mode using both the primary (CS1) and secondary (CS2) filaments. The ratios have been divided by the numbers in parentheses. The increase in the ratio evident above a count rate of 1–2 MHz reflects a loss of counts at high count rates (the "pile up" effect).

ionized smear around mass 65. In addition, peaks for some residual gases from earlier runs are evident near mass 28.

Distinguishing between species—whether parent species or cracking products—with the same molecular mass poses a challenging problem for the INMS investigation. This difficulty is illustrated by the simulated mass spectrum shown in Figure 32, which shows the spectral position and concentration of the primary species and the various cracking products resulting from ionization in the open source (concentrations based on densities at 950 km derived from the model of Toublanc *et al.*, 1995). N_2 is present at higher concentrations than C_2H_4 (both mass 28) and masks the latter, for example, while a cracking product of C_2H_4 (C_2H_3) masks HCN at mass 27. As noted above (Section 2.2.1), however, the masking problem can be addressed by using two different electron energies (70 and 25 eV) in the INMS ion sources. Because of the energy dependence of the efficiencies and cross-sections for dissociative ionization, different electron energies yield different spectra, with different peak heights and different fractionation patterns, for the same gas sample, as can be seen in Figure 33, a spectrum based on closed source calibration data for both high and low electron energies. Comparative analysis of such laboratory calibration data and of the mass spectra acquired in the Saturn system can be performed using standard techniques (Kiser, 1965) to deconvolve the contributions of the various species to the mass spectra.

Figures 34 and 35 illustrate the differences in sensitivity between the two detectors and between the closed and open ion sources. Figure 34 shows a plot of the pulse counter output from detectors 1 and 2 as a function of ion source density (=main chamber number density) for N_2 and CH_4, the major neutral gases in the upper atmosphere of Titan. Data presented in this figure were acquired in the closed source mode using the primary filament. In the left-hand panel (N_2), mass 28 is the parent peak; in the right-hand panel (CH_4), mass 16 is the parent peak, and the remaining peaks are those of the cracking products. As noted above (Section 2.2.4), the use of two detectors that differ in signal detection level by about a factor of 1500 yields the needed dynamic range of $\sim 10^8$.

Count rates as a function of ion source density for each of the neutral gas species used in the characterization of the closed and open sources (using the primary filaments) are shown in Figure 35. Only the counts for the parent species are given. The higher sensitivity of the closed source is evident from a comparison of the two panels. (The lower sensitivity of the open source is due to the geometry of the open source ionization region and the complexity of the ion optics.) With the exception of H_2 and He, most species have about the same sensitivity in a given source. The lower count rates for H_2 and He result from the lower ionization cross sections for these gases. Note: only three data points instead of the usual 10 were acquired per run for C_2H_2 and C_2H_4 because of time constraints.

As can be seen in Figures 34 and 35, the count rate as a function of ion density increases linearly up to a rate of about 4×10^6 counts sec^{-1} (4 MHz). Above this value, however, the secondary electron multiplier pulses begin to "pile up" and several ion pulses are counted as a single pulse by the pulse counter and discriminator circuits. This loss of counts at the higher count rates changes the slope of the mass 28, 16 and 15 peak data in Figure 34 and, most noticeably, in the closed source data for H_2 in Figure 35. The pile up effect is also evident in Figure 36, which shows the normalized ratio of a minor peak to the parent peak for several gases in the closed source (both filaments) plotted as a function of the count rate of the major peak. Above a count rate of 1–2 MHz, this ratio, which should be a constant based on the ratio of the ionization cross sections, increases. The increase occurs because there is a loss of counts in the major peak and very little, if any, in the minor peaks. The pile up effect—i.e., the loss of counts at high count rates—can, in principle, be mitigated by using the output from the second detector once the count rate from the primary detector is larger than 1–2 MHz. A secondary detector count rate of 10^3 counts sec^{-1} is equivalent to 31 counts per sample period and the statistical error is $(31)^{1/2}/31$ or $\sim$0.2. Outputs from both detectors are put into the telemetry packet so that the ratio of the output of the primary detector to that of the secondary detector can be tracked over a Titan pass and that ratio used to predict the primary value based on the secondary detector output.

4.3. Open Source Ion/Neutral Beaming Mode Characterization

An ion beam system was used to characterize the performance of the INMS in the open source ion mode and, in the absence of a neutral beam, in the open source neutral beaming mode as well. An extranuclear ion source in a separately pumped chamber supplied the ions, which were created from He (mass 4), Ar (mass 40), and Kr (mass 84). These particular gases were selected because ion transmission over the INMS mass range (1–99 Daltons) was of primary interest. The ions were focused electrostatically and transmitted from the extranuclear source into the main vacuum chamber, where they were deflected through a quadrupole switching lens into the INMS open source (cf. Figure 30). No magnetic field was used to separate mass-to-charge ratios; so all ions generated in the extranuclear source from a particular gas were present, including fragmentation ions and background gas ions. During the characterization runs, considerable difficulty was encountered with drift effects in both energy and angle for low-energy ions. These most likely resulted from charging effects in the ion source and at the main chamber quadrupole deflector electrodes.

During initial characterization using laboratory electronics, data were taken on the operation of the INMS switching lens in the open source ion mode, with the entrance lens (OL4) set at −30 V, and in the open source neutral beaming mode, with OL4 set at −5 V. Ten ion energies (nominally 2, 3, 4, 6, 8, 10, 12, 14, 17 and 20 eV) plus "0 eV" (thermal gas in the ion source) were used for the initial tests. The ion beam was centered in the INMS aperture by moving the carriage on which the INMS was mounted. The ion flux into the aperture was determined by measuring the current on the top plane lens (TPL). Energy scans were performed by varying the potential on QL3 and setting the other switching lens voltages to QL4 = QL3 and QL1 = QL2 = $-(\text{QL3} + 2K)$, where K is the absolute value of the voltage applied to the entrance lens. QL3 was set to the voltage corresponding to the maximum count rate in the energy scan, and the quad bias voltage was set at -1 V. The ion energy was determined by scanning the quad-bias voltage and determining the voltage value (mean energy value) for which the signal dropped by half its value (the "pseudo-RPA cutoff curve"). (Plots of the results of the energy scans and quad bias scans are shown in Figure 37.) The energy scan was repeated with the quad bias set at -0.5 V below the half-amplitude point as determined by the quad bias scan. Only Ar^+ was used for the initial characterization tests. Spectra were obtained with 1/8-Dalton survey scans.

Characterization tests with the flight electronics were performed using only three energies (3, 8, and 14 eV for Ar^+ and Kr^+ and 3 eV for He^+) at three different flux levels. In addition to test runs with the ion beam centered in the instrument aperture, angle scans around an axis in the horizontal direction and the vertical direction were performed in 0.5° steps out to the maximum range (±5°) of the carriage system on which the INMS was mounted. For each ion flux and energy level, the switching lens was scanned, and the value for QL3 obtained. The switching lens was frequently re-set if the maximum signal did not occur at the anticipated point as determined from earlier laboratory studies. Quad-bias scans were performed, and, if needed, the voltage was re-set at -1 V below the half-amplitude value. Once the correct voltages had been set and verified, normal mass scan and dwell data were taken. A representative spectrum from a 1/8-Dalton mass scan is shown in Figure 38.

During the laboratory characterization tests, the INMS switching lens and quad bias voltages were set to transmit ions of known, fixed energy. For actual measurements at Titan and elsewhere in the Saturn system, however, the energy of the ions (both ambient and those created from neutrals in the open source ionization region) will vary as a function of mass and spacecraft velocity. (The latter will be the same for all masses.) The switching lens and quad bias settings will thus have to vary with ion energy. The switching lens voltages for a given ion energy can be determined from the relationship, established during instrument characterization, between the QL3 voltage corresponding to the maximum signal obtained in an energy scan and the quad bias scan half-amplitude point. As shown by the plot in Figure 39, the QL3 voltage at signal maximum is a linear function of the voltage at the half-amplitude point of the quad bias scans. In Figure 39, the solid and dashed lines represent a linear fit to the data points (circles and squares). The lines are used to predict the appropriate switching lens potential QL3 needed to track the incoming species' kinetic energy as represented by the quad bias half amplitude voltage. For the neutral beaming mode QL3 = $(-9.8754 - 0.90428 \times \text{KE})$ volts. For the ion mode QL3 = $(-54.296 - 0.90646 \times \text{KE})$ volts. KE is the species'

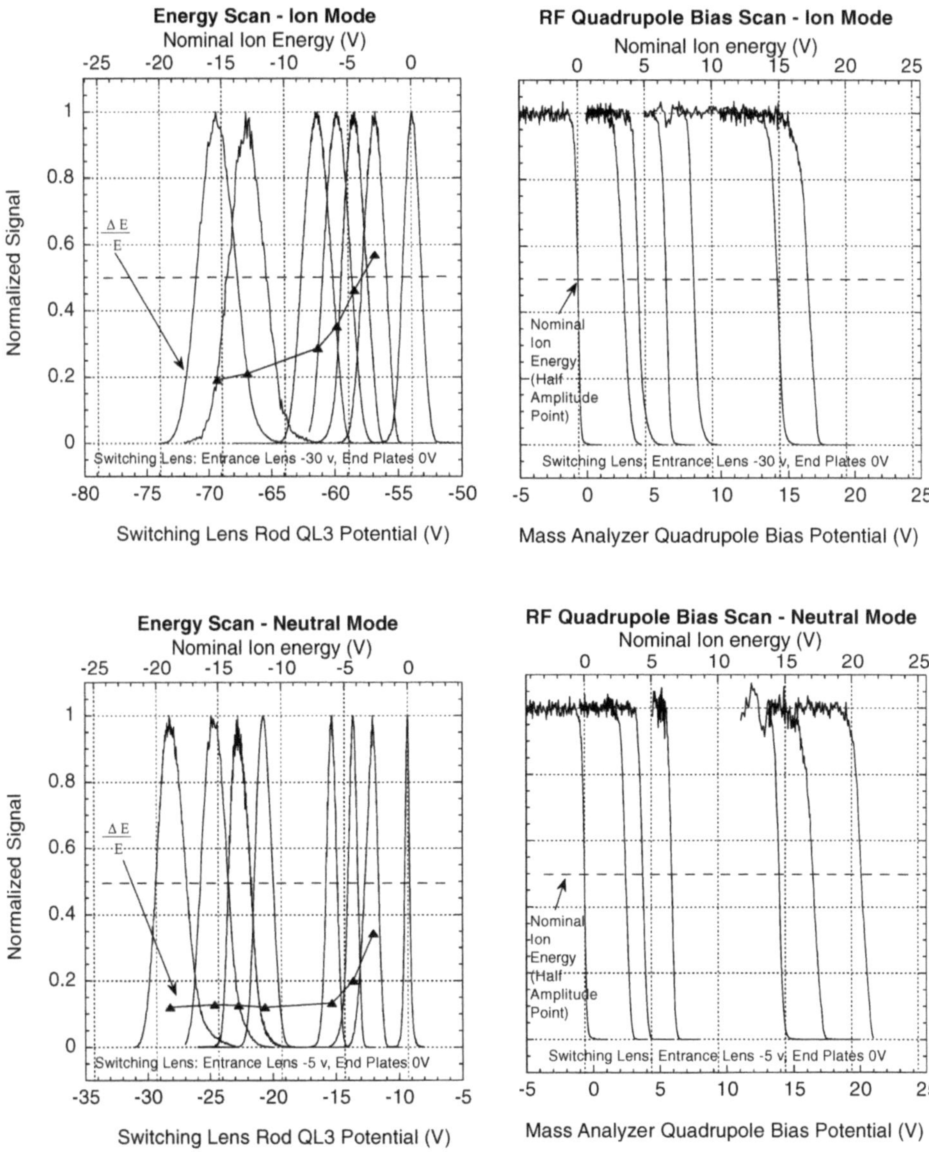

Figure 37. Energy and quadrupole bias scans for various ion energies with INMS operating in the open source ion mode (top) and neutral beaming mode (bottom). The energy scans are obtained by varying the voltage on QL3 and determining the potentials on the other switching lens rods according to the following equations: QL4 = QL3 and QL1 = QL2 = −(QL3 + 2K), where K is the absolute value of the OL4 (and exit lens L1) voltage (i.e., either 5 or 30 V). The half-amplitude value obtained in the quad bias scans is used to establish mean ion energy. The end plate (OL5/6) potentials are set at 0 V rather than at the mean ion energy (=OL4) as in Mahaffy and Li (1990). The INMS $\Delta E/E$ values shown are not constant but decrease with higher ion energy.

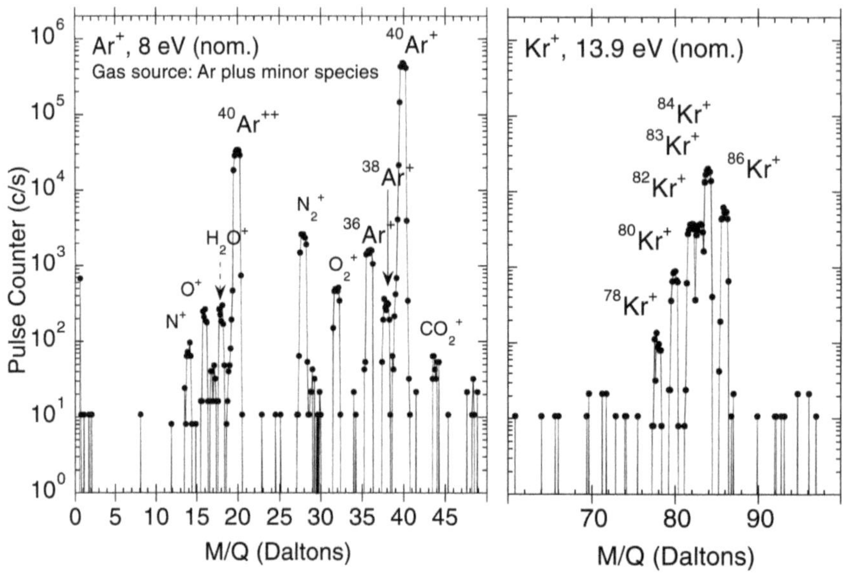

Figure 38. Spectra for Ar^+ and Kr^+ from a 1/8-amu scan in the open source ion mode. Peaks for minor species can be seen in the Ar^+ spectrum and for isotopes in both spectra. A quadrupole bias voltage was applied to retard the mass-analyzed ions and reduce the peak widths (cf. Section 2.2.3).

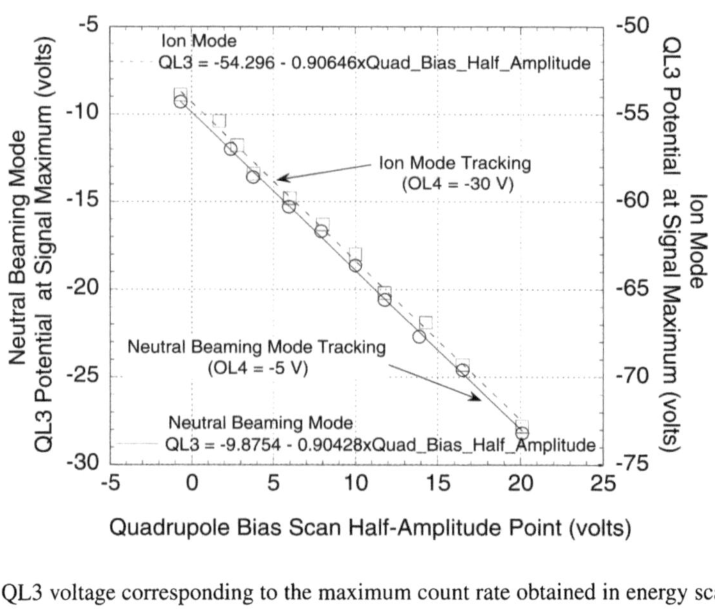

Figure 39. QL3 voltage corresponding to the maximum count rate obtained in energy scans in both the open source ion and neutral beaming modes plotted as a function of the voltage of the quadrupole bias half-amplitude point (=mean ion energy). Data used in generating this plot were taken from Figure 37.

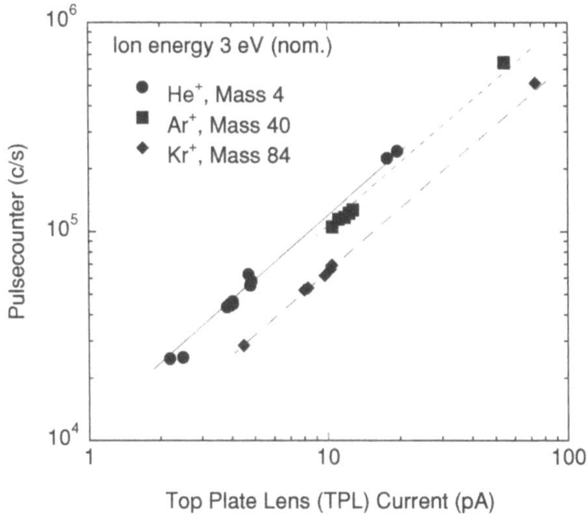

Figure 40. Counts for He^+, Ar^+, and Kr^+ as a function of the top plate lens current, which serves as a proxy for the ion flux into the instrument. Kr data have been corrected to remove isotopic contributions; only the mass 84 contribution is shown. The slope of the lines shown is 1; however, the fits of the data imply slopes ranging from 1.03 to 1.1. It is believed that this discrepancy is due to yet unknown experimental difficulties.

kinetic energy in electron volts as observed in the spacecraft reference frame. For a Titan flyby, with a nominal spacecraft speed of 6 km/sec, KE = (0.191 eV/Dalton) × (species' mass in Daltons). The previous equations are suitable for singly ionized species. In the case of double ionized species, the tracking voltages need to be set to a KE value equivalent to half the species' mass.

The ion flux into the INMS was determined by measuring the TPL plate current at 0° angle. Figure 40 is a plot of the count rates for the three ion species as a function of TPL current. The current to the deflectors was not measured during characterization because earlier measurements had shown them to be negligible for small beam angles.

The rejection of ions by the ion deflector/trap was tested during ion mode characterization. No ion counts were detected for a 20-eV ion beam with the switching lens set to transmit 20 eV ions (OL4 = −30 V, OL1 = OL2 = 0 V) and the deflectors set at D1 = D4 = +30 V and D2 = D3 = −30 V or D2 = D3 = 0 V. Lowering the voltage to D1 = D4 = +5 V and D2 = D3 = −5 V yielded about 1 count per integration period. With D1 = D4 = +1 V and D2 = D3 = −1 V about 530 counts sec^{-1} were observed (compared with 1.2×10^6 counts sec^{-1} with the deflectors and TPL set at ground). With ±30 V on the deflectors, only the thermal gas peak was present when the switching lens was set for open source neutral thermal mode; no ion beam ions were transmitted. For neutral beam conditions (OL1 = −1 V, OL2 = +1 V) and a 20 eV ion beam, no signal was detected. Tests (at ion beam energies of 2.8 eV) indicated that secondary ions were not being created on the deflectors and contributing to the signal.

4.4. Secondary Electron Multiplier Characterization

During characterization, tests using both ion sources were performed usually once a day at a nominal N_2 chamber pressure of 10^{-7} hPa to track possible changes in secondary electron multiplier (SEM) sensitivity. Pulse-height distributions were obtained at reasonably low count rates using low electron energy. A pseudo-secondary electron multiplier gain (SEM) was computed from measurements of the multiplier analog current, and the corresponding count rate using the open source. Three SEM high-voltage levels were used: (1) low, used during the characterization period; (2) medium, corresponding to a SEM voltage that would be needed following pulse-height distribution changes in thermal vacuum and vibration qualification testing; and (3) high, corresponding to possible future changes in the pulse-height distribution during the orbital tour. For the primary detector, the SEM voltages are 2500, 2700, and 2900 V, while for detector 2 they are 2800, 2900, and 3000 V. There are 16 discriminator levels; level 4 is used as the nominal setting. Figure 41 shows the mass-28 peak sensitivity referenced relative to an ion gauge on the main chamber for both the open and closed sources (filament 1), and the computed

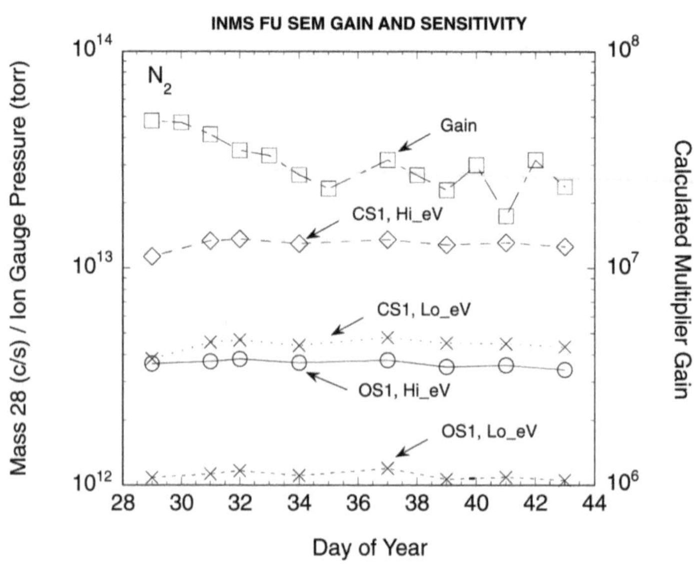

Figure 41. The mass-28 sensitivity relative to reference ion gauge for N_2 at nominally 10^{-7} hPa. OS1: open source, filament 1; CS1: closed source, filament 1; Hi_eV: high electron energy; Lo_eV: low electron energy. Day of year refers to 1997. Multiplier gain is computed from the accumulated SEM counts and the analog current.

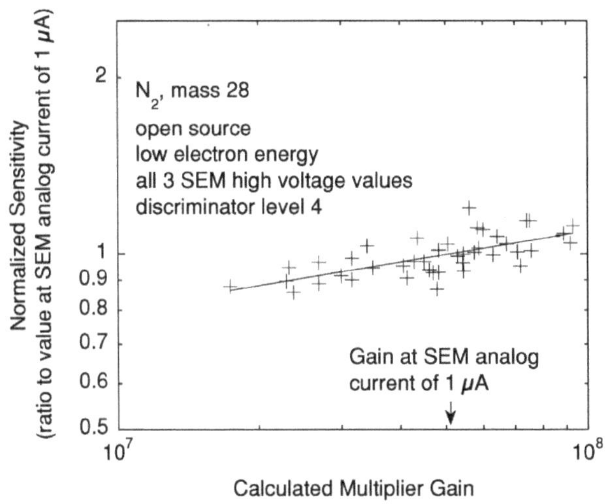

Figure 42. The relative change in sensitivity for a change in multiplier gain.

SEM gain for the low high-voltage setting and discriminator level 4. Figure 42 shows the relative sensitivity change for all three SEM high-voltage levels at discriminator level 4. A factor of 5 change in gain results in about 20% change in sensitivity.

Figure 4.4 shows a plot the integrated pulse-height distributions measured during characterization, after vibration and thermal vacuum testing, and during one of the last spacecraft baseline tests for the three SEM high-voltage settings. Data were taken for each of the discriminator levels except for level 0. As noted above, discriminator level 4 is the normal level used; it was originally chosen to discriminate against the noise counts (upturn in the curve below discriminator level 3 for day 970130), while not severely curtailing the real ion counts above this level. After characterization, SEM voltage was raised to the medium level for the instrument environmental tests. It can be seen from the plot that the slope of the curves above discriminator level 4 changed with time, with the maximum change occurring for the SEM low high-voltage level and the least change for SEM high high-voltage level. It is anticipated that the high high-voltage level will be used for the orbital tour at Saturn. The computed SEM gain during characterization, environmental testing, and baseline testing on the spacecraft is plotted in Figure 44. During characterization, gain values were computed using mass 28 (N_2) data; afterwards, once the sensor was sealed and the getter was activated, gain was computed using mass 40 and mass 20 from Ar. No independent absolute pressure data were available after characterization to reference a possible sensitivity change.

The computed SEM gain can be used as a proxy for estimating the change in ion source sensitivity when a detailed pulse height distribution is not available.

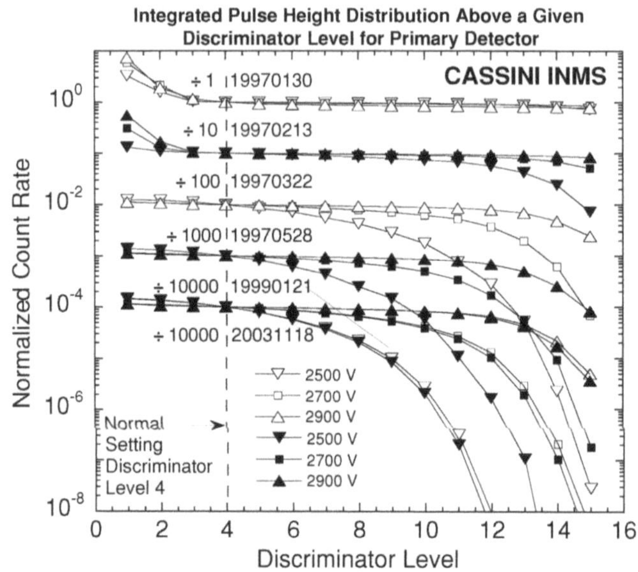

Figure 43. Plot of the integrated pulse-height distributions for the three SEM high-voltage levels (low, medium, high) obtained on four different dates: 970130 (beginning of the characterization period); 970213 (end of the characterization period); 970322 (after environmental testing); and 970528 (testing on the spacecraft). Discriminator level 4 is the normal setting used. Data taken at the beginning and end of the characterization period are for N_2 (mass 28); data taken following environmental testing and on the spacecraft are for Ar (mass 40). The curves have been normalized to discriminator level 4, and the three sets of data for each date have been divided by the values shown just below the curves, to the left of the dashed line. Test dates are given in YYMMDD format. Data are for the primary detector only. Open and filled symbols are used to distinguish data points for alternating dates when the curves overlap.

4.5. Engineering Model and Post-Launch Characterization

The prototype of the Flight Model (FM) instrument, the Engineering Model (EM), was refurbished to more closely emulate the finalized FM unit. This will allow the EM to be used for further laboratory testing, extending the amount of information available for interpretation of the FM results. The EM can also be used to test command sequences and flight computer software changes to be uploaded to the FM. Characterization of the EM in thermal gas mode was done with the same gases used for the FM (see Section 4.2) plus C_6H_6, CO_2, CO, C_2H_6, C_3H_4 (propyne), C_3H_4 (allene), Ne, O_2 and C_3H_8. The vacuum test station was modified to provide an ion source that dropped down in the middle of the main chamber. This eliminated the previous ion source and large quadrupole-switching lens shown in Figure 30. Ion energies similar to those used for the FM were used to compare the flux sensitivity and angular response of the EM to the FM. The open source neutral beaming mode was tested with a low-energy neutral beam. It revealed that lens QL3, which was

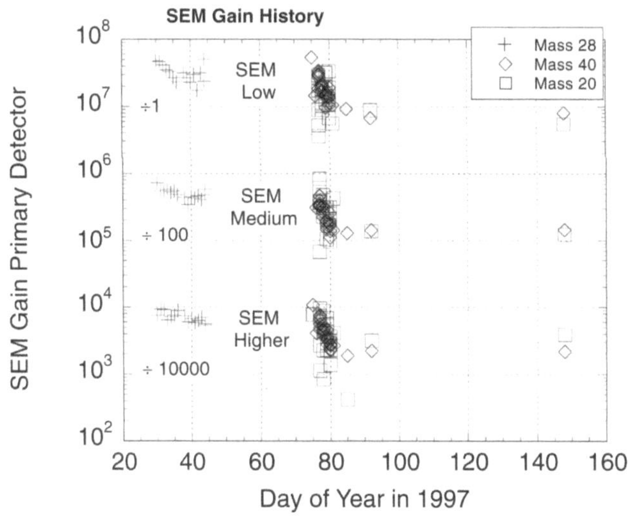

Figure 44. Multiplier gain for the three SEM high-voltage levels, as measured during characterization, after environmental testing, and on the spacecraft. Gain curves have been divided by the values shown immediately beneath the curves. During characterization, gain values were computed using mass 28 (N_2) data; afterwards, with the instrument sealed and the getter activated, the values were computed using mass 40 and mass 20 from Ar.

initially set to be −5 V, needed to be closer to −17 V based on the angular response in that mode when compared to the angular response in the ion mode. Continued laboratory testing with the EM in thermal gas mode, ion mode and neutral beaming mode will be required to provide better operational parameters for the FM as well as to interpret the data collected.

5. INMS Operations Network (ION)

5.1. ION Overview

ION is the ground support system (GSS) that provides complete downlink, uplink, and analysis capabilities for INMS. Most of ION can be accessed through any web browser using a standard point and click hyperlink-based interface. The architecture of ION consists of four primary components—a web server and authentication component, an application component, a database access component and a database component. The main functions of ION are divided into several different components, most of which are web-based applications accessed through the ION web interface, but a few of which are stand-alone applications. Additionally, ION is being developed in three phases that will support downlink, uplink and analysis functions in versions 1.0, 2.0 and 3.0, respectively.

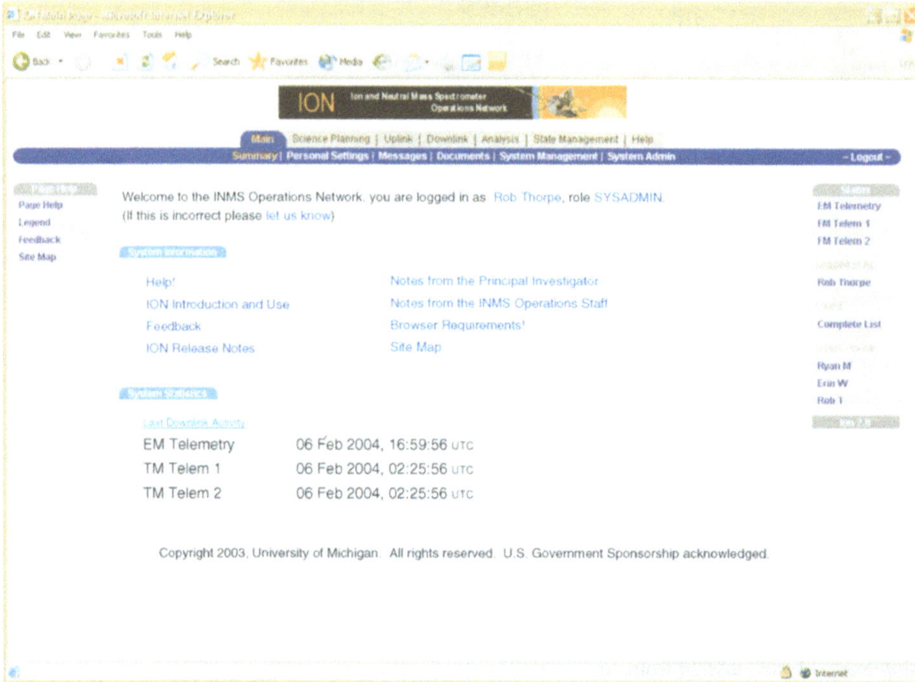

Figure 45. INMS Operations Network screen shot.

ION is accessed through a standard web browser. There is a simple hierarchical menu system for navigation that provides an easy way to access various functions of the system (see Figure 45). While web pages are carefully laid out to provide the right amount of information and color balancing, they are also kept closely aligned to basic HTML standards so many different types of browsers can access the system. This adherence to standards is an essential tenet of ION, supporting the goal of web-based platform-independent access to the INMS ground support system by virtually any web browser.

Architecturally, ION is divided into four components. The Web Server and Authentication Component authenticates a user to the system and provides most of the graphical user interface (GUI) for ION by delivering and processing ION web pages. The Application Components are integrated into the web part of ION and consist of Java Server Pages (JSPs), Java servlets and Java classes. A few applications are part of ION and are not Java-based, and interact with ION through files or direct database connections to the INMS database. The Database Access Component (commonly referred to as the Java Access Layer) consists of a series of Java classes used by all the application components and provides standardized access to the INMS database. Finally, the Database Component is a high-end relational database that stores all INMS GSS data. It is composed of a series of tables, sequences, indices and other data structures that form a performance-oriented and relational series of data sets for INMS operations and analysis.

Functionally, ION is divided into 14 components, which support a combination of uplink, downlink and analysis operations (see Figure 46). Most of these components are web-based and are part of the ION web interface. The Thermal Model, Table Software, and Spacecraft Simulator are stand-alone applications that were produced prior to the ION development effort, or by different organizations. Some of the Analysis Software are also stand-alone applications but may be incorporated into the web-based part of ION.

ION is being developed in three major releases. The first release is focused on downlink capabilities, and will support the storing of INMS housekeeping and science data in the INMS database, real-time and historical telemetry display and review, science operations analysis capabilities, system management capabilities, a help system, a data dictionary, and a messaging system. The second release is focused on uplink capabilities and will contain sequence generation, operations table management, Spacecraft Activity Sequence File (SASF) management, state management and memory management functions. The final release focuses on analysis capabilities, supporting the analysis of science data, health and safety analysis, online charting, data download capabilities for Co-Investigators, extended help capabilities, and other functions as necessary.

5.2. Uplink Operations

INMS orbital sequences are designed using the INMS Operations Table Software and then uploaded to ION. The Sequence Generation functions in ION are then used to create the SASF, which is a combination of INMS commands, INMS memory loads (including INMS Science Sequences) and other spacecraft commands and formats that allow INMS commands to be loaded into spacecraft memory. The Sequence Generation function in ION also accesses an INMS instrument command library to produce the commands appropriate to manage INMS and trigger science sequences. The result of the Sequence Generation function in ION is an SASF file that is stored in the Sequence Library in the INMS Database.

In the case of the INMS Flight Model, this SASF file is sent to the appropriate JPL Cassini office for integration with other SASF command files from other instruments, and ultimately uploaded to Cassini through the Deep Space Network. In the case of the INMS Engineering Model, the SASF file is transferred to the spacecraft simulator, which commands the EM through the RTIU/1553 interface.

The Sequence Library is also a resource for the ION Memory Comparison, State Prediction and State Comparison functions. Memory Comparison functions are used to validate INMS memory prior to proceeding with orbital sequences. The State Prediction and State Comparison functions are used to verify that planned orbital sequences will not violate INMS health and safety operational limits, and

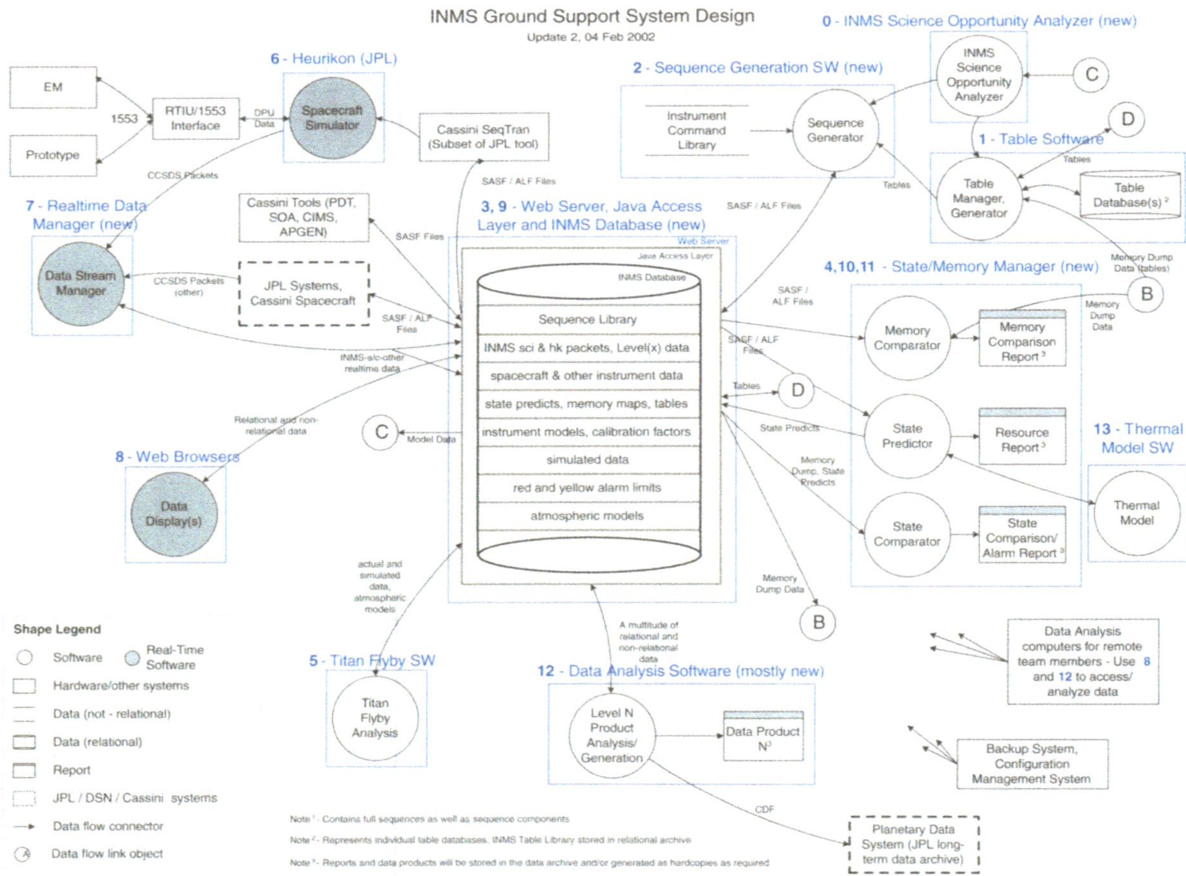

Figure 46. INMS Operations Network design.

to analyze the ongoing operational performance of the instrument. The Sequence Library is also a resource for Cassini modeling and planning tools such as the Cassini Information Management System (CIMS), the Science Opportunity Analyzer (SOA), the Pointing Design Tool (PDT), and others.

5.3. Downlink Operations

INMS science and housekeeping data will be downlinked from the Solid State Recorder (SSR) in the Orbiter's Command and Data Subsystem (CDS) to the Deep Space Network and then transmitted to the Jet Propulsion Laboratory, where they will be stored in the Telemetry Delivery System (TDS). The data is also transferred by the TDS via a Virtual Private Network (VPN) to the INMS Science and Operations Center (ISOC) at the University of Michigan. The data will be received at the ISOC by the Science Operations and Planning Computer (SOPC), a secure communications computer, and then forwarded to the ION data manager, which will store the raw data packets in the ION database. An automatic procedure in the database converts and scales the raw packets (level 0 data) into engineering data (level 1 data) that can be viewed in ION (real-time or historical) via any web browser. Analysis functions in ION will convert the engineering data to science-adjusted data.

In addition to INMS science and housekeeping data, ancillary data (e.g., spacecraft ephemeris, attitude, etc.) will also come into the ISOC via the SOPC and be stored in the ION database. Queries will be designed to cross-reference the spacecraft data with the engineering and analysis data. All appropriate data sets will undergo real-time analysis for health and safety purposes as well as science analysis.

Alternately, science and housekeeping data generated from the INMS engineering model, during calibration and operational testing activities, will be sent directly to the ION data manager and stored in the INMS database. This data will go through the same automatic conversion process and be available for review and analysis by all INMS Team Members using all the functions in ION available for flight model data.

5.4. Data Analysis and Archiving

Data products will be produced from the INMS database using tools in ION under the oversight of the Science Team Leader and Team Members. Some data products will be dynamically available using database queries and others will be stored in the INMS database. Any range and type of data sets will also be available for download by INMS Team Members and other appropriate personnel via the Internet. Level 1 data products will be delivered to the Planetary Data System (PDS) for archiving. The INMS Team Leader will be responsible for validating the science content of the Level 1 products before release to the PDS. Higher-level data products will be developed as specified by the Science Team Members. The appropriate analysis and display tools will be created in ION (and, as appropriate, as remote applications for Co-Investigators) by the Data Analysis Team in consultation with the Science Team. Higher-level products will be stored in the INMS database and, as appropriate, will be archived in the PDS as well.

Exploitation of the data from the INMS investigation will involve both data analysis and modeling studies of the neutral and ionized atmospheres of Titan and of the neutral and plasma environments of the rings and icy satellites.

Analysis of data acquired during the Titan passes will include the computation of neutral and ion densities as a function of altitude and the subsequent computation of the scale height and temperatures. Direct comparison of the INMS thermal ion measurements with CAPS ion and RPWS electron measurements as well as iterative use of the three data sets in the analysis process will be a standard part of the data reduction process. The flyby data sets will also be combined with UVIS measurements to provide information on latitudinal, local time, and possible temporal changes in Titan atmospheric parameters. Of particular importance to the mission's investigation of Titan's atmosphere will be the correlative analysis of INMS high-altitude data and low-altitude Probe data. Analysis of data acquired in the inner magnetosphere—during the ring overflight and ring plane crossing following Saturn Orbit Insertion and during targeted encounters with the inner moons Enceladus, Dione, and Rhea—will yield species densities as a function of Orbiter position. Iterative reduction with CAPS data will be required to determine ion densities and flows.

Extensive use will be made of models in the INMS investigation, both in operations planning and in the interpretation of INMS data. Models of Titan's upper atmosphere will, for example, provide a global context for data acquired at particular local times or locations. In turn, data acquired during the tour—e.g., ion and neutral altitude profiles at Titan at different local times, latitudes, and ram angles—can be used to refine existing models and to develop three-dimensional models of the Titan upper atmosphere. Such three-dimensional models can then be incorporated in global MHD modeling studies of Titan's interaction with the external plasma flow (such as will be conducted as part of the Cassini interdisciplinary science investigation of plasma environment of Saturn's magnetosphere). Models to be used during the INMS investigation will be based on models already developed (or under development) by the INMS Science Team Members and others. These models are listed in Table X.

INMS data analysis will be coordinated with the analysis of complementary data from other Cassini experiments to ensure the maximum scientific yield from the investigation (cf. Sections 1.1.4 and 1.2.3). Data sets from other Cassini experiments relevant to the INMS investigation include: (1) reduced RSS radio occultation profiles of Titan and Saturn; (2) digitized cut-off frequencies of the Saturn Electrostatic Discharges from the RPWS investigation; (3) ion and electron distribution function data from CAPS in the ring ionosphere, from the Titan interaction region, from

TABLE X
Models available for use in the INMS investigation.

Model	Reference
Titan neutral atmosphere	Yung *et al.* (1984), Yung (1987), Toublanc *et al.* (1995), Lara *et al.* (1996)
Titan thermal structure	Yelle (1990)
Titan ionosphere	Fox and Yelle (1997), Cravens *et al.* (1997), Keller *et al.* (1998)
Titan ionospheric dynamics	Cravens *et al.* (1998), Ledvina and Cravens (1998)
Titan TGCM	Müller-Wodarg *et al.* (2001)
Ring atmosphere/ionosphere	Wilson and Waite (1989)*, Ip (1984; 1995)
Icy satellite tori/neutral cloud	To be produced by R.E. Johnson of the CAPS team (cf. Johnson *et al.*, 1989)/(Ip, 1997)
Saturn atmosphere/ionosphere	Waite (1981)*
Saturn TGCM	To be adapted from Jupiter TGCM being developed under Planetary Atmospheres NRA

*To be updated.

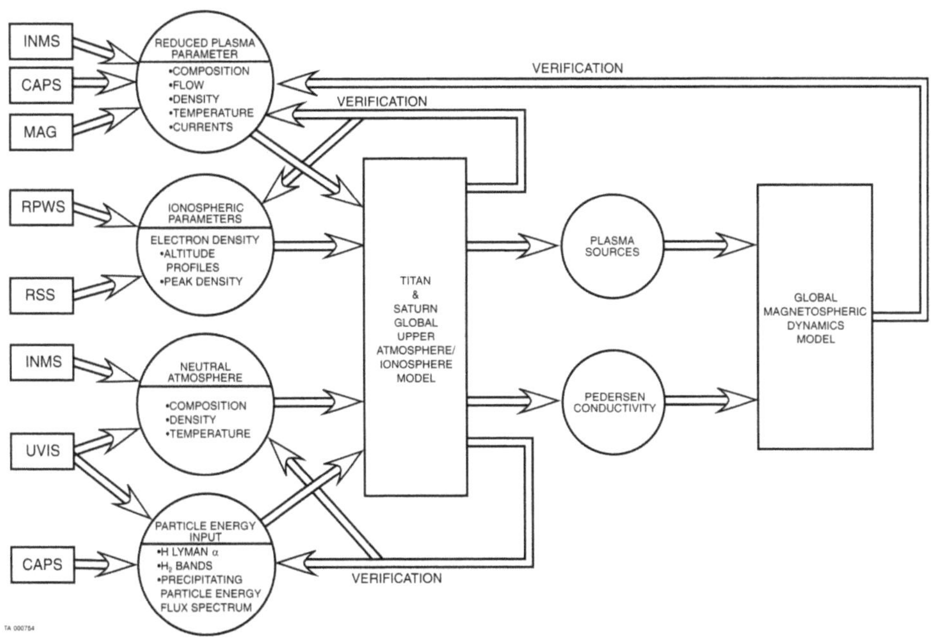

Figure 47. As illustrated in this schematic, INMS data will be integrated with the results of other Cassini experiments to address the science objectives of the Magnetosphere and Plasma Science (MAPS) working group. The working group was formed to coordinate the Cassini investigations relating to magnetospheric and plasma physical processes in the Saturn system, including two inter-disciplinary theoretical investigations (Plasma Circulation and Magnetosphere–Ionosphere Coupling and Plasma Environment in Saturn's Magnetosphere). Modeling studies will play a crucial role in the interpretation of INMS data and in achieving the MAPS science goals.

icy satellite/magnetosphere interaction regions, and in Saturn's magnetosphere; (4) reduced UVIS data on the composition and energetics of the Titan and Saturn upper atmospheres; (5) VIMS data on stratospheric structure and composition; (6) Probe data on neutral composition, aerosol formation, atmospheric dynamics, and radiative transfer in Titan's lower atmosphere; (7) MIMI images of neutral tori; and (8) MAG data. The data from other Cassini experiments will be accessed via the SOPC and will be used as inputs to models and INMS reduction routines with the agreement and collaboration of the responsible Principal Investigators.

The schematic in Figure 47 illustrates how INMS data will be used together with data from other experiments and theoretical models to address key mission science objectives—in this case, those of the Magnetosphere and Plasma Science (MAPS) Working Group.

Acknowledgements

The INMS investigation is a collaborative effort involving many talented and dedicated individuals. The authors of this paper would like to express their deep appreciation to the following people in particular for their contributions to the project: David Anderson, Dana Burket, Butch Carruth, Kristie Frick, Morgan Orcyere, Eddie Weigle, Charles Zinsmeyer (Southwest Research Institute); Robert Abell, Robert Arvey, Steven Cagiano, Dan Harpold (NASA Goddard Space Flight Center); Bette Bilodeau and Florence Tan (IDEA, Inc.); Michael Paulkovich (Jackson & Tull); Mirl Bendt, Christina Carlson, Paul Cursey, Sharad Dixit, Edward Patrick, Eric Raaen, John Westberg (NYMA); Mark McQuaid, Doug Hawk, Rick Bitzell, Anthony Melak, Michael Barciniak (Swales & Associates); Kenneth Arnett, Steven Battel, Bruce Block, John Maurer, Ryan Miller, John Parejko, Steve Sheppard, Erin Walter, Julia Xu (University of Michigan).

References

Anicich, V. G., and McEwan, M. J.: 1997, 'Ion–molecule chemistry in Titan's ionosphere', *Planet. Space. Sci.* **45**, 897.

Banaszkiewicz, M., Lara, L. M., Rodrigo, R., López-Moreno, J. J., and Molina-Cuberos, G. J.: 2000, 'A coupled model of Titan's atmosphere and ionosphere', *Icarus* **147**, 386.

Barbosa, D. D.: 1987, 'Titan's atomic nitrogen torus: Inferred properties and consequences for the Saturnian aurora', *Icarus* **72**, 53.

Barbosa, D. D.: 1990, 'Radial diffusion in Saturn's magnetosphere', *J. Geophys. Res.* **95**, 17167.

Bird, M. K., *et al.*: 1997, 'Detection of Titan's ionosphere from Voyager 1 radio occultation observations,' *Icarus* **130**, 426.

Brecht, S. H., Luhmann, J. G., and Larson, D. J.: 2000, 'Simulation of the Saturnian magnetospheric interaction with Titan', *J. Geophys. Res.* **105**, 13119.

Bridge, H. A., *et al.*: 1981, 'Plasma observations near Saturn: Initial results from Voyager 1', *Science* **212**, 217.

Bridge, H. A., *et al.*: 1982, 'Plasma observations near Saturn: Initial results from Voyager 2', *Science* **215**, 563.

Broadfoot, A. L., *et al.*: 1981, 'Extreme ultraviolet observations from Voyager 1 encounter with Saturn', *Science* **212**, 206.

Carlson, R. W.: 1980, 'Photo-sputtering of ice and hydrogen around Saturn's rings', *Nature* **283**, 461.

Capone, L. A., Dubach, J., Prasad, S. S., and Whitten, R. C.: 1983, 'Galactic cosmic rays and N_2 dissociation on Titan', *Icarus* **55**, 73.

Chappell, C. R., Moore, T. E., and Waite, J. H., Jr.: 1987, 'The ionosphere as a fully adequate source of plasma for the Earth's magnetosphere', *J. Geophys. Res.* **92**.

Chassefière, E., and Cabane, M.: 1995, 'Two formation regions for Titan's hazes: Indirect clues and possible synthesis mechanisms', *Planet. Space Sci.* **43**, 91.

Clarke, J. T., Trauger, J., and Waite, J. H., Jr.: 1989, 'Doppler-shifted H Ly α-emission from Jupiter's aurora', *Geophys. Res. Lett.* **16**, 587.

Comas Solá, J.: 1908, 'Observations des satellites principeaux de Jupiter et de Titan', *Astron. Nach.* **179**, 289.

Connerney, J. E. P. and Waite, J. H., Jr.: 1984, 'New model of Saturn's ionosphere with an influx of water from the rings', *Nature* **312**, 136.

Courtin, R., Gautier, D., and McKay, C. P.: 1995, 'Titan's thermal emission spectrum: Reanalysis of the Voyager infrared measurements', *Icarus* **114**, 144.

Coustenis, A., *et al.*: 1989, 'Titan's atmosphere from Voyager infrared observations', *Icarus* **80**, 54.

Coustenis, A., *et al.*: 1991, 'Titan's atmosphere from Voyager infrared observations. III. Vertical distributions of hydrocarbons and nitriles near Titan's north pole', *Icarus* **89**, 152.

Coustenis, A., *et al.*: 1998, 'Evidence for water vapor in Titan's atmosphere from ISO/SWS data', *Astron. Astrophys.* **336**, L85.

Coustenis, A., *et al.*: 2003, 'Titan's atmosphere from ISO mid-infrared spectroscopy', *Icarus* **161**, 383.

Cravens, T. E., Keller, C. N., and Ray, B.: 1997, 'Photochemical sources of non-thermal neutrals for the exosphere of Titan', *Planet. Space Sci.* **45**, 889.

Cravens, T. E., Lindgren, C. J., and Ledvina, S. A.: 1998, 'A two-dimensional MHD model of Titan's plasma environment', *Planet. Space Sci.* **46**, 1193.

Cravens, T. E., Vann, J., Clark, J., Yu, J., Keller, C. N., and Brull, C.: 2004, 'The ionosphere of Titan: An updated theoretical model, *Adv. Space Res.* **33**, 212.

Edgington, S. G., *et al.*: 1998, 'On the latitude variation of ammonia, acetylene, and phosphine altitude profiles on Jupiter from HST Faint Object Spectrograph observations', *Icarus* **133**, 192–209.

Eviatar, A. and Podolak, M.: 1983, 'Titan's gas and plasma torus', *J. Geophys. Res.* **88**, 833.

Eviatar, A. and Richardson, J. D.: 1990, 'Water group plasma in the magnetosphere of Saturn', *Ann. Geophys.* **8**, 725.

Eviatar, A. and Richardson, J. D.: 1992, 'Thermal plasma in the inner kronian magnetosphere', *Ann. Geophys.* **10**, 511.

Fox, J. L. and Yelle, R. V.: 1997, 'Hydrocarbon ions in the ionosphere of Titan', *Geophys. Res. Lett.* **24**, 2179.

Frank, L. A., *et al.*: 1980, 'Plasmas in Saturn's magnetosphere', *J. Geophys. Res.* **85**, 5695.

Friedson, A. J. and Yung, Y. L.: 1984, 'The thermosphere of Titan', *J. Geophys. Res.* **89**, 85.

Galand, M., *et al.*: 1999, 'The ionosphere of Titan: Ideal diurnal and nocturnal cases', *Icarus* **140**, 92.

Gan, L., Cravens, T. E., and Keller, C. N.: 1992, 'Electrons in the ionosphere of Titan', *J. Geophys. Res.* **97**, 12137.

Gan-Baruch, Z., *et al.*: 1994, 'Plasma observations in the ring plane of saturn', *J. Geophys. Res.* **99**, 11063.
Gautier, D. and Raulin, F.: 1997, 'Chemical composition of Titan's atmosphere', in *Huygens: Science, Payload, and Mission*, ESA Publication SP-1177, European Space Agency, Noordwijk, The Netherlands.
Gurnett, D. A., Kurth, W. S., and Scarf, F. L.: 1981, 'Plasma waves near Saturn: Initial results from Voyager 1', *Science* **212**, 235.
Gurnett, D. A., Scarf, F. L., and Kurth, W. S.: 1982, 'The structure of Titan's wake from plasma wave observation', *J. Geophys. Res.* **87**, 1395.
Hall, D. T., *et al.*: 1996, 'Fluorescent hydroxyl emissions from Saturn's ring atmosphere', *Science* **272**, 516.
Hamilton, D. C. and Burns, J. A.: 1993, 'OH in Saturn's rings', *Nature* **365**, 550.
Hamilton, D. C., *et al.*: 1981, 'Composition of nonthermal ions in the Jovian magnetosphere', *J. Geophys. Res.* **86**, 8301.
Hamilton, D. C., *et al.*: 1983, 'Energetic atomic and molecular ions in Saturn's magnetosphere', *J. Geophys. Res.* **88**, 8905.
Hanel, R., *et al.*: 1981, 'Infrared observations of the Saturnian system from Voyager 1', *Science* **212**, 192.
Hartle, R. E., *et al.*: 1982, 'Titan's ion exosphere observed from Voyager 1', *J. Geophys. Res.* **87**, 1383.
Hidayat, T., *et al.*: 1997, 'Millimeter and submillimeter heterodyne observations of Titan: Retrieval of the vertical profile of HCN and the $^{12}C/^{13}C$ ratio', *Icarus* **126**, 170.
Hidayat, T., *et al.*: 1998, 'Millimeter and submillimeter heterodyne observations of Titan: The vertical profile of carbon monoxide in its stratosphere', *Icarus* **133**, 109.
Hilton, D. A. and Hunten, D. M.: 1988, 'A partially collisional model of the Titan hydrogen torus', *Icarus* **73**, 248.
Hunten, D. M.: 1972, 'The atmosphere of Titan', *Comments Astrophys. Space Phys.* **4**, 149.
Hunten, D. M., *et al.*: 1984, 'Titan', in T. Gehrels and M. S. Matthews (eds.), *Saturn*, University of Arizona Press, Tucson, AZ, pp. 671–759.
Ip, W.-H.: 1984, 'The ring atmosphere of Saturn: Monte Carlo simulation of ring source models', *J. Geophys. Res.* **89**, 8843.
Ip, W.-H.: 1990, 'Titan's upper ionosphere', *Astrophys. J.* **362**, 354.
Ip, W.-H.: 1995, 'The exospheric systems of Saturn's rings', *Icarus* **115**, 295.
Ip, W.-H.: 1997, 'On the neutral cloud distribution in the saturnian magnetosphere', *Icarus* **126**, 42.
Johnson, R. E.: 1998, 'Sputtering and desorption from icy surfaces', in B. Schmitt *et al.* (eds.), *Solar System Ices*, Kluwer Academic Publishers, The Netherlands, pp. 303–334.
Johnson, R. E. and Sittler, E. C.: 1990, 'Sputter-produced plasma as a measure of satellite surface composition—The Cassini mission', *Geophys. Res. Lett.* **17**, 1729.
Johnson, R. E., *et al.*: 1989, 'The neutral cloud and heavy ion inner torus at Saturn', *Icarus* **77**, 311.
Judge, D. L., Wu, F.-M., and Carlson, R. W.: 1980, 'Ultraviolet photometer observations of the Saturnian system', *Science* **207**, 431.
Jurac, S., Johnson, R. E., and Richardson, J. D.: 2001a, 'Saturn's E ring and the production of the neutral torus', *Icarus* **149**, 384.
Jurac, S., Johnson, R. E., Richardson, J. D., and Paranicas, C.: 2001b, 'Satellite sputtering in Saturn's magnetosphere', *Planet. Space Sci.* **49**, 319.
Kabin, K., *et al.*: 1999, 'Interaction of the Saturnian magnetosphere with Titan: Results of a three-dimensional MHD simulation', *J. Geophys. Res.* **104**, 2451.
Kasprzak, W. T., Niemann, H. B., and Mahaffy, P.: 1987, 'Observations of energetic ions on the nightside of Venus', *J. Geophys. Res.* **92**, 291.

Kasprzak, W. T., *et al.*: 1996, 'Cassini orbiter ion and neutral mass spectrometer instrument', *SPIE Proc.* **2803**, 129.
Keller, C. N. and Cravens, T. E.: 1994, 'One-dimensional multispecies hydrodynamic models of the wakeside ionosphere of Titan', *J. Geophys. Res.* **99**, 6527.
Keller, C. N., Cravens, T. E., and Gan, L.: 1992, 'A model of the ionosphere of Titan', *J. Geophys. Res.* **97**, 12117.
Keller, C. N., Cravens, T. E., and Gan, L.: 1994, 'One-dimensional multispecies magnetohydrodynamic models of the ramside ionosphere of Titan', *J. Geophys. Res.* **99**, 6511.
Keller, C. M., Anicich, V. G., and Cravens, T. E.: 1998, 'Model of Titan's ionosphere with detailed hydrocarbon chemistry', *Planet. Space Sci.* **46**. 1157.
Khurana, K. K., Kivelson, M. G., and Russell, C. T.: 1997, 'Interaction of Io with its torus: Does Io have an internal magnetic field?', *Geophys. Res. Lett.* **34**, 2391.
Kiser, R. W.: 1965, *Introduction to Mass Spectrometry and Its Applications*, Prentice-Hall, Inc., Englewood Cliffs, NJ.
Kivelson, M. G. and Russell, C. T.: 1983, 'The interaction of flowing plasmas with planetary ionospheres: A Titan–Venus comparison', *J. Geophys. Res.* **88**, 49.
Kivelson, M. G., *et al.*: 1996, 'Discovery of Ganymede's magnetic field by the Galileo spacecraft', *Nature* **384**, 537.
Kostiuk, T., *et al.*: 1997, 'Ethane abundance on Titan', *Planet. Space Sci.* **45**, 931.
Kuiper, G. P.: 1944, 'Titan: A satellite with an atmosphere', *Astrophys. J.* **100**, 378.
Kunde, V. G., *et al.*: 1981, 'C_4H_2, HC_3N, and C_2N_2 in Titan's atmosphere', *Nature* **292**, 686.
Lammer, H. and Bauer, S. J.: 1991, 'Nonthermal atmospheric escape from Mars and Titan', *J. Geophys. Res.* **96**, 1819.
Lammer, H. and Bauer, S. J.: 1993, 'Atmospheric mass loss from Titan by sputtering', *Planet. Space Sci.* **41**, 657.
Lanzerotti, L. J., *et al.*: 1983, 'Implications of Voyager data for energetic ion erosion of the icy satellites of Saturn', *J. Geophys. Res.* **88**, 8765.
Lara, L. M., *et al.*: 1996, 'Vertical distribution of Titan's atmospheric neutral constituents', *J. Geophys. Res.* **101**, 23261.
Lazarus, A. J. and McNutt, R. L., Jr.: 1983, 'Low energy plasma ion observations in saturn's magnetosphere', *J. Geophys. Res.* **88**, 8831.
Ledvina, S. A. and Cravens, T. E.: 1998, 'A three-dimensional MHD model of plasma flow around Titan', *Planetary and Space Sci.* **46**, 1175.
Lellouch, E., *et al.*: 1989, 'Titan's atmosphere and hypothesized ocean: A reanalysis of the Voyager 1 radio-occultation and IRIS 7.7-μm data', *Icarus* **79**, 328.
Lellouch, E., *et al.*: 1990, 'Titan's thermosphere profile', *Icarus* **83**, 308.
Lewis, J. S.: 1971, 'Satellites of the outer planets: Their physical and chemical nature', *Icarus* **15**, 174.
Lindal, G. F., *et al.*: 1983, 'The atmosphere of Titan: An analysis of the Voyager 1 radio occultation measurements', *Icarus* **53**, 348.
Luhmann, J. G.: 1996, 'Titan's ion exosphere wake: A natural ion mass spectrometer?', *J. Geophys. Res.* **101**, 29387.
Luhmann, J. G. and Walker, R. J.: 1981, 'Model exospheres of the ringed planets', *Geophys. Res. Lett.* **8**, 107.
Luhmann, J. G., *et al.*: 1991, 'A comparison of induced magnetotails of planetary bodies: Venus, Mars, and Titan', *J. Geophys. Res.* **96**, 11199.
Lunine, J. I.: 1993, 'Does Titan have an ocean? A review of current understanding of Titan's surface', *Rev. Geophys.* **31**, 133.
Lunine, J. I.: 1994, 'Does Titan have oceans?', *Am. Scientist* **82**, 136.
Lunine, J. I., Stevenson, D. J., and Yung, Y. L.: 1983, 'Ethane ocean on Titan', *Science* **222**, 1229.

Lutz, B. L., deBergh, C., and Owen, T.: 1983, 'Titan: The discovery of carbon monoxide in its atmosphere', *Science* **220**, 1374.

Maguire *et al.*: 1981, 'C_3H_8 and C_3H_4 in Titan's atmosphere', *Nature* **292**, 683.

Mahaffy, P. M. and Lai, K.: 1990, 'An electrostatic quadrupole deflector for mass spectrometer applications', *J. Vac. Sci.* **A8**, 3244.

Matheson, P. L. and Shemansky, D. E.: 1996, 'Magnetospheric neutral clouds from Saturn's icy satellites' (abstract), *Bull. Am. Astron. Soc.*, Division of Planetary Sciences meeting.

McNutt, R. L., Jr. and Richardson, J. D.: 1988, 'Constraints on Titan's ionsphere', *Geophys. Res. Lett.* **15**, 709.

Morfill, G. E., *et al.*: 1983, 'Some consequences of meteoroid impacts on Saturn's rings', *Icarus* **55**, 439.

Müller-Wodarg, I. C. F., and Yelle, R. V.: 2002, 'The effect of dynamics on the composition of Titan's upper atmosphere', *Geophys. Res. Lett.* **29**, 54-1, doi 10.1029/2002GL016100.

Müller-Wodarg, I. C. F., Yelle, R. V., Mendillo, M., Young, L. A., and Aylward, A. D.: 2000, 'The thermosphere of Titan simulated by a global three-dimensional time-dependent model', *J. Geophys. Res.* **105**, 20833.

Nagy, A. F. and Cravens, T. E.: 1998, 'Titan's ionosphere: A review', *Planet. Space Sci.* **46**, 1149.

Nagy, A. F., Barakat, A. R., and Schunk, R. W.: 1986, 'Is Jupiter's ionosphere a significant plasma source for its magnetosphere?', *J. Geophys. Res.* **91**, 351.

Ness, N. F., Acuna, M. H., Behannon, K. W., and Neubauer, F. M.: 1982, 'The induced magnetosphere of Titan', *J. Geophys. Res.* **87**, 1369.

Neubauer, F. M., *et al.*: 1984, 'Titan's magnetospheric interaction', in T. Gehrels and M. S. Matthews (eds.), *Saturn*, University of Arizona Press, Tucson, AZ, pp. 760–787.

Niemann, H. B., *et al.*: 1997, 'The gas chromatograph mass spectrometer aboard Huygens', in *Huygens: Science, Payload and Mission* ESA SP 1177, p. 85.

Owen, T.: 1982, 'The composition and origin of Titan's atmosphere', *Planet. Space Sci.* **30**, 833.

Pospieszalska, M. K. and Johnson, R. E.: 1989, 'Magnetospheric ion bombardment profiles of satellites—Europa and Dione', *Icarus* **78**, 1.

Pospieszalska, M. K. and Johnson, R. E.: 1991, 'Micrometeorite erosion of the main rings as a source of plasma in the inner Saturnian plasma torus', *Icarus* **93**, 45.

Richardson, J. D.: 1986, 'Thermal ions at Saturn: Plasma parameters and implications', *J. Geophys. Res.* **91**, 1381.

Richardson, J. D.: 1998, 'Thermal plasma and neutral gas in Saturn's magnetosphere', *Rev. Geophys.* **36**, 501.

Richardson, J. D. and Sittler, E. C., Jr.: 1990, 'A plasma density model for Saturn based on Voyager observations', *J. Geophys. Res.* **95**, 12019.

Richardson, J. D., Eviatar, A., and Siscoe, G. L.: 1986, 'Satellite tori at Saturn', *J. Geophys. Res.* **91**, 8749.

Richardson, J. D., *et al.*: 1998, 'OH in Saturn's magnetosphere: Observations and implications', *J. Geophys. Res.* **103**, 20245.

Rishbeth, H., Yelle, R. V., and Mendillo, M.: 2000, 'Dynamics of Titan's thermosphere', *Planet. Space Sci.* **48**, 51.

Roboz, A. and Nagy, A. F.: 1994, 'The energetics of Titan's ionosphere', *J. Geophys. Res.* **99**, 2087.

Samuelson, R. E., *et al.*: 1981, 'Mean molecular weight and hydrogen abundance in Titan's atmosphere', *Nature* **292**, 688.

Samuelson, R. E., *et al.*: 1983, 'CO_2 on Titan', *J. Geophys. Res.* **88**, 8709.

Samuelson, R. E., Nath, N. R., and Borysow, A.: 1997, 'Gaseous abundances and methane supersaturation in Titan's atmosphere', *Planet. Space. Sci.* **45**, 959.

Sandel, B. R., *et al.*: 1982, 'Extreme ultraviolet observations from the Voyager 2 encounter with Saturn', *Science* **215**, 548.
Schardt, A. W., *et al.*: 1984, 'The outer magnetosphere', in T. Gehrels and M. S. Matthews (eds.), *Saturn*, University of Arizona Press, Tucson, AZ, pp. 416–459.
Shemansky, D. E. and Hall, D. T.: 1992, 'The distribution of atomic hydrogen in the magnetosphere of Saturn', *J. Geophys. Res.* **97**, 4143.
Shemansky, D. E., *et al.*: 1993, 'Detection of the hydroxyl radical in the Saturn magnetosphere', *Nature* **363**, 329.
Shi, M., *et al.*: 1995, 'Sputtering of water ice surfaces and the production of extended neutral atmospheres', *J. Geophys. Res.* **100**, 26387.
Smith, G. R., *et al.*: 1982, 'Titan's upper atmosphere: Composition and temperature from the EUV solar occultation results', *J. Geophys. Res.* **87**, 1351.
Smyth, W. H. and Marconi, M. L.: 1993, 'The nature of the hydrogen tori of Titan and Triton', *Icarus* **101**, 18.
Strobel, D. F. and Shemansky, D. E.: 1982, 'EUV emission from Titan's upper atmosphere: Voyager 1 encounter', *J. Geophys. Res.* **87**, 1361.
Strobel, D. F., Meier, R. R., Summers, M. E., and Strickland, D. J.: 1991, 'Nitrogen airglow sources: Comparison of Triton, Titan, and Earth', *Geophys. Res. Lett.* **18**, 689.
Strobel, D. F., Summers, M., and Zhu, X.: 1992, 'Titan's upper atmosphere: Structure and ultraviolet emissions', *Icarus* **100**, 512.
Strobel, D. F., *et al.*: 1993, 'Upper limit on Titan's atmospheric argon abundance', *Icarus* **103**, 333.
Swaminathan, V., Alig, R., Murray, W., and Sarnoff, D.: 1996, 'Design of an improved miniature ion neutral mass spectrometer for NASA applications', NASA Contract NAS5-32823.
Taylor, F. W. and Coustenis, A.: 1998, 'Titan in the solar system', *Planet. Space Sci.* **46**, 1085.
Thompson, W. R., McDonald, G. D., and Sagan, C.: 1994, 'The Titan haze revisited: Magnetospheric energy sources and quantitative tholin yields', *Icarus* **112**, 376.
Toublanc, D.: 1995, 'Photochemical modeling of Titan's atmosphere', *Icarus* **113**, 2.
Trafton, L. M.: 1972, 'On the possible detection of H_2 in Titan's atmosphere', *Astrophys. J.* **175**, 285.
Vervack, R. J., Jr.: 1997, 'Titan's upper atmospheric structure derived from Voyager ultraviolet spectrometer observations', Ph.D. dissertation, The University of Arizona, Tucson, AZ.
Vervack, R. J., Jr., Sandel, B. R., and Strobel, D. F.: 2004, 'New perspectives on Titan's upper atmosphere from a reanalysis of the Voyager 1 uvs solar occultations by Titan', *Icarus*, **170**, 91.
Waite, J. H., Jr.: 1981, 'The ionosphere of Saturn', Ph.D. dissertation, University of Michigan, Ann Arbor, MI.
Weiser, H., Vitz, C., and Moos, H. W.: 1977, 'Detection of Lyman alpha emission from the Saturnian disk and from the ring system', *Science* **197**, 755.
Wilson, E. H.: 2002, 'Investigations into the photochemistry of the current and primordial atmosphere of Titan', Ph.D. Thesis, University of Michigan, Ann Arbor, MI.
Wilson, G. R. and Waite, J. H., Jr.: 1989, 'Kinetic modeling of the Saturn ring-ionosphere plasma environment', *J. Geophys. Res.* **94**, 17287.
Wolf, D. A. and Neubauer, F. M.: 1982, 'Titan's highly variable plasma environment', *J. Geophys. Res.* **87**, 881.
Yelle, R. V.: 1991, 'Non-LTE models of Titan's upper atmosphere', *Astrophys. J* **383**, 380.
Yelle, R. V., *et al.*: 1997, 'Engineering models for Titan's atmosphere', in *Huygens: Science, Payload, and Mission*, ESA Publication SP-1177, European Space Agency, Noordwijk, The Netherlands.
Yung, Y. L.: 1987, 'An update of nitrile photochemistry on Titan', *Icarus* **72**, 468.
Yung, Y. L., Allen, M., and Pinto, J. P.: 1984, 'Photochemistry of the atmosphere of Titan: Comparison between model and observations', *Astrophys. J. Suppl.* **55**, 465.

MAGNETOSPHERE IMAGING INSTRUMENT (MIMI) ON THE CASSINI MISSION TO SATURN/TITAN

S. M. KRIMIGIS[1,*], D. G. MITCHELL[1], D. C. HAMILTON[2], S. LIVI[1], J. DANDOURAS[4], S. JASKULEK[1], T. P. ARMSTRONG[5], J. D. BOLDT[1], A. F. CHENG[1], G. GLOECKLER[2], J. R. HAYES[1], K. C. HSIEH[6], W.-H. IP[3], E. P. KEATH[1], E. KIRSCH[3], N. KRUPP[3], L. J. LANZEROTTI[7], R. LUNDGREN[2], B. H. MAUK[1], R. W. McENTIRE[1], E. C. ROELOF[1], C. E. SCHLEMM[1], B. E. TOSSMAN[1], B. WILKEN[3] and D. J. WILLIAMS[1]

[1]*Applied Physics Laboratory, The Johns Hopkins University, 11100 Johns Hopkins Road, Laurel, MD 20723-6099, U.S.A.*

[2]*Department of Physics, University of Maryland, College Park, MD 20742, U.S.A.*

[3]*Max-Planck Institut für Aeronomie, D-37191, Katlenburg-Lindau, Germany*

[4]*Centre d'Etude Spatiale des Rayonnements, CNRS/Université Paul Sabatier, F-31028 Toulouse Cedex 4, France*

[5]*Department of Physics and Astronomy, University of Kansas, Lawrence, Kansas 66044*

[6]*Department of Physics, University of Arizona, Tucson, AZ 85721, U.S.A.*

[7]*Bell Laboratories, Lucent Technologies, 700 Mountain Avenue, Murry Hill, NJ 07974, U.S.A.*

*(*Author for correspondence: E-mail: Tom.Krimigis@jhuapl.edu)*

(Received 16 September 1998; Accepted in final form 7 November 2002)

Abstract. The magnetospheric imaging instrument (MIMI) is a neutral and charged particle detection system on the Cassini orbiter spacecraft designed to perform both global imaging and in-situ measurements to study the overall configuration and dynamics of Saturn's magnetosphere and its interactions with the solar wind, Saturn's atmosphere, Titan, and the icy satellites. The processes responsible for Saturn's aurora will be investigated; a search will be performed for substorms at Saturn; and the origins of magnetospheric hot plasmas will be determined. Further, the Jovian magnetosphere and Io torus will be imaged during Jupiter flyby. The investigative approach is twofold. (1) Perform remote sensing of the magnetospheric energetic ($E > 7$ keV) ion plasmas by detecting and imaging charge-exchange neutrals, created when magnetospheric ions capture electrons from ambient neutral gas. Such escaping neutrals were detected by the Voyager l spacecraft outside Saturn's magnetosphere and can be used like photons to form images of the emitting regions, as has been demonstrated at Earth. (2) Determine through in-situ measurements the 3-D particle distribution functions including ion composition and charge states ($E > 3$ keV/e). The combination of in-situ measurements with global images, together with analysis and interpretation techniques that include direct "forward modeling" and deconvolution by tomography, is expected to yield a global assessment of magnetospheric structure and dynamics, including (a) magnetospheric ring currents and hot plasma populations, (b) magnetic field distortions, (c) electric field configuration, (d) particle injection boundaries associated with magnetic storms and substorms, and (e) the connection of the magnetosphere to ionospheric altitudes. Titan and its torus will stand out in energetic neutral images throughout the Cassini orbit, and thus serve as a continuous remote probe of ion flux variations near $20R_S$ (e.g., magnetopause crossings and substorm plasma injections). The Titan exosphere and its cometary interaction with magnetospheric plasmas will be imaged in detail on each flyby. The three principal sensors of MIMI consists of an ion and neutral camera (INCA), a charge–energy–mass-spectrometer (CHEMS) essentially identical to our instrument flown on the ISTP/Geotail spacecraft, and the low energy magnetospheric measurements system (LEMMS), an advanced design of one of our sensors flown on the Galileo spacecraft. The INCA head is a large geometry factor ($G \sim 2.4$ cm^2 sr) foil time-of-flight (TOF)

Space Science Reviews **114:** 233–329, 2004.

camera that separately registers the incident direction of either energetic neutral atoms (ENA) or ion species (≥5° full width half maximum) over the range 7 keV/nuc < E < 3 MeV/nuc. CHEMS uses electrostatic deflection, TOF, and energy measurement to determine ion energy, charge state, mass, and 3-D anisotropy in the range $3 \leq E \leq 220$ keV/e with good (~0.05 cm^2 sr) sensitivity. LEMMS is a two-ended telescope that measures ions in the range $0.03 \leq E \leq 18$ MeV and electrons $0.015 \leq E \leq 0.884$ MeV in the forward direction ($G \sim 0.02$ cm^2 sr), while high energy electrons (0.1–5 MeV) and ions (1.6–160 MeV) are measured from the back direction ($G \sim 0.4$ cm^2 sr). The latter are relevant to inner magnetosphere studies of diffusion processes and satellite microsignatures as well as cosmic ray albedo neutron decay (CRAND). Our analyses of Voyager energetic neutral particle and Lyman-α measurements show that INCA will provide statistically significant global magnetospheric images from a distance of ~60 R_S every 2–3 h (every ~10 min from ~20 R_S). Moreover, during Titan flybys, INCA will provide images of the interaction of the Titan exosphere with the Saturn magnetosphere every 1.5 min. Time resolution for charged particle measurements can be <0.1 s, which is more than adequate for microsignature studies. Data obtained during Venus-2 flyby and Earth swingby in June and August 1999, respectively, and Jupiter flyby in December 2000 to January 2001 show that the instrument is performing well, has made important and heretofore unobtainable measurements in interplanetary space at Jupiter, and will likely obtain high-quality data throughout each orbit of the Cassini mission at Saturn. Sample data from each of the three sensors during the August 18 Earth swingby are shown, including the first ENA image of part of the ring current obtained by an instrument specifically designed for this purpose. Similarily, measurements in cis-Jovian space include the first detailed charge state determination of Iogenic ions and several ENA images of that planet's magnetosphere.

Keywords: energetic neutral atoms, gas – plasma interaction, hot plasma composition, magnetospheric imaging, planetary magnetosphere, Saturn magnetosphere, space instrumentation, trapped energetic charge particles

1. Introduction

Saturn, its atmosphere, rings, moons, and plasma envelope or magnetosphere, are all closely coupled and interact through the exchange of matter and energy. A comprehensive study and understanding of Saturn's plasma environment is a central objective for the Cassini mission. The Voyager and Pioneer encounters with Jupiter and Saturn, the Ulysses encounter with Jupiter, the Voyager 2 encounters with Uranus and Neptune, and the Galileo Jupiter orbiter have provided important new insights into magnetospheric processes.

(a) Planetary magnetospheres are mostly populated by energized ionospheric and satellite material.
(b) Planetary magnetospheres inject fast neutral atoms, plasmas and energetic particles into the interplanetary medium.
(c) Energetic particles and plasmas modify the atmospheres and surfaces of natural satellites.
(d) Solar wind plasma pressure on planetary magnetic fields establishes the bow shock, magnetopause, magnetotail, and apparently energizes magnetospheric substorms, but it apparently is not the dominant source for the magnetospheric energetic particle populations.

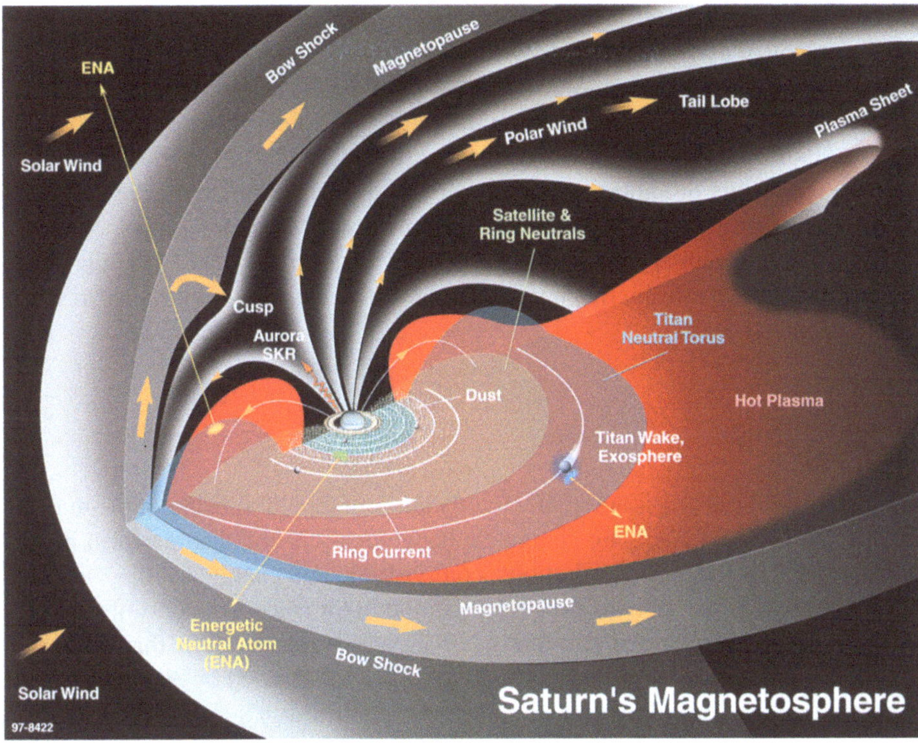

Figure 1. Saturn's magnetosphere.

In the case of Saturn, Titan presents a dense atmosphere within Saturn's magnetosphere where escape of atmospheric constituents contributes to a neutral gas cloud that is an important (perhaps dominant) source of plasma for the magnetosphere.

The Pioneer 11 and Voyager 1 and 2 encounters with the Saturn system have established the existence of a large and complicated magnetosphere, whose global configuration and dynamics are shaped not only by interactions with the solar wind and the planetary atmosphere, but also by interactions with the planetary rings and satellites (Figure 1). At Jupiter, interactions with planetary satellites, particularly Io, are of dominant importance for the magnetospheric mass and energy budgets, but it is not yet clear whether satellite–magnetosphere interactions are comparably important at Saturn.

The Cassini mission to the Saturn system will provide a unique opportunity for exploration of the magnetosphere and investigation of magnetospheric processes. Saturn's aurora may be powered by a solar wind interaction or a Titan interaction, but it is not related to interactions with the icy satellites of the inner magnetosphere. Saturn's kilometric radio emissions are strongly affected by solar wind interaction. On the other hand, the plasmas in the inner magnetosphere are dominated by heavy ions, most likely from the magnetospheric interaction with the icy moons. Nevertheless, the cometary interaction between the magnetosphere and Titan as well as the interaction with Saturn's atmosphere both yield important plasma sources. A fundamental objective of the Cassini mission will be to clarify the relative importance of these interactions for the global mass and energy budgets. Another important objective of the Cassini mission will be to determine if there are substorms in Saturn's magnetosphere. Evidence for Earth-like substorm activity has so far been found in the magnetospheres of Mercury, Uranus, Neptune, and most recently Jupiter (Mauk *et al.*, 1997a). The question of whether substorms are a universal magnetospheric phenomenon is of fundamental importance in space plasma physics.

1.1. Measurements Approach

To address these science issues, MIMI will carry out remote sensing of the magnetosphere by imaging energetic charge-exchange neutrals and also perform in-situ measurements of ion 3-D distributions, composition, and charge state.

The combination of imaging observations and in-situ measurements of the magnetosphere can be expected to yield a far greater science return than either of these types of measurement alone. This is amply demonstrated by experience with combined remote sensing and local measurements of Earth's aurora (DE, DMSP, Viking, Freja, Polar, etc.) as well as the Io torus of Jupiter (Voyager, Galileo). In-situ measurements provide a "ground truth" that validates inferences from remote sensing; furthermore, they provide important information and constraints to be used in deconvolving structures along the line-of-sight. The imaging observations, on the other hand, provide the global context for the local measurements, revealing global patterns of activity that would be missed or that could not be interpreted unambiguously on the basis of in-situ measurements alone.

Remote sensing of energetic charge-exchange neutrals has been used by Roelof *et al.* (1985) to measure the global decay of Earth's ring current and by Roelof (1987) to obtain images of Earth's ring current. Charge-exchange neutrals have been detected by Voyager from Saturn's magnetosphere out to $140R_S$ (Kirsch *et al.*, 1981b) and have been the subject of extensive theoretical modeling (Ip, 1984; Cheng, 1986; Hsieh and Curtis, 1988, 1989; Cheng and Krimigis, 1989b). The far more capable MIMI sensors will image the magnetosphere from apoapsis of the Saturn orbiter tour into the inner magnetosphere. If a substorm or some other large-scale dynamic process should occur in Saturn's radiation belts, it would be detected by MIMI even during the extended periods when the orbiter is outside the magnetosphere (this will be a significant fraction of the time). Imaging and synoptic monitoring of the ring current/radiation belt ion populations by MIMI, in conjunction with imaging of the Saturnian aurora and neutral hydrogen clouds by the ultraviolet imaging spectrograph (UVIS) instrument, should greatly clarify the nature of Saturn's aurora and magnetospheric coupling to Saturn's atmosphere. Imaging and synoptic monitoring of Titan's cometary interaction with the magnetosphere by MIMI should likewise clarify the importance of Titan interactions for the magnetosphere and should determine whether Titan is as important for Saturn as Io is for Jupiter.

In addition to the global imaging of the magnetosphere, MIMI will also carry out in-situ measurements of the energetic ion ($\sim$7 keV/nuc to >8 MeV/nuc) and electron (15 keV to >11 MeV) populations. Ion and electron energy spectra, ion composition, and ion charge state will be measured in order to address fundamental issues concerning the structure and dynamics of the Saturn magnetosphere. Among these are the following: What is the origin of the plasmas in the Saturn system? How and where are the plasmas transported? What processes power Saturn's aurora? What forms of global magnetospheric activity occur, how are they powered, and, specifically, do Earth-like substorms occur? Ion composition and charge state will be measured for the first time in the 10–265 keV/e range, allowing detection and tracking of tracer species from the solar wind, from Saturn's ionosphere, from Titan, and from the icy satellites, not only in the magnetosphere, but also in the upstream solar wind.

2. Scientific Objectives

MIMI science objectives may be summarized as follows:

Saturn

- Determine the global configuration and dynamics of hot plasma in the magnetosphere of Saturn through energetic neutral particle imaging of ring current, radiation belts, and neutral clouds.
- Study the sources of plasmas and energetic ions through in-situ measurements of energetic ion composition, spectra, charge state, and angular distributions.
- Search for, monitor, and analyze magnetospheric substorm-like activity at Saturn.
- Determine through the imaging and composition studies the magnetosphere–satellite interactions at Saturn and understand the formation of clouds of neutral hydrogen, nitrogen, and water products.
- Investigate the modification of satellite surfaces and atmospheres through plasma and radiation bombardment.
- Study Titan's cometary interaction with Saturn's magnetosphere (and the solar wind) via high-resolution imaging and in-situ ion and electron measurements.
- Measure the high energy ($E_e > 1$ MeV, $E_p > 15$ MeV) particle component in the inner ($L < 5R_S$) magnetosphere to assess cosmic ray albedo neutron decay (CRAND) source characteristics.
- Investigate the absorption of energetic ions and electrons by the satellites and rings in order to determine particle losses and diffusion processes within the magnetosphere.

- Study magnetosphere–ionosphere coupling through remote sensing of aurora and in-situ measurements of precipitating energetic ions and electrons.

Jupiter

- Study ring current(s), plasma sheet, and neutral clouds in the magnetosphere and magnetotail of Jupiter during Cassini flyby, using global imaging and in-situ measurements.

Interplanetary

- Determine elemental and isotopic composition of local interstellar medium through measurements of interstellar pickup ions.
- Study the compositional evolution at low energies of shock-accelerated ions in the interplanetary medium.
- Monitor cosmic ray intensity variations ($E_p > 70$ MeV/nuc) and anomalous cosmic rays, and compare with similar measurement in both the inner heliosphere (advanced composition explorer, ACE) and the outer heliosphere (Voyagers 1 and 2).

2.1. Global Imaging

The most fundamental relationship in physics is that of cause and effect, and the study of the magnetospheres of the giant planets has been a valiant struggle with the inseparability of temporal and spatial variations along the flyby trajectories of Pioneers 10/11, Voyagers 1/2, and Ulysses. The exciting new observations from the Galileo orbiter are adding another dimension to our understanding, just as the Cassini orbiter certainly will. However, the most dramatic and unambiguous signatures of magnetospheric electrodynamics will come from imaging the ion populations across the huge expanses of the Saturnian magnetosphere. The ion and neutral camera (INCA) will image energetic neutral atoms (ENAs) that are produced when energetic, singly charged ions undergo charge-exchange collisions with the cold neutral gas that pervades the ion populations. Because the ENA trajectories are straight lines originating at the location of the charge-exchange collision, the counting rate in each INCA pixel of the sky is proportional to the line-of-sight integral of the ion intensity (in the direction of the spacecraft) weighted by the cold neutral densities n_i and the corresponding charge-exchange rate σ_I: $j_{ENA} = \sum \sigma_i \int ds n_i j_{ION}$. Because INCA can resolve ENA hydrogen from ENA oxygen, these two species can be analyzed separately.

Nonetheless, even with resolution of H from O, the unfolding of the ion intensities from the INCA images will probably require sophisticated minimization algorithms in which parametric models of the ion intensities (and the neutral densities) are optimized so that the difference between the measured INCA pixel rates and those predicted by the models is a minimum (in a χ-squared sense).

TABLE I

Measured neutral atom fluxes from planetary magnetospheres near 40 keV energy at large distance d.

Earth	$\sim 100\left(\frac{d}{R_E}\right)^{-2}$ cm^{-2} s^{-1} keV^{-1} (quiet time)
Jupiter	$\sim 440\left(\frac{d}{R_J}\right)^{-2}$ cm^{-2} s^{-1} keV^{-1}
Saturn	$\sim 240\left(\frac{d}{R_S}\right)^{-2}$ cm^{-2} s^{-1} keV^{-1}
Uranus	$< 12\left(\frac{d}{R_U}\right)^{-2}$ cm^{-2} s^{-1} keV^{-1}
Neptune	$< 4\left(\frac{d}{R_N}\right)^{-2}$ cm^{-2} s^{-1} keV^{-1}

Energetic charge-exchange neutral emissions have been detected from the magnetospheres of Earth, Jupiter, and Saturn. Table I compares the measured neutral atom source strengths near 40 keV of five planetary magnetospheres (McEntire and Mitchell, 1989). The fluxes given in Table I are integrated over the source and apply for large distances compared to the source size. The source strengths agree with theoretical expectation (e.g., Ip, 1984; Cheng, 1986; Barbosa and Eviatar, 1986; Krimigis *et al.*, 1988).

At Earth, where ENA imaging was first developed and demonstrated, the cold neutral density takes the comparatively simple form of a nearly spherically symmetrical Chamberlin exosphere of H atoms. Not so at Saturn. Reinterpretation of the Voyagers 1/2 UVS Lyman-α emission demands an H-atom distribution that is more disk-like than spherical, and moreover cannot even be axially symmetric about the planet's rotation axis (Shemansky and Hall, 1992; Ip, 1996). The presence of a large amount of H in the inner system ($r < 10R_S$) implies that Saturn's exosphere is a significant source, but in the outer system, Ip argued that Titan should produce significant densities of H ($10 < n_H < 10^2$ cm^{-3}) extending up to $8R_S$. Furthermore, Ip (1997) has recomputed the original estimates of Johnson *et al.* (1989) for the contrisbutions of the icy moons (Enceladus, Tethys, Dione, and Rhea). He finds concentrations of O, OH, and H_2O in the range 10^2 to 10^4 cm^{-3} in the inner system (3–6R_S), although they are strongly confined to the equatorial plane.

Using the early models for the neutrals distribution, we have simulated idealized ENA images for INCA in Figure 2. "Fish-eye" lens projections of ENA fluxes (>35 keV H) are shown for four locations of Cassini. The bright object in the three more distant images is Titan's exosphere, and the putative Titan torus would be seen through most of the orbit. The color bar is logarithmic, ranging from 1 to 100% of the brightest ENA lines in each image.

Images from 50R_S. Figure 3 shows Saturn and the Titan torus as viewed from the equatorial dawn meridian; distorted dipole field lines are drawn to show compression of the magnetosphere on the front side with the magnetopause at (a) $25R_S$, and (b) $20R_S$. Panel (c) shows the accumulated counts for (a) and (b) (upper

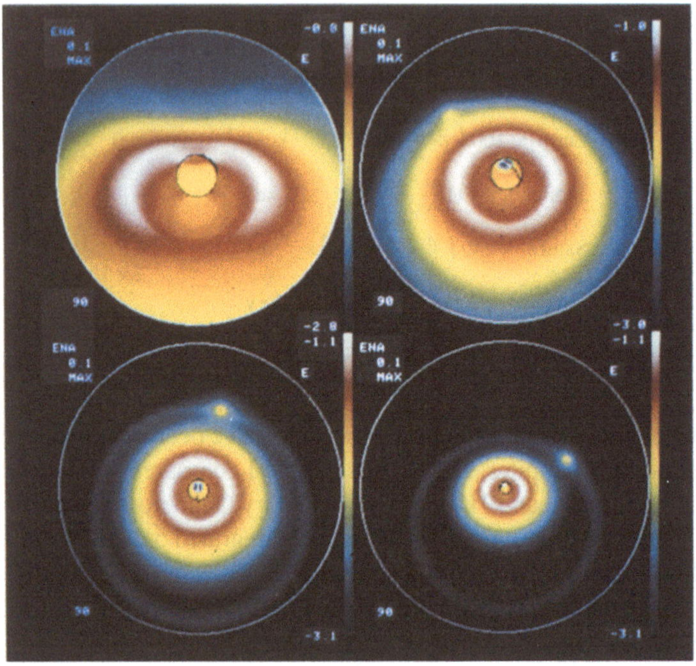

Figure 2. Four idealized ENA images ($1 \times 1°$ angular resolution, no Poisson statistical fluctuations in pixel counts) of the Saturn/Titan system, based on Voyager >35 keV CPME ion measurements and early interpretation of UV observations. "Fish-eye" projection is hemisphere centered on planet (terminator indicated in black). Logarithmic color bar ranges from 1 to 100% of brightest emission in each image. Magnetosphere extends to $25R_S$, and Titan exosphere is shown with putative plasma torus. Cassini positions are every 30° of mean anomaly for inclined orbit with radius, solar longitude, magnetic latitude: (a) $4.4R_S$, 21°, 26°; (b) $5.9R_S$, 39°, 55°; (c) $9.0R_S$, 100°, 71°; and (d) $15.5R_S$, 161°, 55°.

and lower curves, respectively) for a 20-min accumulation. The complete image contains 1163 counts (~10% statistics in inner ring current, 30% at $L = 8$). Most importantly, ENA emission from the dayside Titan torus is significantly diminished as the magnetopause moves inward, thus enabling continuous remote sensing of such compressions, which is not possible through any other technique.

The implications of this complicated neutral distribution for ENA imaging can be looked at two ways. On one hand, it makes the problem of extracting absolute ion intensities from global ENA images extremely challenging. Clearly the Cassini UV measurements of the neutral environment and the INCA ENA intensities must be correlated wherever possible. On the other hand, the concentration of the neutrals close to the equatorial plane actually simplifies the geometry for ion injection events on localized magnetic flux tubes. At Earth, ion injection events are often difficult to analyze precisely because of the spherical distribution of the H atoms – the entire flux tube "lights up" in ENA, making it hard to determine the region of L and mean local time (MLT) containing the injection. However, at Saturn, it will be mainly the equatorial region that will emit the ENA, thus giving directly the L–MLT

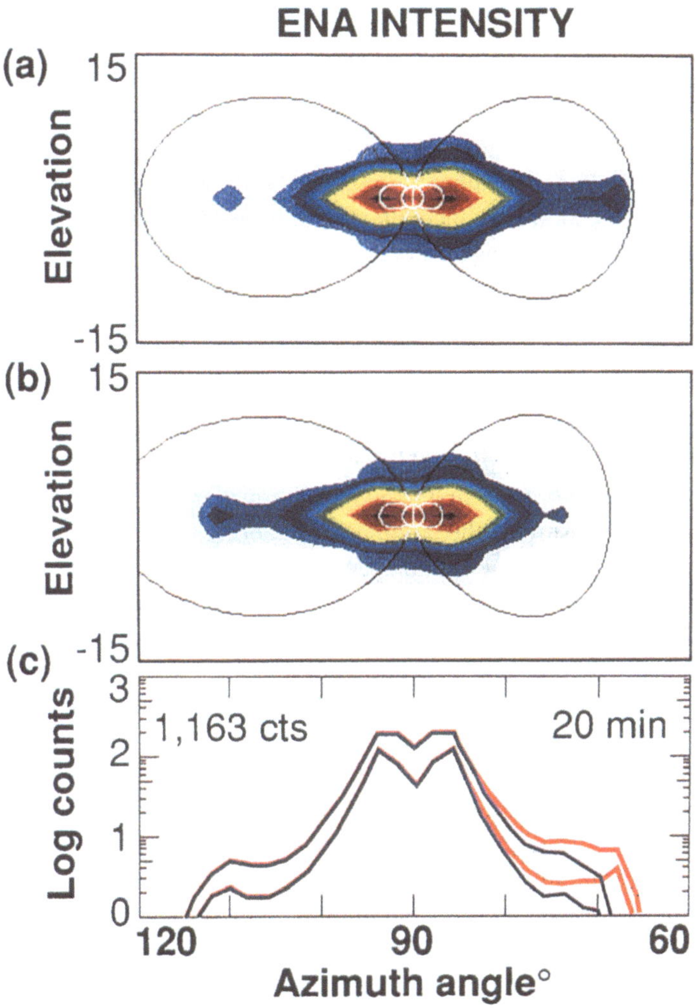

Figure 3. Simulated ENA images ($Y = -50R_S$) of magnetopause compression from (a) $25R_S$ to (b) $20R_S$ detected by (c) change in Titan torus within 20 min (for 30–60 keV H).

location of the ion injection (without complicated unfolding of the line-of-sight integrals).

Consequently we anticipate that INCA, in combination with CHEMS and LEMMS, will produce immediate and quantitative identification of the regions of ion acceleration in Saturn's magnetosphere, some of which we can expect by analogy to the Earth-like ion injection events identified at Jupiter by Mauk *et al.* (1997a) using the Galileo energetic particle detector. It will be the causal relationships – what ions were accelerated where and when – that we look forward to discovering with INCA's global imaging.

2.2. Satellite Interactions and Aurora

2.2.1. *Magnetosphere–Satellite Interactions*

A major magnetospheric goal of the Cassini mission is the measurement of the environments of the Saturnian moons and the determination of their interaction with the Saturnian magnetosphere. To this end the Cassini spacecraft will conduct numerous close (<2000 km) encounters with the moons Titan, Iapetus, Hyperion, Rhea, Dione, and Enceladus. Observations from such close encounters provide the measurements required to determine the environment of the moon and its magnetospheric interaction. In particular, energetic particle measurementswith the LEMMS sensor during close flybys provide vital and often unique results that identify the character of the moon–magnetosphere interaction. For example, as has been shown at Jupiter, energetic particle observations provide a measure of the moon's surface magnetic field (Williams *et al.*, 1997b); directly measure the effect of the moon and its environment on the convecting magnetospheric plasma (Williams *et al.*, 1997a); provide a direct measure of the amount of scattering occurring on planetary field lines intersecting the moon (Williams and Mauk, 1997); provide a measure of the conductivity of the moon and its environment (Paranicas *et al.*, 1998); provide a direct measure of the effectiveness of sputtering in the formation of the moon's atmosphere, ionosphere, and associated gas torus as well as for surface mass redistribution effects (Ip *et al.*, 1997, 1998; Lagg *et al.*, 1998); and for moons with magnetic fields can identify the existence of trapped particles within that field (Williams *et al.*, 1997a).

On a broader scale, simultaneous observations by MIMI and the ultraviolet spectrometer will further elucidate the effects of magnetospheric interactions with Titan and the icy moons. The ultraviolet spectrometer observations will provide the H column density along the same lines of sight viewed by MIMI, enabling the energetic ion distributions to be obtained from the ENA images. The ENA images, obtained separately for each species, will allow a determination of the composition of the distant energetic ion population, thus providing a measure of the strengths of the various moons as sources of magnetospheric ions.

2.2.2. *Saturn's Aurora and Magnetospheric Activity*

Saturn's auroral energetics and morphology are Earth-like rather than Jovian, suggesting that the aurora is powered by a solar wind interaction (Sandel and Broadfoot, 1981). Alternatively, Barbosa (1987) has argued that Saturn's aurora may be powered by a Titan interaction with the magnetosphere. MIMI measurements during the many flux tube crossings at Titan will directly test this possibility. For example, at Jupiter's moon Io, intense energetic electron beams were measured flowing in both directions along Jovian field lines adjacent to Io (Williams *et al.*, 1996). An Io flux tube full of these beams would lead to aurorae at its footprint in the Jovian atmosphere. The flux tube measurements to be made with LEMMS at Titan will yield the energy content of the particles flowing into Saturn's atmosphere and thus measure the auroral producing capability of the moon. The existence or absence of field-aligned beams at Titan will assess the effectiveness of Titan (and its environment) in converting the induced electric field resulting from its motion through the Saturnian magnetosphere into a magnetic field-aligned potential drop. Finally MIMI will search for field-aligned particle flows in the high latitude regions of Saturn's magnetosphere and directly test for the existence of field-aligned auroral zone potentials above ~10 keV. Measurements from MIMI will determine whether the resulting energetic particle precipitation is sufficient to excite aurorae.

The existence of aurorae implies the existence of substorms. At Earth the solar wind interaction with the magnetosphere provides the dominant energy source for substorms. It is not clear what the mechanism for substorm phenomena would be at Jupiter and Saturn and, until the recent systematic study of the Jovian environment by the energetic particle instrument on Galileo by Krupp *et al.* (1998) and by Woch *et al.* (1998, 1999), it was not known whether Jupiter and Saturn could be expected to display substorm phenomena. Energetic ion and electron measurements in the Jovian magnetosphere show signatures that are basically identical to substorm injection events observed at Earth (Mauk *et al.*, 1997a). Measurements of the energy dispersion of both ions and electrons show particle injection over a narrow longitudinal band and subsequent magnetic gradient drift superimposed on magnetospheric corotation. Because of the overwhelming effect of corotation, the dispersion is best observed in the energetic particle population and is very difficult to observe at low energies. Thus, with the existence of aurorae at Saturn, it is reasonable to expect to observe substorm signatures. MIMI will provide the best opportunity to observe such signatures.

2.2.3. *Ring Interactions*

An important goal of the MIMI investigation is to understand the interactions between the planetary magnetospheric environments and Saturn's planetary rings. From the perspective of planetary ring science, charged particle environments can electrically charge ring and other particulates, thereby affecting the transport of the particulates, and perhaps of larger ring members, via collisions with the particulates (Mendis *et al.*, 1984; Goertz, 1989). The "spokes" of Saturn are the best-known example, and the braiding of the F-ring may also have an electromagnetic origin (Mendis *et al.*, 1984). Energetic charged particles can modify, sputter, and erode ring particle surfaces (Johnson, 1990). In addition, charged particle distributions can provide powerful diagnostics on the presence, character, and distributions of sparse ring materials (Krimigis and Armstrong, 1982; Esposito *et al.*, 1984). From the magnetospheric science perspective, rings provide important sources and sinks of the neutral and charged particles of planetary environments (e.g., Krimigis and Armstrong, 1982; Van Allen, 1984). Also, ring/charged-particle interaction features provide powerful diagnostics on the transport of charged particles within the magnetospheric environments (e.g., Van Allen, 1984; Mauk *et al.*, 1994). Finally, the charged particle and ring particulate environments of planetary magnetospheres

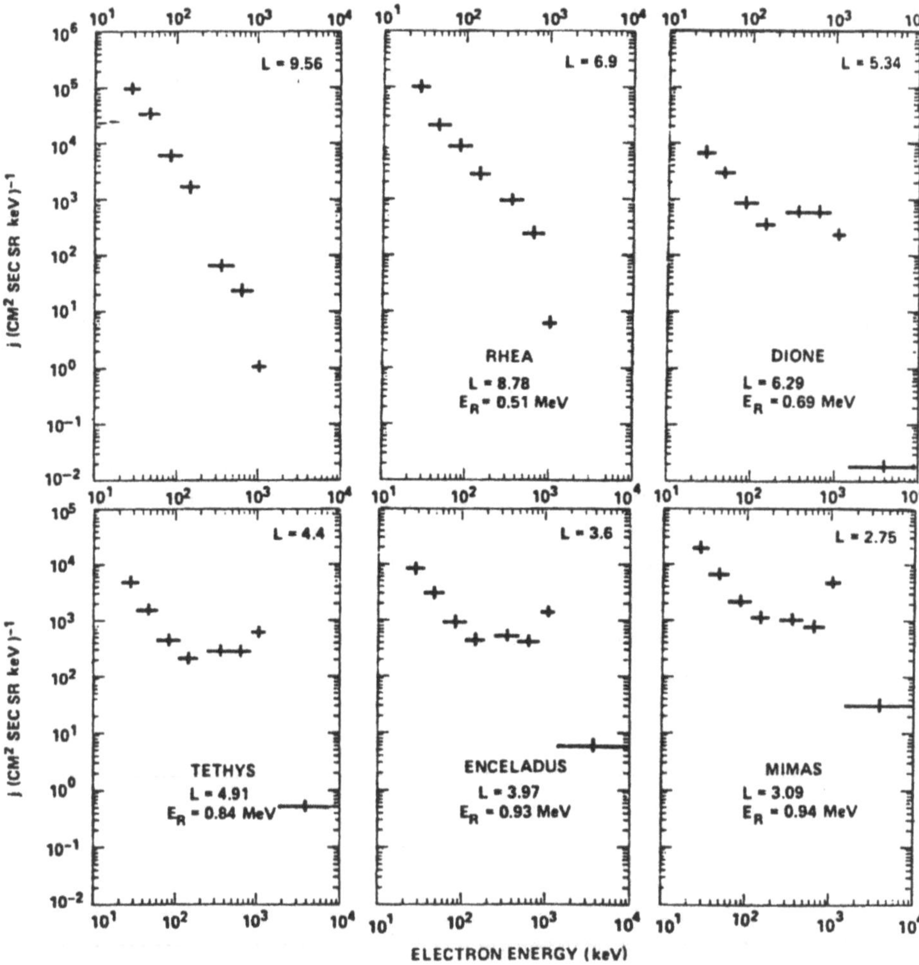

Figure 4. The evolution of the electron energy spectrum at various L values, upper left to lower right, as registered by LECP on Voyager 2. The L values are indicated in each panel, and the resonant electron energy E_R at each satellite is shown in successive panels. From Krimigis *et al.* (1982a).

are perhaps the best laboratories for studying certain aspects and phases of the formation of planetary nebula (Burns *et al.*, 1994).

Numerous signatures of the interaction between energetic charged particles and ring particles were observed by Pioneer 11 and the Voyagers 1 and 2. Figures 4 and 5 show, respectively, examples of explicit radial profile signatures and bite-outs in pitch angle distributions attributed to the effect of ring dust materials. However, the ENA imaging capabilities of INCA, combined with the information of energy spectra, angular distribution, and charge state obtained with LEMMS and CHEMS, provides a powerful new technique for diagnosing ring-particulate/energetic-particle interactions, as described below.

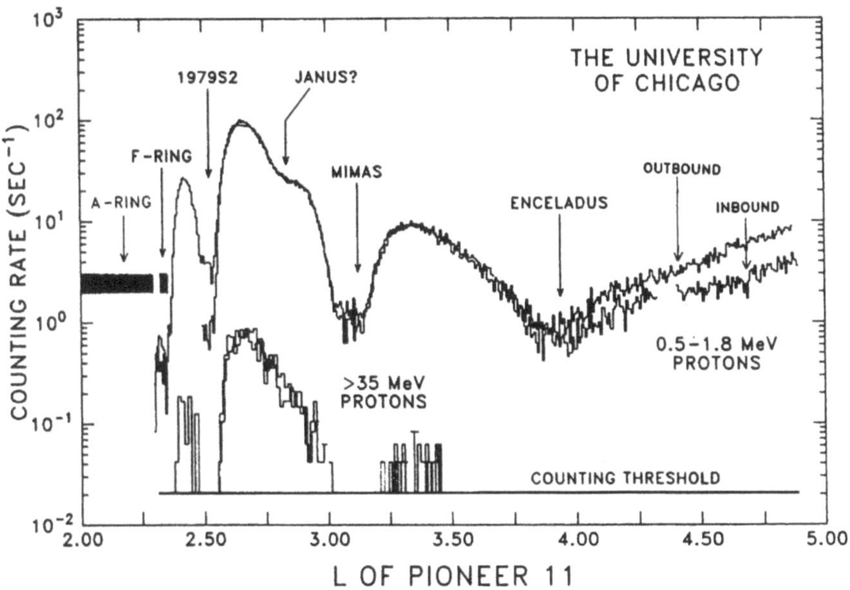

Figure 5. Energetic ions measured by Pioneer 11 in Saturn's inner magnetosphere (after Simpson *et al.*, 1980).

The intense populations of magnetically trapped, energetic charged particles that constitute the radiation belts of Saturn's inner magnetosphere are slowly transported towards the planet by radial diffusion processes. Many of them interact with the rings of Saturn and ultimately are lost to the magnetospheric system. Energetic protons with energies greater than ~50 keV will completely penetrate ring particulates with diameters in the sub-micron regime, such as those that are key constituents of the F, G, and E rings of Saturn. A substantial fraction of those penetrating protons (~60% at 50 keV) will emerge neutralized by the interaction, ending up as hydrogen ENAs. Thus, the interactions between the trapped energetic particles and the ring particulates may be observed remotely by an ENA camera that measures the energy, mass species, and the arrival direction of ENAs. The ability of the INCA sensor of the MIMI instrument to image energetic-particle/ring-particulate interactions within Saturn's inner magnetosphere during the SaturN orbit insertion (SOI) phase of the Cassini mission was recently documented by Mauk *et al.* (1998). A simulated ENA image of Saturn's F-ring from that work is shown in Figure 6. Mauk *et al.* (1998) showed that with such images we can obtain powerful new diagnostics of magnetospheric radial transport of energetic charged particles. Also, the impact rates for the consideration of sputtering and erosion will be better constrained, and the relative importance of the rings as a sink of radiation belt particles will be determined. Finally, the energy spectra of the ENA emissions will provide a new type of constraint on the size distribution of the ring particulates. Figure 7 shows a sample of the modeled sensitivity of the ENA spectra to the dust particulate sizes.

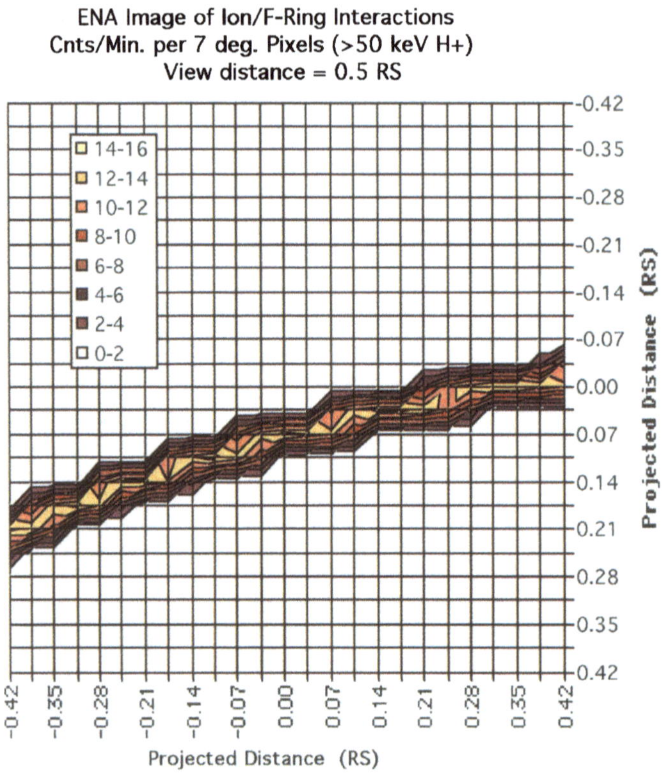

Figure 6. Simulated ENA image of a portion of the F-ring as it might appear to the INCA MIMI ENA camera on the Cassini spacecraft at a distance from the ring segment near the center of the image of $0.5R_S$ during the Saturn orbit insertion phase of the mission. The units are counts per minute. The F-ring dust particulates are assumed to have a size of 0.5 mm. A lower limit assumption has been made concerning the radial transport of energetic protons (~0.01 km/s). After Mauk *et al.* (1998).

2.3. Saturn Charged Particle Environment

Saturn's magnetosphere is a prodigious accelerator of charged particles, as apparently are the magnetospheres of all strongly magnetized planets. There is no clear understanding why planetary magnetospheres have this characteristic. Saturn's magnetosphere has unique properties that can help unravel the relative importance of various proposed mechanisms for particle acceleration and transport. Saturn's energetic particle population is also unique in the degree of interaction between the charged particle populations and the neutral and solid materials such as planetary gases, dust and ring materials, and satellite surfaces. Thus, Saturn's magnetosphere may come closest to emulating the conditions that existed within the Sun's planetary nebula and during later stages in the formation of the solar system.

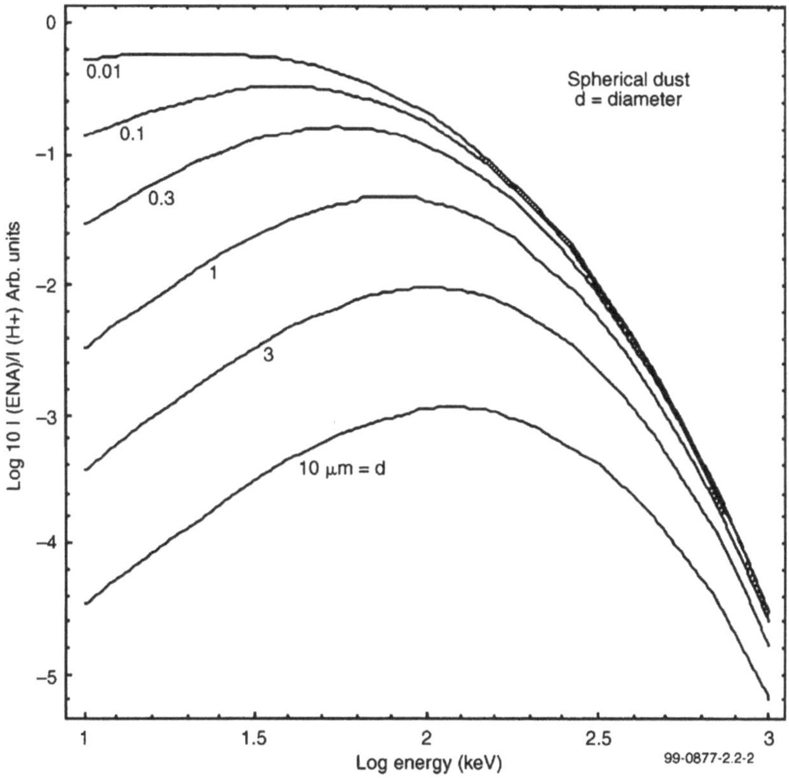

Figure 7. Model calculations of ENA emission energy-spectra from the dust particulates as a function of particulate size and shape. The spectra are normalized by the parent ion spectra. It was assumed the dust particulates are spherical. Other simulations given by Mauk *et al.* (1998) test for sensitivity to particulate shape.

Figure 8 shows Saturn's energetic charged particle environment as characterized with the low energy charged particle (LECP) instrument on Voyager 2 (Krimigis *et al.*, 1983). Energetic particles fill the magnetosphere to the dayside magnetopause and participate in the force balance of the magnetosphere with the interplanetary environment. Energetic particles clearly leak out of the magnetospheric boundary. Strong inbound/outbound (dayside/predusk) asymmetries have been interpreted as being the result of temporal dynamics. Shorter-scale temporal dynamics are also apparent in the so-called "mantle" region (inbound-dayside, $r > 10R_S$), undoubtedly related to dense, detached, cold plasma regions, apparently shed from the interior plasma sheet (Sittler *et al.*, 1983). Large-scale modulations observed in the outbound regions, apparently related to the Saturn rotational modulations reported from the Voyager 1 encounter (Carbary and Krimigis, 1982), are surprising, given the near-exact alignment of Saturn's magnetic axis with its spin axis. Energetic particles in the inner regions show the clear signatures of strong interactions with materials, including the natural satellite surfaces. The very peculiar electron spectra

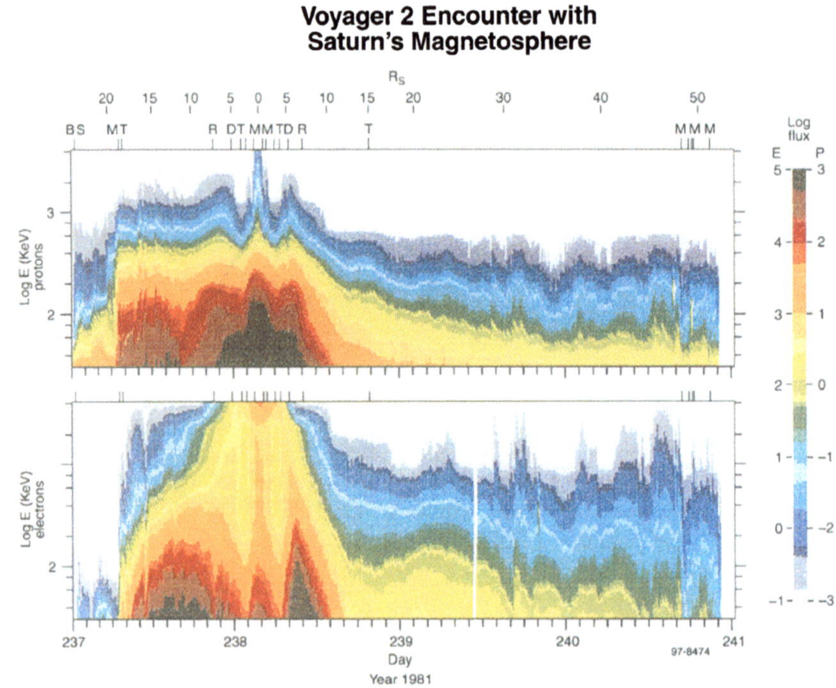

Figure 8. Color spectrogram of energetic ions (top) and electrons (bottom) measured by low energy charged particle (LECP) instrument on Voyager 2 during its encounter with Saturn. The color is coded according to the $\log_{10}$ of intensity ((cm s sr keV) − 1). Radial distance (in R_S) is given below each color panel, and various events are given above each panel. The events are: BS = bow shock, M (first and last three) = magnetopause, T (first and last) = Titan L-shell, R = Rhea L-shell, D = Dione L-shell, T (middle 2) = Tethys L-shell, and M (middle 2) = Mimas L-shell. The unlabeled tick marks between Tethys and Mimas represent the Enceladus L-shell. After Krimigis *et al.* (1983).

near closest approach (Krimigis *et al.*, 1982a), showing the intensities rising at the highest energies, are interpreted as being the signature of a drift period resonance with orbital period of the satellites (see Figure 4).

The temporal dynamics mentioned above are of particular interest for the MIMI investigation of the Saturnian system. Saturn's magnetosphere is thought to occupy a middle ground between magnetospheres such as that of Earth, whose dynamics and energetics are determined by a strong interaction with the interplanetary medium, and magnetospheres such as that of Jupiter, thought to be controlled predominantly by the rapid rotations of the central planet. By comparing the energetic particle characteristics, and their temporal dynamics among Earth, Jupiter, and Saturn, we will be able to sort out the relative influences of rotational and interplanetary environment on the energization of charged particles to high energy. Temporal dynamic processes at Earth are well characterized, and Galileo is now shedding substantial light on the energetic particle dynamics at Jupiter. MIMI will provide for an unprecedented characterization of the energetic particle dynamics within the Saturnian magnetosphere.

The critical influence of such materials as dust, rings, and satellite surfaces on the character of the energetic particle populations is shown most clearly in Figure 5 (Simpson *et al.*, 1980). Dust particulates of the F, G, and E-rings clearly have roles to play, as do both the larger particles of the A and F rings and the satellites. Ring interaction signatures are highlighted further in Figure 9 with the 90° bite-outs observed in energetic particle pitch angle distributions measured near the G-ring (Krimigis and Armstrong, 1982). Also, the very high energy component of the spectra shown (the bump at ~100 MeV) has been interpreted as resulting from CRAND from impacts on the rings. LEMMS will be able to measure these features at all local times, various distances, and latitudes over the entire mission of at least 4 years.

Energetic particle composition is another clear signature of the importance of materials interactions for establishing the character of the energetic particle populations. Figure 10 shows that both Jupiter's atmosphere (H_2, H_3) and the solar wind (He) are contributors to the energetic particle populations, and the presence of heavier ions (C, N, O) are possible tracers of satellite and ring sources (Krimigis *et al.*, 1982b). Evidence not shown here (Krimigis *et al.*, 1983) suggests that oxygen ions dominate the energy density of the ion populations planetward of $\sim 6 R_S$, indicating that sputtering off the icy satellites may be a dominant source of energetic particles in these regions. The relative strengths of all sources, particularly of Titan versus the icy satellites, is not known. MIMI, with compositional and charge state discrimination capabilities that dwarf previous measurements at Saturn, will establish where the energetic particles come from.

Dust, ring particles, satellites surfaces, and gas distributions help determine the characteristics of the energetic particles, but these materials are in turn modified by the energetic particles. Energetic particles process and change the surfaces of materials (e.g., Johnson, 1990), and they affect the transport of dust, and perhaps larger ring particles (Mendis *et al.*, 1984; Goertz, 1989). The use of MIMI in concert with the comprehensive suite of Cassini instruments (e.g., UVIS measurements of gas distributions, imaging science subsystems (ISS) images of dust distributions, etc.) will allow for an epochal characterization of the solar nebula-like interactions that are now occurring within the Saturnian system.

MIMI will also provide unprecedented new information about several other aspects of the Saturnian energetic particle populations that have remained ambiguous since the Voyager encounters. Previous limitations in ion compositional analysis and angular resolution and coverage have prevented us from achieving closure in the analysis of force balance between magnetic fields and plasmas (Mauk *et al.*, 1985). Saturn appears to be unique in the interplay that exists between the contributions of cold and hot plasmas in balancing magnetic stresses. Pressure gradient and anisotropy effects appear to dominate ring current generation in the inner and middle regions, as at Earth, whereas corotational stresses dominate in the outer regions, as was once thought to be true at Jupiter. We will now have the opportunity of

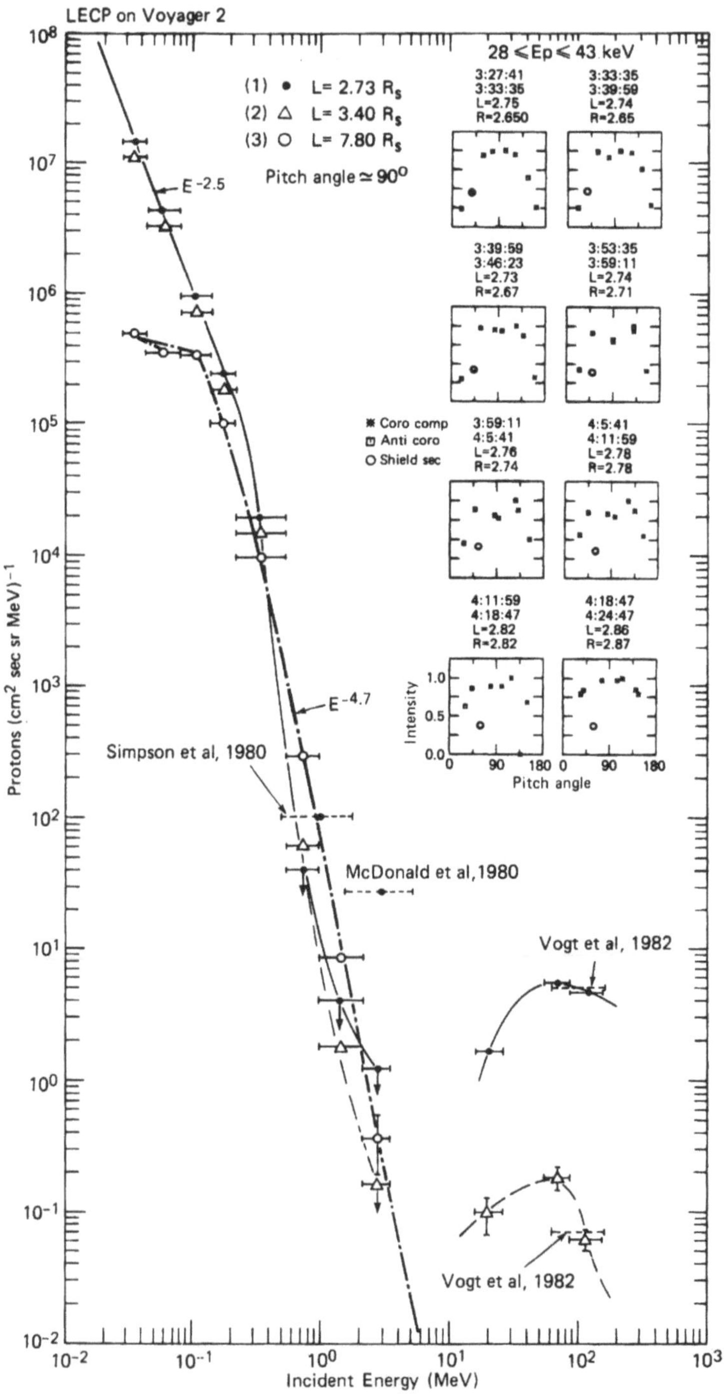

Figure 9. Spectra and pitch angle distributions of energetic ions measured in Saturn's inner magnetosphere by Voyager 2 and comparison with Pioneer-11. After Krimigis and Armstrong (1982).

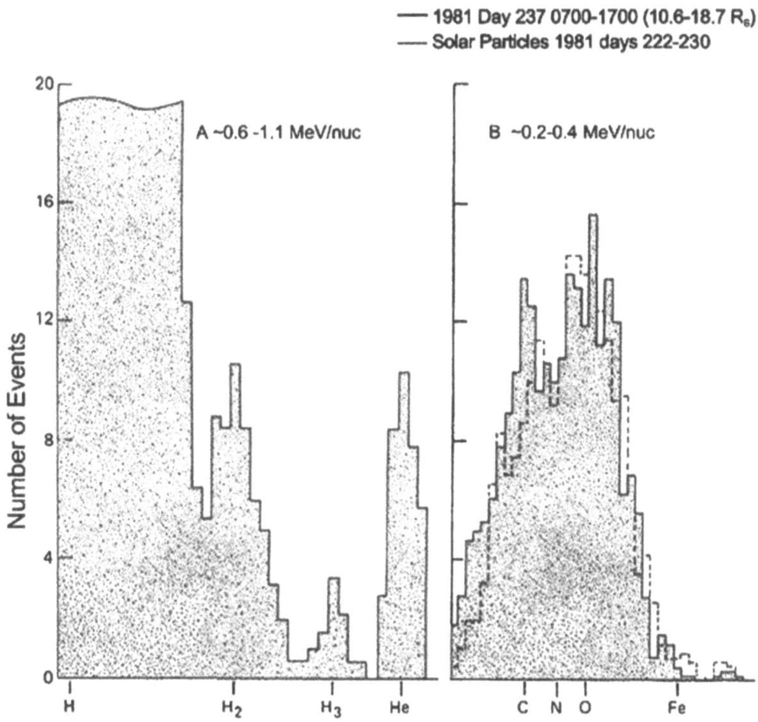

Figure 10. Composition distribution of energetic ions measured by Voyager 2 in Saturn's middle magnetosphere. After Krimigis *et al.* (1982b).

obtaining information about stress balance at Saturn that will tell us a great deal about how such balance is achieved in planetary magnetospheres in general. These same measurement limitations (composition, angular information) have also resulted in very wide uncertainties in our knowledge of radial transport rates of energetic particles at Saturn (Paonessa and Cheng, 1986). Phase space density analyses of the energetic particle populations, in the context of radial diffusion theory (Schulz and Lanzerotti, 1974), should constrain the radial transport and loss rates to a much higher precision than previously available once accurate charged particle invariant parameters are established (with improved angular coverage and resolution) and once the compositional state of the energetic particles can also be established.

2.4. Titan Interactions

Titan's nitrogen-rich atmosphere is subject to direct magnetospheric interaction due to its lack of a significant intrinsic magnetic field. To some extent, the dynamical process will be similar to the solar wind interaction with Mars or comets, with the major difference that no bow shock will form because the local value of the magnetosonic Mach number is less than one. The Voyager 1 plasma measurements at Titan flyby showed that a well-defined magnetotail was formed in the wake region (Ness *et al.*, 1982). However, a four-lobe configuration can also be deduced from both the magnetometer and plasma wave observations (Ness *et al.*, 1982; Gurnett *et al.*, 1982; Neubauer, 1992). This unique phenomenon is caused by the finite gyroradius effect of the exospheric ions created in the vicinity of Titan. This also suggests that Titan's exospheric and ionospheric structures should be highly asymmetric because of the preferential sputtering of the pickup ions on the hemisphere opposite to Saturn. The immediate consequence for the MIMI instrument is that such exospheric structures could be revealed by the ENA images in detail (see Amsif *et al.*, 1997). (We expect a more extensive structure of Titan's exosphere on the anti-Saturn side because of the reimpact effect of the pickup ions.) The MIMI experiment can also be used to study the possible occurrence of particle acceleration in the magnetotail of Titan.

On a global scale, the injection of nitrogen and carbon ions from Titan's exosphere into the Saturnian magnetosphere also provides an ideal set of tracer particles to follow the transport process of energetic ions. The possible detection of molecular ions such as CH_3^+, CH_5^+, and H_2CN^+ created in Titan's ionosphere (Ip, 1990) could also be used to diagnose time-dependent mechanisms since their life-times are limited by electron dissociative recombination.

As indicated, the very weak or nonexistent intrinsic magnetic field of Titan gives the Kronian magnetospheric plasma direct access to the upper Titan atmosphere, where a variety of complex phenomena take place (Eviatar, 1992; Ip, 1992). Energetic ions in the magnetosphere occasionally will undergo a charge-exchange collision with cold neutral atoms from the upper Titan atmosphere, giving rise to the production of ENAs. The energy of the incident ions is almost entirely transferred to the charge-exchange-produced ENAs, which then propagate along nearly rectilinear ballistic trajectories. The coexistence of energetic ions and cold tenuous gas in the Saturn/Titan system thus makes this system particularly suitable for magnetospheric imaging via ENAs (Cheng and Krimigis, 1989b; Curtis and Hsieh, 1989).

The first simulated ENA images of the Saturn/Titan system are due to Hsieh and Curtis (1988), who modelled the Titan torus around Saturn. Later, Roelof and Williams (1990) included Titan itself, and further examples were presented by Cheng *et al.* (1993). ENAs emitted from the Saturn magnetosphere were detected (without imaging) by the LECP instrument onboard Voyager 1, which was designed to detect mainly charged particles, so there was no system to reject the ions (Kirsch *et al.*, 1981a; Cheng, 1986). The observations were made during periods when Voyager 1 was on open field lines, where there were no significant ion fluxes. The emission was modeled by Hsieh and Curtis (1988).

Amsif *et al.* (1997) presented a study of the energetic neutral atom production, resulting from the interaction of Titan's exosphere with Saturn's magnetosphere. This necessitated the development of a model for the altitude density profile and composition of both the inner and the outer exosphere of Titan. They used a Chamberlain model for the inner exosphere (altitude <6000 km), for particles with ballistic and escaping orbits, and included the five major species: H, H_2, N, N_2, and CH_4. The exospheric density and composition altitude profiles obtained from this model were then used in order to calculate the fluxes and the energy spectra of the ENAs produced in the "inner exosphere" of Titan, considering a proton spectrum measured by Voyager in the Saturnian magnetosphere as the parent ion population (Krimigis *et al.*, 1983). The results show how the exospheric species contribute in a different way, as a function of altitude, to the ENA flux spectra. For look directions that go from the Cassini orbiter almost all the way down to the exobase, the contribution to ENA production comes mostly from charge exchanges with CH_4 and N_2. This not only is due to the high integral column density of these molecules, but also is mainly the result of their high charge-exchange cross sections. H_2 has also high integral column densities, but its lower charge-exchange cross sections significantly reduce its contribution to the ENA fluxes. This sensitivity of ENA production to the composition of the neutral atmosphere means that the ENA directional fluxes and spectra, which MIMI will measure, will constitute a powerful diagnostic tool for evaluating the abundance of the different species in the layers of the Titan exosphere. The calculated spectra of the ENAs are also much softer than the spectrum of the parent energetic ion population. This is due to the filter effect of the charge-exchange cross sections, which decreases the efficiency of ENA production at high energies.

In the outer exosphere (altitude >6000 km), molecular hydrogen is the dominant species, and a Chamberlain model that included H_2 particles with ballistic, satellite, and escaping orbits was utilized to evaluate the altitude density profile. This permits an estimate of the ENA intensities several Titan radii from the moon and was used to simulate a flyby of Titan, from 1 day before closest approach to 1 day after, just as will occur about 40 times during the Cassini orbiter tour. Figure 11 (Amsif *et al.*, 1997), shows such a simulated flyby for the 24 h following Titan closest approach. Times are given in hours (H) or minutes (M). The images in each panel are Mercator projections of the entire sky as viewed from the Cassini spacecraft. ENA intensity is shown in the left panels, and equivalent neutral hydrogen column density in the right panels. Near Titan, the molecular H_2 densities were treated as atomic H densities of twice the value. The coordinate system is Sun–Saturn-ecliptic, with the azimuth angle plotted horizontally (in values decreasing from approximately 360° on the left of the panel to 0° on the right, both 0 and 360° being the sunward viewing direction), and the elevation angle plotted vertically (increasing upwards from $-90°$ to $+90°$ relative to Saturn). Coordinate labels are suppressed because of the small size of each panel. A logarithmic color bar has been used, covering a range from the maximum value to 1/100 of that value, for both the ENA unidirectional intensity j_{ENA} (left panels) and column density h (right panels). The maximum values for each panel are given in dimensionless units scaled for convenience so that the maxima from the numerical computation would be of order unity. The Titan exosphere is nearly as bright in ENA as the inner magnetosphere of Saturn, making it possible to image the exosphere out to at least five Titan radii, and even to resolve it when the spacecraft is several Saturn radii distant from the moon.

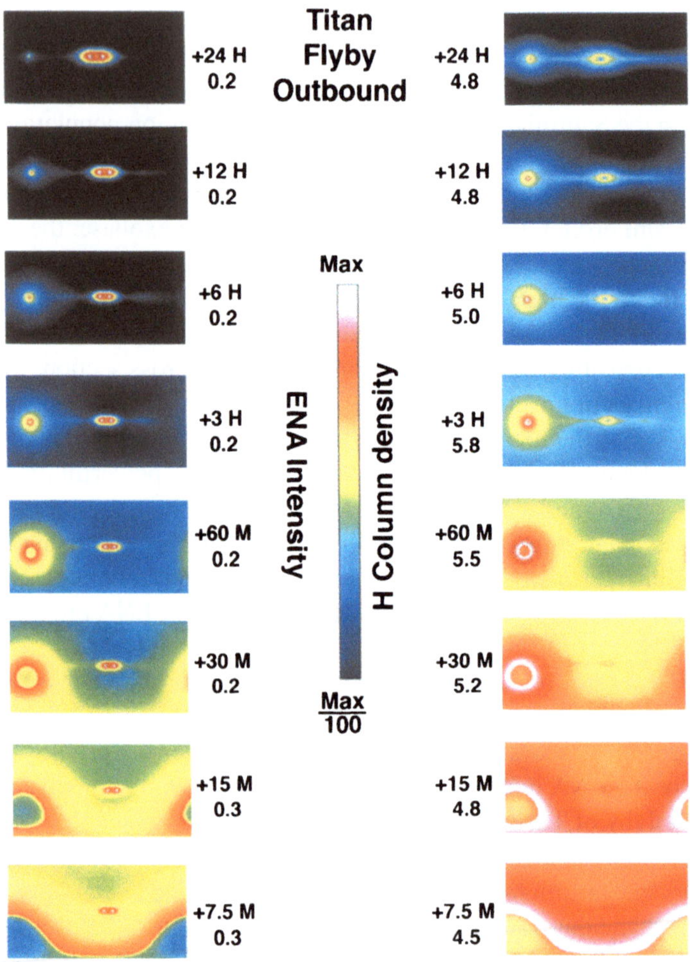

Figure 11. ENA intensity (left panels) and equivalent neutral hydrogen column density (right panels) for a simulated outbound Titan flyby of Cassini orbiter. Where hydrogen is molecular (Titan exosphere), density is $2[H_2]$. Mercator projections of the entire sky (-90 to $+90°$ in elevation centered on Saturn, and $360°$ in Saturn equatorial azimuth, centered for each panel so as to include both Saturn and Titan). Bold numbers are times relative to closest approach (M = minutes, H = hours). Logarithmic color bar (factor of 100) normalized to brightest pixel in each panel. Maximum values in each panel given as dimensionless numbers. Multiply dimensionless column densities by 1.2×10^{13} cm^{-2} (cf. Amsif *et al.*, 1997). Distant Saturn magnetosphere (brightest points are ring current at $L \cong 3$) is comparable in ENA brightness to Titan exosphere, while Titan hydrogen torus (idealized) is much weaker. Note ENA image of Titan exosphere is distinct from Saturn magnetosphere (except within an hour of closest approach).

In Figure 11 it is assumed that the energetic ion population in the vicinity of Titan is uniform and that its pitch angle distribution is isotropic. Well away from the moon these simplifications are the most justifiable. Near Titan, however, and possibly in the vicinity of its wake, finite Larmor radius effects will be important. As an example, a 20 keV proton in a 5 nT magnetic field, as was measured by Voyager near Titan (Neubauer, 1992), has a ~4000 km gyroradius. This is comparable to the Titan exobase radius, which is about 4175 km. Some dark areas in the ENA images are thus expected, in the one side of Titan, due to the screening effect from Titan and its lower atmosphere. In order to simulate these effects, a 3-D model of the interaction of Titan's exosphere with Saturn's magnetosphere has been developed (Amsif, 1996; Dandouras and Amsif, 1999). This necessitated 3-D trajectory tracing techniques for each ion. The results are shown in Figure 12, where a Titan flyby image sequence is simulated, for selected altitude values. Limb brightening effects above the Titan exobase are very clear in these images, and are due to the integrated optical depth in an optically tenuous medium, at certain grazing angles. The other salient feature is shadow effects in the ENA images of the Titan exosphere, which as explained above are due to the finite gyroradii of the parent ion population and the screening effect of Titan. These shadows depend on the ENA energy, on the relative geometry of Titan, and on the magnetic field and the location of the Cassini spacecraft. The INCA images will thus allow remote sensing, not only of the ion fluxes and spectra, but also of the magnetic field in the Titan vicinity. The magnetospheric imaging technique via energetic neutral atom detection is thus a powerful diagnostic tool for the interaction of Saturn's magnetosphere with Titan.

Typical ENA fluxes over Titan are generally of the order of 200–3000 $cm^{-2} \cdot s^{-1} \cdot sr^{-1}$. By taking into account the geometrical factor of the INCA ENA imager, which is ~2.4 $cm^2 \cdot sr$ for the full 90° × 120° field of view, and considering 32 × 32 pixel images with 86-s exposure (high time resolution images), we get on the order of 40–600 counts/pixel in the 10–50 keV energy range. This corresponds to Poisson uncertainties in the 4–15% range (at 1σ), showing the capabilities for performing magnetospheric imaging studies of the Titan–Saturn interaction with good resolution.

Note that the spacecraft velocity effects on this type of imaging are minimal. For a typical Titan flyby, the Cassini relative velocity is ~5.9 km/s, which corresponds to a ~507 km interval along the spacecraft trajectory during the 86 s exposure. If we take as an example a 2781-km minimum altitude flyby, the spacecraft altitude evolves from 5000 to ~4640 km, which corresponds to an evolution of the angle φ, for the exobase, from 33.5 to 35.4°. The resulting $\Delta\varphi = 1.9°$ is smaller than the 2.8° pixel resolution of the ENA images in the high time resolution images (90° × 120° field of view).

2.5. Auxiliary Science: Jupiter and Interplanetary Particle Environment

2.5.1. *MIMI Measurements at Jupiter*

In December 2000, the Cassini spacecraft obtained a gravity assist from Jupiter as it heads toward Saturn. It encountered the Jovian system with a closest approach distance of ~139R_J at ~1600 local time (LT). Several aspects of the Cassini

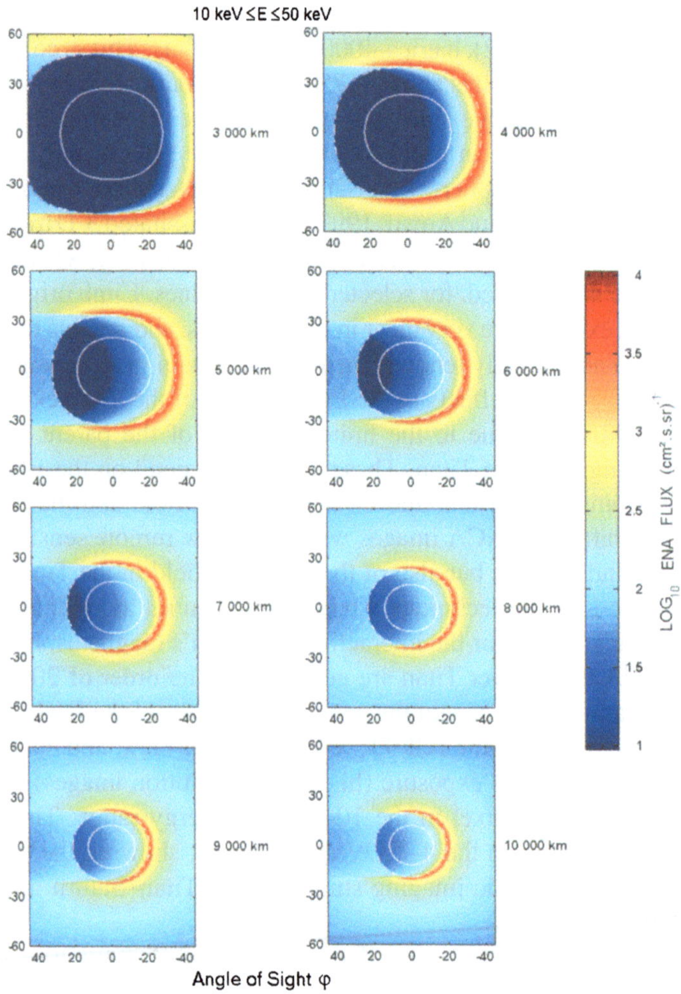

Figure 12. ENA flux as a function of the azimuth view angle ϕ and the elevation view angle θ, for eight altitude values of the Cassini orbiter above the Titan surface, ranging from 3000 to 10,000 km. The ENA energy range is 10–50 keV, and the images correspond to the INCA field of view (90° × 120°). The continuous white circle represents the Titan surface, and the dashed white circle represents the Titan exobase. (Dandouras and Amsif, 1999).

encounter with Jupiter have made it possible to address key science questions about Jupiter's space environment that other missions, including Galileo, have not been able to address as effectively or at all. Among these aspects are the unique capabilities of the MIMI instrument. In particular, at Jupiter the MIMI instrument has (Krimigis *et al.*, 2002):

(1) Provided the first global magnetospheric images of Jupiter using the ENA imaging capabilities of INCA head. The ability of MIMI to obtain high-quality

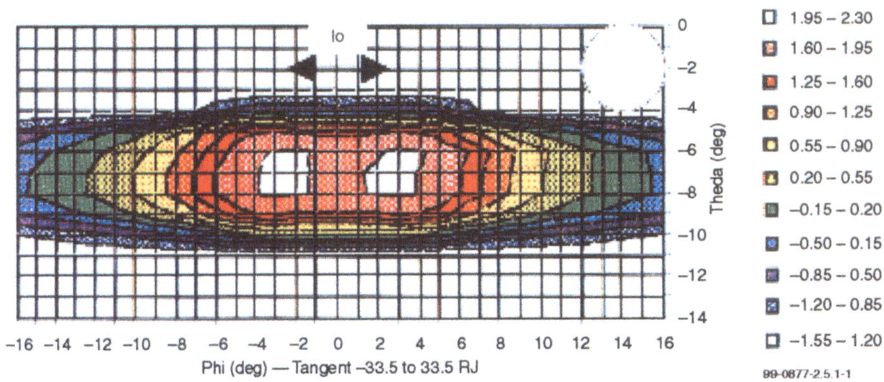

Figure 13. Simulation of an ENA (hydrogen) image of Jupiter's magnetosphere as it would appear to the MIMI/INCA head from $139R_J$ and 5° magnetic latitude. Color is coded according to the log 10 of counts of 60–120 keV H atoms as observed within 4° angular pixels for a 3-h accumulation. Deconvolution would result in a much thinner (more pancake) structure. The circle shows angular resolution capabilities of INCA after deconvolution.

images of Jupiter's inner and middle magnetosphere ($r < 30R_J$) during Cassini's gravity-assist flyby was documented by Mauk *et al.* (1998). A simulated raw image of Jupiter as it would be viewed by INCA from a distance of $\sim 139R_J$ is shown in Figure 13. The color is coded according to log 10 of the counts received in 3 h. Approximately 160 of these images have been obtained over a 20-day period, even while Cassini's main communications antenna remained pointed toward the Earth. The images will be sorted and summed according to the variations in systems drivers, including solar wind variations, Jupiter's rotation, the orbital position of Io, etc. The circle shows the anticipated resolution of the INCA measurements after deconvolution procedures have been applied. We expect to separately characterize the regions planetward of $\sim 10R_J$, the transition region between dipolar and magnetodisc configurations (10–$20R_J$), and the magnetodisc regions out to beyond $\sim 30R_J$. This analysis is currently in progress.

(2) Provided the first measurement of the relative concentrations of singly to multiply charged energetic heavy ions using the Cassini MIMI/CHEMS head. The Ulysses SWICS instrument made the first measurements of energetic ion charge states in Jupiter's magnetosphere (Geiss *et al.*, 1992), but was unable to uniquely identify singly charged oxygen and was essentially blind to singly charged sulfur (Geiss *et al.*, 1992). MIMI/CHEMS is the first instrument at Jupiter capable of making this discrimination. More sensitive than the SWICS instrument by a factor of ~ 30, CHEMS has been able to sense the charge state distributions of the outer magnetospheric regions remotely by characterizing the charge state distributions of pickup ions (PUIs) and other upstream ion events. These events are common in the upstream solar wind regions near Jupiter and are heavily

laden with heavy ions from Jupiter's outer magnetospheric regions and the Io torus (Zwickl *et al.*, 1981; Krimigis *et al.*, 1985; Barbosa and Eviatar, 1986).

With these capabilities the Cassini MIMI instrument will play a critical role in addressing several outstanding questions regarding the Jovian system including:

(a) *What is the influence of the interplanetary environment, relative to other drivers, on the structure and dynamics of Jupiter's magnetosphere and what are the mechanisms by which such influence is exercised?*

The dynamics of Jupiter's magnetosphere are driven predominantly by planetary rotation (Gehrels, 1976; Dessler, 1983). However, the Jovian magnetosphere is much more dynamic than might be anticipated in a system controlled largely by "rock-steady" planetary rotation, and there is increasing evidence that the interaction with the interplanetary medium can exercise an important influence on magnetospheric processes (Desch and Barrow, 1984; Barrow and Desch, 1989; Reiner *et al.*, 1993; Kaiser, 1993; Prangé *et al.*, 1993; Schneider and Trauger, 1995; Baron *et al.*, 1996; Clarke *et al.*, 1996; Satoh *et al.*, 1996; Mauk *et al.*, 1999). The MIMI/INCA head has obtained ENA images that could illuminate the dynamics of Jupiter's inner and middle magnetosphere while other instruments (dual technique magnetometer (MAG), Cassini plasma spectrometer (CAPS)) are monitoring the conditions of the interplanetary environment. These images may discriminate structures that are influenced by interplanetary conditions from those that are not.

(b) *How are hot plasmas transported and energized to generate and support Jupiter's unique magnetodisc configuration?*

It is now known that the magnetodisc is predominantly supported by pressures and pressure anisotropies in the sparse, hot plasma components (Mauk and Krimigis, 1987; Paranicas *et al.*, 1991). We do not know how these hot plasmas are energized and transported, and therefore we cannot begin to understand how these populations are able to form the magnetodisc configuration. According to the "neutral wind" model, corotating Io torus ions are neutralized by charge exchange, reionized in the middle and outer magnetosphere by photoionization, energized by pickup and other processes, and then transported back towards the planet by global radial diffusion processes (Eviatar *et al.*, 1976; Barbosa *et al.*, 1984). Alternatively, Io torus plasmas may be transported outward as a cold component by plasma processes, energized, and then transported inward as a hot component. The charge state of the hot heavy ions provided by the MIMI/CHEMS head, and the angular distribution together with the energy spectral shape provided by MIMI/LEMMS, yield a unique finger-print of the neutral wind model, because according to this model, singly charged ions would be very strongly favored when compared with the charge states measured in the Io torus (Belcher, 1983). Once the transport mechanism is established, researchers can then address mechanisms of magnetodisc formation.

Additionally, the Cassini spacecraft has been the first spacecraft to visit the dusk flanks of the Jovian magnetosphere near the equatorial plane, complementing Ulysses' high-latitude pass. This unique opportunity allows for the completion of the "exploration" of Jupiter's broad magnetospheric region. Of particular interest are the indications of solar wind and boundary layer flow, and the associated draping of the magnetic field configuration inside the dusk flank magnetosphere, as a signature of the strength of interactions between the solar wind flow environment and the corotating magnetospheric plasmas (Cheng and Krimigis, 1989a). Observations of detached Jovian plasmas may provide signatures of the plasma shedding processes in the outer Jovian magnetosphere.

2.5.2. *Pickup Ions*

Before arriving at Saturn, the Cassini spacecraft will spend half a solar cycle in interplanetary space, providing a unique opportunity to study in detail a variety of interplanetary particle populations and acceleration processes. MIMI/CHEMS with its large geometry factor and its ability to measure charge states and resolve ion species (e.g., N from O), will provide detailed information on the variations of these populations with radial distance and solar activity. Of particular interest will be (a) interstellar pickup ions (and the low energy portions of the anomalous component); (b) ion acceleration, especially at high-Mach number shocks and in corotating interaction regions (CIRs); and (c) bow shock processes upstream of Saturn and Jupiter (including leakage of energetic ions from the magnetospheres of these planets).

The presence of pickup ions in the inner solar system (~1–5 AU) is now well established (e.g., Gloeckler *et al.*, 1993; Geiss *et al.*, 1994, 1995; Gloeckler, 1996; Mall *et al.*, 1998; Gloeckler and Geiss, 1998). Investigation of the characteristics of this important heliospheric ion population has already revealed fundamental information about the acceleration of these suprathermal ions (Gloeckler *et al.*, 1994; Fisk *et al.*, 1997), as well as properties of our local interstellar cloud (Gloeckler and Geiss, 1996, 1998; Gloeckler *et al.*, 1997) that could not otherwise be obtained.

The Cassini orbit and the excellent capabilities of CHEMS will make it feasible to gain new information on pickup ions and to measure pickup Ar for the first time. With continuing pickup ion measurements at 1 AU on ACE and from ~1.5 to ~5.5 AU on Ulysses, together with the recent results from Voyager 1 (Krimigis *et al.*, 2000), the Cassini MIMI observations are essential for providing us with a good global, 3-D view of the spatial distribution of neutrals in the heliosphere. Examples of some of the anticipated results are described below.

2.5.2.1. Composition of the Interstellar Cloud. Because the density of pickup ions (except for He) increases almost exponentially with increasing distance from the Sun, Cassini MIMI measurements at ~10 AU with a more sensitive instrument (CHEMS) than those on Ulysses and ACE will give us the most precise information on the composition of pickup ions. In Figure 14 the pickup O^+ spectrum

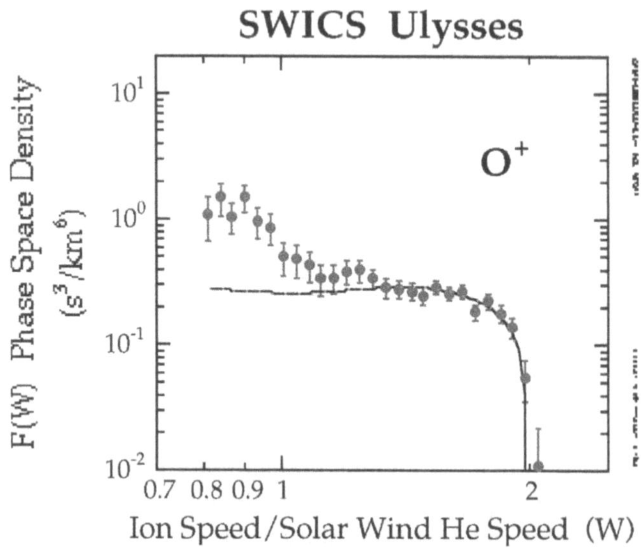

Figure 14. Ulysses SWICS observations of the phase-space density of O^+ averaged over 135 days from from 17 February to 1 July 1992. The average solar wind speed was relatively steady at 442 ± 36 km/s and the distance from the Sun remained close to 5.4 AU during the entire time period of these observations. The curve is the model calculation of interstellar pickup O^+. While the agreement between the model and data is excellent above W of about 1.3, a significant amount of O^+ from another source is evident below $W \sim 1.2$. Adapted from Figure 10 of Gloeckler and Geiss (1998).

obtained with Ulysses at ~5 AU is shown. The data used were averaged over a 5-month period in order to reveal the spectral shape in sufficient detail to identify both the interstellar and inner source for oxygen. With CHEMS at 10 AU the statistical accuracy of pickup ion measurements will improve by about a factor of 10. This will lead directly to establishing accurate elemental (and in some cases isotopic) abundances in the local interstellar cloud surrounding our solar system, information of considerable importance for models of stellar evolution. It will also provide powerful constraints on acceleration models for anomalous cosmic rays.

2.5.2.2. Ionization Rates of the Neutral Gas. Analysis of the distribution functions of pickup ions measured by CHEMS over a range of heliocentric distances ~3–10 AU) combined with similar measurements by SWICS on Ulysses, will give us our best determination of ionization loss rates of the various interstellar neutral species (e.g., H, N, O, Ne) inside the heliosphere. Knowledge of these ionization rates is essential for deriving accurate interstellar abundances.

2.5.2.3. Temperature of the Interstellar Gas. Measurements of the spatial variations of pickup ions with Cassini MIMI as its orbit crosses the so-called gravitational focusing cone (e.g., Holzer, 1977) gives us the only chance to determine the temperature of several atomic species (e.g., O, N, Ne) in the local interstellar cloud. Combining this with knowledge of the atomic He temperature (Witte *et al.*, 1996) will give clues to how this warm cloud is heated.

2.5.3. *Upstream Particle Events: Bow Shock Processes and Magnetospheric Leakage*

Cassini will spend most of its first 15 orbits upstream of the Saturnian bow shock. MIMI/INCA with its large geometry factor in the ion mode is ideal for studies of upstream ion activity, originally observed by Voyagers 1 and 2 (Krimigis *et al.*, 1983; Krimigis, 1986) out to $\sim 200 R_S$. The early Cassini apoapses lie in the dawn sector, where the spiral interplanetary magnetic field is most likely to connect to the bow shock. Substantial upstream particle activity is therefore expected and will be detected by MIMI/LEMMS. During the Voyager 2 flyby, upstream ion bursts were present at least half the time (Krimigis *et al.*, 1985).

At a planetary bow shock, solar wind ions are reflected into the upstream region. Theories of first-order Fermi acceleration (e.g., Lee, 1982) suggest that such ions should act as a seed population for diffusive acceleration by ion-generated irregularities upstream of the bow shock. Cassini measurements offer the opportunity to perform extensive tests of this model at a planetary bow shock with higher Mach number and much larger dimensions than at Earth. The acceleration region upstream of Saturn is at least ten times larger, so that the cutoff energy in the spectrum should be substantially larger than the ~200 keV/e calculated for the case of Earth. MIMI is ideally suited for making the required measurements of energy spectrum, composition, and anisotropy to test diffusive shock acceleration models.

Another source for upstream particle events is leakage of energetic magnetospheric ions (and electrons) into the interplanetary medium. Observations at Earth, Jupiter, Saturn, and Uranus have been interpreted as favoring this hypothesis (Krimigis, 1986, 1992). MIMI/INCA and MIMI/CHEMS will measure ion composition and charge state down to ~7 keV/nuc and ~3 keV/e, respectively, unambiguously separating accelerated solar wind ions (high charge states) from ions of magnetospheric origin (low charge states). MIMI/LEMMS will measure the anisotropy of these events at energies above 30 keV.

2.5.4. *MIMI/INCA Remote Sensing of Heliospheric Shocks*

ENAs are produced whenever energetic singly charged ions undergo charge-exchange (electron capture) collisions with a background population of neutral atoms. These conditions occur not only within planetary magnetospheres, but also in the interplanetary medium throughout the heliosphere. The neutral atoms therein are those of the interstellar gas that flows through the heliosphere and thereby generate the pickup ions, discussed in Section 2.5.2. Those ionization processes systematically diminish the density of interstellar neutrals in the inner heliosphere. For hydrogen, the dominant atom, the attenuation is the most severe, setting in at ~3 AU and reducing the H atom intensity to ~95% of the interstellar value by 1 AU. The energetic, singly charged ions, again predominantly hydrogen, are protons accelerated by processes associated with the formation and propagation of heliospheric shocks.

Three types of shocks can accelerate protons to energies >10 keV and thus be imaged in ENA by MIMI/INCA: (1) the forward and reverse shocks produced by corotating interaction regions (CIRs); (2) the driven shocks associated with large solar flare ejecta and coronal mass ejections (CMEs); and (3) the termination shock (TS) of the heliosphere itself, where the supersonic solar wind makes a transition to the subsonic interstellar plasma. Roelof (1992) estimated the ENA intensities >10 keV as they would be viewed by Cassini at Saturn for the first two types of interplanetary shocks, while Hsieh *et al.* (1992b) provided estimates from 1 AU. A topic of concern in both studies was whether the ENA emission from interplanetary shocks of either type would obscure the ENA emission from the heliospheric termination shock. Hsieh *et al.* (1992a) had proposed that ENA would be produced at the TS by low-energy anomalous cosmic-ray protons that are eventually accelerated to ~100 MeV. Hilchenbach *et al.* (1998) claimed to have identified ENA hydrogen at 55–80 keV coming from the anti-apex direction using the CELIAS/HSTOF ion telescope on the SOHO spacecraft during solar minimum conditions.

While the construction of ENA global images of the energetic proton population associated with the heliospheric termination shock is an exciting possibility for MIMI/INCA, the imaging of interplanetary shocks is practically expected. Based on the Voyagers 1 and 2 measurements of differential proton intensities j_{ion} for $E > 35$ keV, Roelof (1992) estimated the global distribution of ENA hydrogen intensities as they would be viewed from 10 AU (Saturn orbit) from the general relation

$$j_{\mathrm{ENA}} = \sigma \int \mathrm{d}s\, n_{\mathrm{H}}\, j_{\mathrm{ion}}.$$

Using Voyager proton spectra representative of the peak values at the shock j_0, (E) the integral intensity (above an energy E) can be written as

$$J_{\mathrm{ENA}} = \int \mathrm{d}E\, j_{\mathrm{ENA}} = n_0 I \int \mathrm{d}E\, \sigma_{16}\, j_0,$$

where the integral over the spatial dependence of n_{H} and j_{ion} is

$$I = (1.5 \times 10^{-3}\,\mathrm{cm}) \int (\mathrm{d}s/a)(n_{\mathrm{H}}/n_0)(j_{\mathrm{ion}}/j_0).$$

Here σ_{16} is the charge-exchange cross section in units of $10^{-16}\,\mathrm{cm}^2$, $a = 1$ AU, and $n \cong 00.1\,\mathrm{cm}^{-3}$ is H atom density in the interstellar medium.

Values of the Voyager peak intensity spectra (normalized to 10 AU) and the integral $\int \mathrm{d}E\sigma_{16} j_0$ are compiled in Table II, while Figure 15 presents the spatial dependence of I as "fish-eye" ENA images of the sunward hemisphere (as viewed

TABLE II

Integral ENA energy coefficient derived from Voyager energetic ions.

E (keV)	10	20	40	70	100	200	400
σ_{16} (cm^2)	7.75	4.45	1.67	0.377	0.101	6.09(3)	1.76(4)
j_{ion} (cm^2 sr s keV)$^{-1}$	[0.70]	[0.55]	0.35	0.20	0.125	0.036	0.010
dE s$_{16}$ j_{ion} (sr s)$^{-1}$	[71.6]	[34.6]	8.81	1.36	0.31	9(3)	1(4)

Ion fluxes in [brackets] are extrapolated downward in energy from 35 keV.

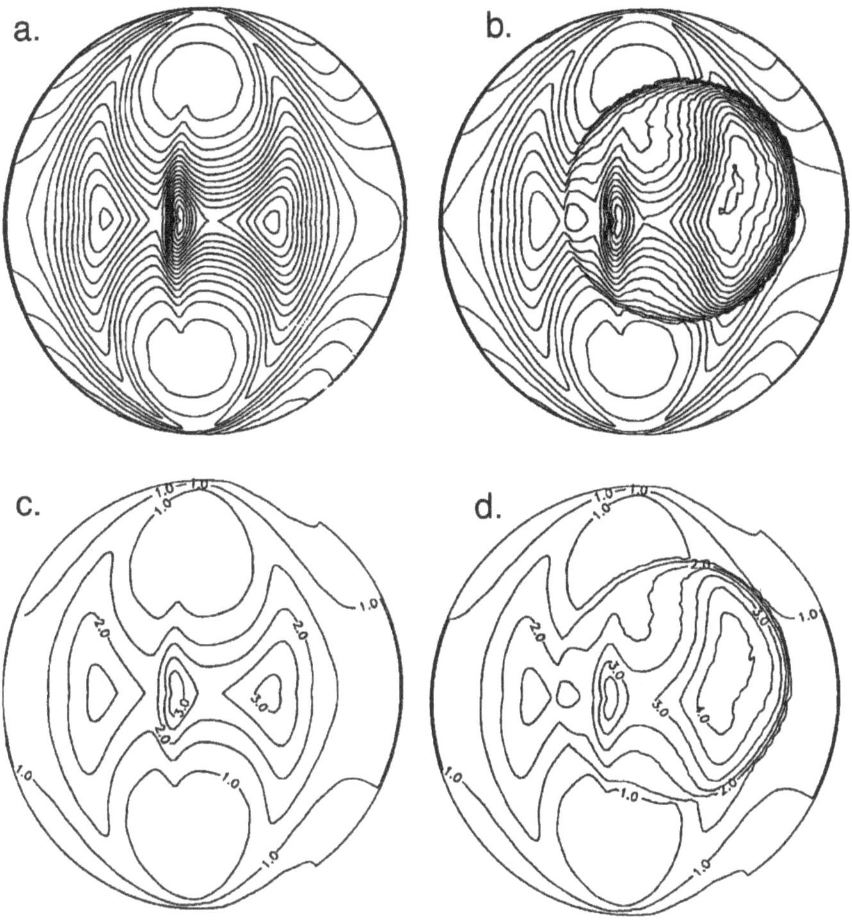

Figure 15. Sunward hemisphere views from Saturn (10 AU) of simulated ENA images of interplanetary shock-associated energetic protons from (a) forward and reverse shocks bounding a corotating interaction region (CIR) and (b) solar transient shock superimposed upon a CIR. ENA flux contours are labeled in (c) and (d) in units of $100I$, as defined in the text, and may be converted into ENA hydrogen flux (as a function of energy) using the values given in Table II.

from Saturn). The ENA images of a large CIR are shown in the left-hand panels, while a large driven shock has been superimposed upon the CIR in the right-hand panels. Contours with $2° \times 2°$ are shown in the upper panels, while a subset of contours of the quantity $100I$ are drawn in the lower panels. The antisunward hemispheres (not shown) exhibit very low contrast, with contours only between 0.5 and 1.0. Based on the instrument response characteristics of MIMI/INCA, Roelof (1992) estimated that images like those shown could be accumulated (in Cassini roll mode) in 3 days with ~10% counting statistics in $6° \times 6°$ pixels. These statistics were deemed adequate for interpreting the image in terms of global topology, particularly in the portions of the image where the contrast is strongest (at the edge of the shock structure). It was also concluded, in agreement with Hsieh *et al.* (1992b), that only the largest CIR or transient solar shocks would obscure the ENA emission from protons associated with the global heliospheric shock.

3. Description of Instrumentation

The MIMI instrument consists of one set of electronics servicing three detector heads which perform a broad variety of measurements. Figure 16 shows the mechanical configuration of the instrument on Cassini with each of the major

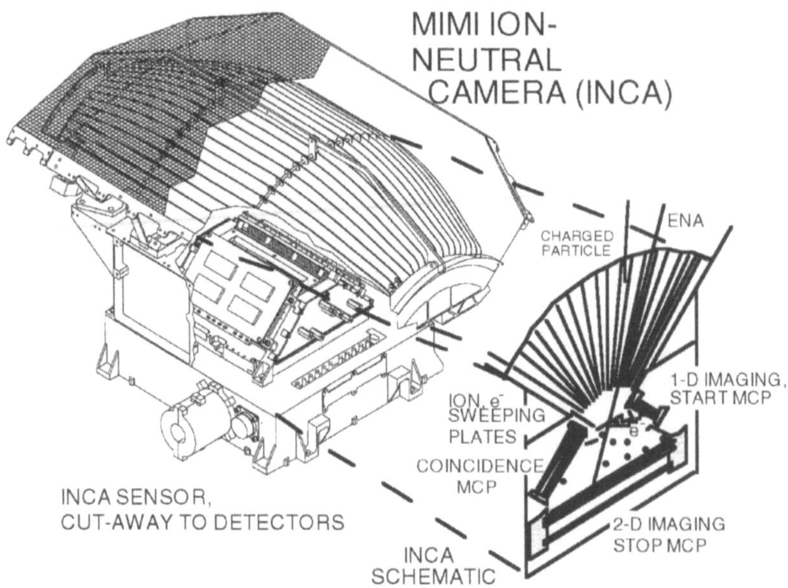

Figure 16. Perspective drawing of the INCA head, with section schematic. Serrated charged collimator plates sweep charged particles out of aperture. ENAs penetrate front foil covering entrance slit, producing secondary electrons for start pulse and 1-D position. ENA travels to back foil and 2-D imaging microchannel plate (MCP), and back-scattered secondary electrons provide coincidence pulse. Dots indicate the locations of wire electrodes for secondary electron steering.

subsystems detailed. The electronics box contains the data processing unit (DPU), as well as all the digital processing electronics for all three detector heads (LEMMS, CHEMS, and INCA). The LEMMS head is double-ended, with oppositely directed 15 and 30° (full angle) conical fields of view (FOV). LEMMS is mounted on a rotation platform, with the spin axis parallel with the spacecraft y-axis, such that when rotating, the LEMMS telescopes sweep through 360° in the spacecraft x–z plane. The CHEMS head sits on the particles and fields instrument pallet, with a field of view of $\sim$160° in latitude (bisected by the spacecraft x–y plane) by 4° in azimuth centered on the spacecraft x-axis. CHEMS measures the charge state and composition of ions from $\sim$3 to 220 keV/e, comprising the most energetically important portion of the Saturnian magnetospheric plasma. The INCA head is separately mounted and nearly co-aligned with the remote sensing instruments, with a field of view $\sim$ 120° in latitude by 90° in azimuth, centered on a vector rotated 9.5° toward the spacecraft $+x$-axis from the $-y$-axis. INCA makes two different types of measurements. It obtains the directional distribution, energy spectra, and crude composition of magnetospheric ions between $\sim$7 keV/nuc and $\sim$3 MeV/nuc; and it makes remote images of the global distribution of the energetic neutral emission of hot plasmas in the Saturnian magnetosphere, measuring the composition and energy spectra of those energetic neutrals for each image pixel. When the spacecraft is rotating, the MIMI sensors obtain measurements over $\sim 4\pi$. The different MIMI sensors share common electronics and provide complementary measurements of energetic plasma distribution, composition, and energy spectrum, and the interaction of that plasma with the extended atmosphere and moons of Saturn.

3.1. The INCA Sensor

INCA is a large ($\varepsilon G \sim 2.4$ cm^2 sr; where G is the geometric geometry factor, and is the efficiency) time-of-flight (TOF) detector that analyzes separately the composition and direction of motion of the incident energetic neutral atom. Elements of the detector assembly can be seen in Figure 16. Sensor characteristics are summarized in Table III. The entrance includes a serrated-plate fan for charged particle deflection, with an FOV of 90° in the nominal (z-axis) spin direction (the azimuthal direction) by 120° in the direction perpendicular to the spin plane (elevation), centered on the spin plane. The calculated angular response in the 120° direction, including efficiencies is shown in Figure 17.

3.1.1. *Deflector*

The magnesium deflector plates are serrated to inhibit forward scattering of incident particles. Commandable potentials of up to $\pm$6 kV are applied to alternate plates to sweep energetic charged particles with energies $\leq$500 keV/e into the plate walls (excluding them from the detector). The shielding effectiveness of the collimator has been measured in the laboratory (Figure 18; also Mitchell *et al.*, 1996) and is

TABLE III

INCA sensor characteristics.

Energetic neutral atoms or ions, chosen by command	
Energy range	<7 keV to 3 MeV/nuc
Velocity resolution	50 km/s (1 ns TOF)
Mass resolution	H, O, Heavies
Field of view	120° × 90°
Angular coverage	0.7π sr (π, if spinning)
Angular resolution	~8° × 4°, $E_H > 50$ keV
(scattering in foils at low E)	>8° × >8°, $E_H \sim 20$ keV
Time resolution	6 s, PHA events
	85 s, low resolution
	6 min, high resolution
	23 min, full sky
$G \times \varepsilon$ (cm^2 sr)	~2.4 for O, ~0.6 for H
$G \times \varepsilon$ for 4° × 4° pixel (O)	0.007, tapered at edges
Dynamic range	$\sim 10^7$

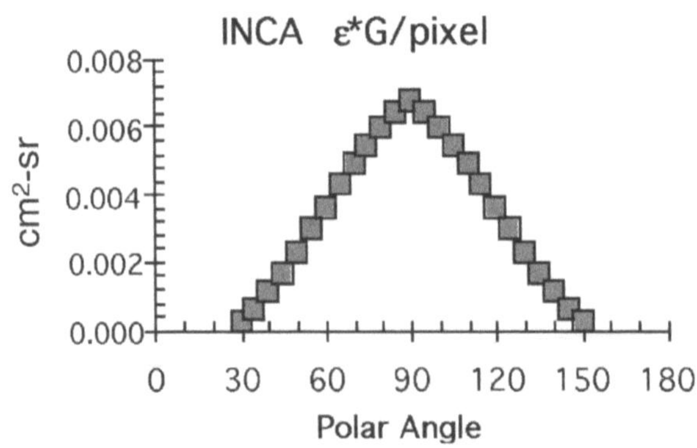

Figure 17. INCA polar angle response function, ε^*G/pixel (oxygen).

adequate to permit neutral particle imaging below the sweeping energy even while the instrument is within moderate magnetospheric plasma environments.

3.1.2. *Measurement Technique*

Incoming neutral particles encounter and penetrate a thin foil, producing secondary electrons (Figure 16). The secondary electrons are first accelerated perpendicular to the foil by the E-field which is locally normal to the equipotential surface of the foil, and then steered electrostatically (using a combination of wires and shaped electrodes at fixed potentials) onto a side, start microchannel plate (MCP) with 1-D

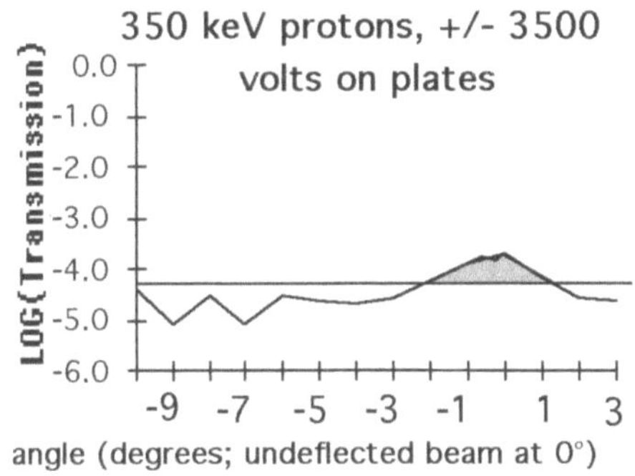

Figure 18. Accelerator test of deflector ion sweeping at an intermediate voltage, indicating charged particle rejection at better than four orders of magnitude. The peak with shading is in the position of the undeflected beam, and results from energetic neutrals created via charge exchange of energetic ions in the Van deGraff beam tube.

position sensitive anode. A start time taken from a capacitively coupled electrode of the start MCP is generated by this event. The original incident particle, after some angular scattering in the front foil, continues through the instrument, striking a second foil just in front of the stop MCP assembly. Secondary electrons produced on the exit surface of the imaging foil are accelerated into the stop MCP and 2-D imaging anode, mapping the position of impact and registering the stop time for the TOF measurement. Figure 19 presents the basic geometry used to determine the ENA trajectory through the sensor. In addition to this TOF and trajectory measurement, secondary electrons produced as the ENA enters the back foil are electrostatically accelerated and guided to the side coincidence MCP (see Figure 16). The electron travel time for these back-scattered electrons is constrained to <40 ns by the steering potentials. The pulse generated in the coincidence plate is used in the valid

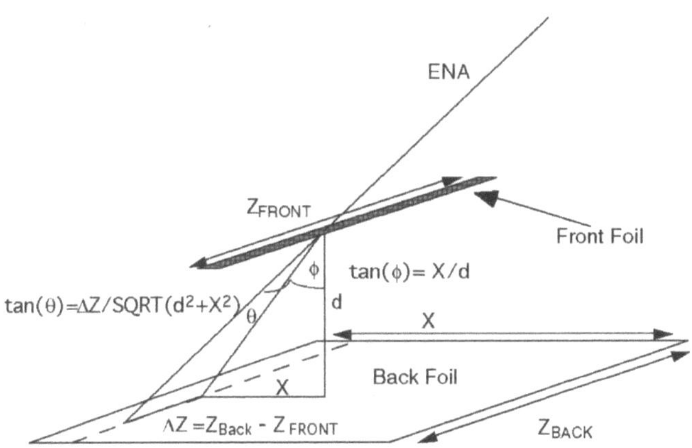

Figure 19. INCA imaging geometry.

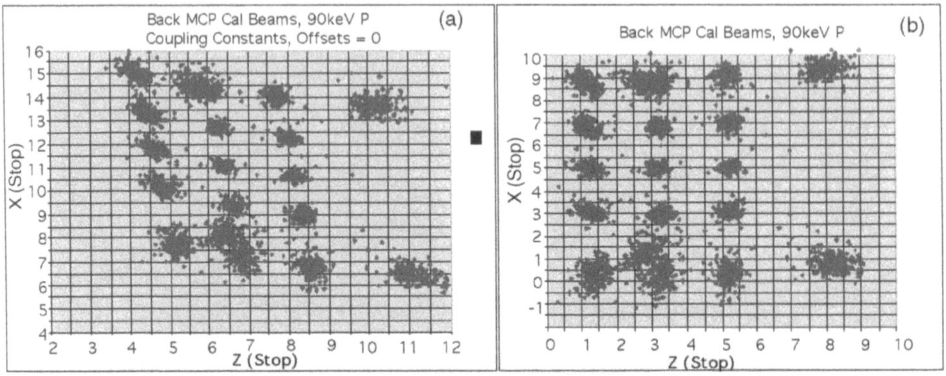

Figure 20. (a) INCA back microchannel plate (MCP) positions (including scattering in front foil) for 90 keV protons, before correction for interelement capacitance. (b) Same data as in (a), with capacitive coupling correction factors applied. Pattern reflects beam positions faithfully.

event logic circuitry as a coincidence check on the measurement, further reducing the probability that uncorrelated background will be falsely counted as a neutral.

The position-sensitive anodes used to register the front foil and back foil positions use proportional charge deposition on shaped elements (wedges and strips) to yield position. Because these elements are in very close physical proximity, there is considerable capacitive coupling between them. The resulting image is skewed, as shown in Figure 20.

3.1.3. *Mass Determination*

The number of secondary electrons produced in each foil is dependent on the atomic number of the neutral atom; for the two most common neutral atoms expected, oxygen has been found to produce several times the number of secondaries that hydrogen will produce. Exploiting this phenomenon, we record the pulse-height of the MCP signal and determine the species based on that measurement (Table IV).

As shown in Figure 21 and Table V (which is based on data shown in Figure 21, as well as higher energy data similar to that in Figures 21 and 22), MCP pulse-height is sufficient to discriminate hydrogen from oxygen. Since hydrogen is expected to

TABLE IV

Algorithm for mass determination based on MCP pulse-height, using the inequality IF[$(\mathrm{PH}_{\mathrm{BACK}}) \cdot \mathrm{GT} \cdot (B - A \times \mathrm{PH}_{\mathrm{FRONT}})$] THEN ⇒ SPECIES = Oxygen.

Energy/nucleon (keV/nuc)	*B*	*A*	H counted as O (%)	O counted as H (%)
31	16,000	4	0.2	17.08
46	19,000	4	0.2	5.34
80	19,000	4	0.2	?

TABLE V
INCA calibration resolution results.

keV/nuc	s_θ H (°)	s_ϕ H (°)	s_θ O (°)	s_ϕO (°)
5	11.6	15.1	7.1	4.6
20	6.75	6.5		
31	5.25	3.5	2.9	1.7
46	3.2[a]	2.6[a]	2.0	1.8
80	3.9[a]	1.8[a]	2.8	1.4
90	4.8	2.2		
200	4.2	1.3		

[a]Large pulse-height events only.

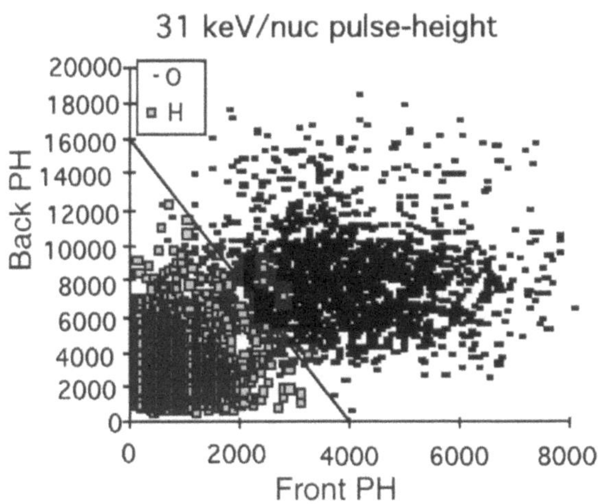

Figure 21. Separation of oxygen and hydrogen using MCP pulse-height signatures. 31 keV/nuc O and H measurements, overplotted. The hydrogen separates as lower MCP pulse-heights. At higher energies, this separation becomes more complete.

dominate the low pulse-height data, the fraction of O mistakenly identified as H is not considered significant. The measured positions of the penetration of the foils determine the particle arrival direction and its path length within the instrument; the TOF associated with this path length (and thus velocity) combined with the species determination based on pulse-height analysis determine particle energy. Although the species are not well resolved, the neutral fluxes at Saturn are expected to consist almost exclusively of hydrogen and oxygen, and these will be relatively well determined.

3.1.4. *Foils and UV Sensitivity*

Electrons are also produced in the foils by photoionization by extreme and far ultraviolet light (EUV and FUV), predominantly in the Ly-α line at Saturn. The

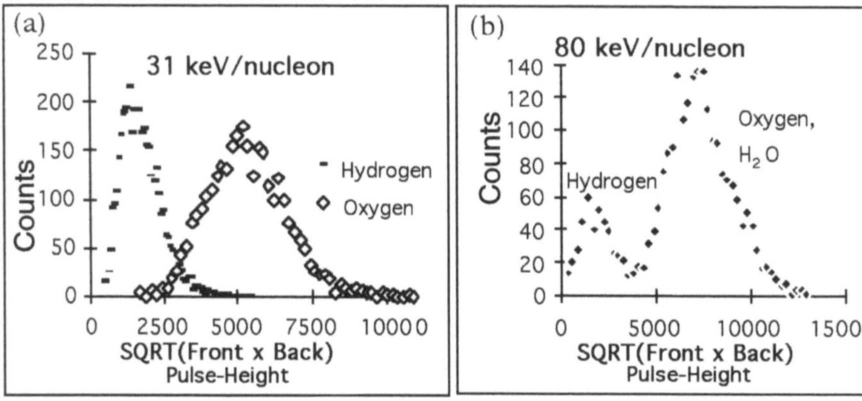

Figure 22. (a) Data shown in Figure 21, re-cast as square root of front pulse-height back pulse-height, showing the separation between hydrogen and oxygen. (b) Same format, for a run at 80 keV/nuc, in which the dominant constituent was water. Some molecules broke up in the beam tube, resulting in a mix of hydrogen and oxygen in the same beam with the same velocity. The separation by pulse-height is clear.

chosen foil must therefore be thick enough to attenuate the Ly-α to an acceptable level, so that the position and TOF circuitry is not swamped by false counts. The TOF measurement is made by recording a start signal followed by a stop signal some time later. Any single photon can produce a start or a stop, but not both. As long as the time between a false Ly-α start and a separate false Ly-α stop is long compared with the maximum valid TOF period ($\sim$100 ns), this background will not produce false events. The electronic design assumes nominal upper limits for Ly-α start events of $\leq 5 \times 10^4$ secondary electrons/s, and for stop events of $\leq$1000 secondary electrons/s. The peak Ly-α flux at Saturn (excluding the Sun itself) is $\sim$1 kR (Broadfoot *et al.*, 1981), which corresponds to $\sim 6 \times 10^8$ photons/s incident on the front foil (which has a geometry factor of $\sim$7 cm^2 sr). The secondary electron production efficiency for Ly-α photons is $\sim$1% (Hsieh *et al.*, 1980). The INCA foil comprises 5.5 mg cm^{-2} Si, 3.6 mg cm^{-2} Lexan, and 1.0 mg cm^{-2} C with a measured 5.0×10^{-3} Ly-α transmittance to reach a conservative design goal of $\leq 3 \times 10^4$ Ly-α induced secondary electrons/s when the brightest portion of the Saturnian atmosphere/exosphere is in view. The choice of Si and C for the foil was based on the work of Hsieh *et al.* (1991). This foil is mounted on a 70 line/inch nickel mesh (82% transmission). The stop foil, composed of Lexan and carbon, will reduce the Ly-α flux to the back, 2-D imaging MCP by about another factor of 100. With the Sun in view at Saturn, because the quantum efficiency for secondary electron production is so high at EUV wavelengths, no practical combination of foil composition and thickness could be identified to sufficiently suppress it. We therefore included a recloseable shutter, designed such that when the Sun is nominally in view, the shutter will reduce the geometry factor enough (by $\sim$30 times) to allow operation.

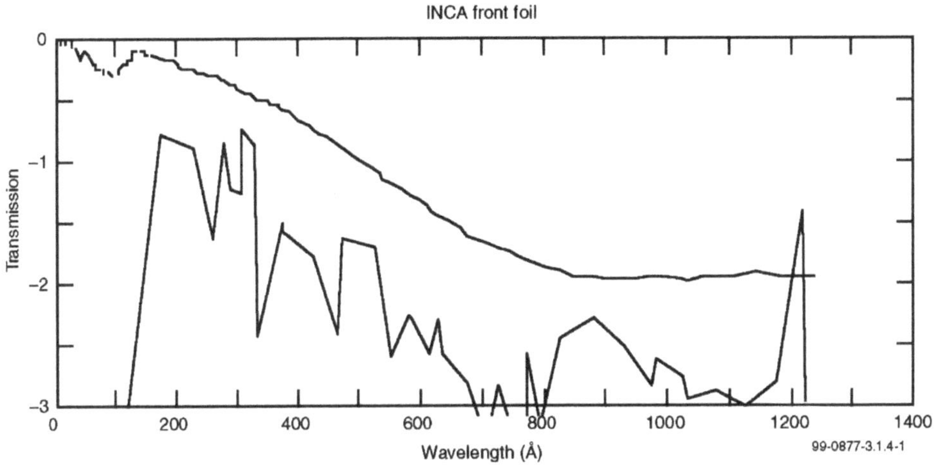

Figure 23. Modeled transmission of INCA front foil as a function of wavelength. Piecewise integrated solar spectrum, after filtering by the foil, is shown in lower curve.

The measured Ly-α front foil transmission, in the lot from which the flight foil was selected, is approximately 5×10^{-3}. The predicted transmission, as a function of warelength, is shown in Figure 23.

3.1.5. *Angular Resolution*

Angular scattering in the foils, along with the angular spread introduced in the electron optics, has been measured during accelerator tests of the complete engineering model INCA sensor. Representative angular standard deviations are given in Table V.

The ϕ resolution is dominated by scattering in the front foil and the entrance slit width in all cases. The θ resolution is dominated by the scattering in the front foil and by the electron optics for the front foil imaging. The latter effect is dominant at higher hydrogen energies, where the number of secondary electrons produced on average is $\sim$1. The position resolution achievable by the start electron optics is modeled in Figure 24.

For the models, data from Meckbach *et al.* (1975) was used to represent the secondary electron energy spectrum. The figure suggests that for single secondary electrons, the spread in the start position will have a full width half maximum (FWHM) of $\sim$2–4 mm. This is consistent with our calibration results for high energy protons. For moderate proton energies, a fraction of the events will produce two or more secondaries; therefore, we will be able to create separate images selected on the basis of start anode pulse height, allowing for higher resolution to be achieved on the two or more secondary events. This has the effect of improving the standard deviations stated above for >45 keV hydrogen by about 1°.

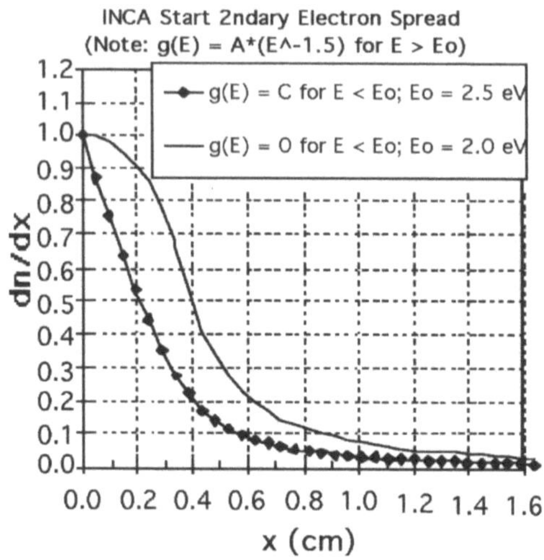

Figure 24. Model start position uncertainty due to random velocities of start secondary electrons. The constant $C = g$ (2.5 eV).

3.1.6. *Sensitivity and Background Rejection*

The INCA head relies upon valid TOF (separate start and stop) measurements, as well as coincidence pulses to differentiate background from foreground events. EUV will produce secondary electrons, but not correlated start, stop, and coincidence pulses. In order to produce false events that appear valid, an EUV start must be followed by EUV (or other) stop and coincidence pulses. Other sources of false signals may come from high-energy penetrating 2 MeV energetic electrons, which should only become a significant background inside $L = 5$–6, as well as ion fluxes above the rejection cutoff energy of the deflection plates. The latter will be discriminated against by their TOF signatures (i.e., too fast to be valid neutrals below the cutoff), although we expect to be able to image neutrals up to ~6 MeV when the ambient energetic ion flux above the cutoff energy is sufficiently low. The expected environment will give foreground count rates ranging from $\sim 10^3\ s^{-1}$ for the inner Saturnian magnetosphere down to $\sim 1\ s^{-1}$ at great distances.

The system has an intrinsic window for valid events of ~100 ns, based on the maximum valid TOF for a ~50 keV oxygen (which is the slowest valid ENA expected to be analyzed), and the coincidence window is ~40 ns. The background rate for false valid events from uncorrelated background rates will therefore be $\sim 1 \times 10^{-7} \cdot \sim 4 \times 10^{-8} \cdot R_{start} \cdot R_{stop} \cdot R_{coincidence}$, where R_{start} (R_{stop}, $R_{coincidence}$) is the uncorrelated singles rate due to penetrating background and EUV on the start (stop, coincidence) MCP. Thus, the false valid event rate is $\sim 4 \times 10^{-15} \cdot R_{start} \cdot R_{stop} \cdot R_{coincidence}$, assuming the accidental rates are uncorrelated. For the EUV generated background rates expected with the design foil thicknesses, i.e., $R_{start} = R_{coincidence} = \sim 30{,}000$ counts/s and $R_{stop} = \sim 300$ counts/s (see discussion in Section 1.3), the maximum false coincidence rate attributable to EUV, will be (4.5 $\times$ 10^{-4} events/s in regions with low foreground rates. If a penetrator hits both MCPs and produces a correlated pulse pair, the probability of a false coincidence event being registered goes up. However, by requiring a minimum TOF of about 1.5 ns (corresponding to a $\sim$6 MeV proton TOF), we will discriminate against some of these very fast particles. Using the coincidence requirements, background rates from penetrators (cosmic rays and magnetospheric energetic particles) combined with uncorrelated EUV, rates are expected to be <0.001 s^{-1} at apoapsis, increasing as Cassini enters the penetrating >2 MeV electron flux in the inner magnetosphere.

Calibrations made with a setup including both a low intensity 50 keV proton beam and a high-intensity UV light source (Figure 25) show that even with an order of magnitude higher UV rates than those expected at Saturn (when the Sun is not in the FOV), the protons can be imaged with a signal-to-noise ratio of between 30 and 100. It should be noted that when MIMI flew past Venus, the accidental rate from the intense UV at Venus was higher by approximately one order of magnitude than would be predicted by the above discussion. While this remains unexplained, the background at Earth from UV was not significant, and the UV background expected at Saturn should be lower than the Earth environment even in the worst case.

Because hydrogen produces only $\sim$1 secondary electron, on average, when it strikes a surface (foil, MCP), any given hydrogen atom may not produce any secondaries at a surface. Therefore, many hydrogen atoms are not counted at all, because they do not produce all three pulses required to be registered as valid events (start, stop, and coincidence pulses). For this reason, protons suffer from an additional (energy dependent) efficiency factor of $\sim$0.2–0.5, relative to oxygen, which produces an average of $\sim$6–8 electrons at a surface.

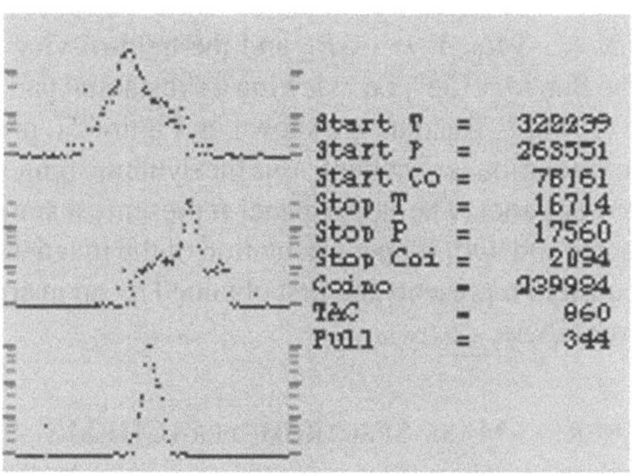

Figure 25. Calibration data for UV rates $\sim$300,000 counts/s. The 50 keV protons in the beam can be seen in the position plots (front foil top, back Z position middle, back X position bottom). Rates are displayed to the right.

3.1.7. *INCA Analog Processing*

Figure 26 shows a block diagram of the INCA analog processing electronics. The INCA sensor generates an MCP start signal, an MCP stop signal, and MCP position signals. The timing signals will be processed in the time-to-amplitude converter (TAC) circuitry over the range of ~2 to ~100 ns with total system resolution of ~1.0 ns. The TAC processing time through the A/D converter is ~10μs. Position measurements are derived from the start and stop MCP anode signals. The timing and angle information from a valid event is latched into an event register, where it is stored until the DPU is ready to process it.

3.1.8. *INCA High Voltage*

Chevron stacks of two MCPs require high bias voltages extending up to about 3500 V to operate properly. While these voltages are high, very little current is drawn (<60 μA), so power consumption is low. These bias voltages are commandable to adjust the gain of the MCP devices, thus correcting any gain shifts due to aging, burn-in, radiation effects, etc. Secondary electron focusing is accomplished with voltages provided by the same supplies. The INCA sensor contains a total of three MCP high-voltage supplies, one to bias the start MCP, one to bias the coincidence MCP, and the third for the stop MCP. The electron steering elements use voltages provided by these same supplies. INCA also has similar high-voltage supplies to bias the charged particle deflection plates. These supplies are commandable, such that the deflection plate voltages can be varied between a low voltage and about 6 kV.

3.1.9. *INCA in-Flight Performance*

The INCA instrument has returned excellent data during the Earth and Jupiter flybys, while its response at Venus was dominated by intense UV interference, as pointed out in Section 3.1.6.

One of the Earth flyby images was obtained while the spacecraft was located at GSE coordinates $X = -9R_E$, $Y = -3R_E$ and the field-of-view spanning a range from ~45° from the Sun to ~135°, i.e., viewing a substantial part of the ring current from duskward to tailward. The data are shown in Figure 27, projected as detailed in the caption. The resolution is excellent, as is the dynamic range of the instrument, as shown in the lower panel. The upper panel represents a smoothed distribution with the Earth to scale and with proper orientation of the magnetic axis. This image of the partial ring current represents the first obtained by an instrument specifically designed to measure ENAs.

3.2. CHARGE–ENERGY–MASS-SPECTROMETER (CHEMS)

The CHEMS instrument will characterize the suprathermal ion populations in the magnetosphere of Saturn and in interplanetary space by measuring the 3-D distribution functions of elemental and molecular ions. CHEMS uses energy

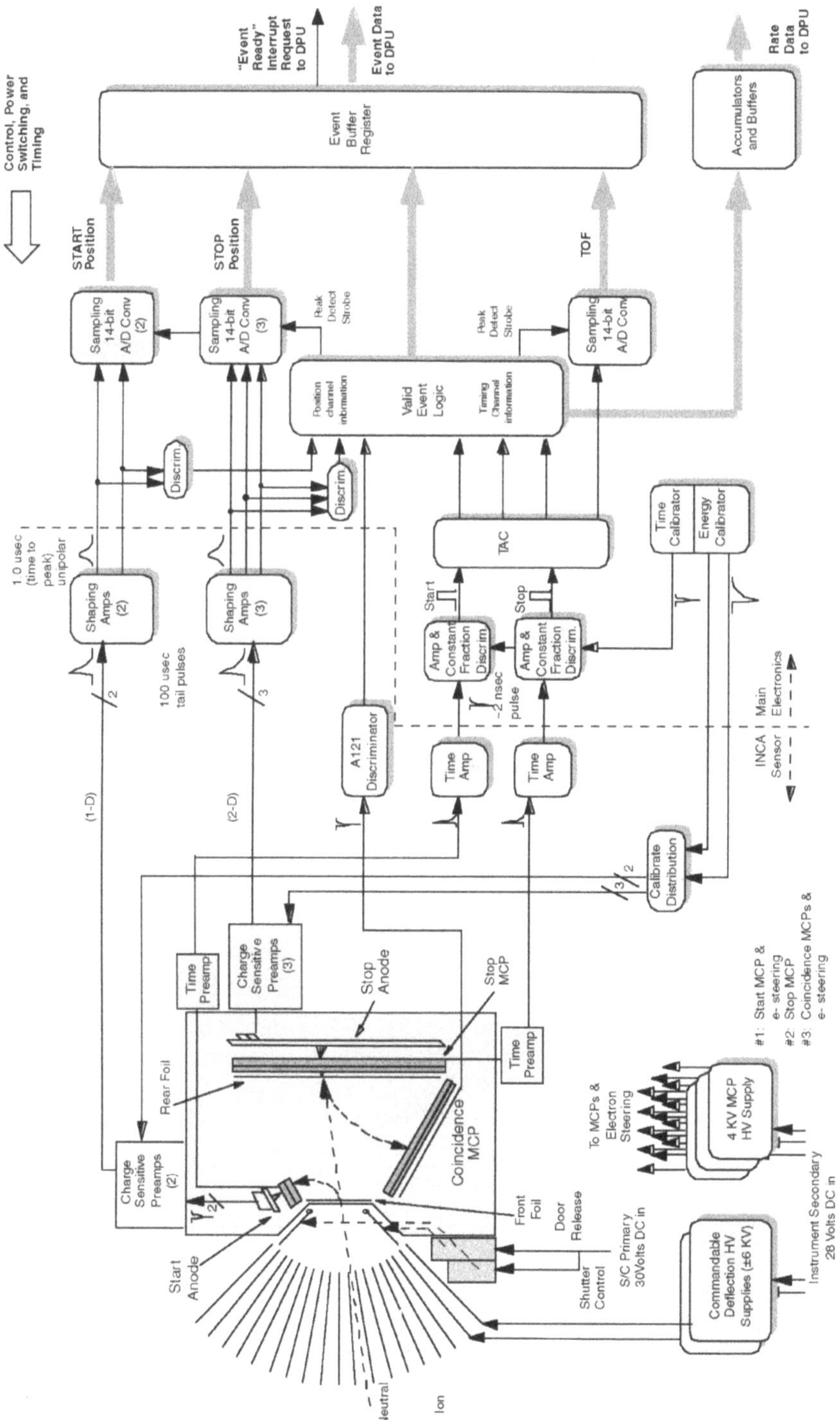

Figure 26. Functional block diagram for INCA.

(a)

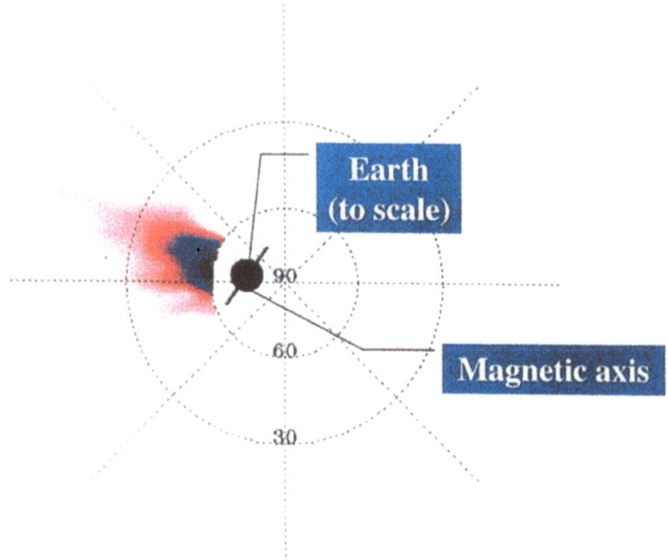

(b)

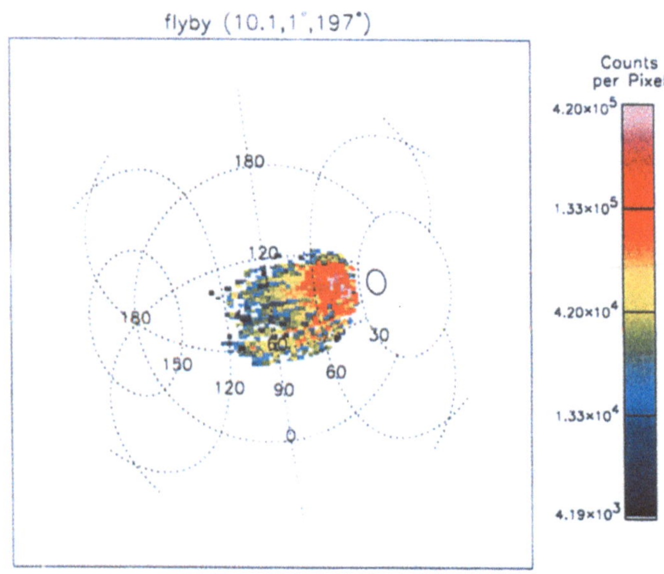

Figure 27. ENA images obtained by the INCA sensor August 18, 1999, 4:17–4:35 UT, during Cassini's Earth flyby. (a) Smoothed INCA image of ring current from time-of-flight channel (27–39 keV, H atoms). Representation is a fish-eye projection centered on the Sun. Cassini is $10.1R_E$ from the Earth on the night side. Region of the dusk side ring current imaged extends between 30° to beyond 60° from the Sun ($5 < L < 10$). (b) INCA image (every pixel) of the Earth's ring current from time-of-flight channel (27–39 keV, H atoms) represented in instrument coordinates (elevation and azimuth). Cassini vantage point is from a radius of $10.1R_E$, 1° latitude and 197° longitude (GSE).

TABLE VI
Capabilities of the CHEMS instrument.

Energy per charge range	3–220 keV/e
Ion species	H–Fe
Mass per charge range	1–80 amu/e
Resolution (FWHM)	
Energy per charge, $\Delta(E/Q)/(E/Q)$	0.03
Mass per charge, $\Delta(M/Q)/(M/Q)$	0.08, He^{2+} at 100 keV/e
	0.07, O^{+} at 100 keV
Mass, $\Delta M/M$	0.15, He^{2+} at 100 keV/e
	0.5, O^{+} at 100 keV
Geometrical factor	0.05 cm^2 sr
Field of view (FWHM)	4.0° by 159°
Dynamic range	10^{10}

per charge analysis followed by the TOF versus energy technique to determine both the mass per charge and mass of ions. CHEMS has a relatively large geometrical factor for a charge state-determining instrument. It also has a nearly 4π viewing geometry when the spacecraft is rolling, affording the capability of measuring 3-D distribution functions. The CHEMS sensor is very similar to the STICS sensor on the GEOTAIL spacecraft (Williams *et al.*, 1994) and the STICS sensor on the Wind spacecraft (Gloeckler *et al.*, 1995). The capabilities of the CHEMS instrument are summarized in Table VI and a photograph is shown in Figure 28.

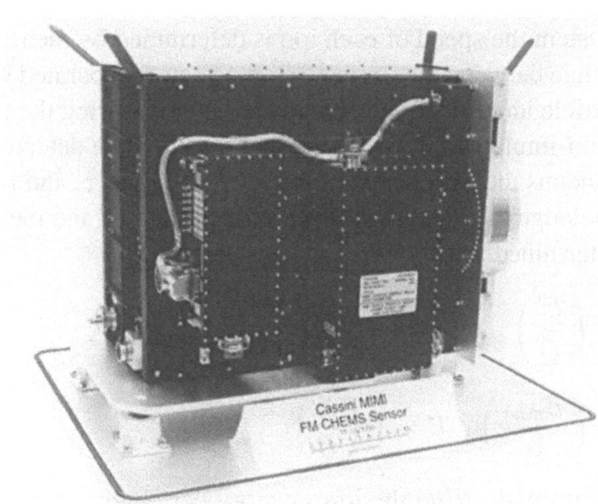

Figure 28. Photograph of the CHEMS instrument.

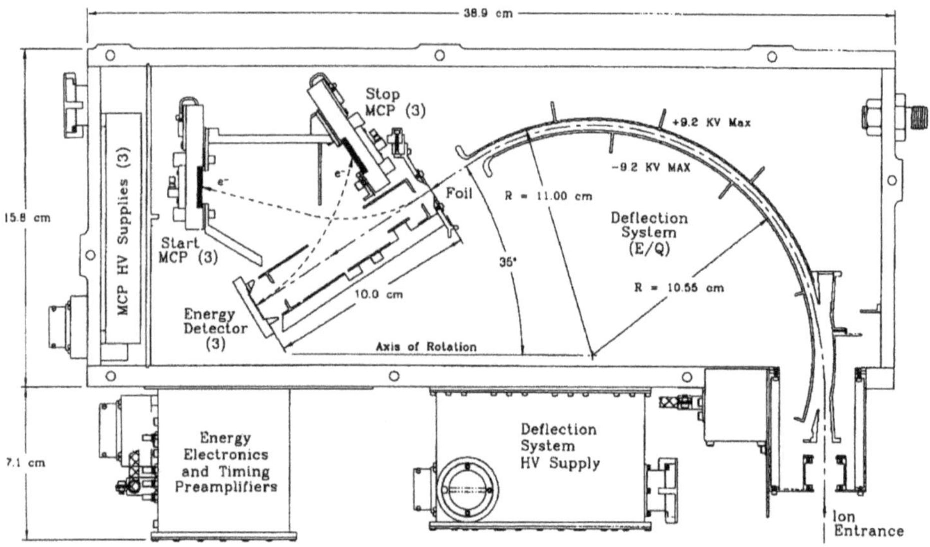

Figure 29. CHEMS mechanical configuration.

3.2.1. *Principal of Operation of CHEMS*

The CHEMS sensor is based on the technique of particle identification using electrostatic deflection combined with TOF and energy measurements (Gloeckler and Hsieh, 1979). The sensor's configuration is shown in Figure 29.

Ions of kinetic energy E, mass M, and charge state Q enter the sensor. The electrostatic deflection analyzer serves both as a UV trap and an energy per charge (E/Q) filter, allowing only ions within a small ($\sim$3%) energy per charge interval centered on a value determined by the stepped deflection voltage to enter the TOF versus energy system.

In the TOF system the speed of each ion is determined by measuring the travel time τ of the particle between the start and stop detectors separated by a distance of 10.0 cm. The particle identification is completed by measuring the residual energy of the ion in an ion-implanted, low-noise, silicon solid-state detector (SSD).

From simultaneous measurements of the time-of-flight, τ, and residual energy, E_{meas}, and a knowledge of E/Q, the mass M, charge state Q, and incident energy E, of each ion is determined as follows:

$$\frac{M}{Q} = 2\left(\frac{\tau}{d}\right)^2\left(\frac{E'}{Q}\right) \quad Q = \left(\frac{E_{\text{meas}}}{\alpha}\right)\left(\frac{E'}{Q}\right)$$

$$M = 2\left(\frac{\tau}{d}\right)^2\left(\frac{E_{\text{meas}}}{\alpha}\right) \quad E = Q\left(\frac{E}{Q}\right)$$

where d is the flight path, E'/Q the ion's energy per charge after the carbon foil, taking into account the small energy loss in the thin foil, and α gives the ratio of measured energy in the SSD to kinetic energy incident at the detector, a number less than 1 due to nuclear defect. The reduction in measured energy due to nuclear defect becomes more important at lower energies and for higher masses (Ipavich *et al.*, 1978).

In general the measurements of τ and E/Q are more accurate than that of E_{meas} at the energies covered by CHEMS. Therefore the determination of M/Q is more precise than that of M or Q individually. For ions with insufficient energy to trigger the 26 keV SSD electronic threshold, only M/Q is determined.

3.2.2. *Description of the CHEMS Instrument*

The CHEMS sensor contains a deflection system, three TOF telescopes, each of which includes an SSD for the energy measurement, the high-voltage power supplies for the deflection analyzer system and the MCPs, analog electronics to produce shaped signals from the SSDs, and six preamplifiers for the fast timing signals. The various detector signals are sent to the MIMI Main Electronics Unit for further processing. Basic instrument characteristics are listed in Table VII, and a block diagram of the CHEMS electronics is given in Figure 30.

3.2.2.1. Deflection Voltage System. Particles enter the CHEMS electrostatic deflection analyzer system through a simple entrance aperture that includes a light trap. The deflection system consists of two concentric spherical segments. The deflection plate power supply (DPPS) supplies symmetric positive and negative voltages to the outer and inner plates, respectively. The DPPS supplies 32 output voltages up to a maximum of ± 9.225 kV as listed in Table VIII. The logarithmic spacing of the steps is about 14.9% and the passband is $\sim$3%. Normally, the DPPS will step through a sequence of 16 steps, consisting of the odd-numbered steps from 31 to 1, but the sequence used is completely commandable. The E/Q values of the steps are used in the MIMI DPU for the mass per charge classification. The E/Q values given in Table VIII are derived from the potential differences between the two DPPS outputs measured before launch, multiplied by the nominal analyzer constant of 11.97. Error in the analyzer constant was found to be $<$1% during accelerator calibrations of the sensor.

3.2.2.2. Time-of-Flight System. The CHEMS TOF versus energy assembly consists of a start and stop detector separated by 10.0 cm. Particles with the correct energy per charge pass through the deflection system and penetrate the thin ($\sim$2.5 μg/cm^2) carbon foil at the entrance of the TOF telescope (see Figure 29). Secondary electrons are emitted from the point where the ion exits the foil. These secondary electrons are deflected by electric fields inside the telescope and strike one of the three start MCPs, generating the start signal for the TOF analysis. At the far end of the TOF telescope, the ion strikes one of the three SSDs. Secondary electrons from the front surface of the SSDs are deflected onto one of three stop MCPs, providing the stop signal for the TOF analysis. The SSDs also provide

TABLE VII

CHEMS instrument characteristics.

Mass	6.84 kg
Dimensions	22.9 cm × 38.9 cm × 28.0 cm (x, y, z)
Carbon foils	Three segments for each of three telescopes supported on 100 line/inch nickel mesh on a conical frame, 80% transmission
	Each segment: 6 mm high, 22.85 mm wide
	Thickness: nominal – 2.0 μg/cm^2; measured – 2.5 μg/cm^2
Microchannel plates (MCPs)	Six borderless chevron pairs from Galileo Electro-Optics Corp. (3 Start, 3 Stop)
	Active area: 15 mm by 37 mm (555 mm^2 each)
	Pore size: 25 m
	Length to diameter ratio: 40
Solid-state detectors (SSDs)	Three specially made, thin window, ion-implanted silicon (PIPS) detectors from Canberra Industries
	Depletion depth: 300 μm
	Entrance window: ∼30 nm Si
	Trapezoidal shape
	Height: 12.0 mm
	Width (top): 12.0 mm
	Width (bottom): 22.2 mm
	Active area: 206 mm^2
Electrostatic deflection system	Two concentric spherical segments with opposite polarities
	Mean radius: 107.75 mm
	Gap: 4.5 mm
	Analyzer constant: 11.97
	Deflection angle: 125°
TOF measurement	Three independent telescopes; 53° FOV each
	Flight path: 10.0 cm
	Minimum TOF: 6 ns
	Maximum TOF: 523 ns
	0.511 ns/channel
	Nominal shielding: 1.57 g/cm 2 Al
Energy measurement	Electronic threshold: 26 keV
	Maximum energy: 2455 keV
	2.46 keV/channel

energy information on the incident particles if the incident energy is sufficiently high to exceed the electronic threshold of 26 keV.

The three TOF telescopes are independent. Each telescope covers 53° in polar angle as shown in Figure 31. The telescopes have a narrow 4° acceptance in the

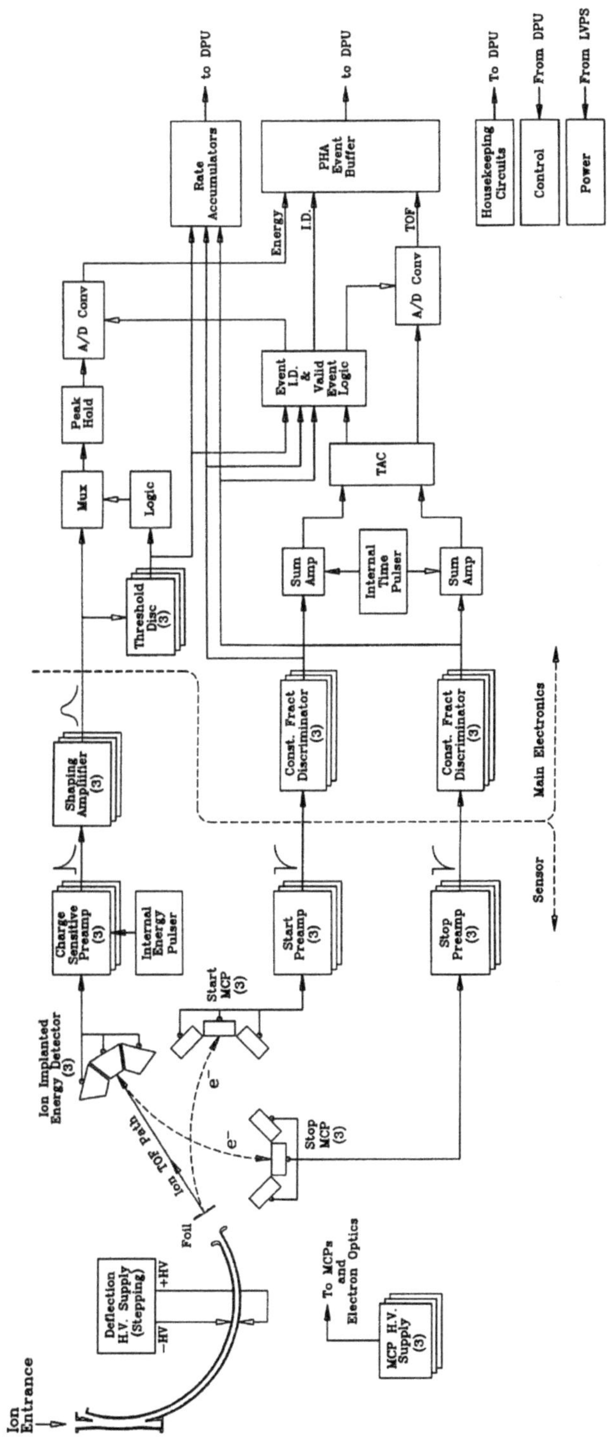

Figure 30. CHEMS analog electronics block diagram.

TABLE VIII
CHEMS deflection system.

Step no.	Energy per charge (keV/e)	Measured deflection voltage (kV)
0	2.81	0.235
1	3.29	0.275
2	3.78	0.316
3	4.40	0.368
4	5.05	0.422
5	5.86	0.490
6	6.75	0.564
7	7.81	0.653
8	8.92	0.745
9	10.3	0.861
10	11.8	0.984
11	13.6	1.141
12	15.7	1.309
13	18.1	1.509
14	20.8	1.734
15	23.9	1.996
16	27.4	2.290
17	31.5	2.635
18	36.2	3.027
19	41.6	3.48
20	47.8	3.99
21	54.9	4.59
22	63.2	5.28
23	72.6	6.07
24	83.4	6.97
25	95.9	8.01
26	110.1	9.20
27	126.6	10.58
28	145.3	12.14
29	166.9	13.95
30	191.6	16.01
31	220.2	18.40

direction perpendicular to that figure. When the spacecraft rolls about the z-axis, CHEMS covers most of the unit sphere. Each of the three telescopes has its own michrochannel plate power supply (MCPPS), which supplies both the start and stop plates. The secondary electron steering voltages are provided by a diode 'OR' of

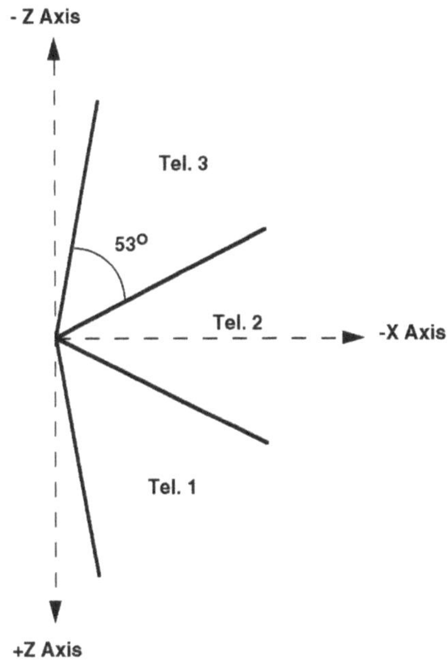

Figure 31. CHEMS telescope view directions relative to the indicated spacecraft axes.

the appropriate outputs of the three supplies. Thus a failure of one MCPPS causes a loss of only one-third of the angular coverage.

Shielding was strategically added in and around the TOF telescopes to reduce the singles counting rates in those portions of Saturn's magnetosphere with high fluxes of relativistic electrons. The MCP housings were made from 1.0-mm tungsten. The cylindrical structure surrounding the TOF region was made from 3.5-mm aluminum. All of the open pockets were removed from the pyramid on which the SSDs are mounted. These measures plus a number of others provide the equivalent of at least 5.8 mm of aluminum (1.57 g/cm^2) in almost all directions, which will protect the detectors from electrons below 2.5 MeV.

3.2.3. *CHEMS Data Types*

CHEMS returns three types of counting rate data along with pulse-height analysis (PHA) events that give complete information on individual ions. Engineering rates (listed in Table IX) are derived from individual detectors or detector coincidences and are useful in assessing instrument health, in the overall normalization of fluxes in high rate periods, and in assessing the extent of any accidental coincidences or pulse pileup in the SSDs.

The other two types of counting rates and the PHA data use the capability of the DPU to rapidly classify individual ion events as to mass per charge and mass (several thousand per second). PHA events contain complete information on a sampling of the incident ions. A PHA event consists of two 10-bit linear channel

TABLE IX

CHEMS engineering rates.

	Name	Coincidence requirement	Description
1	START1	None	Start MCP, Telescope 1
2	START2	None	Start MCP, Telescope 2
3	START3	None	Start MCP, Telescope 3
4	STOP1	None	Stop MCP, Telescope 1
5	STOP2	None	Stop MCP, Telescope 2
6	STOP3	None	Stop MCP, Telescope 3
7	ENERGY1	None	Solid-State Detector, Telescope 1
8	ENERGY2	None	Solid-State Detector, Telescope 2
9	ENERGY3	None	Solid-State Detector, Telescope 3
10	DCR1	Double	Start-Stop Coincidence, Tel. 1
11	DCR2	Double	Start-Stop Coincidence, Tel. 2
12	DCR3	Double	Start-Stop Coincidence, Tel. 3
13	TCR1	Triple	Start-Stop-SSD Coincidence, Tel. 1
14	TCR2	Triple	Start-Stop-SSD Coincidence, Tel. 2
15	TCR3	Triple	Start-Stop-SSD Coincidence, Tel. 3
16	UFSR	None	Universal Start MCP, Any Telescope
17	URSR	None	Universal Stop MCP, Any Telescope

TABLE X

CHEMS pulse height analysis events.

Item	No. of bits
Energy channel number	10
TOF channel number	10
SSD ID	2
Start MCP ID	2
Range ID	3
DPPS step number	5

numbers representing time-of-flight and measured energy in an SSD (Table X), two ID numbers specifying which Start MCP and SSD were triggered, the DPPS step number (equivalent to energy per charge), and a range ID determined by the DPU based on its calculation of mass and mass per charge. For improved accuracy, the DPU uses two-segement linear approximations in the conversion of time and energy channel numbers to physical units.

The seven range IDs are indicated in Figure 32 and Table XI. Each range corresponds to a range of mass per charge and mass as indicated. The DPU calculates

TABLE XI

CHEMS PHA ranges and basic rates.

Range	Number of rates	Rate FOV (°)	Coincidence requirement	NM (inclusive)	NQ (inclusive)	Mass (amu)	Mass per charge (amu/e)
0	3	53	Double	0–0	1–22	0–0	0.70–1.60
1	3	53	Double	0–0	23–73	0–0	1.60–10.9
2	3	53	Double	0–0	74–126	0–0	10.9–80.0
3	3	53	Triple	1–62	1–22	0.50–95.0	0.70–1.60
4	3	53	Triple	1–32	23–73	0.50–7.50	1.60–10.9
5	3	53	Triple	1–62	74–126	0.50–95.0	10.9–80.0
6	3	53	Triple	33–62	23–73	7.50–95.0	1.60–10.9

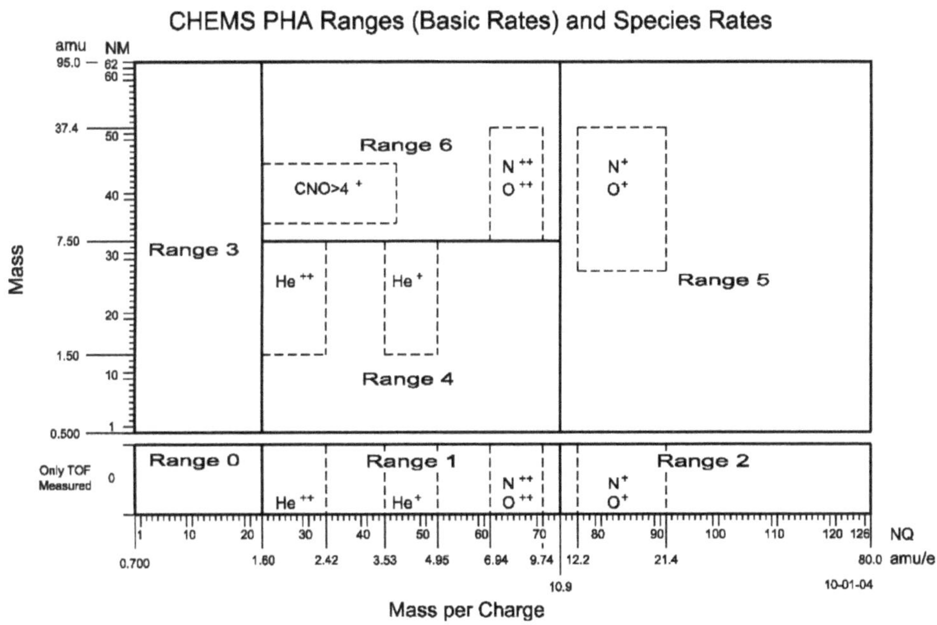

Figure 32. CHEMS PHA ranges, basic rates, and species rates.

two integers, *NQ* (1–126) and *NM* (0–62), logarithmically related to mass per charge and mass, respectively. Ranges 0–2 correspond to $NM = 0$, indicating that there was no SSD trigger, in which case, no mass is determined. Ranges 3–6 correspond to triple coincidences that include an energy measurement as well as a time-of-flight. A basic rate is associated with each range. The basic rates record the total number of events of each range separately for the three TOF telescopes.

There is insufficient telemetry to send down PHA events for all particles, so the DPU selects a random sample of events, giving each range equal weight. Each range

TABLE XII

CHEMS species rates.

Species	Number of rates	Rate FOV (°)	Coincidence requirement	NM (inclusive)	NQ (inclusive)	Mass (amu)	Mass per charge (amu/e)
He^+	1	159	Double	0–0	44–52	0–0	3.53–4.95
He^+	1	159	Triple	14–32	44–52	1.50–7.50	3.53–4.95
He^{2+}	1	159	Double	0–0	23–33	0–0	1.60–2.42
He^{2+}	1	159	Triple	14–32	23–33	1.50–7.50	1.60–2.42
N^+, O^+	3	53	Double	0–0	77–91	0–0	12.2–21.4
N^+, O^+	3	53	Triple	28–51	77–91	4.91–37.4	12.2–21.4
N^{2+}, O^{2+}	1	159	Double	0–0	62–70	0–0	6.94–9.74
N^{2+}, O^{2+}	1	159	Triple	33–51	62–70	7.50–37.4	6.94–9.74
CNO > 4^+	1	159	Triple	36–45	23–45	9.67–22.5	1.60–3.80

purposely contains only double coincidence or triple coincidence events. In high background regions of the Saturnian radiation belts, the double coincidence events will become dominated by accidental coincidences before the triple coincidence events do. By placing the double coincidences in separate ranges, they are never able to displace the more reliable triple coincidences. By command, the PHA events from any range can be excluded from the telemetry.

Nine smaller regions of M versus M/Q space are defined, corresponding to species expected to be common in Saturn's magnetosphere. Events falling in those regions are counted by species rates, as listed in Table XII and shown in Figure 32. Only double and triple coincidence O^+ and N^+ events are separately counted for each telescope. There are no species rates dedicated to H^+ since basic rates 0 and 3 will respond almost entirely to H^+.

3.2.4. *CHEMS Modes of Operation*

CHEMS has relatively few modes. When the Cassini spacecraft is rolling, a complete energy spectrum will be obtained during each 16th of the roll (22.5°) by commanding the CHEMS DPPS through 16 steps. At a nominal roll period of 23 min, CHEMS spends 5.39 s at each energy per charge with 0.150 s after each transition blocked from data accumulation by the DPU to allow the DPPS output voltage to settle. When the spacecraft is not rolling, CHEMS will be commanded through its stepping sequence on a time basis with a similar amount of time spent at each step. There is an in-flight calibrator that can stimulate the CHEMS electronics with timing pulses of known separation (20–300 ns) and energy pulses of known amplitude.

CHEMS has eight trigger modes for the initiation of DPU classification and preparation of PHA events. The standard mode will accept double or triple coincidence events. A triple coincidence only mode exists that may be useful during high background periods. A mode requiring a calibrate strobe is used during in-flight-calibrate periods. There is a mode requiring only an energy pulse that could be used in the event of a TOF failure. The other four trigger modes were used for diagnostic purposes before launch.

3.2.5. *CHEMS Calibration*

The CHEMS flight unit was calibrated before launch at the Goddard Space Flight Center accelerator facility. The final calibration was performed there in April 1997 using the MIMI flight unit DPU. The sensor was exposed to beams of various ion species (H^+, H_2^+, He^+, C^+, N^+, O^+, H_2O^+, Ne^+, Ar^+, CO_2^+, and Kr^+) at energies between 3.3 and 96 keV. These data were used to determine carbon foil thickness, SSD nuclear defect, TOF and energy resolution, and detection efficiencies as a function of energy and species.

3.2.5.1. Time-of-Flight Resolution. It is very important that the CHEMS instrument be able to distinguish between the suggested candidates for the heavy ion plasma in Saturn's magnetosphere. These ions include N^+, O^+, OH^+, and H_2O^+. The TOF distributions for three of those ions plus C^+ are shown in Figure 33 for 96-keV beams. The distributions are well separated. At lower energies, the TOF

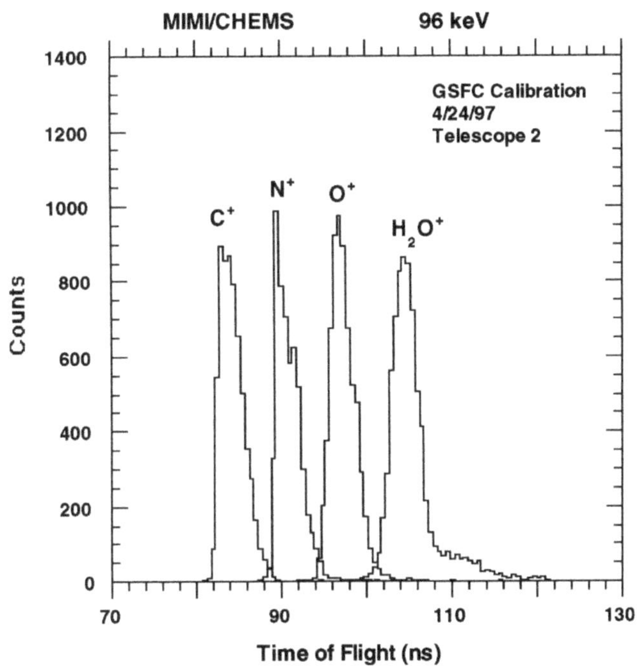

Figure 33. CHEMS time-of-flight distributions for 96 keV ions.

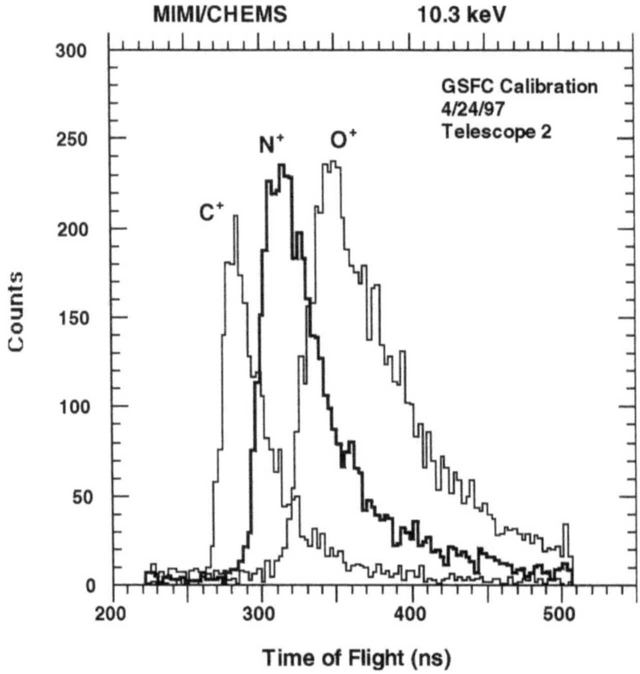

Figure 34. CHEMS time-of-flight distributions for 10 keV ions.

resolution worsens due to energy straggle in the carbon foil. Figure 34 presents distributions from 10.3 keV beams. The means of the three distributions are still well separated but the distributions are much broader.

3.2.5.2. Efficiencies. In general, the TOF detection efficiency increases with increasing energy in the CHEMS energy range. In Figure 35 the TOF start efficiency is shown as a function of beam energy for eight different ion species. The start efficiency was determined by dividing the double coincidence counting rate by the stop singles rate. The variation largely tracks the change in secondary electron emission.

The stop efficiency is displayed in Figure 36. The stop efficiency was determined by dividing the double coincidence counting rate by the start singles rate. The stop efficiency shows a larger variation than the start because it includes not only the variation in secondary electron production, but also the effect of scattering in the carbon foil. Ions that do not hit an SSD do not produce secondary electrons that can be detected. Since the SSD subtends a relatively small angle, the large decrease in stop efficiency at low energies and for heavier ions is largely due to increased scattering in the foil. The total TOF efficiency is obtained by multiplying the start and stop efficiencies.

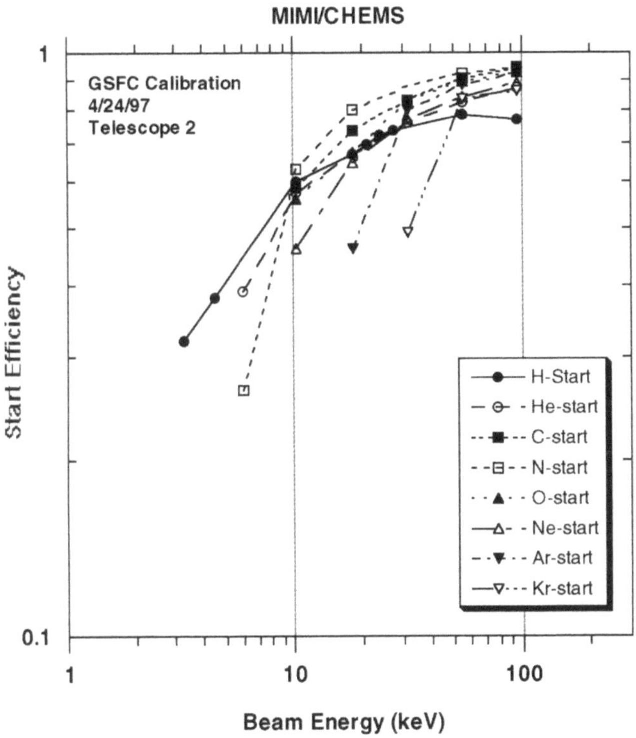

Figure 35. CHEMS TOF telescope start efficiencies. Start efficiencies include the effects of production and detection of secondary electrons from ions passing through the carbon foil.

3.2.5.3. Comparison with the Wind/STICS Instrument. The CHEMS sensor is very similar to the STICS sensor on GEOTAIL (Williams *et al.*, 1994) and the STICS sensor on the Wind spacecraft (Gloeckler *et al.*, 1995). However, a major effort was made to increase the TOF efficiency compared to those instruments. Two major changes were made. The CHEMS solid-state detectors are 56% larger in area than the STICS detectors and extensive 3-D modeling was performed to optimize the steering potentials to improve secondary electron collection. The results were a significant efficiency improvement as shown in Figure 37 for hydrogen and oxygen.

3.2.6. *CHEMS in-Flight Performance*

The CHEMS sensor returned excellent data from the Venus 2 and Earth flybys. Figure 38 illustrates its performance with triple coincidence PHA data from the outer portion of the Earth's ring current. Major species are labeled, and boxes for the four triple coincidence species rate implemented at the time of Earth flyby are shown. Observed resolution is consistent with that expected from prelaunch calibration. In June 2000, small changes were made to the CHEMS Range and Species definitions in the MIMI flight software. These later definitions are shown in

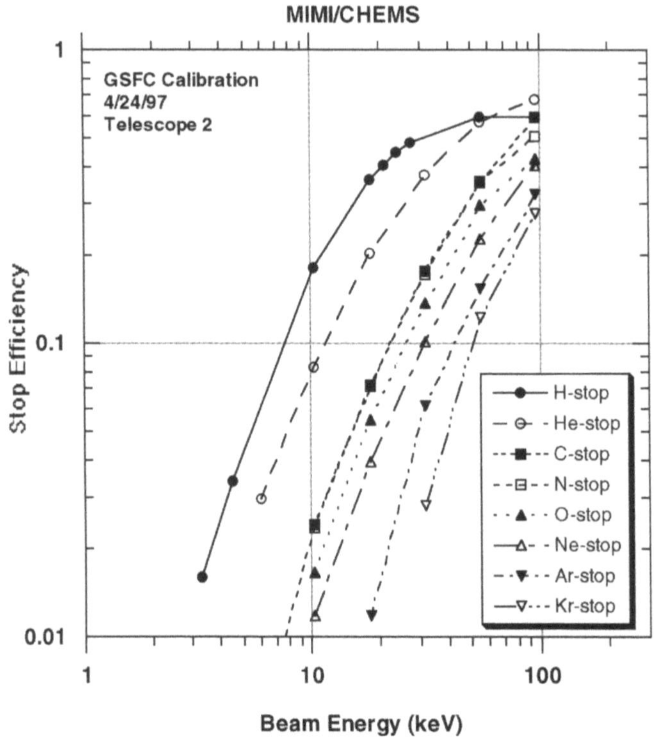

Figure 36. CHEMS TOF telescope stop efficiencies. Stop efficiencies include the effects of particle scattering in the carbon foil as well as the production and detection of secondary electrons from ions that hit the solid state detectors.

Figure 32 and listed in Tables XI and XIII including an additional triple coincidence species rate for high charge state CNO.

3.3. Low Energy Magnetospheric Measurements System (LEMMS)

3.3.1. *Detector Description*

The LEMMS instrument is designed to measure the three-dimensional distribution of energetic ion and electron fluxes. The instrument head consists of a double-ended telescope (low energy and high energy end) heavily shielded by a platinum cover around the whole assembly to avoid penetrating particles with energies $E < 30$ MeV through the sides of the instrument. The instrument head and the electronic boxes are mounted on top of a rotating platform. A picture of LEMMS is shown in Figure 39.

LEMMS rotates about the $-y$-axis of the spacecraft, co-aligned with the remote sensing and optical instruments, and therefore measures angular distributions of ions and electrons within a scan plane, defined by the x–z plane in the spacecraft system of reference (see Figure 40). Three-dimensional particle

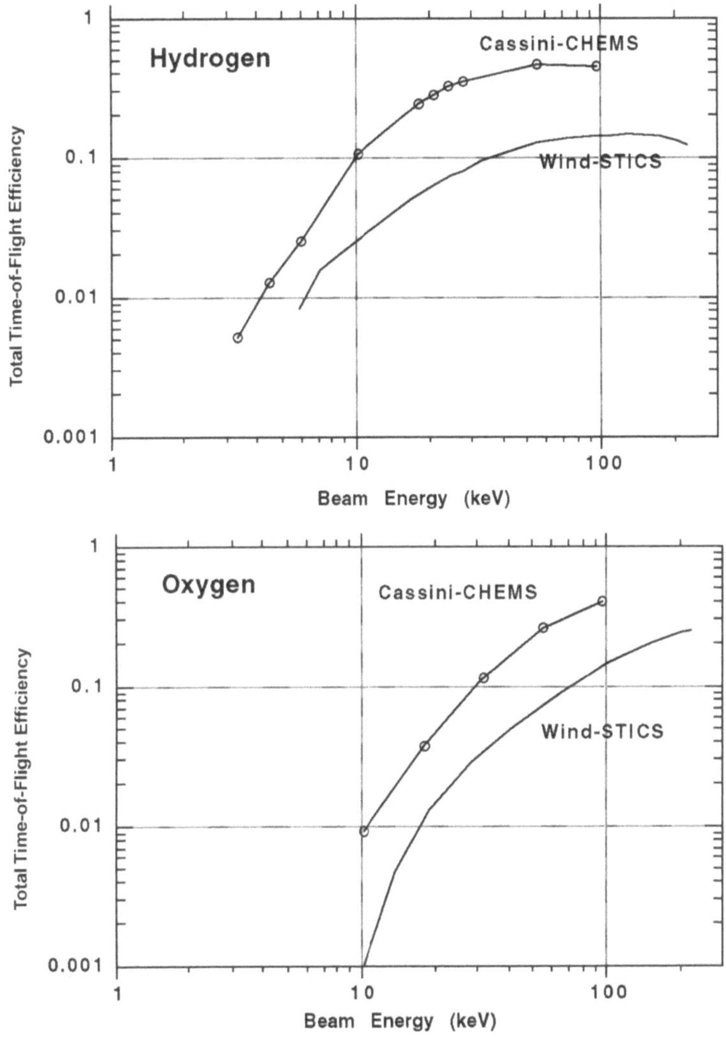

Figure 37. Comparison between Cassini-CHEMS and Wind-STICS TOF efficiencies for hydrogen and oxygen. Total efficiencies are the product of the start and stop efficiencies.

distributions can only be obtained if the spacecraft itself is rotating about the z-axis.

During one complete motor rotation, which is performed in 86 s, 16 subsectors are defined for all 57 counters (see Tables XIII and XIV). In addition six priority counters are defined for which each subsector is divided in eight microsectors as indicated in Figure 40. The low or high energy ends may be positioned behind the calibration shield for background measurements. Collimators define 7 and 19 hexagonal-shaped aperture entrance channels for the low and high energy end, respectively. This geometry is also used on the particle spectrometer EPD-LEMMS

TABLE XIII

Parameter of the semiconductor detectors used in the LEMMS instrument.

Detector	L/H end	Thickness (μm)	Active area (mm^2)
E1	Low	300 ± 15	90
E2	Low	300 ± 15	90
F1	Low	700 ± 15	90
F2	Low	300 ± 15	90
A	Low	150 ± 15	35
B	Low	700 ± 25	35
D1	High	150 ± 15	100
D2	High	700 ± 15	150
D3a	High	700 ± 15	150
D3b	High	700 ± 25	150
D4	High	700 ± 25	150

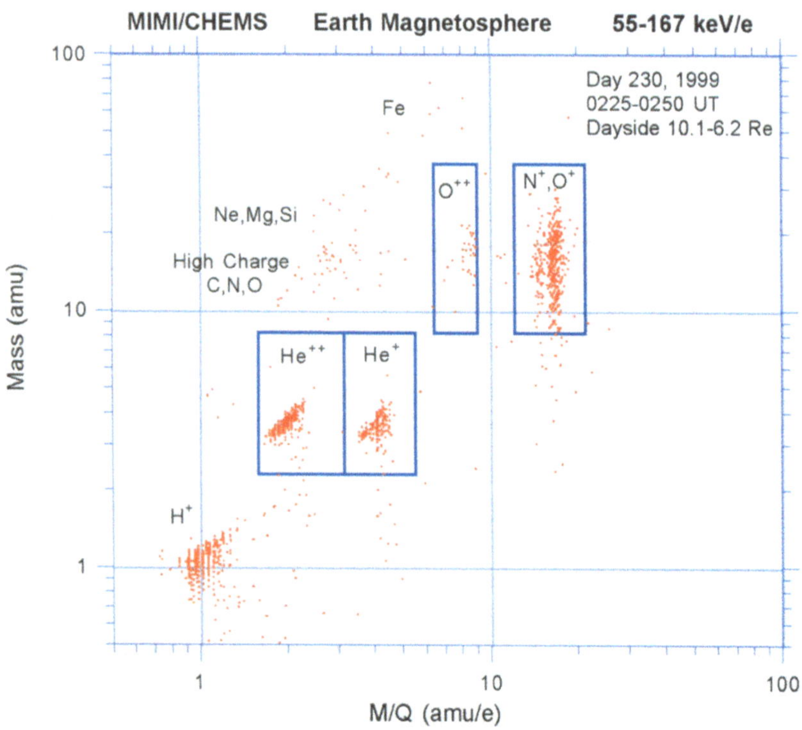

Figure 38. CHEMS PHA data from the outer portion of the Earth's ring current.

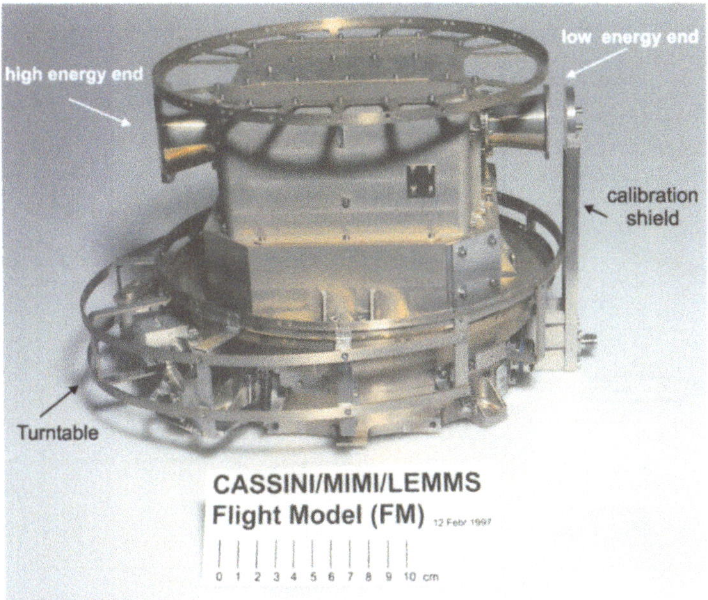

Figure 39. Photograph of the low energy magnetospheric measurement system (LEMMS) flight unit. The double-ended detector head (low energy end and high energy end) is surrounded by the electronic boxes and mounted on the programmable turntable platform. The calibration shield is used for background measurements.

on board the Galileo spacecraft and provides narrow acceptance angles for focused deflection of electrons inside the instrument. The new position of the instrument on the spacecraft (after the removal of the spacecraft scan platform) is different from originally planned. The LEMMS field of view is partly obscured by parts of the spacecraft itself. As shown in Figure 40, the cover of one of the radioisotope thermal generators (RTG), one of the thrusters, and the high-gain antenna are partly blocking incoming particles. This obscuration limits the three-dimensional data coverage and has to be taken into account in the data analysis. Reflected sunlight from the cover of the RTG and the thruster is being removed for software routines comparing intensities in adjacent subsectors and by including energy spectral information.

3.3.2. *Measurement Technique*

The measurement principle of LEMMS is based on energy loss of incidenting particles in semiconductors. The position of the detectors inside the assembly determines the species and energy of the incidenting particle. The overall configuration of LEMMS and the positions of all the 11 semiconductor silicon detectors are shown in Figure 41. Table XIII summarizes the parameters of the LEMMS detectors.

Low Energy End. The low energy end of LEMMS is designed to measure low-energy ions ($E \geq 30$ keV) and electrons (15 keV–1 MeV). An inhomogeneous

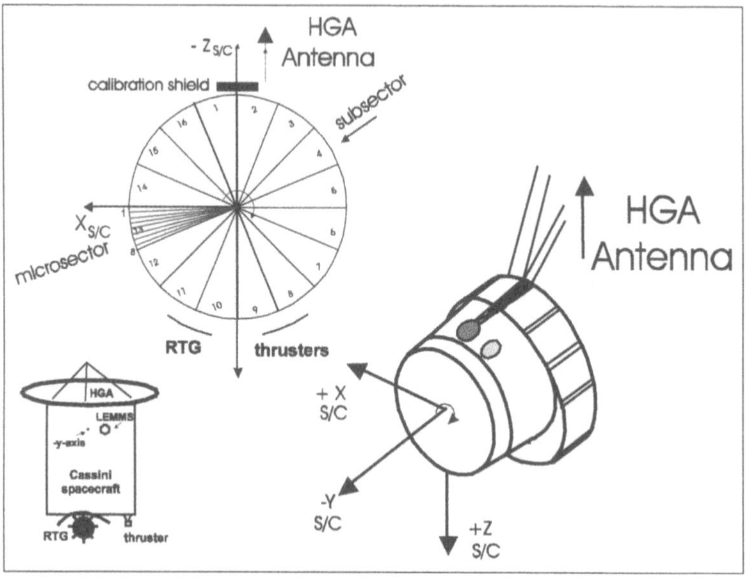

Figure 40. Upper left: LEMMS scan plane defined by the x–z plane in the spacecraft frame of reference; lower left: LEMMS position on the spacecraft. The high-gain antenna (HGA), the cover of one of the radioactive thermal generators (RTG), and one of the thrusters are partly obscuring the field of view of the instrument; right: LEMMS orientation in the spacecraft frame of reference with the scan platform rotation axis along the $-y$-axis.

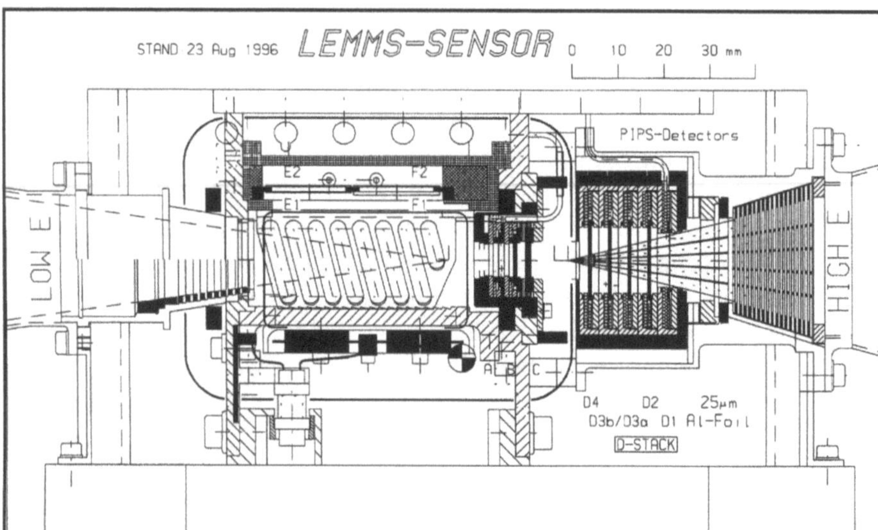

Figure 41. Configuration of the LEMMS detector head. Low energy particles entering hte low energy end (LOW E) are measured with detectors E1, E2, F1, F2 (for electrons) and A, B (for ions). Particles with higher energies entering the high energy end (HIGH E) are detected in a stack of detectors (D1, D2, D3a, D3b, D4). Between detectors B and D4 a gold absorber C is inserted. The whole assembly is shielded with a platinum box.

TABLE XIV

LEMMS channels description (low energy end) from caluclations and calibration measurements.

					p		He		C		O		Fe		Electrons		
Readout N	LEMMS channels	Logic	L/H end	Species	E_{min} (MeV)	E_{max} (MeV)	E_{min} (MeV/N)	E_{max} (MeV/N)	E_{min} (MeV/N)	E_{max} (MeV/N)	E_{min} (MeV/N)	E_{max} (MeV/N)	E_{min} (MeV/N)	E_{max} (MeV/N)	E_{min} (MeV)	E_{max} (MeV)	Geometry factor cm^2-sr
0	A0	A1(A2)(B1)	Low	$Z \geq 1$	0.027	0.035	0.008	0.010	0.004	0.0048333	0.004	0.004	0.003	0.003			0.0112 A0
1	A1	A2(A3)(B1)	Low	$Z \geq 1$	0.035	0.056	0.01	0.016	0.0048333	0.007	0.004	0.006	0.003	0.004			0.0112 A1
2	A2	A3(A4)(B1)	Low	$Z \geq 1$	0.056	0.106	0.01575	0.029	0.007	0.012	0.006	0.010	0.004	0.006			0.0112 A2
3	A3	A4(A5)(B1)	Low	$Z \geq 1$	0.106	0.255	0.02925	0.068	0.012	0.0260833	0.010	0.021	0.006	0.011			0.0112 A3
4	A4	A5(A6)(B1)	Low	$Z \geq 1$	0.255	0.506	0.0675	0.130	0.0260833	0.0483333	0.021	0.038	0.011	0.018			0.0112 A4
5	A5	A6(A7)(B1)	Low	$Z \geq 1$	0.506	0.805	0.13	0.205	0.0483333	0.075	0.038	0.058	0.018	0.025			0.0112 A5
6	A6	A7(A8)(B1)	Low	$Z \geq 1$	0.805	1.6	0.205	0.400	0.075	0.1416667	0.058	0.106	0.025	0.043			0.0112 A6
7	A7	A8(A9)(B1)	Low	$Z \geq 1$	1.6	3.5	0.4	0.875	0.1416667	0.3	0.106	0.231	0.043	0.080			0.0112 A7
8	A8	A9(B1)	Low	$Z \geq 1$	3.5	4	0.875	4.000	0.3	7.25	0.231	8.313	0.080				0.0112 A8
9	B0	A8(A9)B1(B5)(D41)	Low	$Z \geq 1$	4.10	7.70	17.5	41.000	250.00								0.0112 B0
10	B1	A6(A8)B4(B5)(D41)	Low	$Z \geq 1$	7.70	14.00	41.3	67.750									0.0112 B1
11	B2	A9B1(B5)(D41)	Low	$Z \geq 1$	4.00	4.10	4.0	5.500	7.25	316.67	8.375	8.625	13.11	13.36			0.0112 B2
12	B3	A9B5(D41)	Low	$Z > 1$			5.5	15.500	7.58	316.67	8.625		13.41				0.0112 B3
13	BE	A2(A7)B1(B3)(D41)	Low	e−											1	2.6	0.0112 BE
14	C0	E11(E12)(E22)	Low	e−											0.020	0.030	0.0075 C0
15	C1	E12(E13)(E22)	Low	e−											0.030	0.045	0.0207 C1
16	C2	E13(E14)(E22)	Low	e−											0.045	0.060	0.0324 C2
17	C3	E14(E15)(E22)	Low	e−											0.060	0.100	0.0321 C3
18	C4	E15(E16)(E22) + F11(F12)(E22)	Low	e−											0.100	0.200	0.0271 C4
19	C5	F12(F13)(E22)	Low	e−											0.170	0.300	0.0102 C5
20	C6	F13(F14)(E22)	Low	e−											0.300	0.528	0.0250 C6
21	C7	F14(F15)(E22)	Low	e−											0.528	0.900	0.0258 C7
22	AS	A1	Low	$Z \geq 1$	0.027		0.008		0.004		0.003		0.003				— AS
23	BS	B1	Low	$Z \geq 1$	4		4.0		7.250		8.375		13.11				— BS
24	ES	E11	Low	e−											0.02		— ES
25	E2/F2S	E22	Low	e−											1		— E2/F2
26	FS	F11	Low	e−											0.1		— FS

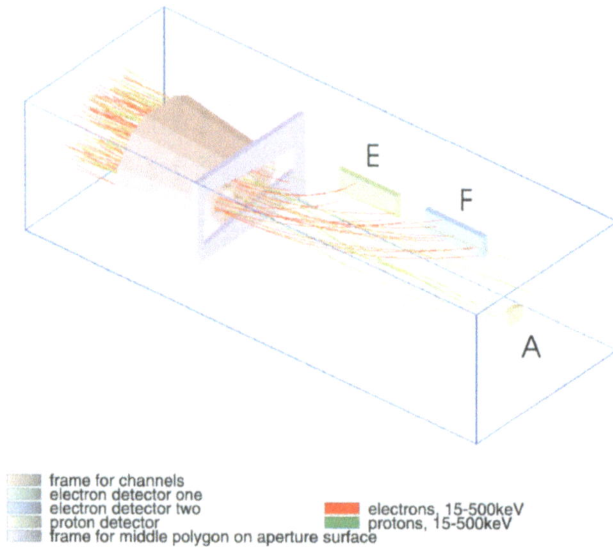

Figure 42. Particle trajectories in the LEMMS low energy end geometry from simulation.

magnetic field produced by an internal permanent magnet separates incidenting ions and electrons which subsequently strike different detectors inside. Figure 42 illustrates sample particle trajectories in the internal magnetic field. Electrons and ions enter the low energy end from the left through the collimator. Low-energy electrons up to 884 keV are deflected by the magnet and measured by the electron detectors E (E1 and E2) and F (F1 and F2) depending on their incident energy. Electrons with energies between 200 and 600 keV are detected in both electron detectors dependent on their collimator entrance angles. Therefore, the calculated geometric factors as shown in Figure 43 overlap in that energy range.

Electrons with energies $E \leq 1$ MeV cannot reach detector A directly. However, low-energy ions, which are less affected by the magnetic field, are measured by the low energy end ion detectors A and B (B not shown in the simulation). The aperture angle of the low energy end collimator is 15°. Between detectors B and the high energy end detector D4, a Gold-absorber (1000 μm, 75 mm^2) is inserted to stop particles penetrating detectors A or B incident from the low energy end. LEMMS data from detectors A, E1, and F1 are also processed through a pulse height analyzer (PHA) that produces 64-channel energy spectra split in two ranges 32 channels each.

In addition to the 27 counters of the low energy end four priority counters (A0, A1, C0, C1) are selected which have microsector angular resolution, as mentioned above. Table XIV summarizes the low energy end electron and ion channels defined by logical conditions between the different electronic detector thresholds. The energy ranges mentioned in the tables are based on energy loss calculations and calibrations for the various detectors.

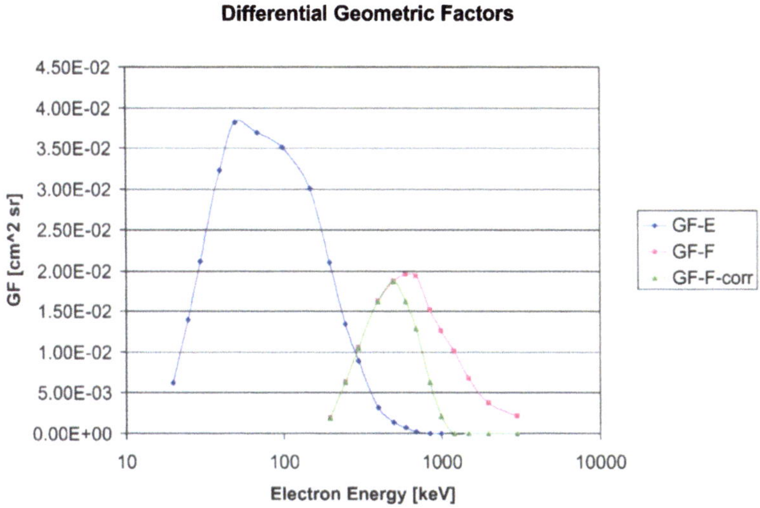

Figure 43. Geometric factors as a function of energy for detectors E and F for electrons calculated from detailed simulation results of the low energy end of LEMMS. The green curve corrects the F geometry factors for penetrating electrons that are rejected by anti coincidence logic.

High Energy End. The high energy end consists of a stack of five detectors D1, D2, D3a, D3b, and D4 to measure high-energy ions (1.5–160 MeV/N) and electrons (0.1–5 MeV). In front of detector D1 a 25 mm Al-foil is inserted to suppress incoming light and the flux of low-energy ions. The opening angle of the high energy end collimator is 30°. The parameters of the high energy end detectors can be found in Table XV, and the calculated and measured electron and ion energy channel boundaries are shown in Table XVI. A summary of the thicknesses and active areas of the detectors used in LEMMS is given in Table XIII.

3.3.3. *LEMMS Turntable*

Description of Features. To be able to measure particles in every direction, the LEMMS experiment is mounted on a rotating platform which was developed by the Finnish company VIT Automation Space Technology. A photograph of the turntable without the LEMMS detector head and electronic is shown in Figure 44.

The turntable (TT) is able to rotate continuously left- or right-handed within 360°. It is designed for a 3 million revolutions lifetime in the temperature range of −35 to +45°C. The rotation speed is selectable from a range of 2° to about 12°/s. This is realized with a stepper motor and custom designed drive electronics with closed-loop control. The accuracy to let the motor rotate to a fixed position is about 0.01°. The TT provides the mechanical platform for the instrument and an attachment ring for the spacecraft (MLI). It is locked during launch by a Marman clamp ring, which is released by a thermal wax actuator. The TT also provides a 21 contact slip ring assembly for the LEMMS electrical signals, and a feedthrough for purge gas. The weight of the TT is 3.2 kg, and the power consumption is 2.5 W.

TABLE XV

LEMMS channels description (high energy end) from caluclations and calibration measurements.

Readout N	LEMMS channels	Logic	L/H end	Species	p E_{min} (MeV)	p E_{max} (MeV)	He E_{min} (MeV/N)	He E_{max} (MeV/N)	C E_{min} (MeV/N)	C E_{max} (MeV/N)	O E_{min} (MeV/N)	O E_{max} (MeV/N)	Fe E_{min} (MeV/N)	Fe E_{max} (MeV/N)	Electrons E_{min} (MeV)	Electrons E_{max} (MeV)	X-rays E_{min} (MeV)	X-rays E_{max} (MeV)	Geometry factor cm^2-sr
27	E0	D11(D12)(D21)	High	Z = 1, e−	1.2	1.4									0.15	0.2	0.08		0.37 E0
28	E1	D11(D12)D21(D31)	High	e−											0.25	1			0.512 E1
29	E2	(D11)D21(D22)(D31)	High	e−											0.4	0.75	0.06		0.512 E2
30	E3	(D12)D21(D22)D31(D32)(D41)	High	e−											0.7	4			0.512 E3
31	E4	(D12)D21(D22)D32(D33)(D41)	High	e−											0.9	3			0.512 E4
32	E5	D21(D22)D32(D33)	High	e−											1.1				0.512 E5
33	E6	D32(D33)D41(D42)(B1)	High	e−											2.7	18			0.512 E6
34	E7	D41(D42)B1(B2)	High	e−											18				0.125 E7
35	G1	(D21)D31(D32)(D41)	High	X-ray													0.05	0.4	0.435 G1
36	P1	D13(D15)(D21)	High	All	1.42	2.28	1.4	1.48	2.17	2.25	2.31	2.44	2.75	2.86	0.3	1	0.4	1.8	0.371 P1
37	P2	D15(D16)(D21)	High	Z = 1, Z > 1	2.28	4.49	1.5	2.00	2.25	2.42	2.44	2.56	2.86	3.00			1.8	5.4	0.371 P2
38	P3	D15(D16)D21(D23)(D31)	High	Z = 1	4.49	5.74													0.371 P3
39	P4	D12(D16)D23(D24)(D31)	High	Z = 1	5.74	7.89													0.371 P4
40	P5	D12(D15)D24(D25)(D31)	High	Z = 1	8.31	11.45													0.371 P5
41	P6	D24(D25)D32(D34)(D41)	High	Z = 1	11.47	13.43													0.512 P6
42	P7	D21(D25)D33(D35)	High	Z = 1	12.10	58.90	42.5	475.00											0.512 P7
43	P8	D12(D15)D22(D23)D33	High	Z = 1	25.19	59.00	142.5	475.00											0.371 P8
44	P9	D22(D23)D32(D33)	High	Z = 1, e−	58.99		475.0								1				0.371 P9
45	H1	D16(D17)(D21)	High	Z > 1			2.0	4.5	2.42	3.08	2.56	3.13	3.04	3.36				5.14	0.371 H1
46	H2	D16(D17)D21(D26)	High	Z > 1			4.5	10.5	83.33	166.67									0.371 H2
47	H3	D25(D26)D32	High	Z > 1			11.3	25	83.33	666.67									0.512 H3
48	H4	D23(D25)D35	High	Z > 1			25.0	42.5	666.67										0.512 H4
49	H5	D22D34(D35)D45	High	Z = 1, Z > 1	20	25	42.5	75											0.512 H5
50	Z1	D17(D21)	High	Z > 6					3.08	8.00	3.13	9.38	3.36	14.64				19	0.371 Z1
51	Z2	D17D21(D31)	High	Z > 6					8.00	20.92	9.38	24.38	14.64	42.86					0.371 Z2
52	Z3	D26D32	High	Z > 6					20.92	83.33	24.38	187.50	42.86						0.512 Z3
53	D1	D11	High	All	1.40		1.3		2.17		2.31		2.73		0.15	0.27			—D1
54	D2	D21	High	All	4.40		4.5		8.00		9.38		14.64		0.18				—D2
55	D3	D31	High	All	11.20		11.3		21.00		24.38		42.86		0.5				—D3
56	D41	D41	High	All	19.00		19.5		36.17		42.50		76.79		1				—D41

TABLE XVI
Nominal electronic thresholds for the LEMMS detectors.

Detector A → PHA				Detector B			
	Thresholds				Threshold		
Energy	Level	Description	Amp.	Energy	Level	Description	Amp.
12 keV*	43.5 mV	A1	2nd	100 keV*	120 mV	B1	2nd
30 keV	108.75 mV	A2	2nd	400 keV	480 mV	B2	2nd
50 keV	181.25 mV	A3	2nd	1 MeV	230.8 mV	B3	1st
100 keV	362.5 mV	A4	2nd	4 MeV	923 mV	B4	1st
250 keV	906.25 mV	A5	2nd	13 MeV	3 V	B5	1st
500 keV	1.813 V	A6	2nd				
800 keV	2.9 V	A7	2nd				
1.6 MeV	0.96 V	A8	1st				
5 MeV	3.0 V	A9	1st				
	Detector D1				Detector D2		
75 keV*	121.4 mV	D11	2nd	50 keV*	90 mV	D21	2nd
200 keV	323.7 mV	D12	2nd	700 keV	1.26 V	D22	2nd
400 keV	647.4 mV	D13	2nd	3 MeV	247.5 mV	D23	1st
760 keV	1.23 V	D14	2nd	6 MeV	495 mV	D24	1st
1.6 MeV	293 mV	D15	1st	12 MeV	0.99 V	D25	1st
5 MeV	916.7 mV	D16	1st	40 MeV	3.3 V	D26	1st
18 MeV	3.3 V	D17	1st				
	Detector D3				Detector D4		
50 keV*	90 mV	D31	2nd	100 keV*	66 mV	D41	1st
400 keV	720 mV	D32	2nd	400 keV	264 mV	D42	1st
3 MeV	618.7 mV	D33	1st	1 MeV	660 mV	D43	1st
6 MeV	1.238 V	D34	1st	2.8 MeV	1.85 V	D44	1st
16 MeV	3.3 V	D35	1st	5 MeV	3.3 V	D45	1st
	Detector E1 → PHA				Detector F1 → PHA		
15 keV*	100 mV	E11	2nd	100 keV*	120.4 mV	F11	1st
30 keV	200 mV	E12	2nd	170 keV	205 mV	F12	1st
45 keV	300 mV	E13	2nd	300 keV	361 mV	F13	1st
60 keV	400 mV	E14	2nd	530 keV	638 mV	F14	1st
100 keV	667 mV	E15	2nd	880 keV	1.06 V	F15	1st
200 keV	1.333 V	E16	2nd				
	Detector E2/F2 (anti)						
$50 keV*	100 mV	E2F2	1st				

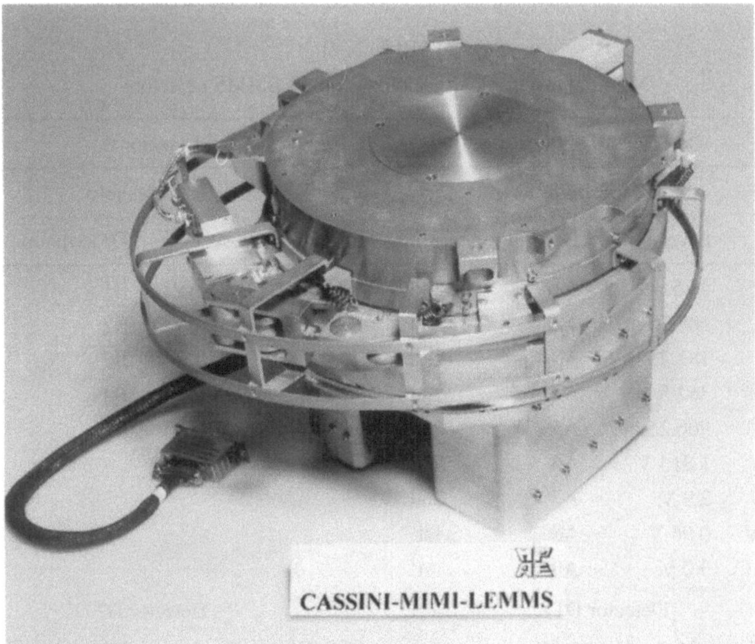

Figure 44. Photograph of the LEMMS turntable, developed by the Finnish company VTT.

3.3.4. *LEMMS Electronics*

The LEMMS sensor electronics block diagram is shown in Figure 45. The signals coming from the silicon detectors are led to nine separate amplifier boards. Each of these boards includes low noise input FET, charge sensitive amplifier (CSA), differentiator, Gaussian filter and linear amplifiers. A power supply filter for each detector is also implemented. The amplifier boards for A, E1, D1 are designed for low-noise contribution. Amplifier board A with FWHM pulse-width of 900 ns generates less than 11 keV (FWHM) noise, E1 with 1900 ns/15 μs as bipolar pulse less than 8 keV, and D1 with 470 ns width less than 15 keV. A power supply generates the detector bias voltages. The first supply provides four different voltages, 3-bit programmable from 0 to approximately 176 V. The second supply has been implemented both for redundancy and detector fine adjustment. It is 3-bit programmable with about 10% higher voltages.

The amplifier output pulses travel to the coincidence board #3, containing discriminators, coincidence logic, and counters. Eighteen 4-bit commandable discriminators and sixteen 24-bit accumulators are integrated in one ASIC each. Together, four accumulator and three discriminator chips are used. The control and communication to the sensor controller has been integrated into one 1020 Actel FPGA. 58 accumulators out of 64 for coincidences and 49 thresholds can be programmed individually from ground.

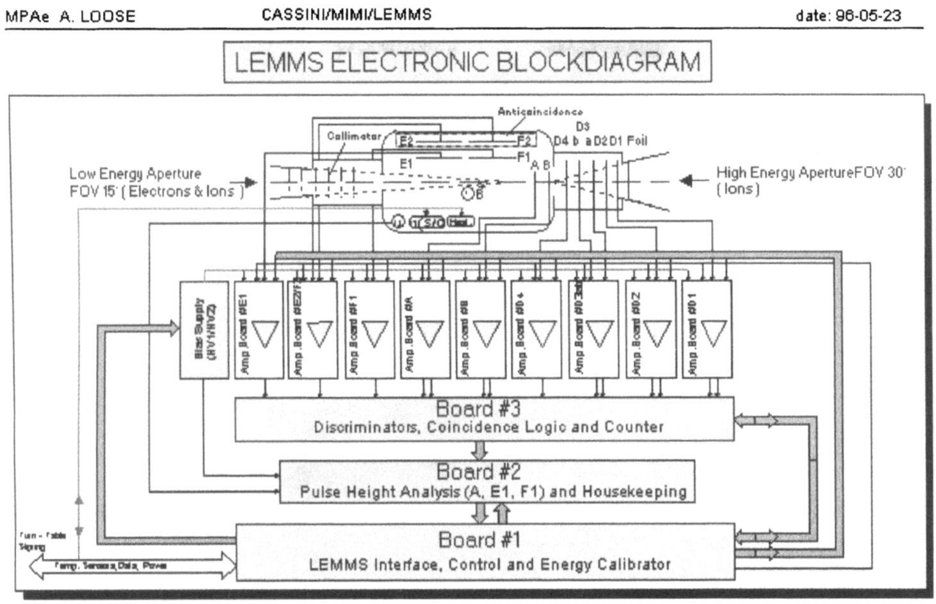

Figure 45. LEMMS electronics block diagram.

For the A, E1, and F1 detectors, a 7-bit pulse height analysis is done. The three channels are serially multiplexed. To cover the whole dynamic range there is a low- and a high-resolution mode for each detector. The corresponding mode is automatically indicated. These electronics are located on board #2, together with the house-keeping electronics that convert the sensor and electronic temperatures, as well as supply voltages into 7-bit digital data.

Board #1 contains the differential interface to the DPU, the sensor controller, integrated into one 1020 Actel FPGA and the energy calibrator. This internal calibrator generates stimulation signals for all amplifier boards from 3.5 keV to 40 MeV in two 8-bit ranges to check the electronics and the logic.

Data, power, and control signals to the sensor connect from the DPU via a 21 contact slip ring, integrated in the turntable. The electronic thresholds of the various LEMMS detectors are summarized in Table XVI.

3.3.5. *LEMMS Calibration*

More than 4600 calibration files have been accumulated over several calibration campaigns which have been performed to investigate the instruments characteristics. We used different beam facilities to calibrate the instrument with ions and electrons in various energy ranges. In addition we used radioactive sources for high-energy electron and gamma calibration.

The complete flight hardware was the subject of extensive tests, including internal calibrator tests, energy scans, motor position scans, vertical and horizontal

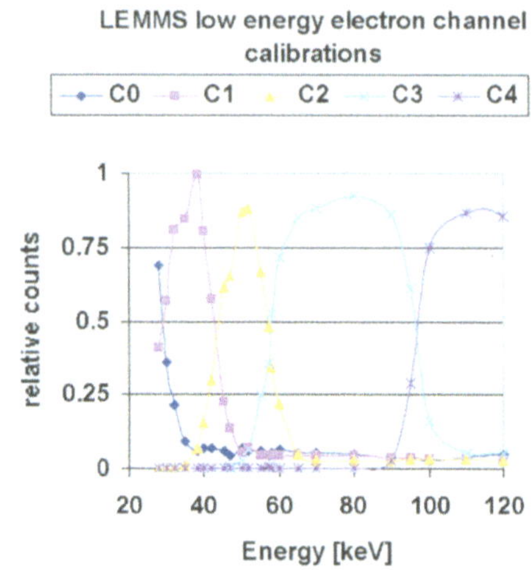

Figure 46. Determination of energy channel limits for electrons for LEMMS channels C0–C4 of the low energy end.

angle scans, detector threshold scans, and scans with and without collimators. As an example we show two energy scans from which the energy channels have been determined. Figure 46 shows the normalized count rates of several energy channels (C0–C4) from detectors E and F as a function of energy. The points where two curves from different channels cross each other were used to determine lower and upper energy limits for each channel. Figure 47 shows a similar energy scan for

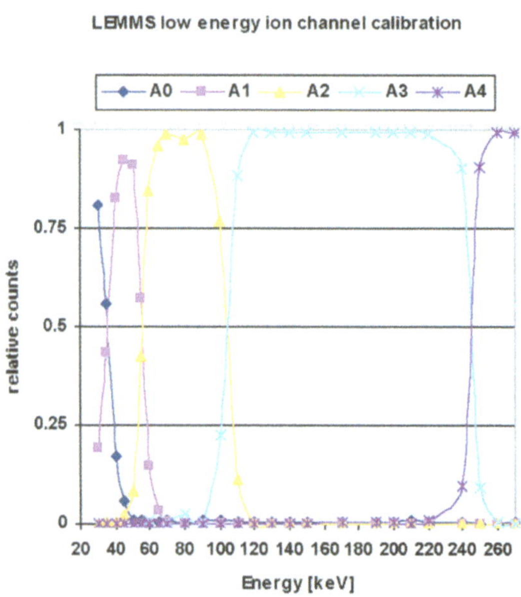

Figure 47. Determination of energy channel limits for ions for LEMMS channels A0–A4 of the low energy end.

ions measured in detector A (A0–A4). The results of these energy scans analyzed so far are summarized in Table XIV for the low energy end and in Table XV for the high energy end of the instrument.

3.3.6. *First in-Flight Results*

After a calibration period in January 1999 and a turn-on of LEMMS during the second Venus flyby in June 1999, the spacecraft returned to Earth in August 1999 to perform a close flyby. During this encounter LEMMS was switched on for nearly 4 weeks (August 17 to September 14, 1999). Cassini passed the Earth's radiation belts and left the Earth through the dawn magnetotail. The flyby was used for calibration in a "known" environment. Figure 48 shows the LEMMS particle measurements obtained during the Earth swingby inside the magnetosphere. The dynamic range of LEMMS covers nearly 6 decades in proton energies, more than 3 decades in electron energies, and 7 decades in intensies. The LEMMS parameters are summarized in Table XVII.

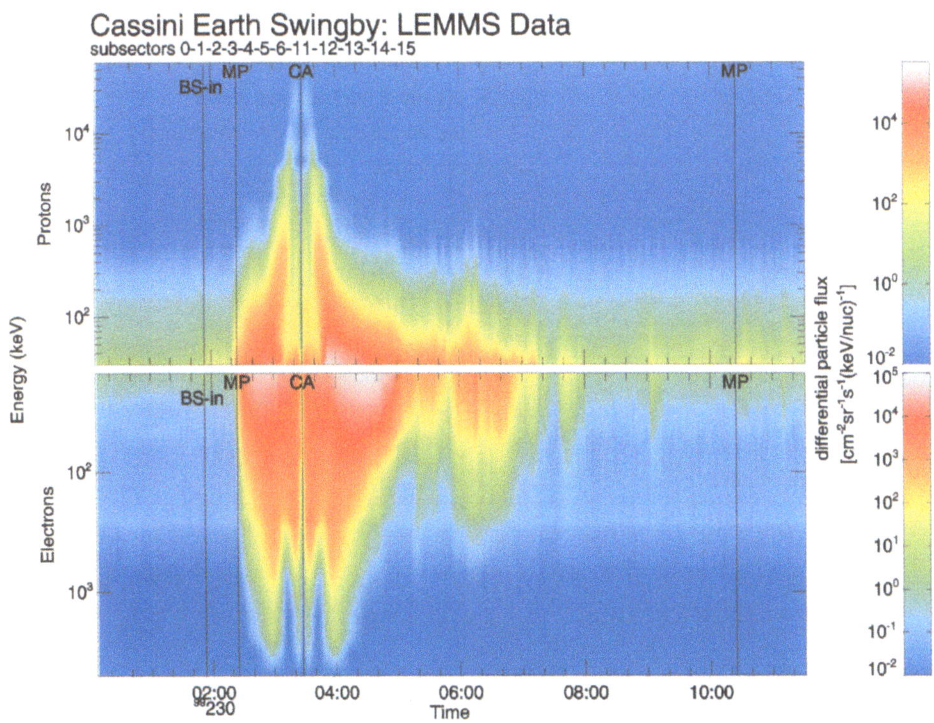

Figure 48. LEMMS energy spectra for protons and electrons in the Earth magnetosphere during the Cassini flyby on August 18, 1999. Solid lines mark the bow shock (BS) and magnetopause (MP) crossings, and the closest approach (CA) of the spacecraft. The highest intensities were measured in the radiation belts.

TABLE XVII
LEMMS parameters summary.

Instrument name	Low energy magnetospheric measurement system (LEMMS)
Scientific objectives	Determination of the angular distribution of energetic particles in Saturn's magnetosphere
Principle	Measurement of energy loss of incident particles in semiconductor detectors
Design	Double-ended sensor head mounted on top of a programmable turntable
Aperture opening angle (full width)	15° (low energy end); 36° (high energy end)
Number of solid-state detectors	11
Species identification	Separation between electrons and ions through magnetic deflection (low energy end) and coincidence logic in a stack of four detectors (high energy end)
Number of channels	57 normal counters
	6 priority counters
	64 PHA channels for detectors A, E1, F1
Rotation axis	$-y$-axis of the S/C system
Scan plane	x–z plane of the S/C system
Angular resolution	22.5° per subsector (16 subsectors)
	2.81° per microsector (128 microsectors)
Time resolution	86 s per rotation
	5.31 per subsector
	0.66 s per microsector
Energy range	Ions: 0.030–160 MeV
	Electrons: 0.015–5.0 MeV
	Priority counters
	Ions: 0.030–0.036 and 0.036–0.053 MeV
	Electrons: 0.015–0.028 and 0.028–0.043 MeV
Weight	6.72 kg
Power consumption	5.2 W

3.4. System Configuration

3.4.1. *Main Electronics Unit and Power Distribution*

The MIMI instrument electronics is spread throughout the four assemblies that comprise the instrument (Figure 49). As seen in the sensor block diagrams (Figure 50), each of the three sensors contains amplifier chains and high voltage/bias supplies to operate detectors and provide steering voltages for the particle optics. This minimal set of electronics is sufficient to detect events and buffer the signals for further

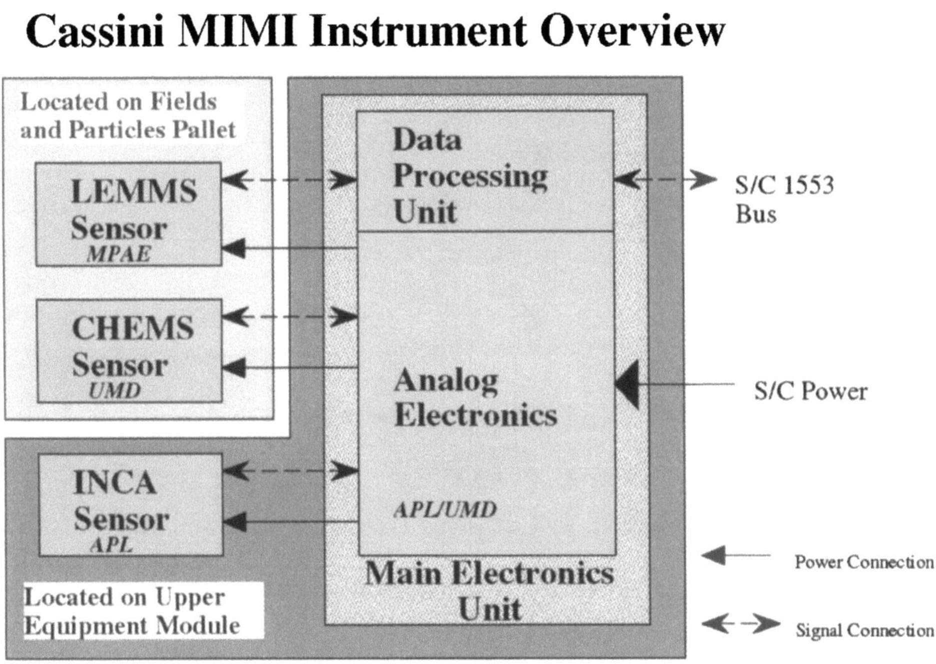

Figure 49. Cassini MIMI instrument overview.

central processing. The overall system parameters are shown in Table XVIII. *MEU*. The majority of the instrument analog and digital processing electronics is located in the Main Electronics Unit (MEU). This subsystem accepts the information from the sensors and performs necessary amplification, processing, and buffering of the event data. The signals from the CHEMS and INCA sensors are sent to the MEU as analog waveforms. The LEMMS events are sent in digital form, however, to reduce the effects of noise pickup in the sensor's rotating slipring interface. This is possible because the LEMMS event processsing is significantly simpler than that

TABLE XVIII
MIMI system overview.

Subsystem	Mass (kg)	Nominal power (W)	Allowable flight operational	Temp limits non-ops
INCA	6.92	3.0	−20/+35	−25/+50
CHEMS	6.66	3.5	−20/+35	−25/+40
LEMMS	6.72	5.2	−20/+35	−25/+40
MEU	7.20	8.6	+5/+50	+5/+50
Coax cables	0.61			
Total	28.11	20.3		

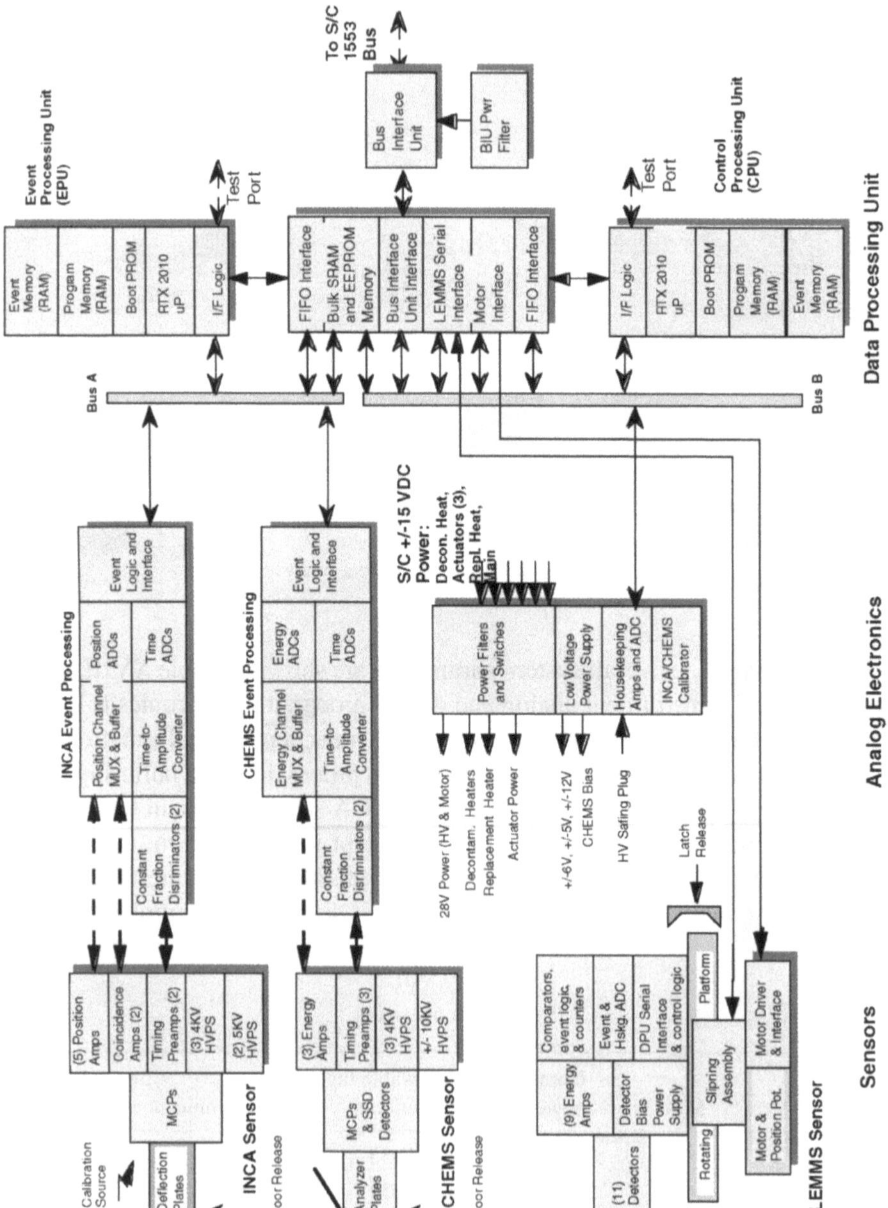

Figure 50. MIMI instrument block diagram.

for INCA or CHEMS, and the necessary digital logic could be co-located with the detector amplifiers.

In addition to processing events, the MEU also handles all spacecraft data and power interfaces, produces all secondary power voltages, and monitors the analog and digital status of the instrument. These functions are split between the two halves of the MEU: all analog processing is performed in the analog electronics (AE) section, and almost all digital processing is performed in the data processing unit (DPU). Both of these are physically located in the MEU enclosure, separated by a thin wall (to reduce EMI).

The analog electronics is comprised of nine PC boards (Figure 51): the low voltage power supply (LVPS), the LEMMS power interface board, the analog electronics motherboard, and six analog processing boards (power interface, housekeeping, CHEMS processing, CHEMS TAC, INCA processing, and INCA TAC).

MIMI has six spacecraft primary power interfaces: the LEMMS replacement heater, the sensor decontamination heaters (all run in parallel), the main instrument electronics power interface (LVPS), and one each for the sensor wax actuators (door/latch releases). The power for each of these interfaces is switched in the spacecraft via a solid-state hybrid. The hybrid provides current measurement, overcurrent limiting and cutoff, and controlled rise time activation.

The LVPS, mounted to the MEU side wall, supplies power to all three sensor assemblies and their processing electronics in the MEU. The converter produces

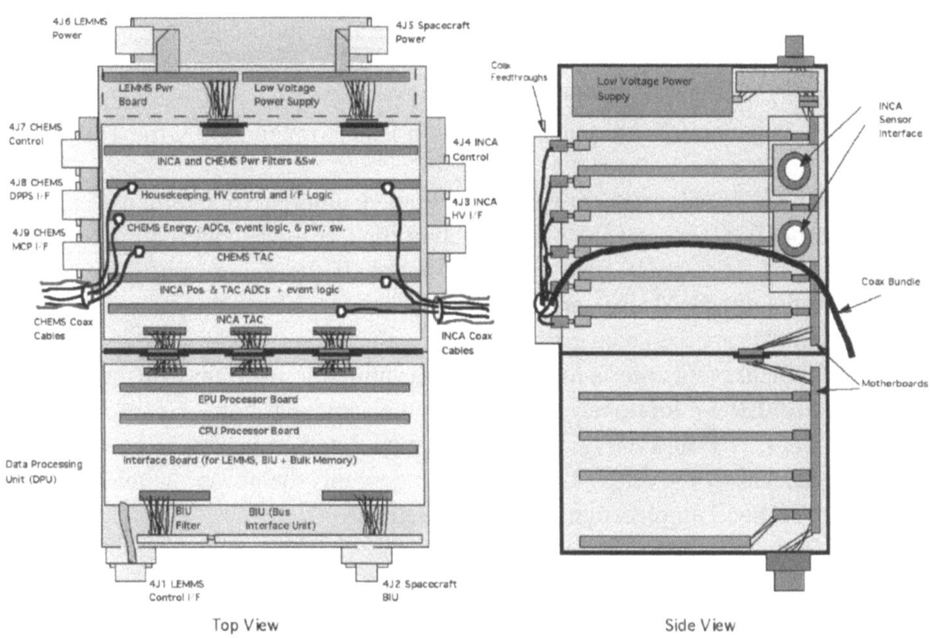

Figure 51. MIMI MEU configuration.

secondary voltages of ±5 V (analog), ±6, ±12, ±5 V (digital), +28, and +150 V (CHEMS bias) with an overall efficiency of 73% (nominal loads). The converter runs at 250 kHz (synchronized to the BIU 1 MHz reference) using push–pull drivers and current mode control. The +5V DPU and CHEMS bias outputs have separate windings, while the other outputs are stacked on a single tapped winding; all outputs share a single return (signal ground), which is tied to the spacecraft chassis ground only within the MEU.

The instrument includes a number of power switches to control the INCA and CHEMS high voltage supplies, the LEMMS motor and processing electronics, the analog processing electronics, and the INCA shutter mechanism. Most of the power switching is accomplished on the LEMMS power interface board and INCA/CHEMS power interface board. Also included on these boards are power filters for each of the spacecraft interfaces; these were included in an effort to isolate the instrument from noise generated elsewhere on the spacecraft. Figure 52 shows the power distribution within the instrument.

3.4.2. *Instrument Processing Boards*

The INCA and CHEMS event data receive similar processing within the MEU. Two boards are dedicated to each sensor; one board to process time-of-flight information and the other to process the energy (CHEMS) or position (INCA) information.

The CHEMS time-to-amplitude (TAC) board accepts "start" and "stop" pulses from the microchannel plate (MCP) anodes in each of the three telescope heads (Figure 53). If the time between correlated "start" and "stop" pulses is measured to be between 9 and 540 ns, a "valid TOF" signal is sent to the CHEMS processing board, along with a TOF pulse whose amplitude is proportional to the measured particle flight time. If no "stop" is detected, the TAC circuitry resets itself at the end of the 540-ns window; the reset is delayed an additional 1000 ns when a "valid TOF" pulse is generated. No pile-up detection is performed during the TOF event analysis.

The CHEMS processing board accepts a shaped, unipolar (600 ns to peak) solid-state detector (SSD) energy pulse from each of the three telescopes in the CHEMS sensor and the TOF pulse from the TAC board. If the energy pulse exceeds a programmable threshold level, and valid event criteria are met, the energy and TOF pulse amplitudes are then converted to digital values and stored for retrieval by the DPU. The typical event criteria for CHEMS is a valid time-of-flight with the START, STOP, and ENERGY (optional) signals from the same telescope. Composite rates and individual TOF and energy rates are accumulated in hardware counters for periodic retrieval by the DPU. The board also contains the power switches for the INCA and CHEMS amplifier and analysis circuitry.

The INCA TAC board is very similar to the CHEMS TAC board, but only one start and one stop input are used (Figure 54). The acceptable TOF range is from 1 to 100 ns; the circuitry will reset itself 120 ns after an uncorrelated "start" pulse is measured, and 1.5 μs after a "start" pulse and correlated "stop" pulse are

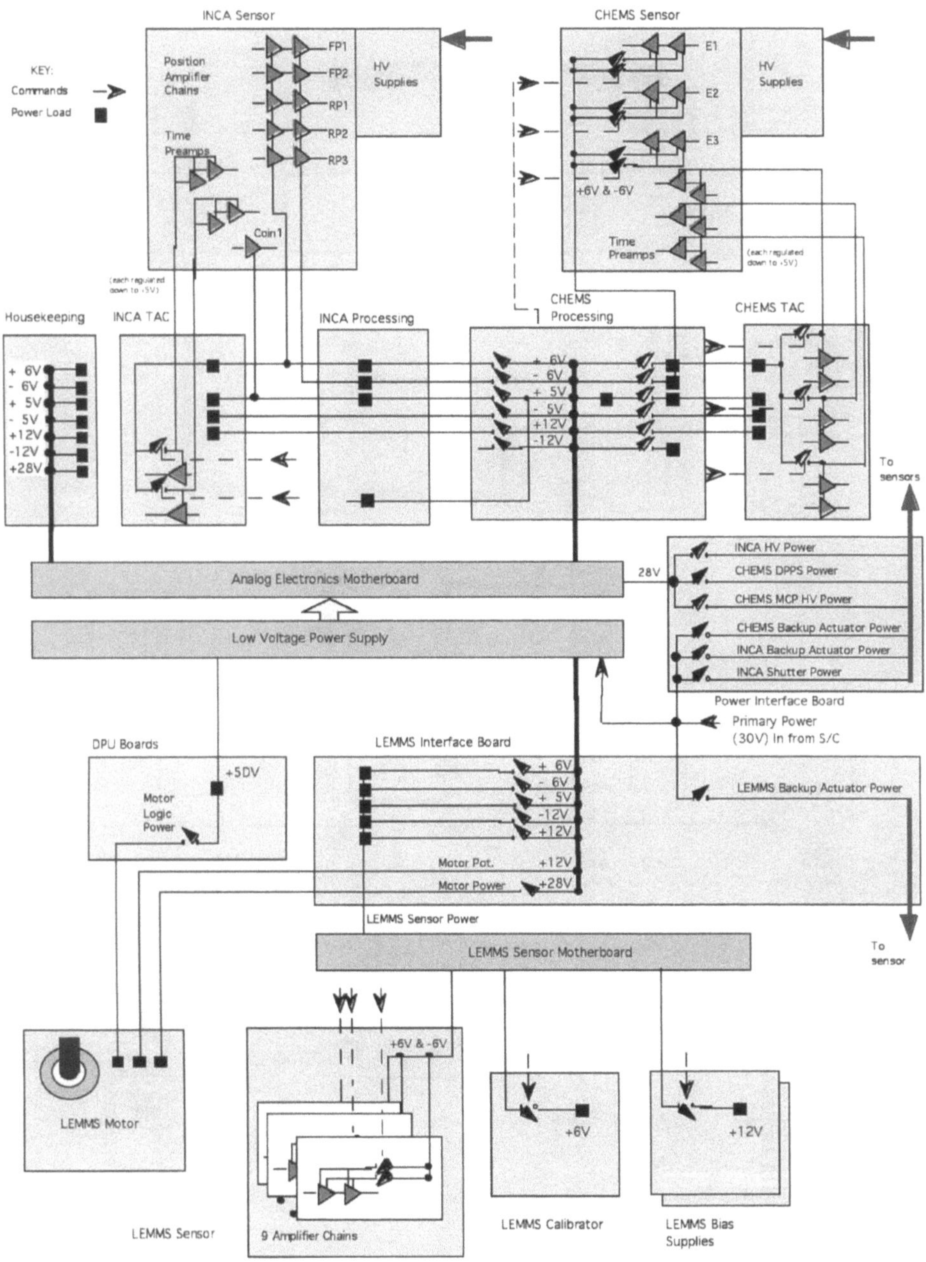

Figure 52. MIMI electronics power distribution.

detected. No pile-up detection is performed during the TOF event analysis. The INCA TAC board additionally has commandable input thresholds to help discriminate against UV-generated counts. The TOF calibrator circuitry is also on this board.

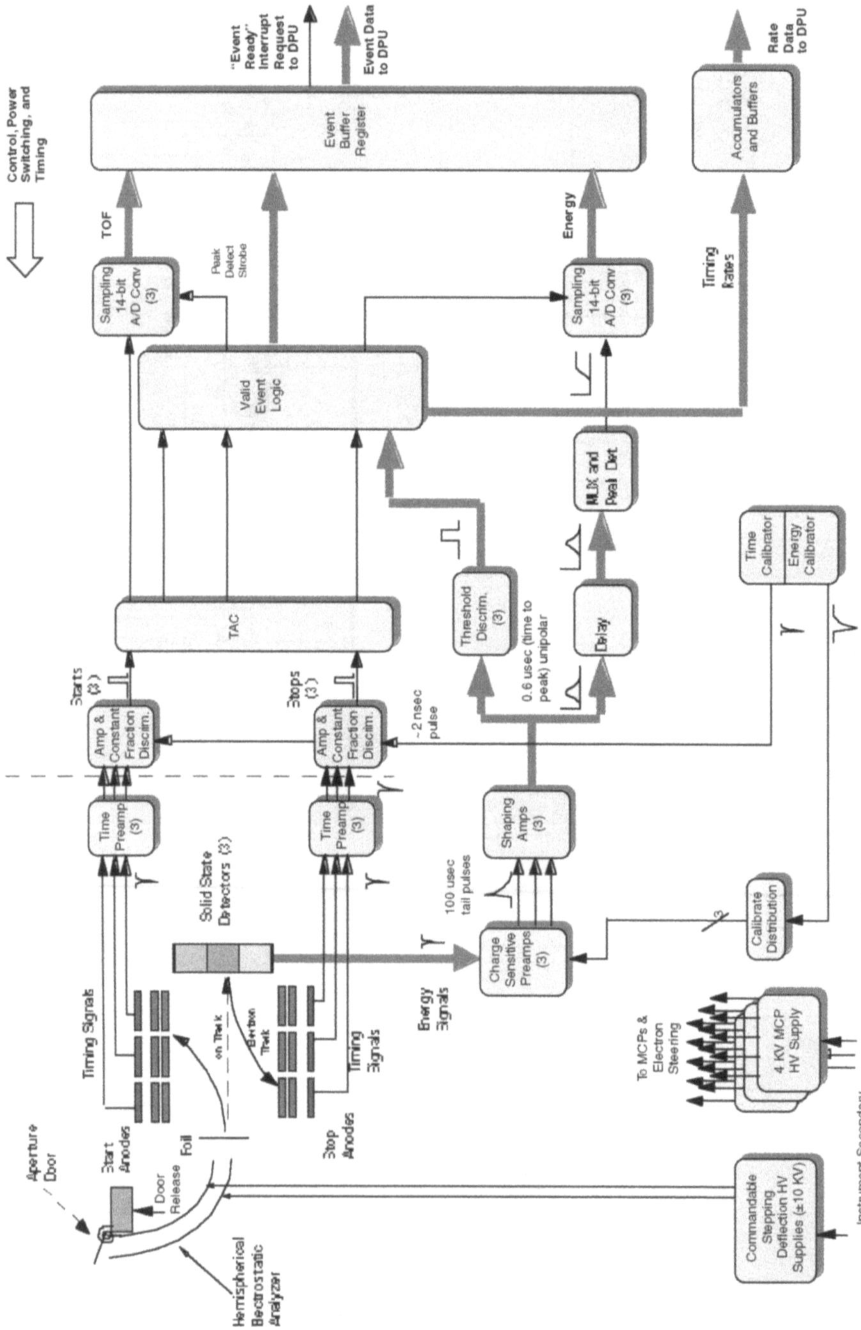

Figure 53. CHEMS subsystem block diagram.

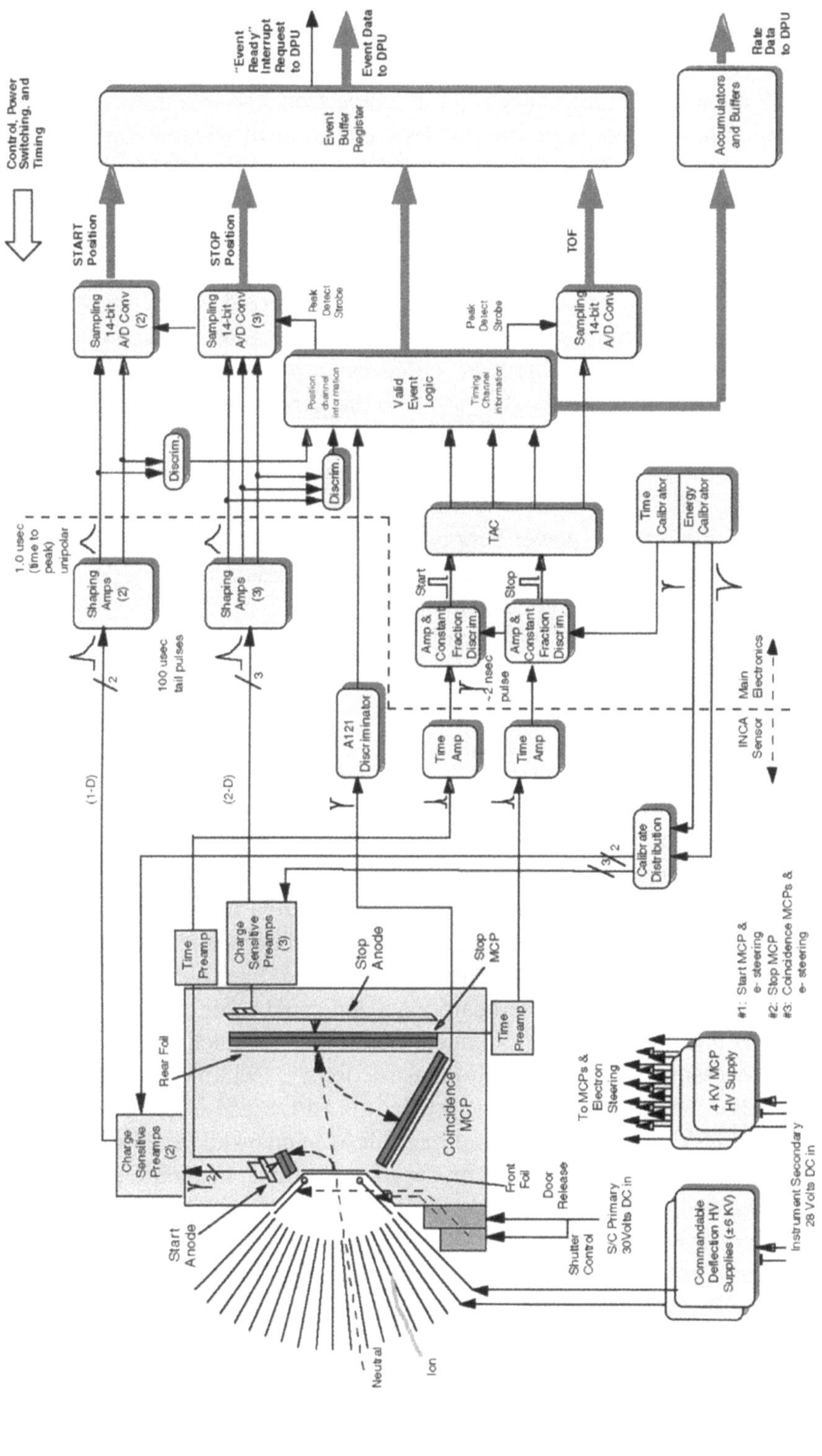

Figure 54. INCA block diagram.

The INCA processing board is very similar to the CHEMS processing board, but with the energy channel electronics replaced with five shaped, unipolar (1000 ns to peak) "position" channels: two shaped pulses come from the one-dimensional Start MCP anode and three shaped pulses come from the two-dimensional Stop MCP anode. These pulses represent the detected charge distribution on the START and STOP position-sensitive anodes, and will be used in the DPU to calculate the particle trajectory in eteration and azimuth angle. If valid event criteria are met, the position and TOF pulse amplitudes are then converted to digital values and stored for retrieval by the DPU. The typical event criteria for INCA are a valid TOF, start position, stop position, and coincidence signal all detected within approximately 100 ns of each other. Composite rates, coincidence MCP count rates, and individual TOF rates are accumulated in hardware counters for periodic retrieval by the DPU.

The housekeeping board handles most of the general status setting and monitoring for MIMI. Analog measurements are made for supply voltages (low- and high-voltage) and currents, instrument temperatures, and the LEMMS motor position. Analog references are provided for setting thresholds, calibrator pulse-heights, and high-voltage power supply (HVPS) levels. Digital outputs provide power switch-enables, HVPS enables, and digital references for the TOF calibrator and the CHEMS deflection plate power supply.

The LEMMS subsystem, as shown in Figure 55, is fairly self-contained; no LEMMS hardware is contained within the MEU. A detailed description of LEMMS can be found in Section 3.3.

3.4.3. *Data Processing Unit (DPU)*

The DPU (right side of Figure 50) is responsible for all instrument control, data analysis, command, and telemetry functions. It is comprised of six boards: the control processing unit (CPU), the event processing unit (EPU), the support board, the bus interface unit (BIU), the BIU filter, and the DPU motherboard. The CPU, EPU, and support board are mounted into the motherboard, and make up the heart of the DPU. The BIU and its power filter are mounted directly to one of the MEU walls; they interface to the motherboard via a dedicated cable.

The CPU and EPU share an identical board design. Each has a Harris RTX 2010RH processor running at 6 million instructions/s, 192 kb of rad-hard SRAM, rad-hard PROM (8 kb for the CPU, 4 kb for the EPU), and 64 kb of rad-tolerant EEPROM. An additional 832 kb of rad-hard SRAM and 64 kb of rad-tolerant EEPROM is provided for each processor on the support board. In addition to providing this bulk memory to the processors, the support board also contains dedicated FIFO circuitry through which the EPU and CPU communicate, and FPGA-based interfaces for the LEMMS motor, LEMMS data, and the BIU. The BIU board, designed and built at JPL, provides redundant physical and logical interfaces to the spacecraft MIL-STD-1553 data bus.

The CPU and EPU are each dedicated to different hardware interfaces and tasks. The EPU only handles processing for the INCA and CHEMS events, and

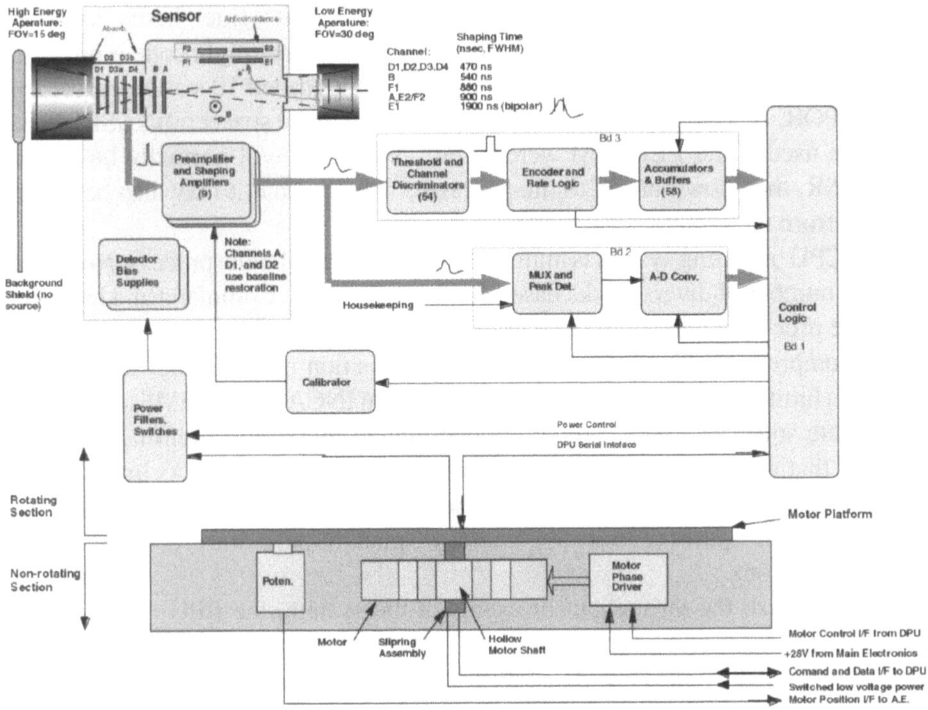

Figure 55. MIMI LEMMS subsystem block diagram.

is slaved to the CPU for all command and timing interfaces via the support board FIFOs and dedicated control lines. The EPU communicates with the INCA and CHEMS TAC and ADC boards and the support board. The CPU handles all other instrument tasks. It communicates with the housekeeping board, the LEMMS serial and motor interfaces, and the BIU interface. Since each processor can access only a portion of the instrument hardware interfaces, both are required to fully operate the instrument.

The BIU handles all 1553 communications, and communicates with the CPU via a dual-port 8 kb SRAM. When command and telemetry data are written to assigned memory locations within this SRAM, hardware flags are automatically set to indicate that data is available for pickup. The BIU has two dedicated digital I/O ports that the spacecraft can directly control/read without the CPU's participation. These lines provide the spacecraft the capability to provide high-level command/control lines to the instrument that are not dependent on the instrument processor itself. In MIMI, these lines are used to enable high voltage operation, reset the processors, enable the watchdog timers, and request the "SLEEP" mode. Status tell-tales also let the BIU monitor the DPU boot process to aid in possible fault analysis. A 9720 μF capacitor bank is connected to the BIU +5 V supply line to filter out the large transients generated when the BIU is transmitting data over the 1553 bus.

DPU Software. Software for both the EPU and CPU is written in the Forth language, which is directly executable by the RTX processors. Flight code is normally stored in the processor's EEPROM memories, and is loaded 10 s after a processor reset or POR. A copy of the flight code is also stored in the spacecraft's SSR, which could be used if the EEPROM were to fail. New code will normally be written to the SSR, then downloaded to the processor memory, but it may also be loaded directly from the ground if necessary.

The CPU operating system is multitasking, which enables the processor to handle its large number of diverse tasks based on a priority basis. Command and hardware interface processing are of highest priority, followed by telemetry processing and image compression. The EPU operating system is non-multitasking, since it only handles a limited number of processes. Events from INCA and CHEMS are buffered in separate software FIFOs, which are then polled in a round-robin fashion, thus insuring that neither sensor can monopolize the EPU processing. As implemented in the software at launch, each INCA event requires (200 μs to process, while each CHEMS event requires (135 μs to process; we plan to speed this processing in later software versions.

To maximize the throughput of events in the system, the EPU is responsible only for the INCA and CHEMS event collection, processing, binning, and image cropping. The CPU takes this data (at spin/sector boudaries) and further processes it. The resulting CHEMS and INCA data is then formed into suitable telemetry packets.

There are several sources of computational dead-time in the EPU. When data is transferred between the EPU and CPU on microsector, subsector, sector, and quadrant boundaries, the normal EPU data collection is halted. The largest dead-times will be on sector and quadrant boundaries when INCA high-time resolution and low-time resolution images are transferred. A more significant source of dead-time is the image motion compensation (IMC) calculations performed by the EPU once every subsector. This results in a total dead-time of up to 0.6 s while in the "Staring Mode," and 0.07 s while in the "Spin Mode." The IMC dead-time applies only to event processing, however; new events will still be added to the software FIFOs (up to their 256 event limit).

Command Functions. The DPU command processing is very flexible. Normal commands and memory loads are transferred to the instrument on an immeditate or time-delayed basis via the spacecraft CDS. Commands for MIMI are sent from the spacecraft over the 1553 bus. Each command is comprised of one or more 16-bit words, where the MSB of the first command word is a unique opcode. All commands have a fixed length (except for IEB_LOAD). A flight rule restricts normal commanding to no more than one command per second to reduce possible command errors; memory loads are transferred at rates up to 1280 16-bit words/s from the SSR.

The instrument has the ability to store and run two different types of command sequences. The first type, known as macros, represent quasi-permanent sequences used for instrument control functions that are not expected to change. Most instrument alarm responses and the LEMMS motor maintenance procedures are stored as macros. The second type, known as instrument expanded blocks, or IEBs, represent temporary sequences that are used for instrument orbital operations. IEBs are created and compiled on the ground using GSE software. A dedicated set of instrument commands is used to load, manage, and trigger IEBs in the DPU memory. Normally the trigger commands will be integrated into the spacecraft activity timeline, thus sychronizing a set of predefined MIMI science operations with the overall spacecraft activity plan.

3.4.4. *MIMI Operational Modes and Controls*

Operational Modes. The instrument has a large number of operating modes, given that the three sensors may be commanded to a number of different configurations and individually enabled or disabled. Of this superset, however, a small subset has been chosen as standard operating modes, as shown in Figure 56:

POR Mode: This is the mode that is entered when the instrument is first turned on or reset. Only the DPU and housekeeping boards will be active; analog and sensor power will be off.

FULL Mode: This is the standard operating mode of the instrument. All three sensors are power and active, high voltages are at their operational levels, and the LEMMS motor is rotating; only the INCA shutter and the electronic calibrators are off.

SLEEP Mode: This mode is used to temporarily reduce the instrument power consumption, thus allowing the spacecraft to time multiplex instrument operations. When activated, this mode will turn off the LEMMS motor, the INCA shutter, all INCA and CHEMS amplifier, calibrator, and processing electronics, and will command the EPU to a slower clock speed. When this mode is exited, the former state of the electronics will be resumed. Note that all high voltages and LEMMS sensor power are unaffected.

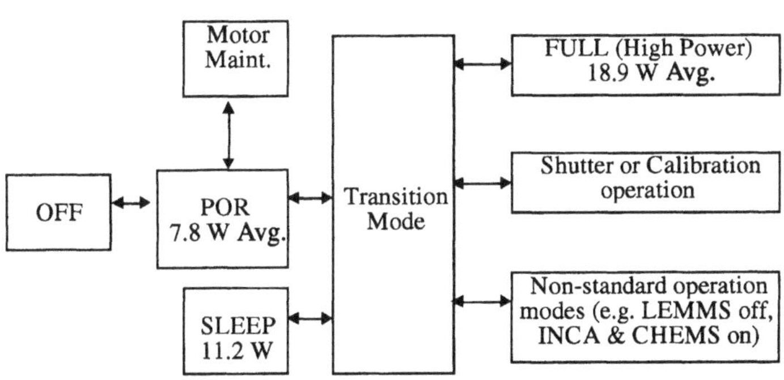

Figure 56. MIMI power states.

MIMI can also be commanded to store several different instrument configurations for later recall. This can be used to switch between standard operating modes during the mission with only a few commands.

Fault monitoring and protection. The DPU monitors the instrument operation via both digital and analog status data. All instrument temperatures, secondary voltages and several secondary currents are measured parameters.
In-flight calibration. The CHEMS and INCA sensors share an electronic calibration circuit, located in the MEU, which provides predefined energy and timing stimuli to the processing electronics.

The LEMMS sensor contains a similar circuit to provide energy stimuli to its processing electronics. The MIMI DPU will control the calibrator levels and sequencing; the calibrators will be powered off during nominal instrument operation. It should be noted that the CHEMS/INCA calibrator use is limited due to a known electonics flaw.

The INCA shutter, thermostatically driven by a coil-wound bimetalic spring heated with a 1.5-W heater, requires approximately 15 min to open or close. When the shutter is closed, the sensor will see counts from two AM-241 sources, each 0.075 μCi, producing 5.4 MeV α particles as well as 60 keV X-rays.

Thermal control. The thermal design of MIMI is tightly coupled to that of the spacecraft. All four subassemblies are conductively mounted to the spacecraft structure, and their bulk thermal performance is largely determined by the overall spacecraft temperatures. The INCA collimator assembly is thermally isolated from the rest of the sensor, however, to limit radiation losses. The rotating section of LEMMS is not thermally coupled well to the spacecraft, due to the limited mechanical contact through its motor bearings. The sensor electronics or a replacement heater is always on to insure that the sensor temperature does not drop too low. Instrument temperatures are independently measured by both the spacecraft and the MEU.

4. Data and Image Processing

Because of the high sensitivity and high angular, energy and time resolution of the three MIMI sensors, continuous transmission of the raw data would require impossibly large bit rates. So, extensive onboard data processing and data compression is a fundamental aspect of the MIMI experiment. The MIMI flight software has been designed to meet the scientific requirements of the mission even in limited data transmission allocation conditions. A variety of data products have been defined, each corresponding to a virtual instrument, and a combination of them is transmitted to the telemetry.

All MIMI data acquisition and processing is synchronized to an internally generated "sector clock," which when the spacecraft is in spinning mode is synchronized to the spacecraft rotation. One spacecraft rotation is divided into 16 sectors, each sector thus corresponding to 22.5° of spacecraft rotation. Each sector is then subdivided into 16 subsectors, and each subsector into 16 microsectors. For the nominal spacecraft spin period, which is ~23 min, a sector corresponds to ~86.2 s, a subsector to ~5.4 s, and a microsector to ~0.34 s. These values are updated at the beginning of each sector, by taking into account the most recent spin rate information transmitted by the spacecraft AACS. This allows the MIMI sectoring scheme to adapt to spacecraft spin-ups and spin-downs. When the spacecraft is in a staring mode, the same MIMI sectoring scheme is still applied, but now the sector duration is a fixed parameter value, that mimics spacecraft rotation. Transition between the two modes, i.e. staring or spinning, is performed automatically, by monitoring the spacecraft spin rate data transmitted by the AACS.

The MIMI instrument incorporates two 16-bit RTX 2010 processors; the control processing unit handles all instrument command, telemetry, and control tasks, as well as the LEMMS data collection. The event processing unit is dedicated to collecting and processing the raw INCA and CHEMS data.

INCA image data are organized in three different image groups, simultaneously generated: high spatial resolution images, acquired over four sectors; high time resolution images, acquired each sector; and high time-of-flight and mass resolution images, acquired over four sectors (Figure 57).

The EPU collects data from the INCA instrument, generates the images, performs image motion compensation (by taking into account the spacecraft attitude data, correcting thus for spacecraft mosaic-like attitude motion), and sends the images to the CPU. The images are then compressed by the CPU and sent to the telemetry. A double buffering process is used in order to acquire an image set in the EPU in parallel with the processing of the previous instrument cycle image set in the CPU. The INCA image types generated in neutral (ENA) mode are also shown in Table XIX.

TABLE XIX

INCA image types generated in neutral (ENA) mode.

	Number of images	Image accumulation period	Pixels/image	Memory pixels/image
High spatial resolution Low TOF	3	4 sectors	64 × 64	128 × 96
High spatial resolution High TOF	3	4 sectors	32 × 32	64 × 48
High time resolution Low TOF	1	1 sector	32 × 32	64 × 48
High time resolution High TOF	1	1 sector	32 × 32	64 × 48
High *m*-TOF resolution: (neutral mode)	5 × 8	4 sectors	16 × 16	32 × 24

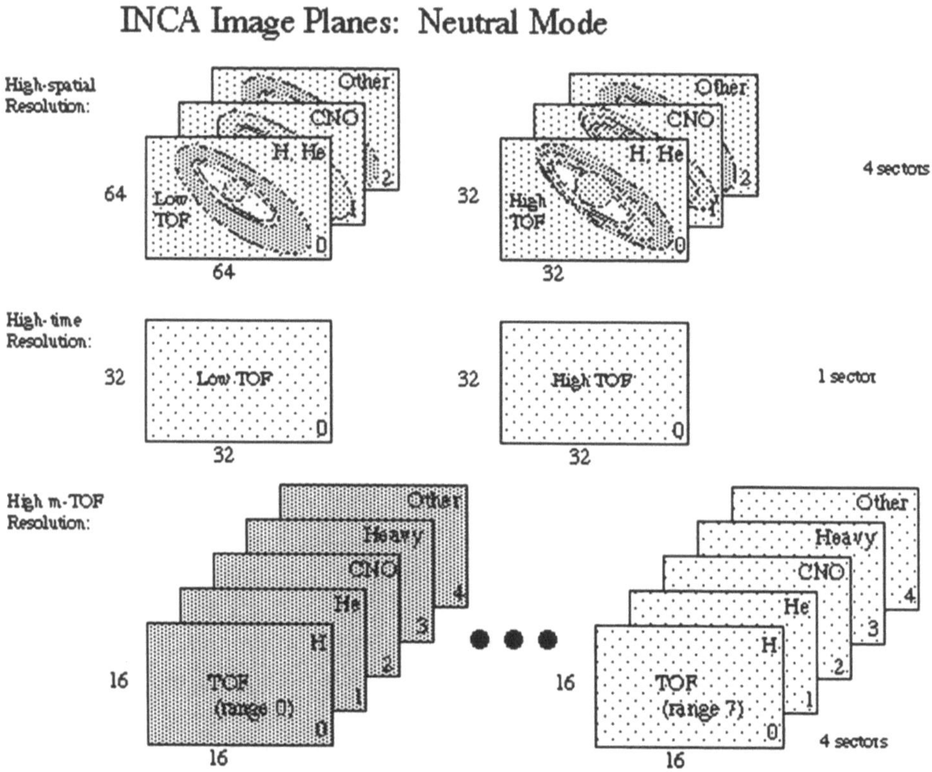

Figure 57. INCA images.

The INCA images must be compressed in order to fit into the available telemetry allocation. The data compression is a four-step process: event binning, selection of the "useful" part of the image, logarithmic compression, and "real" image compression.

Event binning, and organization of the image data into the scheme shown in Table XIX, results in a preliminary data compression by a factor of $\sim$20. As shown in the table, every four sectors INCA generates three 64×64 pixel images (high spatial resolution), eleven 32×32 pixel images (eight high time resolution and three high spatial resolution images) and forty 16×16 pixel images (high TOF/mass resolution). This corresponds to an average of $\sim$98 pixels/s. Without the implementation of such a binning scheme, a single set of images would be generated, providing the full spatial, time and m-TOF resolution, i.e. 5×8 images, 64×64 pixels each, generated every one sector. This would have required an average of $\sim$1900 pixels/s.

Selection of the "useful" part of the image results in the reduction of the initial $180^\circ \times 180^\circ$ equivalent FOV, allocated in the DPU memory, to the $90^\circ \times 120^\circ$ image sent to the telemetry. At the start of image integration each $180^\circ \times 180^\circ$

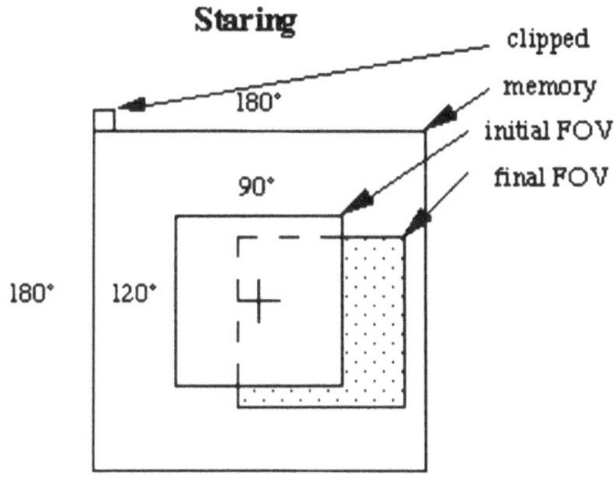

Figure 58. INCA image integration in stare mode.

image buffer is centered with INCA's FOV (Figure 58). As the spacecraft moves, the FOV can move and rotate within the image buffer, image motion compensation taking into account for that. Nevertheless, at the end of the image accumulation, only the 90° × 120° part of the image, i.e. equal to the INCA instantaneous field of view (corresponding to the most exposed part of the 180° × 180° image) is sent to the telemetry. This corresponds to a data reduction factor of 3.

Logarithmic compression of the accumulated counts is then performed by the CPU on each pixel (except during the very high bit rate), and results in a 16–8-bit compression. This is lossless only for the first 32 values, but the maximum relative error does not exceed 3.1%.

"Real" image compression is then performed by the CPU, and it is a lossless Rice or Fast compression. The compression algorithm is selected via uplink command, and its use is flagged in the downlink data stream as a header variable. The Fast algorithm is an encoding scheme that examines a block of J pixels and finds the maximum pixel value in the block. It then calculates the minimum number of bits required to code this maximum value. Each pixel of the block is then represented using only this smaller number of bits. A block header is required for each block, that contains the value used for encoding the pixels in the block. The Rice algorithm is based on using a Comma code. The Comma code for any positive integer, k, is simply k leading zeroes followed by an ending "1." As in the Fast encoder, the data are processed in a 4 × 4 pixel block at a time. The Rice algorithm considers the block values as containing a certain number, m, of incompressible least significant bits (lsbs). These lsbs are stripped off of each block value and saved. The Rice algorithm must determine the best value for m, and the number of lsbs to strip off for each block of data, so that the overall code length of the encoded block is minimized. The remaining most significant bit values (msbs) are then Comma encoded. Both Rice and Fast algorithms have been tested on a test image set, generated from simulated INCA images, as the one shown in Figure 12, in which instrument noise was then introduced in order to examine the noise effects on the image entropy and the compression factor.

Compressed images are then organized in telemetry subpackets, each subpacket corresponding to a self-contained part of the parent image, including the necessary header information. It should be noted that image compression is performed within the image part corresponding to each subpacket. This allows decompression of each subpacket independently, so as to allow reconstruction of correct partial images in case of loss of some subpackets. INCA image subpackets (and all other MIMI science data product subpackets) are inserted into the fixed-length MIMI science packets, and are transmitted to the spacecraft solid-state recorder (SSR).

Other MIMI data products include:

- INCA and CHEMS accumulator rates, which are periodic readouts of the hardware accumulators in these instruments (9 INCA accumulators and 17 CHEMS accumulators). These are subject to a 24–10 bits logarithmic compression.
- INCA and CHEMS PHA events, which correspond to the complete information on the subset of the particles detected by these two instruments. For INCA this includes: TOF, pulse-height of the signal output from the front and rear MCPs, MCP identification (coincidence, start/stop), calculated azimuth and elevation, and mass range. For CHEMS it includes: TOF, pulse-height of the signal output from the SSD, MCP, and SSD identification, deflection voltage (DPPS) step, and calculated range (one of seven basic rates, each one being a subdivision of the $m - m/q$ space).
- CHEMS basic rates and science rates, i.e., total number of detected ions corresponding to each of the 7 basic rates and 34 science rates, measured for each DPPS step. These are subject to a 16–8 bits logarithmic compression.
- LEMMS accumulator rates, which are periodic readouts of the 57 normal counters and the 4 priority counters, each corresponding to a different particle, energy, or species range. These are subject to a 24–10 bits logarithmic compression.
- LEMMS PHA data, corresponding to the periodic readout of 64 energy channels $\times 3$ detectors and are subject to a 16–8 bits logarithmic compression.

Table XX gives an overview of the MIMI science data products and of the periodicity of their transmission to the telemetry stream. Note that during spinning mode, INCA 16-sector images, corresponding to a 360° azimuthal FOV, are transmitted to the telemetry as four separate 90° azimuthal FOV images, each one accumulated over a 4-sector period.

In addition to the above INCA, CHEMS and LEMMS sensor data products, the MIMI science telemetry stream also includes some auxiliary data, such as LEMMS

TABLE XX
Overview of MIMI science data products.

	Staring		Spinning	
	Neutral	Ion	Neutral	Ion
INCA				
High spatial resolution images	4 sectors	–	16 sectors*	–
High time resolution images	1 sector	1 sector	1 sector	1 sector
High *m*-TOF resolution Images	4 sectors	1 sector	16 sectors*	1 sector
PHA events	4 subsectors			
Accumulator rates	1 subsector (# of events varies with telementry mode)			
CHEMS				
Science rates	1 subsector (2 subsectors in *Very Low Bit Rate*)			
Accumulator rates	1 subsector			
Basic rates	1 subsector (2 subsectors in *Very Low Bit Rate*)			
PHA events	1 sector (# of events varies with telementry mode)			
LEMMS				
Accumulator rates	4 "priority" counters: 2 microsectors 60 remaining counters: 16 microsectors (32 in *Very Low Bit Rate*)			
PHA events	1 subsector in *Very High Bit Rate* and *High Bit Rate* 2 subsectors in *Medium Bit Rate* 4 subsectors in *Low Bit Rate* 8 subsectors in *Very Low Bit Rate*			

turntable rotation data, spacecraft attitude data, MIMI instrument status data, etc.

The MIMI science packet and subpacket headers have been designed to include all the necessary information for data and mode identification, decompression, and time tagging, while minimizing the overhead to the telemetry bit rate. This overhead is about 2.4%, including the standard CCSDS and Cassini project packet headers.

A flexible scheme has been implemented in order to optimally match the limited SSR allocation and the variable compression rates, while transmitting the maximum amount of information possible. Five telemetry rates have been defined for each of the three MIMI sensors: very high bit rate, high bit rate, medium bit rate, low bit rate, and very low bit rate (cf. also Table XX). Each 24-h period, corresponding to an SSR allocation, has been divided into a series of "observation periods," and a default telemetry rate is programmed by uplink command for each of the three sensors and for each observation period. The higher telemetry rates can thus be programmed for the orbit sections corresponding to satellite flybys, flux tube crossings, etc. The MIMI DPU monitors the science packet data volume sent to the spacecraft SSR during each observation period. If this volume exceeds a predefined threshold, the sensors automatically switch to a lower than the default programmed bit rate for the next observation period, so as to avoid an early saturation of the SSR partition allocated to MIMI. But if the monitored data volume is lower than anticipated, then for the next observation period the sensors switch to a higher than the default programmed bit rate, thus making optimum use of the SSR partition.

5. Toward Accomplishment of Science Goals

The unprecedented capabilities of the MIMI investigation, coupled with the comprehensive nature of the overall Cassini mission, will result in an epochal change in our understanding of the space environment of the Saturnian system. A summary of the measurement advances that the MIMI investigation will make over previous measurements would include: (1) the first global magnetospheric imaging of the hot ion populations of the Saturnian magnetosphere, Titan and possibly icy satellite exospheres, and dust interactions within the inner magnetospheric regions; (2) unambiguous composition analysis of hot ions, critical for ascertaining plasma sources, and the first measurement of charge state distributions, critical for the study of processes, such as charge exchange, that fractionate the compositional distributions; and (3) much more comprehensive in-situ angular coverage and finer angular resolutions, critical for calculating the charge particle invariant quantities that order the transport of hot plasma populations. These instrumental advances must be added to the advances that will accrue simply from the fact that the Cassini is an orbital mission, allowing a comprehensive in-situ survey of all of the critical regions of the Saturnian system. We have emphasized in this paper that the energetic charged particle populations interact strongly with other components of the Saturnian space environment, including Saturn's atmosphere, the rings and dust particulates, neutral gas populations, icy satellite surfaces, and Titan's atmosphere. When the advanced capabilities of the MIMI investigation are combined with other measurements, our understanding of these interactions will also increase dramatically.

Table XXI summarizes the anticipated accomplishments of the MIMI investigation. It lists the major science objectives, the observations required for those objectives, and the results expected. Needless to say, such a table must be limited in scope. There is not the space here to discuss the rich array of possible multi-instrument studies involving MIMI and all of the other advanced instrumentation that is flying with the Cassini mission.

TABLE XXI

Expected accomplishments of the MIMI investigation.

Objective	MIMI product	Interpretation
Global morphology of Saturn's magnetosphere	ENA images (remote), orbital intensity survey (in-situ)	Direct views of remote images; image inversion; combine with UVIS neutral gas images; orbital data assimilation statistics and models
Plasma source(s)	Composition and charge state distributions, energy spectra	Species identification; ionization states; source process modeling; analysis of CRAND spectral features
Plasma energization	Energy spectra (in-situ); pitch angle distributions (in-situ); ENA images (remote)	Location and evolution of energization events; spatial and temporal evolution of spectra
Global dynamics and possible substorms	ENA images (remote); orbital intensity survey (in-situ)	Direct views of remote images; image inversion; orbital data assimilation and orbit-to-orbit comparisons
Plasma convection	Ion angular distributions (in-situ)	Quantitative plasma flow analysis
Hot plasma and radiations transport	Phase space densities (in-situ); intensity profiles (in-situ); ENA dust interaction image (remote)	Diffusion theory analysis of phase space densities; material interaction feature evolution analysis (e.g. satellite macro/micro signatures), dust interaction modeling
Saturnian aurora energy source	ENA images (remote); orbital survey of electron and ion spectrum moments (in-situ), phase space densities	Correlate hot plasma dynamics with auroral dynamics; quantitative analysis of electron and ion energy precipitation limits; pitch angle diffusion analysis.
Hot plasma/Saturnian atmosphere interactions	Composition spectra (in-situ)	Identification of atmospheric species by location
Hot plasma/satellite and ring interactions	ENA images (remote); intensity profiles, phase space densities, composition spectra (in-situ)	Quantitative ENA intensity analysis of dust emissions; micro/macrosignature transport analysis; species identification; quantitative surface deposition sputtering calculations
Titan remote monitoring of magnetospheric dynamics	Remote ENA images of Titan	Analysis of position of magnetopause based on the intensity of Titan ENA emissions
Jovian science	ENA images (remote), ion composition and charge state distributions (in-situ)	Images analysis and sorting by various system drivers (solar wind, Jupiter rotation, Io position, etc.); composition source analysis; analysis of charge state in context of hot plasma transport models.

6. Summary

It is beyond doubt that the Cassini orbiter mission to Saturn will increase manyfold our knowledge of the overall Kronian system, beginning with the rings, Titan, icy moons, magnetosphere, atmosphere, aurora, dust environment, etc. If any proof were needed, it has been provided by the spectacular Galileo results from the Jovian system and the preliminary results from Cassini's flyby at Jupiter.

The MIMI investigation is designed to make key contributions to each of the subjects enumerated above, using an array of sensors that will provide comprehensive measurements of the hot plasmas in Saturn's environment with unprecedented resolution in energy and species. Furthermore, MIMI's ability to image plasmas indirectly, through their ENA emissions, represents a heretofore unreachable objective of space plasma research that has long eluded instrument designers in this field. It has been amply demonstrated on the IMAGE mission in Earth orbit (Mitchell *et al.*, 2001). This new capability opens up a large class of studies on plasma dynamics that may enable us to not only address long-standing unsolved questions (e.g., substorm initiation and development) in magnetospheric physics, but also lead to radically new insights on the large-scale structures and evolution of such systems.

The in-flight performance of the instrument has been nominal so far, and we expect that to continue through Saturn orbit insertion and the 4-year encounter operations. The comprehensive orbiter payload will make possible a large array of collaborative studies that guarantee an exciting harvest of scientific results.

Acknowledgments

We are grateful to a great number of people who contributed to the hardware effort for MIMI at the institutions central to the activity, namely, the Johns Hopkins University Applied Physics Laboratory (JHU/APL), the Department of Physics at the University of Maryland, the Max-Plank-Institute für Aeronomie (MPAe), and Centre d' Etude Spatiale des Rayonnements (CESR), in Toulouse. At JHU/APL the assistance of personnel in the Technical Services Department made possible the timely layout and fabrication of INCA and the MEU. In the Space Department key contributions were made by several people, among whom are J. Heiss, Dr. N.P. Paschalidis, R.E. Thompson, and S.R. Vernon. The CHEMS sensor was developed by the space physics group in the Department of Physics at the University of Maryland. Many individuals contributed to its design, development, fabrication, testing, and calibration over the years 1991–1997. We are particularly grateful to the following individuals: Paul Wilson IV, Jeff Rizza, Edvin Tums, Richard Pappalardo, Maurice Pairel, Pota Floros, Scott Lasley, Wayne Shanks, George Ho, and Joseph Zanelotti. The success of the LEMMS sensor at MPAe owes much to the excellent work of W. Boeker, B. Goll, K. Heerlein, H.G. Kellner, Dr. A. Lagg, A. Loose, O. Matuschek, H. Sommer, and W. Tappert. We would also like to thank the team from VTT Automation Space Technology under the lead of K. Viherkanto for an excellent job in building the turntable. Many thanks also to the staff from the Institut für Geophysik und Meteorologie at the Technical University of Braunschweig for LEMMS internal magnetic field measurements. Thanks to Prof. Treusch, Dr. R. Maier, Dr. F. Dworschak, W. Brautigam and Dr. Protic and their teams from the Kernforschungszentrum Jülich, Germany; the people from the 88-inch cyclotron of the Lawrence Berkeley Laboratory; Drs. B. Blake and M. Looper from the Aerospace Corporation in providing excellent beam facilities and help during instrument calibration. We especially acknowledge the work in the design of LEMMS by our colleague B. Wilken who unfortunately died in 2001, 3 years before Cassini's arrival at Saturn. We will miss him as a very knowledgeable scientist and friend. The contributions to the DPU of personnel at CESR, including those of M. Gangloff and N. Casajus, with key assistance from T. Psomoulis of Democritos University of Thrace are appreciated with thanks. Important contributions were also made by personnel at the University of Arizona (foils) and the University of Kansas (LEMMS design). The dedicated work of S. Brown and C. Smith in operating the Goddard Space Flight Center accelerator was instrumental to the successful calibrations of all three MIMI detector systems. Finally, we are most grateful to the JPL project personnel, including R. Spehalski, T. Gavin, W.G. Fawcett, D. Matson, J. Feynman, M. Burton, and S. Bolton for their assistance in various phases of the program. The MIMI program at JHU/APL was supported by the NASA office of Space Science under Task I of contract N00024-97-C-8119 between the Johns Hopkins University and the Department of the Navy, and by subcontract at the Universities of Maryland, Kansas, and Arizona. Mission operations and data analysis are supported under Task Order 003 of contract NAS5-97271 between the NASA/GFSC and the Johns Hopkins University. The German part of LEMMS at MPAe was financed by the Bundesministerium für Bildung und Forschung (BMBF) through the Deutsches Zentrum für Luft und Raumfart e. V. (DLR) under contract no. 50 OH 93028. The work at CESR was supported by CNES.

References

Amsif, A.: 1996, *Etude et modélisation de la production d'atomes énergétiques neutres dans l'exosphère de Titan*, Ph. D. Thesis, P. Sabatier University, Toulouse, France.

Amsif, A., Dandouras, J., and Roelof, E. C.: 1997, *J. Geophys. Res.* **102**, 22169.

Barbosa, D. D.: 1987, *Icarus* **72**, 53.

Barbosa, D. D. and Eviatar, A.: 1986, *Astrophys. J.* **310**, 927.

Barbosa, D. D., Eviatar, A., and Siscoe, G. L.: 1984, *J. Geophys. Res.* **89**, 3789.

Baron, R. L., Owen, T., Connerney, J. E. P., Satoh, T., and Harrington, J.: 1996, *Icarus* **120/122**, 437.

Barrow, C. H. and Desch, M. D.: 1989, *Astron. Astro. Phys.* **213**, 495.

Belcher, J. W.: 1983, in: Dessler, A. J. (ed.), *Physics of the Jovian Magnetosphere*, Cambridge University Press, Cambridge, UK and New York, p. 68.

Broadfoot, A. L., Sandel, B. R., Shemansky, D. E., Holberg, J. B., Smith, G. R., Strobel, D. F., McConnell, J. C., Kumar, S., Hunten, D. M., Atreya, S. K., Donahue, T. M., Moos, H. W., Bertaux, L., Blamont, J. E., Pomphrey, R. B., and Linick, S.: 1981, *Science* **212**, 206.

Burns, J. A., et al.: 1994, *An Integrated Strategy for the Planetary Sciences: 1995–2010*, Report of the Committee on Planetary and Lunar Exploration, Space Studies Board, National Research Council, Washington, DC.

Carbary, J. F. and Krimigis, S. M.: 1982, *Geophys. Res. Lett.* **9**, 420.

Cheng, A. F.: 1986, *J. Geophys. Res.* **91**, 4524.

Cheng, A. F. and Krimigis, S. M.: 1989a, *J. Geophys. Res.* **94**, 12003.

Cheng, A. F. and Krimigis, S. M.: 1989b, in: Waite, J. H., Burch, J., and Moore, R. (eds.), *AGU Solar System Plasma Physics*, p. 253.

Cheng, A. F., Keath, E. P., Krimigis, S. M., Mauk, B. H., McEntire, R. W., Mitchell, D. G., Roelof, E. C., and Williams, D. J.: 1993, *Remote Sens. Rev.* **8**, 101.

Clarke, J. T., et al.: 1996, *Science* **274**, 404.

Curtis, C. C. and Hsieh, K. C.: 1989, *AGU Solar Syst. Plasma Phys., Geophys. Monogr. Ser.* **54**, 247.

Dandouras, J. and Amsif, A.: 1999, *Planet. Space Sci.* **47**, 1355.

Desch, M. D. and Barrow, C. H.: 1984, *J. Geophys. Res.* **89**, 6819.

Dessler, A. J.: 1983, *Physics of the Jovian Magnetosphere*, Cambridge University Press.

Esposito, L. W., Cuzzi, J. N., Holberg, J. B., Marouf, E. A., Tyler, G. L., and Porco, C. C.: 1984, in: Gehrels, T., and Matthews, M. S. (eds.), *Saturn*, The University of Arizona Press, Tucson, p. 463.

Eviatar, A.: 1992, *Adv. Space Res.* **12**(8), 367.

Eviatar, A., Mekler, Y., and Coroniti, F. V.: 1976, *Astrophys. J.* **205**, 622.

Fisk, L. A., Schwadron, N. A., and Gloeckler, G.: 1997, *Geophys. Res. Lett.* **24**, 93.

Gehrels, T.: 1976, *Jupiter*, University of Arizona Press, Tucson, Arizona.

Geiss, J., Gloeckler, G., Fist, L. A., and von Steiger, R.: 1995, *J. Geophys. Res.* **100**, 23373.

Geiss, J., Gloeckler, G., Mall, U., von Steiger, R., Galvin, A. B., and Ogilvie, K. W.: 1994, *Astron. Astrophys.* **282**, 924.

Geiss, J., et al.: 1992, *Science* **257**, 1535.

Gloeckler, G.: 1996, *Space Sci. Rev.* **78**, 335.

Gloeckler, G. and Geiss, J.: 1998, *Space Sci. Rev.* **86**(1–2), 127–159.

Gloeckler, G. and Geiss, J.: 1996, *Nature* **381**, 210.

Gloeckler, G. and Hsieh, K. C.: 1979, *Nucl. Inst. Methods* **165**, 537.

Gloeckler, G., Fisk, L. A., and Geiss, J.: 1997, *Nature* **386**, 374.

Gloeckler, G., Balsiger, H., Bürgi, A., Bochsler, P., Fisk, L. A., Galvin, A. B., Geiss, J., Gliem, F., Hamilton, D. C., Holzer, T. E., Hovestadt, D., Ipavich, F. M., Kirsch, E., Lundgren, R. A., Ogilvie, K. W., Sheldon, R. B., and Wilken, B.: 1995, *Space Sci. Rev.* **71**, 79.

Gloeckler, G., Jokipii, J. R., Giacalone, J., and Geiss, J.: 1994, *Geophys. Res. Lett.* **21**, 1565.

Gloeckler, G., Geiss, J., Balsiger, H., Fisk, L. A., Galvin, A. B., Ipavich, F. M., Ogilvie, K. W., von Steiger, R., and Wilken, B.: 1993, *Science* **261**, 70.

Goertz, C. K.: 1989, in: Waite, J. H., Burch, J. L., and Moore, R. L. (eds.), *AGU Solar System Plasma Physics, Geophysical Monograph Series* Vol. 54, p. 427.

Gurnett, D. A., et al.: 1982, *J. Geophys. Res.* **87**, 1395.

Hilchenbach, M. et al.: 1998, *Astophys. J.* **503**, 916.

Holzer, T. E.: 1977, *Rev. Geophys. Space Phys.* **15**, 467.

Hsieh, K. C. and Curtis, C. C.: 1989, in: Waite, J., Burch, J., and Moore, R. L. (eds.), *AGU Solar System Plasma Physics*, p. 159.

Hsieh, K. C. and Curtis, C. C.: 1988, *Geophys. Res. Lett.* **15**, 772.

Hsieh, K. C., Shih, K. L., Jokipii, J. R., and Gruntman, M. A.: 1992a, *Astophys. J.* **393**, 756.

Hsieh, K. C., Shih, K. L., Jokipii, J. R., and Gruntman, M. A.: 1992b, in: Marsch, E., and Schwenn, R. (eds.), *Proceedings of the 3rd COSPAR Colloquium*, p. 365.

Hsieh, K. C., Sandel, B. R., Drake, V. A., and King, R. S.: 1991, *Nucl. Inst. Methods* **B61**, 187.
Hsieh, K. C., Keppler, E., and Schmidtke, G.: 1980, *J. Appl. Phys.* **51**, 2242.
Ip, W. H.: 1997, *Icarus* **126**, 42.
Ip, W. H.: 1996, *Astrophys. J.* **457**, 922.
Ip, W. H.: 1992, in: *Proceedings of the Symposium on Titan*, Toulouse, France, ESA SP-338, p. 243.
Ip, W. H.: 1990, *Astrophys. J.* **362**, 354.
Ip, W. H.: 1984, *J. Geophys. Res.* **89**, 2377.
Ip, W. H., Williams, D. J., McEntire, R. W., and Mauk, B. H.: 1998, *Geophys. Res. Lett.* **25**, 829.
Ip, W. H., Williams, D. J., McEntire, R. W., and Mauk, B. H.: 1997, *Geophys. Res. Lett.* **24**, 2631.
Ipavich, F. M., Lundgren, R. A., Lambird, B. A., and Gloeckler, G.: 1978, *Nucl. Inst. Methods* **154**, 291.
Johnson, R. E.: 1990, *Energetic Charged Particle Interactions with Atmospheres and Surfaces*, Springer-Verlag, New York.
Johnson, R. E., Pospieszalska, M., Sittler, E., Cheng, A. F., Lanzerotti, L. J., and Sieveka, E. M.: 1989, *Icarus* **77**, 311.
Kaiser, M. L.: 1993, *J. Geophys. Res.* **98**, 18757.
Kirsch, E., Krimigis, S. M., Ip, W. H., and Gloeckler, G.: 1981a, *Nature* **292**, 718.
Kirsch, E., Krimigis, S. M., Kohl, J. W., and Keath, E. P.: 1981b, *Geophys. Res. Lett.* **8**, 169.
Krimigis, S. M.: 1992, *Space Sci. Rev.* **59**, 167.
Krimigis, S. M.: 1986, *Comparative Study of Magnetospheric Systems*, CNES, LEPADUE Editions, Toulouse, France, Vol. 99.
Krimigis, S. M. and Armstrong, T. P.: 1982, *Geophys. Res. Lett.* **9**, 1143.
Krimigis, S. M., Carbary, J. F., Keath, E. P., and Armstrong, T. P.: 1982a, *EOS* 1068.
Krimigis, S. M., Armstrong, T. P., Axford, W. I., Bostrom, C. O., Gloeckler, G., Keath, E. P., Lanzerotti, L. J., Carbary, J. F., Hamilton, D. C., and Roelof, E. C.: 1982b, *Science* **215**, 571.
Krimigis, S. M., Carbary, J. F., Keath, E. P., Armstrong, T. P., Lanzerotti, L. J., and Gloeckler, G.: 1983, *J. Geophys. Res.* **88**, 8871.
Krimigis, S. M., Decker, R. B., Hamilton, D., and Gloeckler, G.: 2000, *AIP Conf. Proc.* **528**, 333–336.
Krimigis, S. M., Zwickl, R. D., and Baker, D. N.: 1985, *J. Geophys. Res.* **90**, 3947.
Krimigis, S. M., Mitchell, D. G., Hamilton, D. C., Dandouras, J., Armstrong, T. P., Bolton, S. J., Cheng, A. F., Gloeckler, G., Hsieh, K. C., Keath, E. P., Krupp, N., Lagg, A., Lanzerotti, L. J., Livi, S., Mauk, B. H., McEntire, R. W., Roelof, E. C., Wilken, B., and Williams, D. J.: 2002, *Nature* **415**, 994.
Krimigis, S. M., et al.: 1988, *Planet. Space Sci.* **36**, 311.
Krupp, N., Woch, J., Lagg, A., Wilken, B., Livi, S., and Williams, D. J.: 1998, *Geophys. Res. Lett.* **25**, 1249–1252.
Lagg, Andreas: 1998, *Energiereiche Teilchen in der inneren Jupitermagnetosphaere: Simulation und Ergebnisse des EPD-Experimentes an Bord der Raumsonde GALILEO*, Dissertation, Max-Planck-Institut fuer Aeronomie, Lindau/Harz, Germany, MPAE-W-807-98-01.
Lee, M. A.: 1982, *J. Geophys. Res.* **87**, 5063.
Mall, U., Fichtner, H., Kirsch, E., Hamilton, D. C., and Rucinski, D.: 1998, *Planet. Space Sci.*, **46**, 1375–1382.
Mauk, B. H. and Krimigis, S. M.: 1987, *J. Geophys. Res.* **92**, 9931.
Mauk, B. H., Krimigis, S. M., Mitchell, D. G., Roelof, E. C., Keath, E. P., and Dandouras, J.: 1998, *Planet. Space Sci.* **46**, 1349.
Mauk, B. H., Williams, D. J., and McEntire, R. W.: 1997a, *Geophys. Res. Lett.* **24**, 2949.
Mauk, B. H., Williams, D. J., McEntire, R. W., Khurana, K. K., and Roederer, J. G.: 1999, *J. Geophys. Res.* **104**, 22759.

Mauk, B. H., Krimigis, S. M., Mitchell, D. G., and Roelof, E. C.: 1998, *Adv. Space Res.* **21**, 1483.
Mauk, B. H., Krimigis, S. M., and Acuña, M. H.: 1994, *J. Geophys. Res.* **99**, 14781.
Mauk, B., Krimigis, S. M., and Lepping, R.: 1985, *J. Geophys. Res.* **90**, 8253.
McEntire, R. W. and Mitchell, D. G.: 1989, in: Burch, J., and Waite, J. (eds.), *Outstanding Problems in Solar System Plasma Physics*, AGU Monograph.
Meckbach, W., Braunstein, G., and Arista, N.: 1975, *J. Phys. B* **8**, L344.
Mendis, D. A., Hill, J. R., Ip, W. H., Goertz, C. K., and Grün, E.: 1984, 'Electrodynamic processes in the ring system of Saturn', in: Gehrels, T., and Matthews, M. S. (eds.), *Saturn*, The University of Arizona Press, Tucson, p. 546.
Mitchell, D. G., Hsieh, K. C., Curtis, C. C., Hamilton, D. C., Voss, H. D., Roelof, E. C., and Brent, P. C.: 2001, *Geophys. Res. Lett.* **28**, 1151.
Mitchell, D. G., Krimigis, S. M., Cheng, A. F., Jaskulek, S. E., Keath, E. P., Mauk, B. H., McEntire, R. W., Roelof, E. C., Schlemm, C. E., Tossman, B. E., and Williams, D. J.: 1996, in: *Proceedings SPIE International Symposium on Optical Science Engineering and Instrumentation*, Mission to the Sun, Vol. **2803**, p. 154.
Ness, N. F., et al.: 1982, *J. Geophys. Res.* **87**, 1369.
Neubauer, F. M.: 1992, in: *Proceedings of the Symposium on Titan*, Toulouse, France, ESA SP-338, p. 267.
Paonessa, M. and Cheng, A. F.: 1986, *J. Geophys. Res.* **91**, 1391.
Paranicas, C., Cheng, A. F., and Williams, D. J.: 1998, *J. Geophys. Res.* **103**, 15001.
Paranicas, C. P., Mauk, B. H., and Krimigis, S. M.: 1991, *J. Geophys. Res.* **96**, 21135.
Prangé, R., Zarka, P., Ballester, G. E., Livengood, T. A., Denis, L., Carr, T., Reyes, F., Bame, S. J., and Moos, H. W.: 1993, *Geophys. Res.* **98**, 18779.
Reiner, M. J., Fainberg, J., Stone, R. G., Kaiser, M. L., Desch, M. D., Manning, R., Zarka, P., and Pedersen, B. M.: 1993, *J. Geophys. Res. Planets* **98**, 13163.
Roelof, E. C.: 1992, in: Marsh, E., and Schwenn, R. (eds.), *Proceedings of the 3rd COSPAR Colloquium*, p. 385.
Roelof, E. C.: 1987, *Geophys. Res. Lett.* **14**, 652.
Roelof, E. C. and Williams, D. J.: 1990, *Johns Hopkins APL Tech. Dig.* **11**, 72.
Roelof, E. C., Mitchell, D. G., and Williams, D. J.: 1985, *J. Geophys. Res.* **90**, 10991.
Sandel, B. R. and Broadfoot, A. L.: 1981, *Nature* **292**, 679.
Satoh, T., Connerney, J. E. P., and Baron, R. L.: 1996, *Icarus* **122**, 1.
Schneider, N. M. and Trauger, J. T.: 1995, *Astrophys. J.* **450**, 450.
Schulz, M. and Lanzerotti, L. J.: 1974, *Particle Diffusion in the Radiation Belts*, Springer-Verlag.
Shemansky, D. E. and Hall, D. T.: 1992, *J. Geophys. Res.* **97**, 4143.
Simpson, J. A., Bastian, T. S., Chenette, D. L., McKibben, R. B., and Pyle, K. R.: 1980, *J. Geophys. Res.* **85**, 5731.
Sittler, E. C., Ogilvie, K. W., and Scudder, J. D.: 1983, *J. Geophys. Res.* **88**, 8847.
Van Allen, J. A.: 1984, in: Gehrels, T., and Matthews, M. S. (eds.), *Saturn*, p. 281.
Williams, D. J. and Mauk, B. H.: 1997, *J. Geophys. Res.* **102**, 24283.
Williams, D. J., Mauk, B., and McEntire, R. W.: 1997a, *Geophys. Res. Lett.* **24**, 2953.
Williams, D. J., Mauk, B. H., McEntire, R. W., Roelof, E. C., Armstrong, T. P., Wilken, B., Roederer, J. G., Krimigis, S. M., Fritz, T. A., Lanzerotti, L. J., and Murphy, N.: 1997b, *Geophys. Res. Lett.* **24**, 2163.
Williams, D. J., Mauk, B. H., McEntire, R. W., Roelof, E. C., Armstrong, T. P., Wilken, B., Roederer, J. G., Krimigis, S. M., Fritz, T. A., and Lanzerotti, L. J.: 1996, *Science* **274**, 401.
Williams, D. J., McEntire, R. W., Schlemm, C. E., Lui, A. T. Y., Gloeckler, G., Christon, S. P., and Gliem, F.: 1994, *J. Geomagn. Geoelectr.* **46**, 39.
Witte, M., Banaszkiewicz, M., and Rosenbauer, H.: 1996, *Space Sci. Rev.* **78**, 289.

Woch, J., Krupp, N., Lagg, A., Wilken, B., Livi, S., and Williams, D. J.: 1998, *Geophys. Res. Lett.* **25**, 1253–1256.

Woch, J., Krupp, N., Khurana, K. K., Kivelson, M. G., Roux, A., Perraut, S., Louarn, P., Lagg, A., Williams, D. J., Livi S., and Wilken, B.: 1999, *Geophys. Res. Lett.* **26**, 2137–2140.

Zwickl, R. D., Krimigis, S. M., Carbary, J. F., Keath, E. P., Armstrong, T. P., Hamilton, D. C., and Gloeckler, G.: 1981, *J. Geophys. Res.* **86**, 8125.

THE CASSINI MAGNETIC FIELD INVESTIGATION

M. K. DOUGHERTY[1,*], S. KELLOCK[1], D. J. SOUTHWOOD[1], A. BALOGH[1], E. J. SMITH[2], B. T. TSURUTANI[2], B. GERLACH[3], K. -H. GLASSMEIER[3], F. GLEIM[3], C. T. RUSSELL[4], G. ERDOS[5], F. M. NEUBAUER[6], S. W. H. COWLEY[7]

[1]*The Blackett Laboratory, Imperial College London, SW7 2BZ, U.K.*

[2]*Jet Propulsion Laboratory, California Institute of Technology, Pasadena, CA 91109, U.S.A.*

[3]*Technische Universitat Braunschweig, D-38106 Braunschweig, Germany*

[4]*UCLA Institute of Geophysics and Planetary Physics, Los Angeles, CA 90024-1567, U.S.A.*

[5]*KFKI Research Institute for Particle and Nuclear Physics, PO Box 49, H-1525 Budapest, Hungary*

[6]*Institute for Geophysics and Meterology, Köln University, 50923 Köln, Germany*

[7]*Department of Physics and Astronomy, Leicester University, Leicester, LE1 7RH, U.K.*

*(*Author for correspondence: e-mail: m.dougherty@imperial.ac.uk)*

(Received 1 December 1997; Accepted in final form 9 December 2002)

Abstract. The dual technique magnetometer system onboard the Cassini orbiter is described. This instrument consists of vector helium and fluxgate magnetometers with the capability to operate the helium device in a scalar mode. This special mode is used near the planet in order to determine with very high accuracy the interior field of the planet. The orbital mission will lead to a detailed understanding of the Saturn/Titan system including measurements of the planetary magnetosphere, and the interactions of Saturn with the solar wind, of Titan with its environments, and of the icy satellites within the magnetosphere.

Keywords: Cassini, magnetometer, saturn

1. Introduction

The ultimate goal of the Cassini–Huygens mission is a detailed understanding of the Saturn system with special emphasis on Titan. The Cassini orbiter provides extended monitoring of the Saturn system over a 4-year period resulting in a detailed description of the dominant physical processes and their variation with time, allowing separation of temporal and spatial features. In particular, repeated flybys of Titan provide a rather complete survey of the environment of this moon. The magnetometer in turn addresses the nature of the internal field, the planetary magnetosphere and its interaction with the solar wind, the magnetic states of the moons and how they interact within their plasma environments.

A magnetometer is an essential instrument for planetary studies and many different designs have been flown. The Cassini magnetometer has been designed and built specifically for measurements within the Saturnian environment. Previous planetary flybys have shown the existence of an internal Saturn magnetic field, a magnetosphere and a strong plasma interaction between Titan and its plasma surroundings. By measuring the planetary field and determining the relative size of

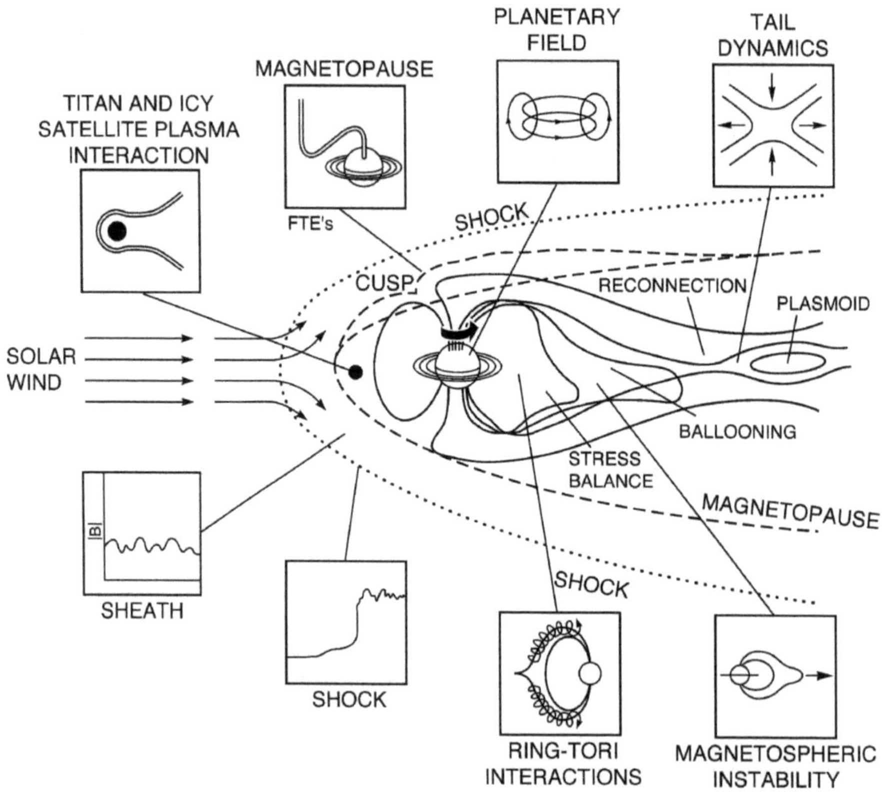

Figure 1. Schematic of Saturn's magnetosphere and magnetometer science objectives.

the major spherical harmonic components, which originate deep inside the planet the magnetometer will provide information on conditions in and near the dynamo region. However, the magnetometer probes not only the deep interior of the planet where the internal planetary field originates but also makes detailed measurements of the planetary environment, its magnetosphere and ionosphere, and of those of the Saturnian moons. This paper describes the science objectives of the magnetometer team, the instrument sensors and electronics, instrument calibration, data processing, as well as early cruise operations and the critical calibration data collected during the swingby past the Earth in August 1999.

The main scientific objectives of the Cassini magnetic field investigation are listed below as well as being shown schematically in Figure 1. They can be divided up into two main areas of research:

Saturn

- Resolve the internal planetary magnetic field to at least fourth order.
- Establish relative contributions to electromagnetic and mechanical stress balance.
- Identify the energy sources for dynamical processes.

- Understand the coupling that occurs between the magnetosphere and ionosphere.
- Characterise the phenomena of the distant dayside/flank planetary environment.
- Survey satellite/dust/ring/torus electromagnetic interactions.
- Determine the tail structure and dynamic processes.
- Provide a three-dimensional global model of the magnetospheric magnetic field.

Titan

- Determine the magnetic state of the body and conditions of the atmosphere.
- Characterise and model the Titan electromagnetic environment.
- Interpret and model all Titan-plasma flow interactions and the variation of the Titan-magnetosphere interaction with respect to the Titan orbital phase.
- Determine the nature of the low frequency waves in the near-Titan plasma environment.

In addition the magnetic state and plasma environments of the Saturnian icy satellites will be probed to the extent possible within the limitations of the finally chosen tour.

The design of the instrument has been driven by a range of requirements; unique to a long-term interplanetary cruise, to the Cassini spacecraft and to the science goals listed above. The extremely long mission duration, from 1997 until at least 2008, requires high reliability and survivability. Due to the long delays in communication between the spacecraft and the ground, both the spacecraft and instruments must operate autonomously. To conform to practical limits on the use of the DSN network, the mission makes use of on-board data storage. The science aims require measurements with high sensitivity over a wide dynamic range (from tens of thousands of nanotesla (nT) during Earth flyby, through a few thousand nT expected during Saturn flybys to the extremely low field strengths in the solar wind). We have met these requirements with a novel instrument, a dual technique magnetometer (such as flown on Ulysses, Balogh *et al.*, 1992) where we use vector helium (VHM) and fluxgate magnetometers (FGM) and also operate the helium device in a scalar mode (SHM). The instrument is capable of determining the *absolute* magnitude of the field to an accuracy of 1 nT. By calibrating FGM measurements with the absolute scalar measurements, determination of the fields in the Saturn environment will be possible with hitherto unachieved accuracy. The combination of V/SHM and FGM allows very sensitive determination of wave fields. The VHM is more sensitive below ~1 Hz, with the FGM being more sensitive at high frequencies. Simultaneous operation of both instruments will assure maximum sensitivity over the broad range of frequencies from 0 to 20 Hz. The Cassini mission will be the first time that the scalar magnetometer has been used on a planetary mission.

The instrument is a result of the co-operation and assistance of many people and science data analysis will involve many more. Team members are drawn from

Imperial College London (Imperial), Leicester University, Jet Propulsion Laboratory (JPL), University of California Los Angeles (UCLA), Boston University, Central Institute for Physics, Budapest (KFKI), Technical University of Braunschweig (TUB) and Köln University.

2. Science Objectives

Saturn's magnetosphere is one of the most interesting planetary plasma environments in the solar system sharing many of the characteristics of the magnetosphere of the Earth (solar wind dominated) and Jupiter (rotation dominated). In many respects, however, the Saturnian magnetosphere is unique due to the diversity of coupling mechanisms that exist between the various components of the Saturn system (such as the solar wind, Saturn's ionosphere, Titan, the icy satellites, the rings and the large clouds of dust and neutrals). Our existing knowledge of Saturn arises from the three planetary flybys (Pioneer 11 and Voyager 1 and 2), remote sensing observations and modeling work. This has allowed us to gain a preliminary understanding of the magnetic field, plasma, energetic particles, waves and auroral signatures. There remain a large number of unanswered questions with regards to our understanding of the Saturn environment, a number of which will be addressed in this paper. Probably one of the most remarkable and enigmatic issues, which remain is the high degree of axisymmetry of the Saturnian internal magnetic field, with its close alignment of dipole and rotation axes.

2.1. Interiors of Saturn, Titan and the Icy Satellites

The primary scientific objective of the magnetometer is a determination of the internal magnetic field of Saturn to at least the fourth order spherical harmonic terms. In addition a more complete characterization of the planetary field will be achieved and possible secular changes since the initial Pioneer and Voyager flybys will be investigated. Following the recent Galileo flybys of the Galilean moons which have revealed the existence of an internal field at Ganymede and induced magnetic fields at Europa and Callisto (see Section 2.1.2) the possibility cannot be ignored that Titan and some of the icy satellites may have internal/induced fields of their own. We describe the potential of the magnetic field investigation to probe the deep interior of Saturn and its satellites, thus enabling us, in effect, to "see inside" these planetary bodies. The Sun and most of the planets have large internal magnetic fields, which are being continuously generated by dynamo action deep in the interior of the planet as a result of highly conducting material which resides there. Dynamo action is known to generate magnetic fields at the different planets, for example, the magnetic field generated by the dynamos of the Earth and the Sun both reverse and the dipole axes for Uranus and Neptune are tilted by large angles away from their rotation axes. The best fit model for the Jupiter field, which is mainly that due to a dipole but with additional quadrupole and octupole terms is tilted by 10° away from the rotation axis and when using a dipole model for the field results in an offset of the axis away from the planetary centre. For Saturn the surprising feature is that there is only less than 1° tilt between the rotation and dipole axes and the field shows a remarkable symmetry about the rotation axis.

2.1.1. *Saturn*

In situ measurements of the magnetic field of Saturn have been obtained from the three previous flyby missions, Pioneer 11 (1979) and Voyagers 1 and 2 (1980 and 1981), each of which consisted of a single flyby through the Saturnian environment. The Cassini mission as an orbiter, has an immense advantage over the previous flyby missions as regards the determination of the internal planetary field, since an orbiter provides a much more complete three-dimensional sampling of the magnetic field measurements over a wide range of latitudes, longitudes and radial distance. The internal field at Saturn is highly unusual, being highly axisymmetric with a close alignment between its dipole and rotational axes. However, auroral observations, periodicities seen in kilometric radiation and the formation of spokes in the B ring as well as energetic particles spectral modulation and spin periodicity observed in magnetic field observations (Carbary and Krimigis, 1982; Kaiser *et al.*, 1984; Grun *et al.*, 1992; Bhardwaj and Gladstone, 1982; Espinosa and Dougherty, 2000) suggest that longitudinal asymmetries do exist close to the planet.

In modelling the internal planetary field of a strongly magnetized planet it can be assumed to first order that outside the planet there are no significant sources and therefore the magnetic field can to a good approximation be derived from a scalar potential function, $B = -\nabla V$, where

$$V = a \sum_{n=1}^{\infty} \left(\frac{a}{r}\right)^{n+1} \sum_{m=0}^{n} P_n^m(\cos\theta) \left[g_n^m \cos(m\phi) + h_n^m \sin(m\phi)\right],$$

where r is the distance to the planet's centre, a is the planetary radius, θ and ϕ are the colatitude and east longitude, respectively, P_n^m are associated Legendre functions with Schmidt normalisation and the g_n^m, h_n^m (Schmidt coefficients) are internal field parameters. The Schmidt coefficients are chosen to minimise, in a least squares sense, the differences between the model field and the observed field. Field modelling at Saturn to date (Smith *et al.*, 1980a; Smith *et al.*, 1980b; Connerney et al., 1982; Acuna *et al.*, 1983; Davis and Smith, 1986, 1990) has revealed that the dipole is axially coincident with the rotation axis, with significant quadrupole and octupole terms. Taking the summation over n beyond the octupole ($n = 3$) term is unwarranted with existing data and the only significant model coefficients generated are those of g_1^0, g_2^0 and g_3^0, shown in Table I, in this instance for the SPV harmonic model (Davis and Smith, 1990), generated from magnetometer data from all three of the previous flyby data sets.

An important question is whether non-axisymmetric coefficients are present in the field and whether they can be inferred from more accurate measurements of the

TABLE I
Model coefficients for the SPV internal Saturn field

n	m	g_n^m	h_n^m
1	0	21,160	
1	1	0	0
2	0	1,560	
2	1	0	0
2	2	0	0
3	0	2,320	
3	1	0	0
3	2	0	0
3	3	0	0

magnetic field or will become evident in the higher degree terms. Suggestions put forward to explain this absence of non-axisymmetric terms include the possibility of a highly conducting shell outside of the dynamo region and rotating about Saturn's dipole axis (Stevenson, 1980) as well as the suggestion that the generation of the magnetic field has ceased at Saturn and that the higher order non-axial terms are decaying more rapidly than the axial lower order ones. The Cassini magnetometer has been designed specifically to detect these very weak higher order terms ($n > 3$) whose contributions outside the planet are significantly smaller than the lower order terms.

Extension of the spherical harmonic sequence describing the internal field above that of the dipole, quadrupole and octupole terms depends on the magnetometer being able to make very accurate measurements at low altitudes, inside of $4R_S$ over a wide range of latitudes and longitudes. Figure 2 (adapted from Connerney, 1993) reveals the close flyby trajectories of the previous spacecraft flybys past Saturn, that of Pioneer 11 and Voyagers 1 and 2, as a function of latitude and longitude. The radial distance of the spacecraft is also denoted at various intervals along the different trajectories. This figure clearly shows the very limited range of latitudes and longitudes at which the Saturn planetary field has already been measured for distances less than $4R_S$ and confirms how crucial measurements which Cassini will make inside of this distance will be (see Figure 3). To date inside of $4R_S$ there are only three short swaths of observations, for Pioneer 11 (periapse distance $1.3R_S$) over latitudes of [$\pm 5°$] and longitudes of [135–190°]; from Voyager 1 (periapse distance $3.1R_S$) over latitudes of [$-15°$ to $-35°$] and longitudes of [170–200°]; and for Voyager 2 (periapse of $2.7R_S$) from latitudes [-20–30°] and longitudes [315–350°].

Table II predicts the expected size of the weak higher order coefficients which the magnetometer will need to measure in order to delineate the fourth and fifth

TABLE II
Expected internal field contributions at Cassini orbit

Harmonic Coefficient	Magnetic Field Magnitude at $1R_S$	$1.5R_S$	$2R_S$	$2.5R_S$	$3R_S$	$4R_S$
g_1^0	21,160	6,270	2,645	1,354	784	331
g_2^0	1,560	308 (4.9%)	98 (3.7%)	40 (2.9%)	19 (2.4%)	6 (1.8%)
g_3^0	2,320	306 (4.9%)	73 (2.6%)	24 (1.8%)	10 (1.3%)	2 (0.6%)
g_4^0	2,000?	176 (2.8%)	31 (1.2%)	8 (0.6%)	3 (0.4%)	0.5 (0.2%)
g_5^0	2,000?	117 (1.9%)	16 (0.6%)	3 (0.2%)	1 (0.1%)	0.1 (0.03%)

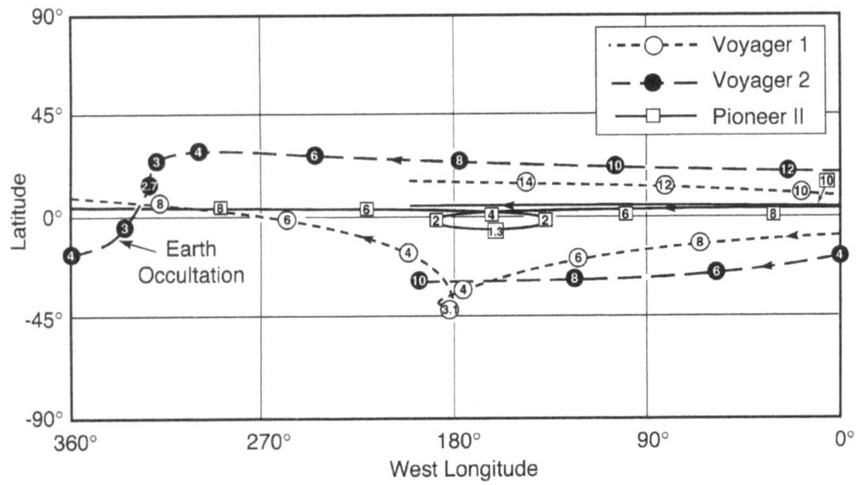

Figure 2. Close flyby trajectories of previous spacecraft.

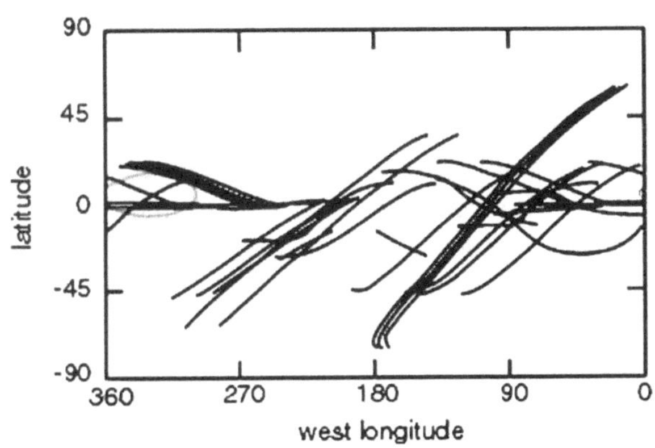

Figure 3. : Planned latitude and longitude coverage inside $4R_S$ by the Cassini spacecraft.

order components of the internal field. These values were estimated assuming that the higher degree terms were similar to the average of the quadrupole and octupole terms. If the field in the dynamo has equal power at all degrees as is true for the Earth (for example Elphic and Russell, 1978) these values will be somewhat reduced. These values stress how important the magnetic cleanliness of the Cassini spacecraft has been in order to carry out the science goals of the magnetometer as well as of the other fields and particles instruments, see Section 6. In addition the values in this table show how important data inside of $2.0R_S$ is for resolution of orders 4 and 5. The only time that this will occur is during Saturn Orbit Insertion confirming the criticality of magnetometer observations during this time for resolution of our prime science objective. The values shown in parentheses in Table II are the percentages of the total dipole field strength contributed by the coefficient at that distance.

Figure 3 shows the close flyby trajectories (inside of $4R_S$) of Cassini during the planned 4-year orbital tour, where ideally as much coverage of latitude and longitude space as possible is required in order to resolve the higher order moments.

It is the capability of the Cassini scalar magnetometer, which will be crucial to the high field operation of the instrument, with the scalar magnetometer allowing for precise absolute calibration of the fluxgate. In the past, a problem with achieving high accuracy from vector measurements has been the precise attitude knowledge required. For example, in a 10,000 nT field an accuracy of 1 nT in any component implies that the appropriate orientation angle must be known to within 10^{-4} or 20.6 arc sec. This can be achieved using special star trackers, for example, but with the SHM the problem is avoided by measuring the field magnitude directly by way of a resonance magnetometer. The fluxgate instrument, which has a large dynamic range has shortcomings in deriving accurate estimates of weak field components in a strong field. The vector helium magnetometer is able to make measurements of 1 part in 10^3 at best for weak field components in a strong background field. However, on operating the helium magnetometer in its scalar mode and thereby directly measuring the Larmor frequency, measurements with an accuracy of 1 part in 10^5 can be achieved. This improved accuracy of the SHM will allow measurements of the 4th and 5th order components of Saturn's internal field at distances further from the planet than those needed for the FGM. With the present mission profile, such an opportunity for the FGM would probably only occur at initial orbit insertion. SHM measurements can also be used to calibrate the analogue vector measurements to better than 1 part in 10^4 (Langel *et al.*, 1981). Vector measurements in high fields near periapsis are critical in separating internal dynamo field components from externally generated magnetospheric currents.

2.1.2. *Titan and Icy Satellites*

The magnetic field measurements taken during the Galileo flybys of the Galilean moons revealed that Ganymede has its own internal field and that an electromagnetically induced magnetic fields arise at Europa and Callisto and possibly Ganymede due to currents on the conducting interior (assumed to be salty water) when the external field changes as Jupiter rotates (Kivelson *et al.*, 1997; Kivelson *et al.*, 2000). The induced magnetic fields point to the probable existence of oceans below the icy crusts of these satellites (Neubauer, 1999; for a review see Kivelson *et al.*, 2002).

There are 44 close flybys of Titan planned during the 4-year orbital tour and the prime Titan-related science objective is to determine the internal magnetic field sources, be they from dynamo action, electromagnetic induction or remanent magnetisation in a "dirty ice" crust; as well as from sources external to it, such as ionospheric and plasma currents. Gaining an understanding of these sources will allow us to gain an understanding of the internal structure and evolution of Titan. The high sensitivity of the magnetometer instrument will ensure that Cassini will measure even a very weak internal Titan field to improve on upper limits and magnetic moments derived in the past (Neubauer *et al.*, 1984; Kabin *et al.*, 2000). The planned multiple flybys will allow measurements to be made at a range of latitudes, longitudes and altitudes enabling a better determination of an internal field to be made. If an internal ocean exists such studies will allow this to be probed.

In addition there will be 12 targeted (close) flybys of some of the icy satellites, including Phoebe, Mimas, Enceladus, Tethys, Dione, Rhea, Hyperion and Iapetus; as well as 28 untargeted (and therefore more distant) flybys as well. The science objectives arising from these flybys include the search for possible evidence of ancient dynamos and crustal remanence in the icy satellites as well as to investigate the icy satellite plasma environments (see Section 2.4). The study of any induced or intrinsic fields associated with satellites is important for a variety of reasons. It enables us to better understand the interior structure, composition and dynamics of the satellite, the way that particles from the magnetosphere impinge on the satellite and how this affects their evolution and the generation of magnetic fields in solid bodies in the solar system.

2.2. The Magnetosphere

The Saturnian environment has to date been visited by three spacecraft, Pioneer 11 and Voyager 1 and 2. The trajectories of these spacecraft flybys are shown in Figure 4 (adapted from Sittler *et al.*, 1983) in a view projected onto the equatorial plane with approximate bow shock and magnetopause boundaries shown. In addition, the specific regions during which the spacecraft observed shock and magnetopause crossings is shown, clearly revealing the dynamic nature of the size of the magnetosphere. The limited spatial and local time coverage of the flybys is clearly evident.

2.2.1. *Electromagnetic and Mechanical Stress Balance*

Through much of the Saturnian magnetosphere the dynamics of the system are believed to be controlled by the rotation of the planet (McNutt, 1983), with the planet appearing to enforce corotation through much of the inner regions of the

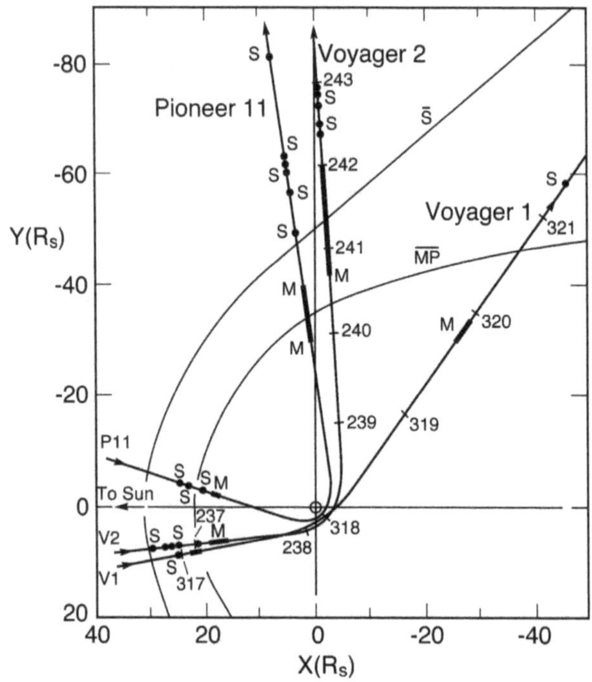

Figure 4. Equatorial view of the three spacecraft flybys of Saturn. On the trajectories the black dots labelled S show observed bow shock crossings, and black rectangles labelled M show observed magnetopause crossings.

magnetosphere. As material is ionised it is picked up in the corotational flow in the equatorial magnetospheric regions and rotational energy in the form of momentum is transferred from the planet to the equatorial plane. An unavoidable feature of any plasma which has been generated by ionisation where there is a background flow is that the thermal speed is at least comparable to the flow speed picked up. Separating the various forces exerted by the different parts of the particle distributions and by different species is a very important question and in a quasi-steady situation since all the mechanical forces are taken up by the magnetic field, the field is a natural area to study in order to aid the particle teams in this endeavour. A natural extension of the characterisation of the magnetospheric Saturnian field (see Section 2.2.7) would be to use the magnetic field data in conjunction with data from the other fields and particles instruments to examine the fundamental questions of what mechanical stresses the magnetic field stresses link to. The magnetic field imposes coherent behaviour on the collision free regions of the planetary environment and hence acts as the sinews of the medium. Stress is communicated by the field and energy is stored in the field and sometimes released by dramatic explosive phenomena like reconnection. Derived field and current patterns will aid in determining the balance of stresses and the relative importance of pressure and centrifugal forces.

2.2.2. *Energy Sources for Dynamical Magnetospheric Processes*

At the Earth it is clear that the dominant energy source for the magnetosphere is that due to the solar wind with the bulk of energy which drives magnetospheric dynamics arising from the coupling between the magnetosphere and the solar wind. At Jupiter, with its dominating internal plasma source Io, very fast rotation rate and very large magnetosphere, it is internal processes which are the dominant energy source, with the bulk of the magnetospheric energy being gained by transfer of the rotational energy from the planet out to the magnetosphere. At Saturn, although there can be no doubt that the outer Saturnian magnetosphere undergoes some form of dynamic interaction with the surrounding solar environment, the importance of energy transfer via the solar wind to the Saturnian system is, as a whole, thought to be of minor importance. The high rotation speed of the planet along with high corotation speeds imply that the centrifugal forces are very large in the outer magnetosphere and may alone dominate steady or sporadic transfer of material, momentum and energy between the planetary environment and the interplanetary medium. This question of which are the dominant processes governing plasma transport as well as the overall time variability of the Saturnian system will be addressed by the Cassini orbits of Saturn. We expect that in the inner magnetosphere, the dominant processes will be centrifugal in origin and hence plasma originating in the inner regions will undergo outward diffusion. Transport processes in the outer regions are more likely to be time dependent and are less well known. An important test of dynamical processes within the magnetosphere is being able to distinguish between the occurrence of substorms governed by solar wind behaviour or the periodic detachment of tubes of plasma being driven by planetary rotation or other time dependent effects.

A surprise discovery during the Ulysses flyby of Jupiter was the presence of what have been called null field events in the outer magnetosphere (Haynes *et al.*, 1994). These nulls are sporadic drop-outs in field strength detected in the outer magnetosphere. It is believed that the spacecraft flew through regions of enhanced plasma pressure which were elongated along the ambient field direction. Following this discovery null fields were also found in all of the previous Jupiter flybys, that of Pioneer 10 and 11 and Voyagers 1 and 2 (Leamon *et al.*, 1995). It has been proposed that these events are formed by the breaking off of material from the outer edge of the Jovian magnetodisk (Southwood *et al.*, 1995). The plasma material is hence moving outwards and the nulls, therefore, play a role in the outward transport of iogenic material. At Saturn a number of similar types of field drop-outs with corresponding density enhancements have been observed, both inward and outward of the Titan orbit in the noon sector. In the Voyager 1 data they are seen on either side of the main Titan encounter plume and the observations are consistent with a general aging of the plume characterised by their dispersal by the centrifugal interchange instability. That is, there could be multiple encounters with an extended Titan plume, which has been wrapped around Saturn by corotation and could last for up to three Saturnian rotations before reaching background levels (Eviatar *et al.*, 1982). In the Voyager 2 data density enhancements have been described as plasma blobs which have become detached from the outer boundary of the nightside extended plasma sheet (Goertz, 1983). During the Voyager 2 flyby Titan was apparently in the magnetosheath or near the magnetopause boundary for a few days before the spacecraft entered the magnetosphere and hence one would not expect to see any plume evidence. The two theories, therefore, do not necessarily contradict one another. The Voyager 1 events translate to spatial dimensions of less than 1,600 km (or $0.027R_S$) if spacecraft motion alone is considered, whereas if it is assumed that the events are convected past the spacecraft at corotational speeds of order 150 km/s then the dimensions become 15,000 km (or $0.27R_S$). It is important to generate a consistent theory which explains these plasma blobs which have been observed on the dayside of fast rotating magnetospheres as well as ascertain whether such signatures are to be seen on the dusk or nightside of Saturn's magnetosphere. If it can be firmly established that the blobs are associated with the breaking away of material from the magnetodisk then the discovery will be important for the light it can throw on the nature of transport processes in the outer Saturnian magnetosphere and in fast rotating magnetospheres in general.

2.2.3. *Magnetosphere Ionosphere Coupling*

At Saturn it is the angular momentum arising from planetary rotation which is the main source of momentum input into the magnetospheric plasma. A central issue arising from this concerns the coupling of the planet's ionosphere to the magnetospheric medium by way of momentum and angular momentum transfer which takes place via field aligned currents. At Earth it is generally directed from the magnetospheric plasma to the ionospheric plasma, with the Earth gaining momentum from the solar wind. For Saturn, however, the flow is usually in the opposite sense with angular momentum being transferred from the ionospheric plasma to the magnetospheric plasma with the magnetosphere thereby gaining momentum from planetary rotation. Auroral emission is direct evidence of magnetosphere/ionosphere/atmosphere coupling which exists for a magnetised planet such as Saturn. The driving mechanism for such emission is charged particle precipitation along field lines. The association of auroral signatures with processes in the terrestrial and Jovian magnetosphere is well documented and accounts for a range of dynamical phenomena coupled via the magnetic field. Although many global properties of the Saturnian magnetosphere are different from those at the Earth and at Jupiter, the magnetic field undoubtedly plays a similar role in connecting magnetospheric processes with the auroral ionospheric events they trigger, and in particular in establishing large scale current systems closed in the ionosphere. There are a number of fundamental questions which remain unanswered about Saturn's aurora; are they due to electron or ion precipitation, are the precipitations due to wave-particle interactions and if so what type of waves are amplified and in what regions does this take place? Observations of Saturn in the near UV (220 nm) with the Faint Object Camera (FOC) on the Hubble Space Telescope (HST) have revealed a dark northern auroral oval (Ben-Jaffel *et al.*, 1995). Auroral activity is usually characterised by light emitted around the poles of the planet. The dark oval observed at Saturn reveals that transport of energy and charged particles from the magnetosphere to the atmosphere of the planet at high latitudes induces auroral activity that not only produces auroral light but also UV dark material near the poles. The first bright aurorae at Saturn's northern and southern poles were observed in the far UV by the wide field planetary camera on HST. (See http://oposite.stsci.edu/pubinfo/pr/1998/05/index.html). These images allowed resolution of a luminous, circular band centred on the north pole where an enormous auroral curtain rose as high as 2,000 km above the cloud tops. Observations of magnetic field measurements in conjunction with observations from the Cassini optical remote sensing instruments during the Cassini orbits as well as with telescope viewing of the aurorae will be able to go a long way to answering some of the questions raised above.

2.2.4. *Characterisation of Phenomena in the Distant Dayside Planetary Environment*

Due to the orbiting nature of the Cassini mission as well as the rather elastic nature of the Saturnian magnetosphere, as evidenced by the range of dayside bow shock crossing distances observed by the previous Saturn missions of Pioneer 11, and the Voyager 1 and 2 spacecraft (Slavin *et al.*, 1983), Cassini is expected to observe a large number of bow shock and magnetopause crossings over a range of local times. This region is the one which holds the key to the solar wind interaction with the planetary environment. In the most distant regime of the planetary environment, beyond the bow shock, upstream waves are generated on field lines which connect to the bow shock by ions which are reflected from the shock. This results in a complex foreshock region with the waves scattering the streaming particles and thereby reflecting and accelerating them. The Saturnian magnetosheath is known to be home to very large amplitude mirror mode waves (Tsurutani *et al.*, 1982, 1993) such as have been observed at both the Earth and Jupiter. The characterisation of the sheath MHD wave spectrum will have important implications for the nature of the solar wind input into the planetary system. Detailed analysis of the magnetopause and its immediate environment is also important since it will enable one to determine whether the field penetrates through to the boundary, that is, whether the flux of the interior planetary field connects to the exterior solar wind field. This will answer the question as to whether solar wind activity directly controls internal magnetospheric activity. A full study of the magnetopause boundary crossings requires combining the field and plasma data to establish whether material is in fact crossing though the boundary and whether acceleration and heating are taking place. The boundary crossing is unlikely to be stationary with the motion being possibly caused by a variety of phenomena among which are: the steady reconnection process, sporadic reconnection, changes in solar wind pressure, Kelvin–Helmholtz instability and the centrifugally driven instability to name but a few. The wave modes and wave particle interactions in the regions both upstream and downstream of the bow shock will be determined. The internal structure of the shock and magnetopause crossing will be examined as well as the occurrence of waves, reconnection and other phenomena at and near the magnetopause. The effect of internal magnetospheric rotation and external solar wind conditions on the shape and position of the magnetopause and bow shock will also be investigated.

2.2.5. *Survey Satellite Dust Ring Torus Electromagnetic Interactions*

The magnetosphere of Saturn has multiple internal plasma sources, with ionised material originating from the hydrogen torus, the Titan torus, Titan itself, the icy satellites, the planetary ionosphere and most uniquely from the rings of the planet. There is additionally the potential external source of the solar wind plasma. The dynamics of the magnetosphere as a whole and in particular that of the inner magnetosphere are controlled by the planetary rotation. As material is ionised within the magnetosphere, large amounts of rotational energy have to be transferred from the planet to the corotating plasma in the equatorial plane in order to bring the freshly ionised material up to corotation. This energy transfer takes place via the magnetic field which is effectively frozen into the flow. This energy and momentum is transferred via stress in the field by large scale field aligned currents which couple the planetary ionosphere and magnetosphere. These current systems generate an azimuthal field component which reverses across the equator and hence the existence of the current systems can be measured by perturbations in the B_ϕ field component. Departures from corotation have been measured at Saturn by the Voyager spacecraft (Richardson, 1986) and more recently by Ulysses at Jupiter (Dougherty *et al.*, 1993) where corresponding auroral signatures were also observed (Dougherty *et al.*, 1998). It is highly unlikely that the coupling between the planet and the newly ionised material occurs smoothly and this results in the presence of Alfvenic oscillations which, therefore, are a diagnostic of this coupling.

2.2.6. *Structure and Time Dependence of the Magnetotail*

One of the central issues to be addressed by the Cassini orbiter is the relative importance of solar wind stresses and planetary rotation stresses in determining the morphology of the near tail structure and, therefore, the overall dynamics of the outer magnetosphere. The time dependent studies made possible by Cassini will enable us to distinguish between substorm like processes, in which solar wind stress is released sporadically in an explosive manner or where centrifugally driven ballooning causes plasmoid formation (see plasma bubbles discussion in Section 2.2.2). From the Cassini magnetometer measurements we will be able to characterise the configuration of the Saturnian magnetic tail in the near planetary environment in terms of lobe field strength, orientation and variations, as well as properties of the magnetic field in the central plasma sheet. One will be able to, in conjunction with particle and plasma measurements, identify the relative contributions of rotational, pressure and field stresses to stress balance and thus determine which are the dominant processes.

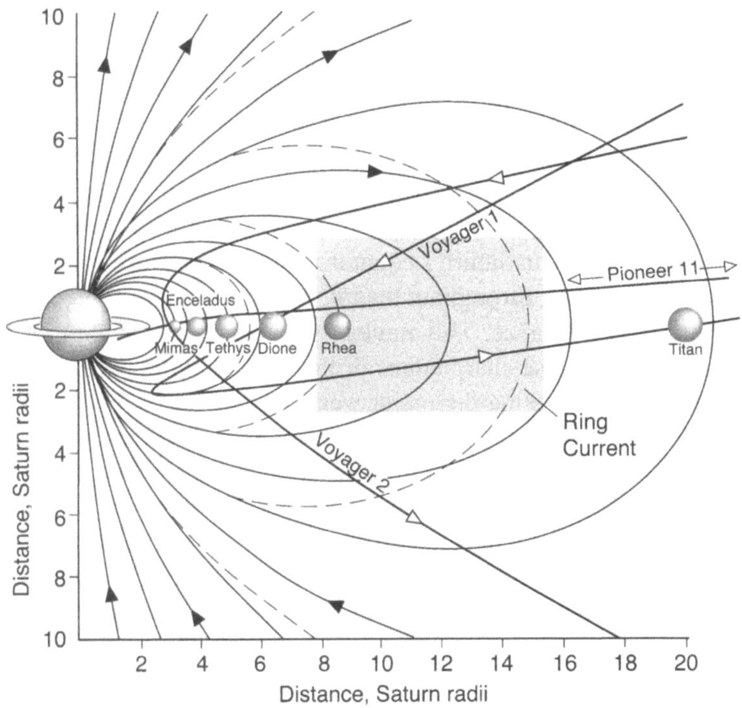

Figure 5. Meridional view of the modelled Saturn magnetic field showing the flyby trajectories overlain.

2.2.7. *Three-Dimensional Magnetospheric Field Model*

Figure 5 shows a meridional view of the modelled magnetic field lines (solid line) at Saturn (adapted from Connerney *et al.*, 1983; Blanc *et al.*, 2002). In the middle magnetosphere the dashed lines denote field lines due to the internal field alone, and the solid lines are those resulting from a combination of both internal field and the effect due to the ring current (which is assumed to flow in the shaded box, azimuthally around the planet). One can see that although the radial stretching of the field lines is clearly evident in this region, this effect is moderate compared to that at Jupiter. Also shown in figure 5 are the trajectories of the three planetary flybys and hence the latitudinal coverage achieved can clearly be seen.

The total magnetic field at Saturn consists not only of the internal planetary field, but also of the sum of contributions from a variety of external sources. For example, the equatorial ring current, magnetopause and magnetotail currents all generate magnetic fields which have an effect on the overall field. The modelling of the external field provides a means of encapsulating the field measurements on a global magnetospheric scale. It allows the magnetic field morphology of the Saturnian system to be concisely defined and will allow extrapolation of the field outside of the region of space which will be explored by the Cassini spacecraft.

The dominant external field component is that due to the equatorial ring current which results from the requirement that the centripetal acceleration which is associated with planetary rotation must be balanced by an electromagnetic J × B force, as well as from pressure gradient effects. These currents then distend the mainly dipolar field in the radial direction. The azimuthal equatorial ring current at Saturn has until now been modelled in a very similar way to the Jovian magnetodisk model (Connerney *et al.*, 1983). This ring current is represented by an annular disk extending from 8 to $15.5R_S$ in Saturn's equatorial plane. The current is assumed to be uniformly distributed in z throughout the total disk of thickness $6R_S$ and decreases inversely with radial distance. This model is a crude representation of the Saturnian plasma sheet with the sharp edges of the sheet and the purely radial fall off of current with distance being the major weaknesses. This is, however, the model which has generally been used in recent research into Saturnian physics although improvements to this model are being sought (Bunce and Cowley, 2002; Giampieri and Dougherty, 2002). There are a number of models which have been generated for the Jovian and terrestrial magnetodisks which describe the physical conditions which arise there in a far superior way. We plan to consider, in collaboration with Cassini particle instrument teams, such alternate approaches at Saturn in order to generate a more physically valid magnetodisk model. Three of these approaches are: (1) A self-consistent model of the magnetodisk including centrifugal force and pressure effects (Caudal, 1986); (2) Analytical models which allow for modelling of current disks of finite thickness flanked by current free regions (Tsyganenko and Peredo, 1994); (3) A model using the Euler potential approach (Khurana, 1997) which allows the model to incorporate the hinging and delay of the current sheet with radial distance, sweep-back of the field lines and have realistic azimuthal current density profiles within the magnetosphere.

The contribution to the magnetic field arising from magnetopause currents depends strongly on the characteristics of the solar wind interaction with the planet. It is the balance between the internal magnetic pressure and the dynamic pressure of the solar wind which establishes the outer shape of the magnetosphere. Since the magnetised solar wind does not penetrate into the magnetosphere, currents develop on the magnetopause surface, which in turn induce a contribution to the magnetic field. The only model available to date which models the interaction of the solar wind with the Saturnian magnetosphere has been developed by Maurice and Engle (1995). This computes the size and shape of the magnetopause as well as the additional contribution to the total field from the surface currents. This has been carried out for a few values of solar wind pressure and only one direction for the incident solar wind. Extensions to this type of model will allow scenarios to be generated for large variety of solar wind pressure values (as observed at Saturn) as well as a variety of incident solar wind geometries with large inclinations to the ring plane.

The magnetotail of Saturn has not yet been traversed by any of the previous spacecraft which have visited the Saturnian magnetosphere. The magnetotail region into which the field lines and plasma are able to expand as they rotate around the planet is thought to be a very important region for energy transport processes within the magnetosphere. The only existing quantitative model of the magnetic tail region was proposed by Behannon *et al.* (1981) where cross-tail currents which extend from 16 to $100R_S$ downtail along the Sun–Saturn line have been incorporated which close onto the magnetopause. One route which we plan to follow is to model the magnetotail contribution along the lines of a general Tsyganenko terrestrial model (Tsyganenko, 1989) to Saturn's dimensions. This model takes into account the effect of warping of the tail current in two dimensions. Any models generated will naturally depend on the physical measurements and processes which are observed during the Cassini orbits.

2.3. Titan

Following the recent orbits of the Galileo spacecraft past the Galilean satellites of Io, Ganymede, Europa and Callisto with their measured internal/induced fields—there is now a renewed interest in the history and evolution of such moons. This holds not only for Jovian satellites but also for the Saturnian moons with the upcoming Cassini observations. The possibility of an internal/induced magnetic field at Titan is discussed in Section 2.1.2 as are the measurements which the Cassini orbiter will be able to make in this regard. Other Titan-related science targets of the Cassini orbiter are described briefly below.

2.3.1. *Empirical Model of the Titan Magnetoplasma Environment*

Titan has by far the densest atmosphere of any known satellite and is the second largest satellite in the solar system (with Ganymede being the largest). It is at a radial distance of $20R_S$ from Saturn and although it spends most of the time within the Saturnian magnetosphere there are times when the solar wind pressure is high enough such that Titan can in fact be outside of the Saturnian bow shock or magnetopause (as probably occurred during the Voyager 2 flyby in 1981; Sittler *et al.*, 1983). Another interesting effect which occurs at Saturn is that there are periods when the planet is located within the Jovian magnetotail. It seems very likely that this occurred during the Voyager 2 flyby past Saturn (Lepping *et al.*, 1983). Depending on its position within the Saturnian environment, Titan constitutes an obstacle to the magnetospheric, magnetosheath or solar wind flow. Following the Voyager 1 pass through the wake of Titan, it is known that a strong interaction occurs with a magnetotail configuration (reviewed in Neubauer *et al.*, 1984) and an associated plasma cavity (Keller *et al.*, 1994). The implications of the relatively small scale size of Titan in relation to the Larmor radius of the upstream ions implies that serious asymmetries could result from the interaction. During the numerous Titan flybys which will take place during the Cassini mission, a description of the Titan environment which quantifies the structures detected in the vicinity of the body will be derived utilising the magnetic field measurements. Examples of such structures are: the size and shape of the magnetic cavity, size and opening angle of the magnetotail, width of the plasma sheet and mantle and critical parameters for any Alfven wing-like structures and wave phenomena in its environment. A bow shock might be observed when Titan is in the solar wind.

2.3.2. *Titan-Plasma Flow Interactions and Variations of Titan/Magnetosphere Interaction with Respect to Titan's Orbital Phase*
As discussed in Section 2.3.1 the Titan orbit can, depending on solar wind conditions, be either inside or outside of the Saturnian magnetosphere. During the periods when Titan is inside of the magnetosphere, encounters will take place at a variety of phases of the Titan orbit, resulting in large stress variations between Titan and the surrounding plasma. Such stresses will be revealed by perturbations in the magnetic field measurements in the vicinity of the moon. Near the noon meridian, exits into the magnetosheath and the solar wind are also possible depending on solar wind conditions at the time. All such encounters will be documented and compared with theoretical models. As Titan orbits Saturn, the angle between the corotational flow and the direction of the ionising solar photons varies through 360°. This leads to a variety of interesting cases at different local times which may prove to be important for the diagnostics of the plasma physical processes of the Titan interaction. For example, around Saturnian local time (LT) 06:00 hours induction effects should be most visible, around (LT) of 18:00 hours the cumulative action of solar photoionization and collisional ionization by magnetospheric electrons will lead to the strongest interaction.

2.4. Icy Satellites Electromagnetic Environment

As described in Section 2.1.2, there will be numerous targeted and untargeted flybys of the icy satellites during the 4-year Cassini orbital tour. In addition to the search for internal/induced magnetic fields at these moons we will also investigate the icy satellite plasma environments. This will allow us to understand the electromagnetic interaction which arises between the various satellites and their surrounding plasma medium, whether it be Saturnian plasma or solar wind plasma. Here, the inner satellites seem to be particularly promising with Enceladus and the E-ring being prominent.

2.5. Nature of the Low Frequency Wave Spectrum

For a multi-ion plasma, such as found in the vicinity of Titan and the other icy satellites, even the presence of small concentrations of heavy ions greatly modifies the properties of the waves. In particular, new critical frequencies are introduced, in the form of resonance and cut-off frequencies (Dougherty and Southwood, 1993; Thompson *et al.*, 1995). Resonant frequencies are important since they reveal the frequency at the boundary of a stop band and in warm plasma theory wave particle interactions occur close to resonant frequencies. The cut-off frequencies separate frequency bands of propagation to ones of evanescence. An additional critical frequency which arises in a multi-ion plasma is that of the cross-over frequency at which the phase velocity of right and left hand circularly polarised waves is the same as that of the transverse mode and in the vicinity of the proton Alfven speed. Since the new critical frequencies depend not only on the gyro-frequencies of each species present but also on the relative abundances of each, this implies that by careful interpretation of the frequency spectrum information can be obtained on which species are present as well as on the relative abundances of these species. Ion cyclotron wave activity at Jupiter has recently been examined in both the Ulysses (Dougherty *et al.*, 1997) and Galileo (Huddleston *et al.*, 1997) magnetic field data sets and observations of such wave activity were observed inside Saturn's magnetosphere by the Pioneer 11 spacecraft (Smith and Tsurutani, 1983). Work on the Galileo Io observations (Huddleston *et al.*, 1998; Huddleston *et al.*, 1999) compares pick-up processes near Io to those in the solar wind as well as showing the wave spectra can be used to infer plasma composition.

There are in fact a number of reasons for interest in ion cyclotron wave activity. Not only can they be used as a diagnostic as to what type of ion species arise in a plasma, but they can also be used as a means of coupling different species via wave particle interactions. Some ultra-low frequency (ULF) pulsations are standing MHD waves in the Earth's magnetosphere, with there being indications that such eigen-oscillations may also exist at Jupiter and Mercury although none have been observed in the Saturnian magnetosphere. Work by Cramm *et al.*, 1998 shows that the Voyager 1 spacecraft crossed an area of resonant mode coupling during its Saturn encounter. The sensitivity and high time resolution of the magnetometer will enable us in conjunction with the Cassini radio wave instrument to better resolve the wave spectrum at Saturn, for example, very close to the planet only the magnetometer has the capability of resolving down to the proton gyro-frequency.

3. Instrument Details

The chosen instrument configuration consists of a fluxgate magnetometer (FGM) and a vector helium magnetometer capable of operating in scalar mode (V/SHM). The instrument is intended to measure small changes in fields spanning four orders of magnitude with extremely high sensitivity. This is achieved in part by mounting the sensors on an 11 m spacecraft boom, the V/SHM at the end of the boom, the FGM halfway along it. The magnetometer boom distances the sensors from the stray magnetic field associated with the spacecraft and its subsystems and especially with spacecraft generated field variations, spacing the sensors at different distances along the boom allows this field to be better characterised and removed from the observations. The efforts taken to constrain the stray spacecraft field are described very briefly in Section 6. Sensitivity is also enhanced by the use of 16 bit analogue-digital converters (with 14 bit values being sent to ground). Mounting the sensors on a boom could result in their orientation with respect to the spacecraft axes changing from time to time, after spacecraft manoeuvres for example. A means of sensor alignment determination has been provided by the Cassini project; the Science Calibration Subsystem, SCAS. This consists of two, perpendicular, coils rigidly mounted on the spacecraft body with a known alignment to the spacecraft axes. These produce well-defined magnetic fields on command which can be detected by the sensors and used to correct for any changes in sensor orientation. SCAS measurements taken to date will be described in Section 4.2.

What are the scientific advantages of the FGM, VHM combination and the scalar mode (SHM)? Both magnetometers are well capable of measuring the magnetic field vector at the location of the respective sensor at rates from DC up to 10 Hz (VHM) or at least 30 Hz (FGM). The VHM optimises low frequency vector measurements in low fields, FGM measurements are best at high frequency and it operates over an extremely wide dynamic range from very low fields ranging up to measurements in Earth's high field. The particular advantages of joint FGM, SHM measurements have been discussed in Section 2.1.1. The twin sensor configuration also contributes to overall instrument reliability; if one sensor fails field measurements can be made with the other sensor, with sufficient performance to achieve many of the major objectives of the investigation. Reliability has been further increased by the provision of redundant instrument power supplies and data processing units and by careful selection of electronic components which can survive the radiation environments encountered during the long cruise phase of the mission and in the Saturnian system. The VHM provides the stability needed to maintain the calibrations obtained in the solar wind while Cassini is inside the Saturnian magnetosphere for long periods during the 4-year tour.

An implicit feature of scalar or resonance magnetometers are null zones which arise if the ambient field falls outside a cone of 45° half angle with respect to the optical axis of the magnetometer. This results in the signal being dramatically weakened and hence the absolute accuracy of the instrument will suffer. As a result of use of the scalar magnetometer on Cassini, a requirement has been placed on the mission to avoid spacecraft orientations inside of $4R_S$ which cause the planetary field to lie within the null zones of the scalar device. This results in the signal being weakened with an associated decrease in absolute accuracy.

Other features of the instrument which have been driven by the characteristics of the mission and by the design of the spacecraft are to be found in the data processing unit (DPU). This contains a bus interface unit (BIU) provided by the Cassini project for interfacing to the onboard data handling subsystem (CDS) bus. In line with the spacecraft design, the DPU is capable of handling Packet Telemetry and Telecommands and features a flexible telemetry storage and generation scheme to support the multiple telemetry modes of the spacecraft. The tour operations concept requires that the DPU is able to handle trigger commands which initiate multiple actions within the instrument (macro commanding). Further, in order to optimise the analysis of discrete events such as shock crossings, a snapshot capability has been implemented by which up to 16 Mbytes of data can be stored for later downlink at higher time resolution than normal. This capability can be initiated by command or triggered by pre-defined events.

Magnetic field information is needed by other investigations on the spacecraft. To this end magnetic field data are made available to onboard users every second. These onboard ancillary data are raw, uncalibrated vectors, the data source being selectable by command between the two sensors.

In total, the instrument consists of the two, boom-mounted, sensors, subchassis No. 1 (an assembly containing electronics for the FGM, VHM and SHM, the heater control electronics, the power supplies and power management system) and subchassis No. 2 (an assembly containing the data processing unit). Both subchassis are mounted in bay 4 of the Orbiter upper equipment module (UEM). Figure 6 shows the location of the magnetometer hardware on the spacecraft together with the spacecraft and sensor axes. An overall block diagram of the instrument showing how the hardware responsibilities for the individual elements have been divided between Imperial, JPL and TUB is shown in Figure 7. The instrument ground support equipment was provided by KFKI and TUB.

Table III lists the main characteristics of the instrument. Power and data rate values vary according to instrument mode. The values given are for the delivered flight model where power values include power drawn by the Cassini-provided BIU.

3.1. The FGM

The FGM sensor is mounted halfway along the magnetometer boom, its associated analog electronics form part of the electronics assembly on Subchassis No. 1. A cable of approximately 6.5 m length runs along the boom between sensor and electronics. A high efficiency, tuned drive design of the electronics has been chosen to reduce power consumption and the effect of cable loading.

The FGM is similar to the Imperial instrument flown on Ulysses and to many others flown on numerous missions. It is based on three single-axis ring core flux-gate sensors mounted orthogonally on a machinable glass ceramic block. Ceramic is chosen for its low thermal expansion coefficient, minimising misalignments between sensors due to temperature changes. In each sensor, a drive coil is wound around a high permeability ring core which is completely enclosed in a sense winding. The drive coil is driven by a crystal-controlled 15.625 kHz square wave which is used to generate a magnetic field driving the core into saturation twice per cycle. The three drive coils are connected in series to simplify the cabling and circuitry. The presence of an ambient magnetic field component parallel to the axis of the sense coil causes the saturation of the core to become asymmetrical. This induces a second harmonic of the drive frequency in the sense coil which is proportional to the magnitude of the magnetic field component along that axis. The signal is processed through a narrow band amplifier tuned to the second harmonic of the

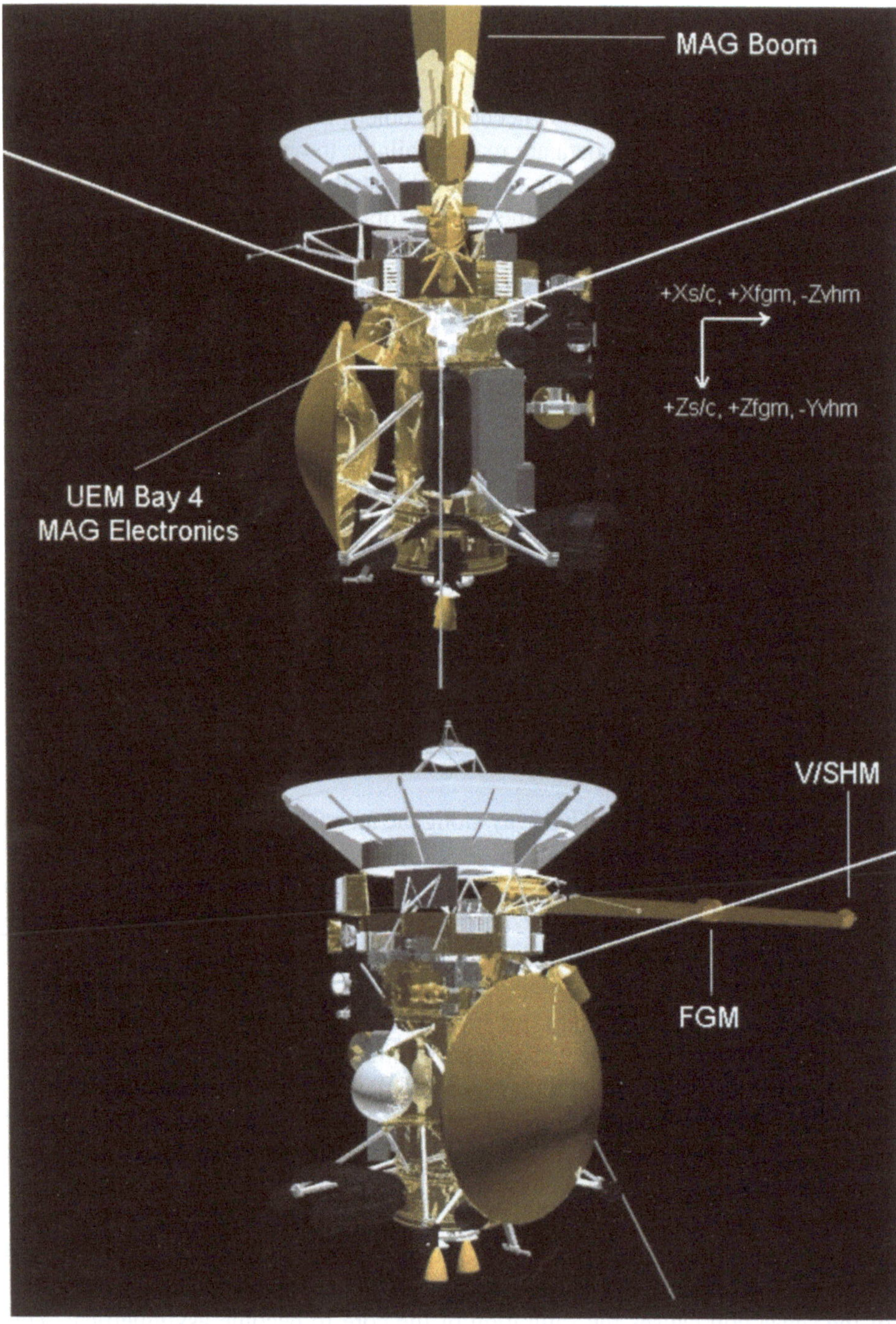

Figure 6. The placement of magnetometer hardware on the spacecraft is shown as well as the axes of the sensors (source: JPL).

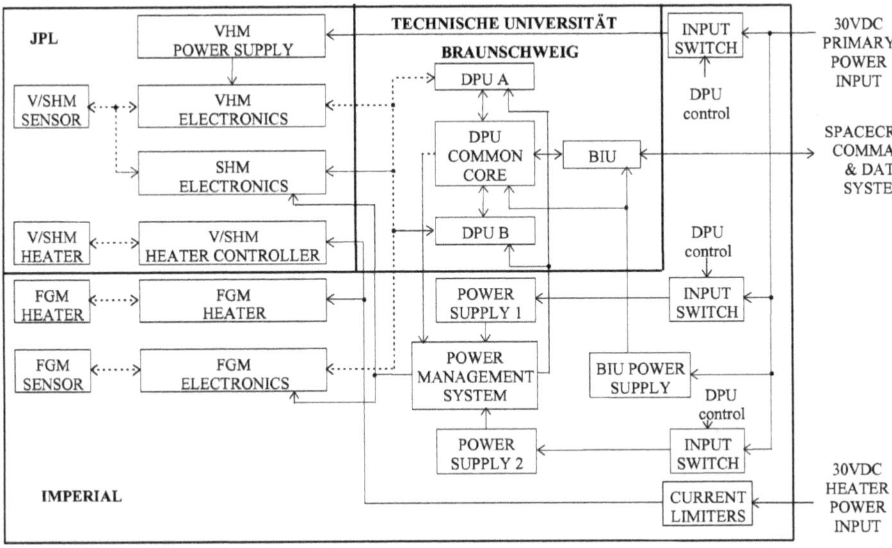

Figure 7. Overall block diagram of the magnetometer instrument which also shows the institutes responsible for hardware. Command and data interfaces are shown as dotted lines and power interfaces are shown by solid lines.

drive frequency, which attenuates harmonics other than the second. The result is integrated, converted to a current and fed back to the sensor coil to null the ambient field. The integrated output voltage, amplified and corrected for scale factor and alignment errors, is proportional to the ambient field. The three analogue vector components are passed to the DPU for analogue to digital conversion and data processing. The noise performance of the FGM, measured on ground at the analogue output of the electronics, is better than 5 pT/$\sqrt{}$Hz at 1 Hz. The electronics can be checked in flight using an in-flight calibration (IFC) capability built into the electronics and controlled by command from the DPU. The IFC applies a fixed offset to each of the three vector outputs corresponding to a signal of approximately 10 nT. The frequency, number of on/off cycles, of the IFC is selectable by command.

Changing the electronics feedback path and the output amplification allows the sensor to be operated in one of four different full scale magnetic field ranges, as listed in Table III. The largest range (±44,000 nT) was included mainly for ground testing in the Earth's field. Switching between ranges in normal operations is automatic, controlled by the DPU. The DPU monitors each FGM component field value, if the magnitude of any of these exceeds an upper threshold for greater than a specified number of samples, the DPU will switch the FGM to a higher range. Similarly, if all three component value magnitudes fall below a lower threshold for more than a specified number of samples, the DPU will switch the FGM to a lower range. All parameters are modifiable by command and autoranging can also be disabled and manual range changes commanded.

TABLE III

Main Instument Characteristics

Mass (kg)	
V/SHM Sensor	0.71
FGM Sensor	0.44
Subchassis#1 (Power Supplies, Sensor Electronics)	5.15
Subchassis#2 (DPU)	2.52
Total	8.82
Power (W)	
Sleep Mode	7.50
Vector/Vector Mode (FGM+VHM)	11.31
Vector/Scalar Mode (FGM+SHM)	12.63
Normal downlink data rate	
FGM	32 vectors/s
VHM	2 vectors/s
SHM	1 value/s
Housekeeping	24 bits/s
Total	2,000 bits/s
Dynamic range, resolution	
FGM	±40 nT, 4.9 pT
	±400 nT, 48.8 pT
	±10,000 nT, 1.2 nT
	±44,000 nT, 5.4 nT
VHM	±32 nT, 3.9 p
	±256 nT, 31.2 pT
SHM	256 –16384 nT, 36 pT

A 1 W heater has been provided to maintain the FGM within its operating temperature range of −30 to +50 °C. The specially designed, non-magnetic unit is mounted on the ceramic sensor block and has control electronics on Subchassis No. 1. Further thermal control is provided by an aluminised mylar covered fibreglass case over the sensor block and by three, project-provided, radioactive heater units mounted equidistant around the sensor base (the latter provide a total of 3 W). Figure 8 is a photograph of the Fluxgate sensor, with its case removed showing the location of the single-axis sensors on the sensor block, together with its associated electronics.

3.2. The V/SHM

The V/SHM sensor is mounted at the end of the 11 meter magnetometer boom. A photograph of the V/SHM sensor and the VHM electronics is shown in Figure 9.

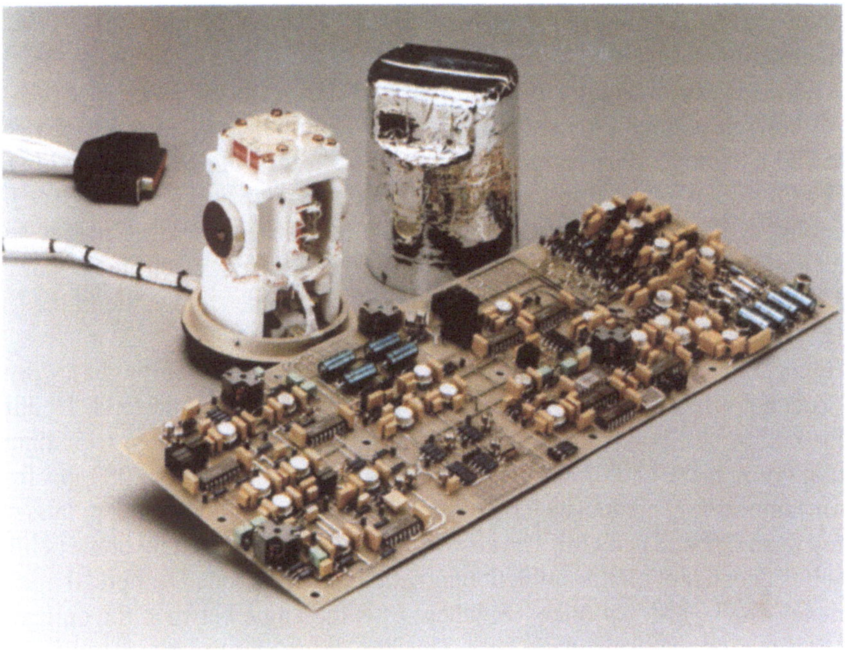

Figure 8. Ultra Electronics photograph of the FGM (with cover off) and electronics board.

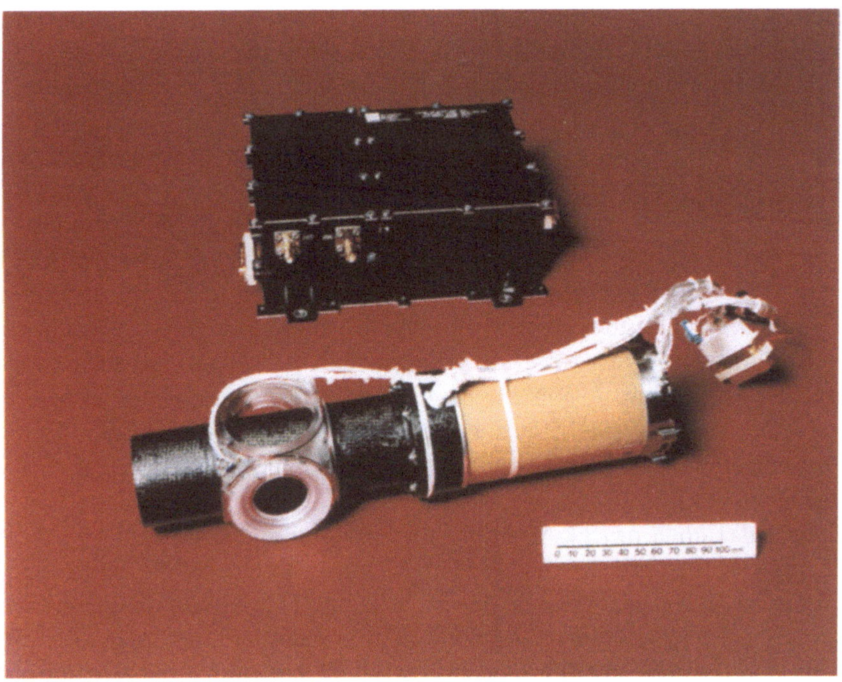

Figure 9. V/SHM sensor and VHM electronics.

A set of cables running the length of the boom connect it to the VHM and SHM electronics on Subchassis No. 1. The flight sensor is actually the flight spare Ulysses vector helium magnetometer sensor with the addition of a small pair of coils nested inside a pair of the larger Helmholtz coils used in the vector mode and a coaxial cable carrying the signal from the voltage controlled oscillator (VCO) to that cable. Similarly, the VHM electronics box is the Ulysses flight spare unit with small modifications to change the sensor operating ranges and to compensate for the different boom cable lengths. A new electronics board has been added to Subchassis No. 1 containing the electronics to operate in the scalar mode.

The operation of the magnetometer is based on field dependent light absorption (the Zeeman effect) and optical pumping to sense the magnetic field. Helium in an absorption cell is excited by a radio frequency (RF) discharge to maintain a population of metastable long-lived atoms. Infrared radiation at 1083 nm from a helium lamp, also generated by RF excitation, passes through a circular polariser and the cell to an infrared detector. The absorption (pumping efficiency) of the helium in the cell is dependent on the ambient magnetic field direction. The optical pumping efficiency is proportional to $\cos^2\Theta$ where Θ is the angle between the optical axis and the direction of the magnetic field. This directional dependence is utilised in the vector mode by applying low frequency sweep fields rotating about the cell, which allow the extraction of the three orthogonal ambient field components. These fields are fedback using a set of triaxial Helmholtz coil mounted on the sensor housing around the cell. In the scalar mode, the directional dependence results in a "field of view" restricted to a cone with half angle approximately 45°, centred on the optical axis detector.

Changing the VHM sweep fields allows the sensor to operate in different ranges. Two VHM ranges have been selected for Cassini (see Table III). As for the FGM, automatic ranging has been implemented in the DPU. The VHM electronics also have an internal autoranging capability (used for the Ulysses instrument). A single range has been implemented for the SHM. Injection of known currents into the Helmholtz coil system provide an in-flight calibration (IFC) capability. The calibration fields apply an offset of approximately 1/8 of the full scale range to each vector component. A non-magnetic proportional heater using up to 2 W is incorporated into the V/SHM sensor and is controlled from electronics built into the VHM electronics box on Subchassis No. 1. The operating temperature range of the sensor is -10 to $+40$ °C.

In the scalar mode, a weak AC field at the Larmor frequency is applied to the cell, which opposes the optical pumping. The decrease is detected by the IR detector as a decrease in transmitted light from the lamp. The Larmor frequency which is proportional to the ambient magnetic field is measured. In order to track the ambient field the applied field is frequency modulated so that the detector output contains a signal component harmonically related to the modulation frequency. The

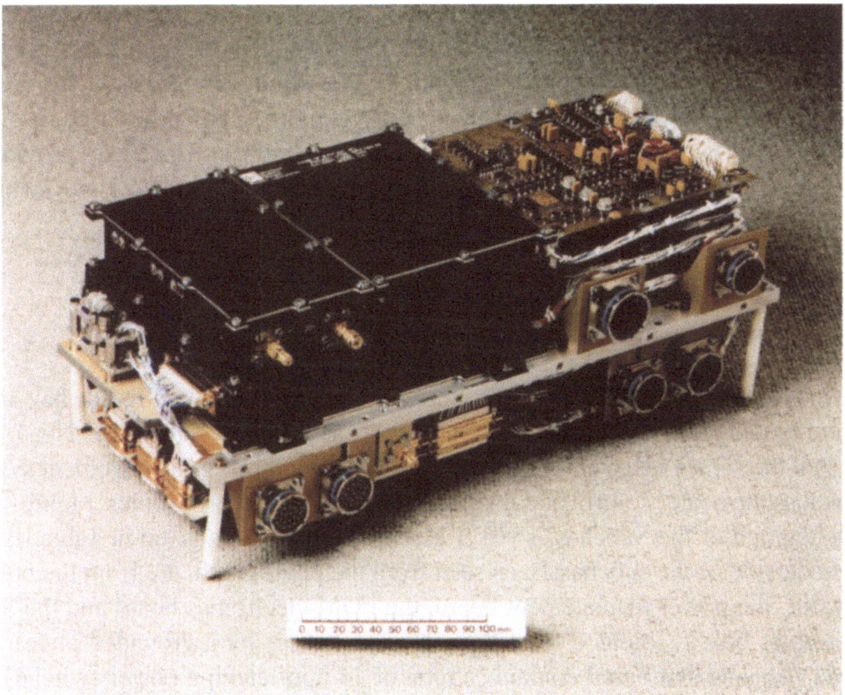

Figure 10. Photograph of Subchassis No. 1 with the VHM sensor on top.

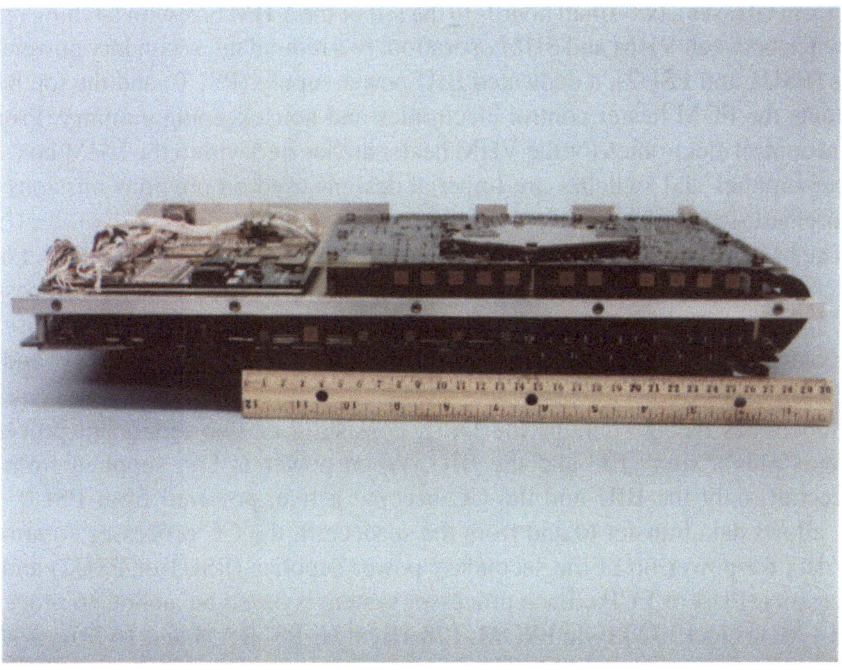

Figure 11. A photograph of Subchassis No. 2.

proportionality constant is the gyromagnetic ratio which for helium is 28.023561 Hz/nT. Detection and measurement of the Larmor frequency lead to a very accurate measurement of the ambient field magnitude. The result is passed as a 20 bit scalar word from the SHM electronics to the DPU. A more detailed description of the V/SHM may be found in Kellock *et al.* (1996). Smith *et al.* (2002) provides a detailed description of the SHM operation and observations from the Earth swingby in August 1999.

3.3. Instrument Electronics

The instrument electronics are all mounted in Upper Equipment Module bay 4 (see Figure 6) and are split between two subchassis assemblies. Subchassis No. 1 contains all the sensor electronics, the power supplies and power management system as well as the heater control and instrument housekeeping electronics. Figure 10 is a photograph of the Subchassis No. 1 assembly, its mass is given in Table III. On the underside of the subchassis, as seen from the photograph, are from the bottom upwards, the power management board, the FGM electronics board and the SHM electronics board (shielded from the other electronics by a grounded plate). The power management board contains a total of 14 non-latching power switches and cross-strapping circuitry for the two redundant secondary power supplies. The top side of the subchassis contains the black VHM electronics box (a modified spare unit from Ulysses), two small boards to the left of the VHM box with latching relays to switch between VHM and SHM operation, two redundant, secondary power supplies (PSU1 and PSU2), a dedicated BIU power supply (PSU0) and the top board contains the FGM heater control electronics and housekeeping circuitry. Proportional control electronics for the VHM heater are located within the VHM box. The power supplies and switches are Imperial designs used on previous missions and feature built-in overcurrent trips.

The basic power distribution scheme is described in Kellock *et al.*, 1996. Power switches for the secondary voltage lines are controlled by the active processing unit and power switches for the power supplies and processing units themselves are controlled by discrete commanding from the spacecraft via the BIU and the Common Core (CC).

Subchassis No. 2 contains the DPU, consisting of two redundant processor systems plus a small CC and the BIU. When power is first supplied from the spacecraft, only the BIU and the CC become active, powered from PSU0. The BIU allows data transfer to and from the spacecraft, the CC processes commands and data for power up of the secondary power supplies (PSU1 or PSU2) and the processors (PUA or PUB). Each processor system is based on an 80C86 processor with 4 MHz clock, 32 kByte PROM, 128 kByte Hi-Rel RAM and 16 MByte state-of-the-art commercial DRAM. The systems normally operate singly but can be operated in parallel. A high accuracy 16 bit analogue to digital converter (ADC) is integrated into each processor system for sensor data collection. Two ADC clock speeds are available; 1 and 2 MHz, the former being the default. Tantalum shielding has been used for the ADCs, DRAMs and Operational Amplifiers to reduce their susceptibility to radiation. The assembled subchassis is shown in Figure 11. The DPU boards are folded around the subchassis. The electronic components face to the subchassis, because the 2.4 mm thick 16-layer boards provide additional radiation shielding. The Sensor Interface Board is on the right hand side. The outer side shielding of the ADCs and Operational Amplifiers can be seen. The flexible connection board goes through subchassis cutouts to the Processor Board on the other side. The JPL-provided BIU plus its associated cabling is located on the left side.

To satisfy the demands of a deep-space mission with limited communications, the DPU has been designed with a large measure of autonomy and sophisticated data handling functions. The functions it performs are as follows: telecommand handling, sensor autoranging and IFC, sensor data collection, sensor data processing, snapshot data handling, telemetry generation, error correction, fault detection and recovery, and onboard ancillary data generation. Autoranging and IFC have been discussed above; data functions are described in the data processing Section 7. The remaining functions are briefly described below.

The DPU is designed to handle both the packet telecommand standard adopted by Cassini, used for normal commanding, and discrete telecommands used when the processor is not active. A large variety of command functions are supported which are discussed later. The DPU must be able to accept telecommands at all times, in all operational modes. Commands may be for immediate execution, or can contain relative or absolute timetags for delayed execution (relative timetags cause execution at a fixed time with respect to reception of the command, absolute timetags cause execution at specific spacecraft times). As noted earlier, a macro commanding capability has been implemented whereby sequences of instrument commands can be stored in the DPU and the sequence started by executing a macro command.

Hamming single bit error correction and double bit error detection is provided for all memory devices (except those in the BIU). Single Event Upsets (SEUs) can change the content of memory cells, cyclic access to every memory cell corrects single bit and reduces the risk of double bit errors. Memory scrubbing is initiated every 64 s in the Hi-Rel RAM and the 16 MByte, multi-snapshot, DRAM. It takes about 1 h to scrub the complete RAM. Additional memory checks can be initiated by command: occurences of single and double bit errors are monitored. The PROMS are checked separately and contain a pre-defined error pattern for detection. If a permanent RAM problem arises, the DPU can be commanded to run its software directly from PROM. In-flight tests have also been implemented for the ADC to check the noise and conversion and settling times on each analogue channel.

Both processor systems contain four separate dual level latch up detectors, one for the processor, one for the multi-snapshot memory and one each for the ADC ± 12 V supply voltages. Detectors of similar design have been flown on the GEOTAIL, WIND and SOHO spacecraft. If a latch up is detected in the processor, it will be immediately switched off and on again and the instrument will be automatically reconfigured into its previous mode. Latch ups in either the memory or the ADC will cause it to be immediately switched off and back on again. There is also a hardware watchdog function in the DPU which will detect problems in the DPU program flow and which initiates a hardware reset followed by an automatic instrument reconfiguration.

The SCAS, provided by the Cassini project for alignment checks of the sensors on the magnetometer boom, consists of two coils, one (Y coil) located in UEM bay 4 and one (X coil) located in UEM bay 7. Its operational use is entirely under the control of the magnetometer team. The standard operational procedure is as follows: Upon electronics power up the SCAS X coil is energised and operates for 34 min 8 s, swapping polarity every 2 s. After this SCAS automatically switches to the$Y$ coil which is operated in the same way. This cycle repeats until SCAS is powered off.

4. Calibration

Instrument calibration is an ongoing activity that begins at instrument assembly and continues throughout the mission. Before launch the sensors were calibrated in highly accurate test facilities; in-flight calibration signals have been regularly applied since launch; and the well characterized magnetic field of the Earth was used as a calibration source during Earth swingby.

4.1. Ground Calibration

4.1.1. *Calibration of the Fluxgate Magnetometer*

The fluxgate magnetometer was calibrated in the 2.5 m Braunbek coil system of the Institute of Geophysics at the Technical University of Braunschweig (TUB) during the period June 24 to July 12, 1996. Tests were performed to calibrate the sensitivity (nT/bit), orthogonality, zero levels, linearity and frequency response. Functional tests of the auto-ranging, in-flight calibration and snapshot event trigger were also performed at this time. Detailed test procedures are documented in TUB reports CA-IGM-TP001 and test results in CA-IGM-TR-0002 and -0003. Key test data were reduced by two independent teams and then cross-checked. Table IV lists the sensitivities of each of the three sensors in each of the four ranges of operation of the magnetometer, normalised by the nominal range.

TABLE IV
Fluxgate magnetometer sensitivities

Range (nT)		Output/Input X sensor	Y sensor	Z sensor
R0	40	0.9966	1.0002	0.9991
R1	400	0.9994	0.9997	0.9997
R2	10,000	0.99931	0.99952	0.99953
R3	44,000	0.99938	0.99950	0.99950

TABLE V
Orthogonalisation matrices

Range	Matrix
R2	(1.0007 −0.00210 0.00304) (0.00202 1.0005 0.00159) (−0.00169 −0.00769 1.0005)
R3	(1.0006 −0.00210 0.00302) (0.00205 1.0005 0.00161) (−0.00169 −0.00771 1.0005)

Table V lists the matrices that convert the sensor measurements into a calibrated orthogonal field measurement. Both tables are for the sensor at room temperature. Only the matrices for the two highest field ranges are given here because the small size of the off-diagonal terms makes these elements difficult to determine in low field strengths. To a high degree of accuracy the corresponding matrices for the low field ranges can be obtained by replacing the diagonal terms in the R2 or R3 matrix by the reciprocal of the corresponding sensitivities in Table IV. Converting the off-diagonal terms to the equivalent angles between the sensors gives the following: XY angle, 89.998°; XZ angle, 90.076° and YZ angle 89.650°.

The frequency response of the basic (analog) fluxgate magnetometer is such as to have negligible impact on the instrument performance. The 5 dB point for the X and Z sensors is about 158 Hz and for Y about 70 Hz. The phase lag is about $-100°$ at 200 Hz on all three sensors. The phase and amplitude response of the magnetometer in the final data stream will be set principally by the high rate sampling of the analog to digital converter (64 samples/s) and the averaging of the data processing unit (nominally 32 samples/s). The linearity test revealed the instrument to be extremely linear. The only non-linearity detected was at the level of $\pm 4 \times 10^{-5}$ of full scale in the x-sensor. This non-linearity appears to be sinusoidal and can be easily removed if necessary in data processing.

4.1.2. *Calibration of the V/SHM*

The calibration of the V/SHM in the vector mode was carried out at JPL in a magnetically-shielded walk-in room that contains a precisely calibrated set of Helmholtz coils. The sensor was mounted on a turntable whose angular position is controlled from outside the room. Means are available to vary the temperature of both the electronics and sensor independently. A standard comprehensive calibration procedure was followed that has been developed for past missions such as Ulysses. It involves accurate measurements of essentially all performance parameters including the triaxial scale factors (voltage out for a given field in) and linearity, intrinsic instrument noise, triaxial offsets (equivalent to the output in zero field), the amplitude and phase response as a function of frequency, response to the in-flight calibration signals (several positive and negative field steps applied to the sensor coils upon command), and orthogonality of the three axes.

Since the Cassini Flight Model is basically the Ulysses spare VHM built in the early 1980s, a long history exists consisting of calibration data taken at approximately 6 month intervals when the instrument was removed from JPL bonded stores for that purpose. Within the instrument field-equivalent-noise level (30 pT^2/Hz independent of frequency/white noise), no changes in any of the parameters with time or temperature have been detected. This stability has been confirmed in-flight by the Ulysses VHM over an interval of 12 years with accompanying excursions in temperature as the spacecraft traveled between 1 and 5.4 AU. In particular, any variation in magnetometer offsets is a few pT per year at most (and even this may be changes in spacecraft field).

Operation in the vector, and especially the scalar mode, was tested and calibrated at a coil facility maintained by the US Coast and Geodetic Survey magnetic observatory located outside Fredricksburg, Virginia. The SHM output was carefully compared with the fields produced by a set of large coils about 10 meters in diameter arranged in a Braunbek array that had previously been calibrated with a proton precession magnetometer. A high precision standard resistor was placed in series with the coils to monitor the coil currents very accurately. Close agreement was found between the SHM output and that based on the currents in the large coils and the corresponding calibration constants.

Special tests were run to measure small departures of the SHM output frequency from the true Larmor frequency having multiple causes. The sensor exhibits small vector offsets when operated in the scalar mode which had to be determined accurately so as to be able to remove them from the data if necessary. An additional small source of error that was measured accurately was the so-called "light shift" that arises because of the perturbation of the quantum energy levels by the incident radiation. Finally, at low fields (less than 1,000 nT), the Bloch-Siegert shift, whose magnitude depends on the ratio of the amplitude of the VCO-generated field to that of the ambient field, was characterized. The procedures used and the data taken have been documented in a test report and a final calibration report, the latter to provide the parameters need to reduce the V/SHM data to accurately known fields.

TABLE VI
Vector helium magnetometer sensitivities

	Output/Input		
Range (nT)	X sensor	Y sensor	Z sensor
R0 32	1.005167	1.005515	1.009559
R1 256	1.008589	0.998337	1.006285

Table VI lists the sensitivity of each axis of the VHM, normalized by the nominal maximum field value divided by the maximum number of bits, for each of its two ranges. The digital output from the SHM consists of the number of cycles in an interval of 0.997126 sec and is stored in a register before being transferred to the DPU. The equivalent frequency is then converted to field strength by using the gyromagnetic ratio.

4.2. In-Flight Calibration

For the vector magnetometers, conversion of raw data to magnetic field values is generally achieved by applying an offset and a gain (nT/bit) to the raw data. Additional corrections have to be applied for the sensor alignment and temperature dependencies. Ground calibration provides initial values for these parameters which will vary after launch. In particular, the stray spacecraft field adds to the intrinsic sensor offsets. Further, the alignment of the sensors with respect to the spacecraft cannot be established until after the magnetometer boom is deployed. Various means are available to re-establish the calibration of the instrument after launch.

The built-in IFC functions, described earlier provide a check of the sensor gains. Changes in alignments, both between sensor axes and with respect to the spacecraft axes can be detected by the use of SCAS. Since the alignment of the SCAS coils with respect to the spacecraft are known (measured before launch) and the alignment of the spacecraft axes with respect to an inertial reference system are known, the relationship of the sensor outputs to the inertial reference system can be established. This relationship was also cross-checked by operation of the sensors in the well known Earth's field during flyby. It is likely that the orientation of the sensors on the boom may change slightly after events such as spacecraft manoeuvres or as the thermal environment changes with distance from the sun. To counter this, periodic SCAS alignment checks are planned for the mission, starting during the Science cruise phase. The approximate SCAS fields (zero to peak) expected to be seen at the sensors are, for the FGM, $X = 25$ nT, $Y = 15$ nT and, for the VHM, $X = 2.25$ nT, $Y = 2$ nT (X and Y are spacecraft axes). The periodic change in direction of the fields will make them distinguishable from the ambient external field and spacecraft field.

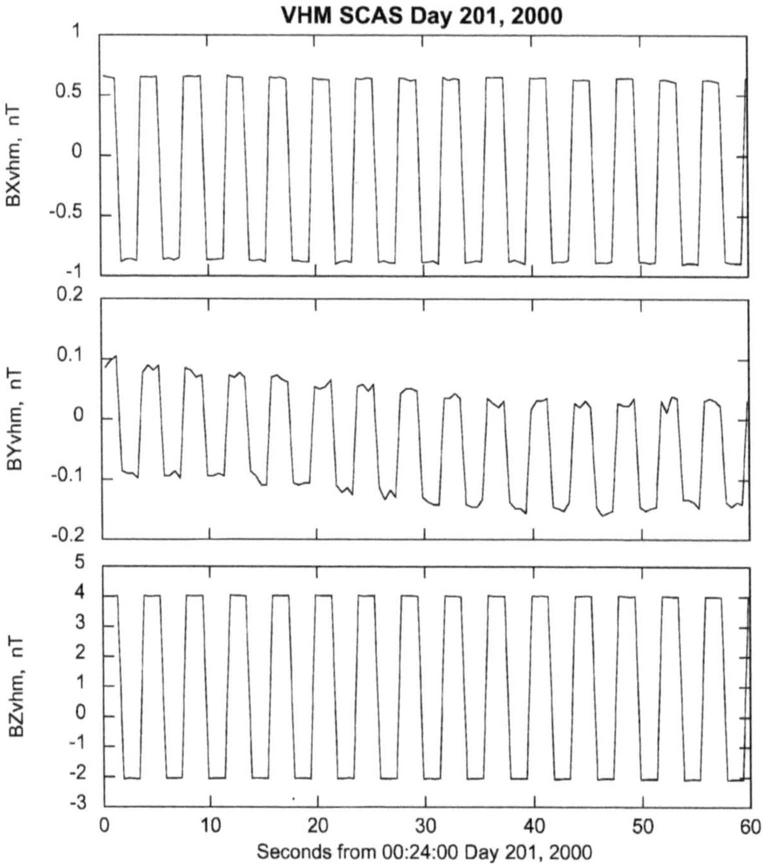

Figure 12. VHM data taken during the SCAS operation of Day 201, 2000.

SCAS was operated three times after boom deployment near Earth in August 1999 and twice more during the following year. Figure 12 shows an example of VHM data taken when the SCAS X coil was on. The data are in VHM sensor coordinates (S/C $+X$ = VHM $-Z$ and S/C $+Y$ = VHM + X). The time resolution is 8 VHM vectors/second.

The SCAS signals are fairly easily estimated, since their peak to peak values are not affected by the sensors' zero levels, and the operation interval includes a large enough number of 4-sec cycles so that the effects due to ambient fields are negligible. The directions of the SCAS fields at both sensors are consistent within 0.1° for all the operation intervals analysed, so the orientation of the sensors has not changed more than this amount over the 1 year from August 1999 through July 2000. Comparison with the predicted SCAS fields at the sensors indicate that the FGM is within 0.3° of alignment with the spacecraft axes and the VHM is within 1°.

Having two independent measurements of the local field at any one time (FGM/VHM or FGM/SHM) provides possibilities for cross-calibration between sensors. In particular, joint operation of the SHM and FGM (scalar/vector mode), for example during Earth flyby, allow the absolute measurements of the scalar sensor to be used to re-calibrate the FGM, see next section. This is done by least squares fitting of the FGM vector measurements to the scalar field measurements, a technique that has been used on previous missions such as MAGSAT (Langel *et al.*, 1981).

Dual sensor measurements also allow estimates of changes in the spacecraft field to be made. The spacecraft field may vary in-flight due to onboard configuration changes. As the mission progresses we expect to be able to establish a rough correspondence between the spacecraft configuration and spacecraft field which will help in subsequent data analysis. The instrument checkout period (see Section 8.3) provided an opportunity for this since the magnetometer stayed on as each instrument performed its checkout operations. Rotation of the spacecraft (around one axis followed by another) allows an estimate of the offsets (sensor plus spacecraft field) to be made. This is because the detected field component due to the ambient external field will be seen to rotate and that due to sensor offsets and spacecraft field will remain static (in the sensor frame of reference). Using thrusters, the spacecraft is capable of rolling at approximately 0.4°/sec. We plan periodic calibration sessions (of 30 rolls about the Z axis followed by the same about X) during science cruise (once per month) and tour (at each apoapsis). While in the solar wind, offsets (sensor and spacecraft) can also be estimated using a variance technique developed by Davis and Smith (1968), Belcher (1973) and Hedgecock (1975) and used on many previous missions. This technique is based on the fact that interplanetary field fluctuations tend to be transverse to the average direction of the field and conserve the field magnitude. Solar wind observations are possible during science cruise and during those parts of the tour when the orbit leaves the Saturn magnetosphere.

4.3. Calibration from Earth Swingby

Just prior to Earth swingby, the magnetometer boom was deployed. Deployment was not allowed within 0.85 AU of the Sun, since the boom might have lost structural integrity if too hot. It could also not be deployed until after the trajectory correction manoeuvre (TCM) which optimised the spacecraft trajectory for Earth flyby, due to the additional attitude control uncertainties during TCM introduced by the 11 meter boom. Both the instrument and SCAS were operated during, and immediately after, deployment. This allowed the deployment to be monitored, the spacecraft field to be measured as the separation increased between the sensors and the spacecraft and the post-deployment sensor alignment to be assessed. This was the first of three SCAS operations which took place during the flyby period. The second took place during the flyby itself and the third shortly afterwards. The sensor alignments were checked after deployment, as discussed in the previous section. A combination of vector/vector and vector/scalar modes were used for these activities.

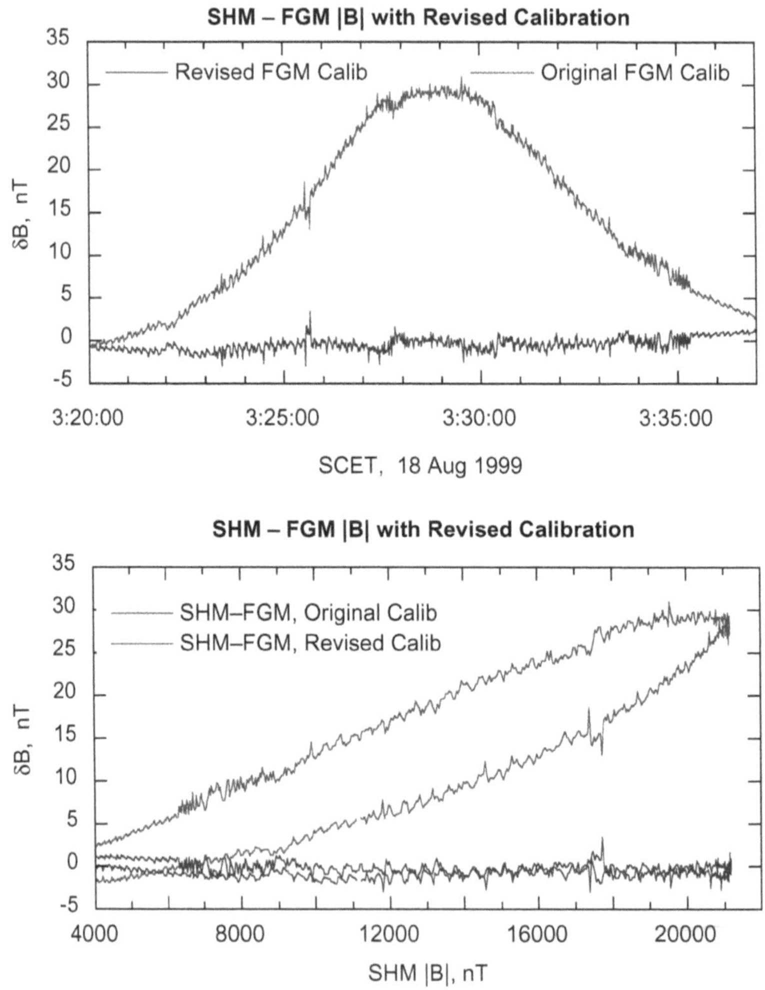

Figure 13. SHM – FGM (|B|) differences with original and revised calibrations plotted as a function of time (*top*) and $|B|$ (*bottom*).

The trajectory during Earth swingby can be seen in Section 9.2, Figure 19. The high resolution data taken during the flyby then provided an important calibration opportunity for comparing the SHM measurements with the well known Earth field (Smith *et al.*, 2002) and also with the magnitudes derived from the FGM vector measurements. The difference between SHM and FGM |B| reached 30 nT at the peak of 21,200 nT, when the FGM sensor was in its highest range (R3). Adjusting the elements in the FGM R3 calibration matrix that depends upon the non-orthogonality of the FGM sensor's axes reduced the difference to less than 2 nT over the 17-min interval during which |B| varied from 4,000 to 21,200 nT (Figure 13). Using the adjusted calibration matrix also resulted in a significant

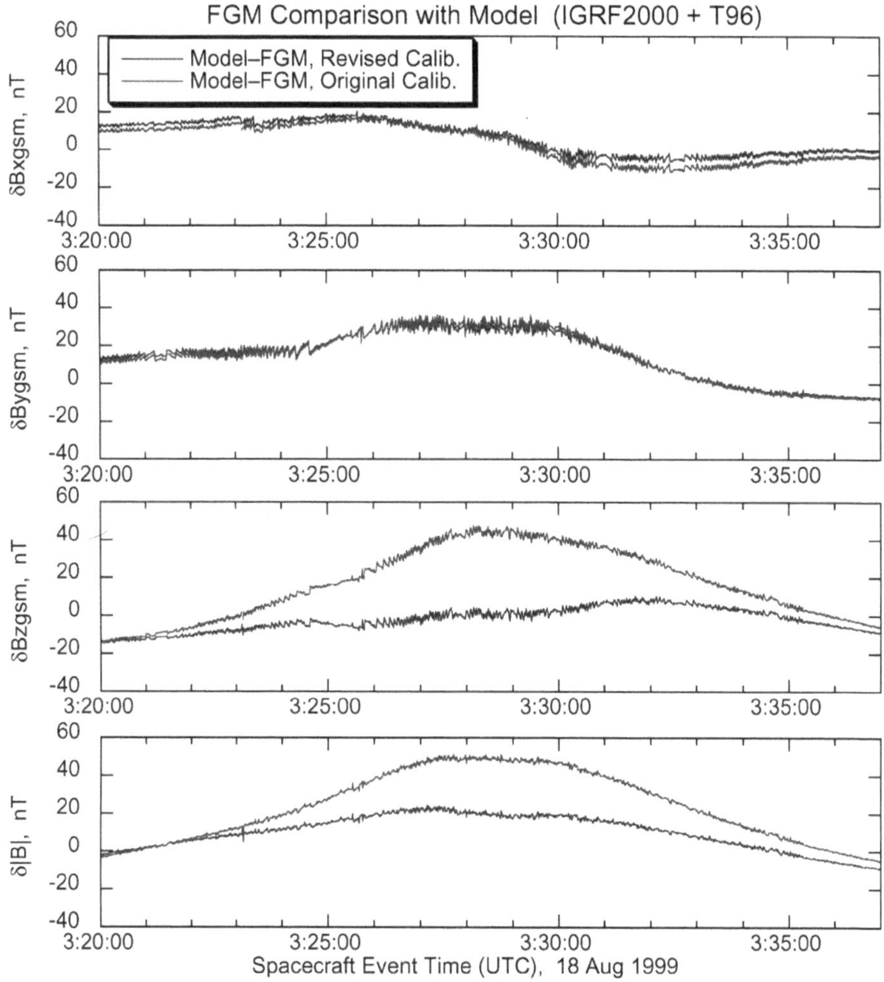

Figure 14. Differences of FGM components and magnitude from field model (IGRF2000 + T96), with revised and original FGM calibrations.

decrease of the residuals of FGM vector components and magnitude from the Earth field model (Figure 14). This adjustment is a good example showing how the SHM data can be used to improve the FGM calibration in-flight.

The method used to calculate the adjustments to the calibration matrix was to minimize $(|Af_i| - s_i)^2$ with respect to the elements a_{21}, a_{31} and a_{32} of the matrix A. The diagonal elements of A are set equal to the X, Y, Z ground calibration sensitivities and the upper triangular elements are zero. Each SHM measurement is indicated by s_i and f_i denotes the FGM average vector corresponding to the interval of the SHM measurement. Values of a_{21}, a_{31}, and a_{32} from ground calibration were −0.00005, 0.00135, and −0.0061, respectively; adjusted values are 0.0022, −0.00115, and −0.0053. The adjustments represent changes in the *XY* orthogonality angle of 0.13°, the *XZ* angle of 0.14°, and the YZ angle of 0.05°.

The SHM mode was exercised between 4.0 and $1.17R_E$ both inbound and outbound. The measurements were compared with the Earth's internal field using the International Geomagnetic Reference Field (IGRF 2000). The differences, representing the disturbance field caused by external currents during an on-going magnetic storm, were compared with two models, the Tsyganenko 96 model (available on the NSSDC web site) and the model derived from observations by the POLAR spacecraft (Tsyganenko *et al.*, 1999). Better agreement was obtained with the POLAR model although small adjustments to the model parameters were necessary in order to arrive at a closer agreement (within a few nT). Since the Disturbance field contributes mainly to the low degree spherical harmonics describing the internal field, a theorem was developed that allowed a comparison between the higher degree moments of the model and the SHM measurements after removing the low degree terms and the Disturbance field. A close correspondence was then found demonstrating the absolute accuracy of the SHM measurements at the very high level that was anticipated.

5. Spacecraft Magnetic Cleanliness

Due to the complex nature of the Cassini spacecraft the potential for magnetic contamination of the FGM and V/SHM magnetic field data were quite high. Since the goal of achieving a maximum allowable spacecraft magnetic field of 0.2 nT at the V/SHM location would be compromised by this, a magnetic cleanliness program was initiated by the Cassini project team to ensure that the goal would be met. To oversee this program a Magnetics Control Review Board (MCRB) was established, consisting of magnetometer team members as well as members of the Electromagnetics Compatibility Group at JPL. The actions and results which arose from the MCRB are described in detail in Narvaez (2003).

A model of the spacecraft field was developed by the MCRB. A semilogplot of the predicted spacecraft field along the magnetometer boom (Figure 15) provides an indication of the inverse cube fall of the spacecraft field and shows the expected *Y* field at the position of the two sensors after boom deployment.

A comparison of the VHM and FGM predicted spacecraft fields, following the Huygens probe deplyment, are made in the polar plot in Figure 16. The inverse cube fall-off relationship between the spacecraft fields at the FGM and V/SHM locations are shown. Note also the nearly dipolar shape of both fields at the two locations. Magnetic field measurements will be taken during the deployment of the Huygens probe in December 2004. This will allow the changing spacecraft field from before to after probe deployment to be measured and comparisons to be made with predictions.

The Cassini magnetics control program has been very successful. Each Cassini spacecraft subsystem has made the significant contributions which were necessary to achieve the goals set out by the MCRB. The resulting efforts have paid off for the

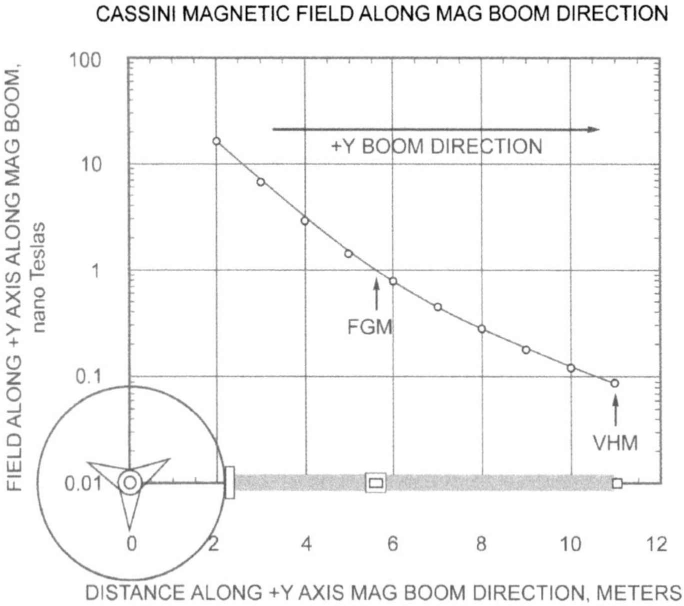

Figure 15. Cassini magnetic field along MAG boom direction.

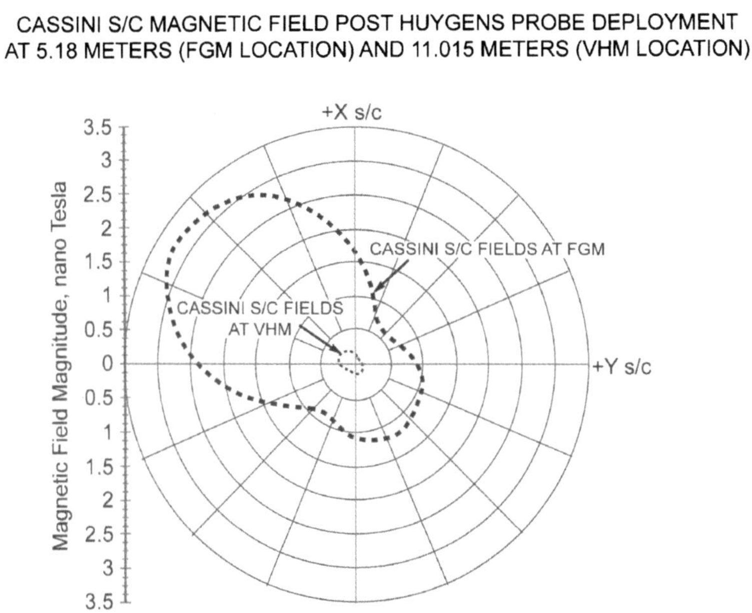

Figure 16. Cassini spacecraft field after Huygens deployment.

science aims of the mission, since if no magnetic cleanliness had been instituted for the Cassini program, the spacecraft fields would have been quite high and variable. The science goals of the magnetometer and other fields and particles instruments would have suffered (see for example the discussion in Section 2.1.1 on the interior of Saturn).

6. Instrument Modes and Control

Due to both the large amount of cross-strapped instrument hardware redundancy and the extensive software functionality built into the DPU, numerous operating states can be defined for the instrument. However, a small set of operational modes have been defined which represent the instrument configuration for normal operations: Off, Sleep, Maintenance, Vector/Vector and Vector/Scalar. Figure 17 shows the allowed mode transitions. The modes are further described below:

Off mode: In this mode the DPU and instruments are unpowered with the exception of the heater control electronics and sensor heaters. Power is required for the heaters throughout the mission, irrespective of the powered state of the instrument itself. Thus, whenever the instrument is not operating it will be in this mode. No telemetry data are generated by the instrument, and sensor temperatures are available using thermistor data which is processed by the spacecraft. If both heaters were fully powered, the average power consumption would be 3.42 W.

Sleep mode: This mode was requested for each instrument by the Cassini project as a minimum power, minimum data rate state that the instrument can be commanded into by single command. Increases in instrument power/data rate are not

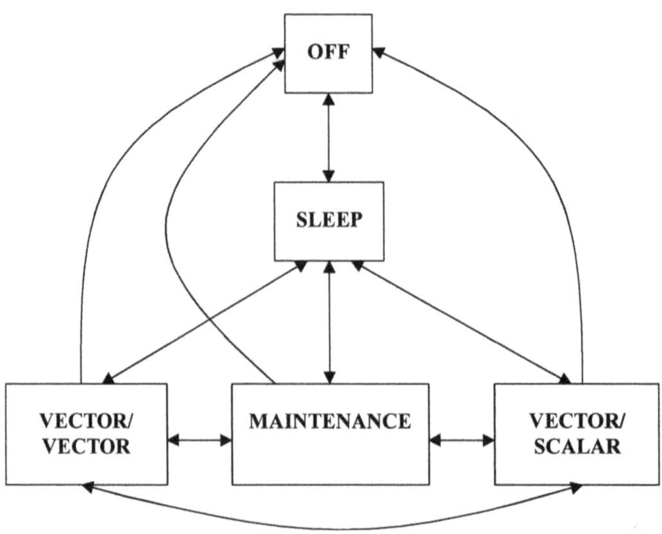

Figure 17. Allowed instrument operational mode transitions.

allowed until the instrument is first set active again and commands that could do so must be rejected by the instrument. Spacecraft operational modes have been defined where a particular subset of instruments have priority, the others being in sleep mode. Transition from one spacecraft mode to another can be achieved very quickly and easily since all instruments are still powered. The DPU is powered and running and will block commands to activate other parts of the instrument. Only housekeeping packets are generated; they will normally be collected by the CDS at a rate of one packet per 64 s. The average power consumption is 7.50 W (given in Table IV) and includes heater power.

Maintenance mode: During long cruise intervals with the instruments off periodic operation of the V/SHM sensor is needed to maintain the health of the helium cell. In this mode both the DPU and VHM are powered. The DPU prepares science and housekeeping packets but only the housekeeping packets are picked up by the CDS. The packet collection rate is 1 packet per 128 s. Average power consumption is 11.31 W (including heaters).

Vector/Vector mode: This mode is used for magnetic field observations in low ambient fields, <256 nT. Both FGM and VHM are operating, providing vector measurements of the field. This permits intercalibration of the sensors which is useful in pinpointing possible changes in the spacecraft field. In normal operations, science packets are collected every 4 sec, housekeeping packets every 64 sec. The average power consumption, including heaters, is 11.31 W.

Vector/Scalar mode: Magnetic field observations in high ambient fields, > 256 nT, will be made in this mode. Here the FGM and SHM are operating together. Packet rates are normally the same as those for vector/vector mode, VHM vectors being replaced by SHM scalar values. Average power consumption (including heaters) is 12.63 W.

Most of the commandable functions of the instrument are referred to either directly or indirectly in other sections of this paper. A summary of these functions is provided in Table VII with an indication of their planned usage for the mission shown.

7. Onboard Data Processing

The DPU collects sensor science and housekeeping data from the ADC. The FGM science data collection rate is commandable from 16 to 500 vectors/sec, VHM data are collected at a fixed ratio of one sixteenth of the FGM rate. The normal collection rate is 64 FGM vectors/sec. An SHM scalar word is collected once per second. In most operational modes, sensor data are collected at a higher rate than they can be downlinked in telemetry, or may be collected when no telemetry downlink is available. The DPU is able to average the collected data until downlink is available. Two averaging schemes have been implemented, a simple fixed window average where consecutive sets of data are averaged down to a single value, and a running average with a commandable averaging window size. These schemes adapt

TABLE VII
The main commandable functions of the instrument

FUNCTION	USAGE
Heater electronics on/off	Always On
Power supply and processor selection	Redundant units only used after failure of primary ones
BIU control	Not needed for normal operations
Sleep/active state selection	Spacecraft operations mode control
Latch up detector and watchdog control	Not needed for normal operations
PROM/RAM software select	Not needed for normal operations
Software control (reset, registers, tables, parameters etc.)	Not needed for normal operations
Science data sampling rate and averaging selection	Science observation dependant
Telemetry packet content selection	Fixed for normal operations
Snapshot data control	Science observation dependant
ADC control/test	Not needed for normal operations
FGM, VHM and SHM control	Science observation dependant
Initiate Macro commands	Used for observations during tour
Memory load/dump/test	Not needed for normal operations

automatically to the data rates applied by the spacecraft. It is also possible to store and downlink sets of unaveraged data (i.e., sets of data sampled at the best achievable time resolution). Such data are called snapshot data and would be used to sample discrete magnetospheric events. Snapshot data collection can be triggered by command or by built-in trigger algorithms. Several algorithms have been implemented to detect changes in magnitude of the field vector or vector component, changes in field direction, changes in the variance of the field vector or vector component. Up to four algorithms with individually assigned priorities for telemetry readout can be run in parallel and algorithm control parameters are commandable. The snapshot data are stored in one of two areas; in one half of the Hi-Rel 128KB RAM for high reliability but short coverage data and in the commercial 16MB RAM for extended coverage. At a sampling rate of 64 vectors/sec, up to 170 sec worth of data can be stored in the Hi-Rel RAM or 12 h of data in the commercial RAM. Up to 33 snapshot events can be stored at any one time.

Cassini uses the international packet telemetry standard. The DPU generates two types of telemetry packets, housekeeping and science. These packets are of fixed length, 192 bytes housekeeping and 988 bytes science. The generation rate of the packets is modified according to the spacecraft telemetry mode. For normal operations, a science packet is generated every 4 sec and a housekeeping packet every 64 sec, leading to a downlink data rate of 2000 bits/sec. Science packets also contain a housekeeping data set but the bulk of the packet data area is set aside for FGM, VHM and SHM science data. Again by command, the science data area can be used for special data requests; whereby the science data are replaced by memory readout, ADC noise test or ADC turbo test data. A portion of the packet data area is set aside for the downlink of subcommutated snapshot data and snapshotting status, such as selected algorithms, trigger status and statistics of events occupying the snapshot memory.

The amount of housekeeping data exceeds the size of a housekeeping packet so some data are subcommutated i.e., the full data set is split up and a portion of it put in each packet. The housekeeping packet contains analogue and digital housekeeping data plus the configuration image (the set of all parameters used by the onboard software) and the set of error counters (providing visibility of all tested fault conditions and their frequency). Analogue housekeeping is subcommutated over four packets, the configuration image over 32 packets and the error counters over 64 packets. A portion of the packet, called the maintenance data area, has been set aside for special data requests. By command, the user can select the contents of this area from the following data: configuration image, error counters, high resolution analogue housekeeping and snapshot. Selected data sets which are normally subcommutated can be obtained much more quickly in this way. Optionally, a small set of highly averaged science data, ADC noise test data, ADC turbo test data or memory readout data can be selected by command.

8. Launch and Early Cruise Operations

8.1. Introduction

Launched on October 15th 1997, the spacecraft will travel 7 years before Saturn Orbit Insertion (SOI) in July 2004 – this period is called the cruise phase. After arrival it will spend at least 4 years orbiting Saturn and Titan performing science operations – the tour phase. Its trajectory during cruise has included two swingbys around Venus, one around Earth and one past Jupiter. In order to reduce operational costs, spacecraft activities during the cruise phase have been held to the minimum required to maintain the trajectory and the health of the spacecraft and instruments, although some unique science data have also been obtained, see Section 9. Two years before SOI, instrument operations have been increased to include calibration and characterisation in preparation for the tour. Implicit in this concept is a low spacecraft operations team staffing level for the majority of the cruise phase, limiting the possibilities for major science operations. Design of the tour phase operations and the development of key tour phase elements have been deferred to the cruise phase. The magnetometer operations team has followed the same approach. An instrument ground segment has been established which will support the essentially engineering nature of instrument cruise operations and the development of a full science planning and data processing system is being pursued during cruise.

The general mission concept is to avoid real-time commanding where possible and to use the CDS command sequencing capability. However, a near-real-time command capability has been provided which allows commands to be prepared and sent at relatively short notice – from a week to a day in advance.

8.2. Cruise Operations Concept

All instrument operations for the magnetometer are based at Imperial in line with the distributed operations concept employed by the Cassini project. The operations team is considered a part of the overall spacecraft operations team and is fully responsible for operations. In addition, the team has been assigned responsibility for operations of SCAS. Mission planning is achieved through 'virtual teams' composed of members of the instrument and spacecraft operations teams. The cruise phase is subdivided into approximately 6 month planning periods which are addressed by the Mission Planning Virtual Team (MPVT). Each period is further subdivided into approximately 60 day sequences. As the MPVT completes the planning for one period, these data are handed over to two Sequence Virtual Teams (SVTs) who produce the detailed commands for each sequence. The team provides instrument planning input to the MPVT, reviews planning products, and provides instrument commands to the SVTs.

The main tool for operations support is the Science Operations and Planning Computer (SOPC), provided by the Cassini project and located at Imperial. This workstation is connected via a router and dedicated voice and data lines to the spacecraft monitor and control system at JPL. It is supplied with a suite of software tools to support operations, the same tool as those used by the spacecraft operations team at JPL. The SOPC provides access to instrument and spacecraft telemetry both real-time and that stored in the telemetry database at JPL. Other tools allow the transfer of files containing planning information from the Cassini central database (CDB), allow the generation and transfer of command files to the CDB and give visibility into the detailed spacecraft activity plan and the command sequences prepared for uplink.

Figure 18 is a block diagram of the operations segment at Imperial. Spacecraft and instrument housekeeping and instrument science data are transferred from JPL to the SOPC. The housekeeping data are displayed and stored on the SOPC for instrument health and safety monitoring. The instrument housekeeping and science data are transferred to and stored on the instrument Ground Support Equipment (GSE) for science quicklook data and additional instrument health checks. Instrument command sequences are generated on the PCs and are tested by transfer to the GSE and execution on the Engineering Model (EM) instrument. The error free sequences are then delivered to JPL via the SOPC. Files generated at JPL during the uplink preparation process are checked on the operational PCs using software developed to filter these files for instrument-related information. If required, instrument flight software patches will be produced by the DPU developers, TUB. The patch

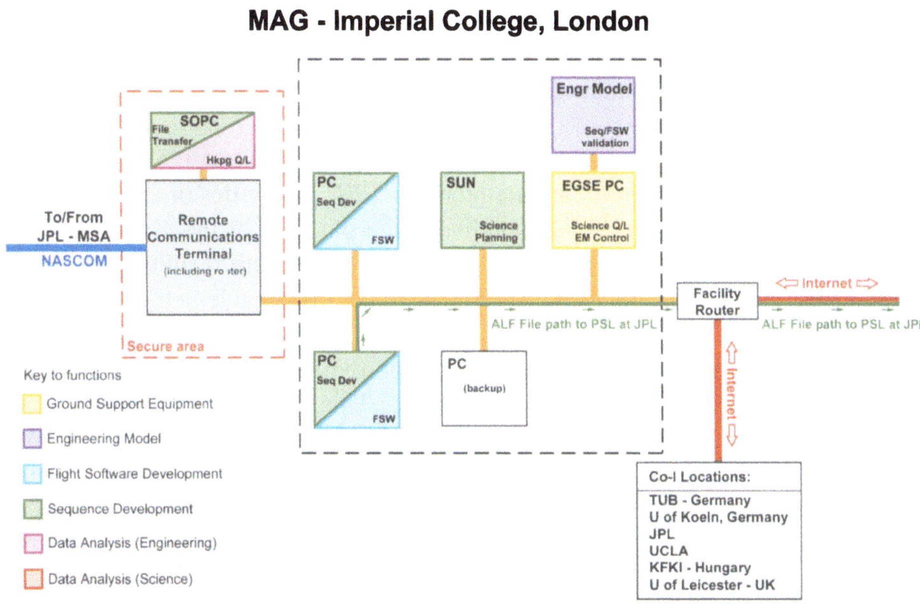

Figure 18. Operations segment at Imperial College.

files will be transferred to the PCs where they will be converted into the load format needed by JPL and then delivered to the JPL software library. The science planning workstation is used to generate predictions of expected magnetic field strengths and directions as a function of time. These predictions will drive the spacecraft attitude and instrument modes requested during the mission planning process.

Reliability in an operational system is essential. To guard against failures, the operations PCs are triple redundant and the software on the science planning workstation is duplicated on the SOPC. Duplication has also been built into the network interfacing the operational machines.

8.3. Activities

Instrument operations during cruise are limited to the following: heater activation, sensor temperature monitoring, periodic VHM maintenance, instrument and SCAS checkout, boom deployment, sensor alignment verification during Earth flyby, in-flight calibration. These activities are briefly described below. In addition, quasi-continuous operation usually at low data rate is allowed for cruise science when not impacting other higher priority spacecraft activities. This has included operation during the spacecraft flyby of Jupiter.

Safe non-operating and operating temperature ranges have been defined for both sensors. The sensor temperatures are being monitored regularly since launch to detect trends outside the safe limits. As described in the modes section, heaters have been provided to help maintain the temperatures within limits. The heaters were activated during the launch sequence, at 75 min after launch. Their function is being monitored by comparing the sensor temperatures with the current drawn by the control electronics.

The VHM sensor must be operated occasionally to maintain the performance of the gas cell. Periodic instrument maintenance (PIM) activities of 2.5 h duration take place approximately every 3 months for the duration of the cruise phase if the VHM is not already powered for cruise science. The instrument is operated in maintenance mode. Only real-time, low rate, housekeeping telemetry is available during PIM.

About 14 months after launch, a period of approximately 25 days was set aside for a comprehensive checkout of each instrument. This was the first opportunity for proper instrument operation since launch. A period of 24 h was allocated for the magnetometer instrument checkout during which all instrument modes and commands were exercised. The magnetometer was then powered for the remainder of the checkout period in order to observe the magnetic signatures of the other instruments (although at this stage the boom was not yet deployed). Checkout of SCAS was also included. The importance of instrument operations during Earth flyby for alignment calibration has been discussed above. In-flight calibration operations will start 2 years before SOI. The activities to be performed have been described earlier in the calibration section.

Boom deployment and instrument calibration and science operations during Earth swingby are discussed elsewhere. More intensive in-flight calibration operations will start 2 years before SOI. The activities to be performed have been described earlier in the calibration section.

9. Cruise Science

Cassini was launched on October 15, 1997 at 08:43 GMT. The instrument heaters were successfully activated at 10:08 GMT and within a day the sensor temperatures had stabilised. Due to the long interplanetary cruise trajectory which the spacecraft had to follow in order to make use of numerous planetary gravity assists to place Cassini into Saturn orbit there have been a number of opportunities for cruise science. These consist of interplanetary cruise science, Jupiter flyby science and Earth swingby science and calibration. The analysis of the cruise science data is in its infancy since the team has been concentrating on science planning for the 4-year orbital Cassini tour.

9.1. Interplanetary Cruise Science

Continuous taking of fields and particles data began in February 2000, following High Gain Antenna Acquisition. This is very important scientifically since it will allow interplanetary observations to be made between 2.3–9.5 AU around the time of solar maximum with an advanced payload of instruments. This will enable a multi-point study to be made with distance of the heliospheric magnetic field signatures of large-scale solar wind flow structures such as stream interaction regions and coronal mass ejections (CMEs). In particular, the interplanetary signatures of CMEs have not previously been studied with such comprehensive instrumentation at distances beyond the orbit of Jupiter. In this distance range they are thought to become less and less distinct from the ambient solar wind. These observations by Cassini are being carried out in conjunction with near-Earth spacecraft at 1 AU, with Ulysses between 1.3–5.4 AU and with Voyager beyond 75 AU, the first time that such a joint opportunity has arisen. There are also plans to study waves and turbulence in the solar wind at a much higher time resolution than has previously been possible at these distances (up to 32 vectors/sec).

9.2. Earth Swingby

The crucial aspect of this flyby for the instrument was the deployment of the magnetometer boom 4 days before closest approach to the Earth and the acquisition of vector and scalar data in the well known Earth's magnetic field allowing calibration of the instrument. This aspect of the flyby was discussed in some detail in Section 4.3. Figure 19 shows the spacecraft trajectory during the flyby (from Southwood *et al.*, 2002). Science analysis of the magnetic field data from the flyby has resulted in numerous publications clearly highlighting the instrument's performance in detection of magnetic phenomena in a planetary environment. Due to the very high flyby speed ($\sim$16 km/sec) and the high time resolution of the magnetic field observations (32 vectors/sec) significant new information was obtained on a variety of topics including possible evidence of the plasmasphere interchange instability (see Khan *et al.*, 2001; Smith *et al.*, 2001; Southwood *et al.*, 2001; Tsurutani *et al.*, 2001).

9.3. Jupiter Flyby

This unique science opportunity during late 2000, consisted of observations from the Cassini spacecraft while in the upstream solar wind on its approach to Jupiter as well as on the previously unexplored equatorial dusk flank where it made numerous entries into the magnetosheath as well as two brief entries into the flank magnetosphere. In addition during this time the Galileo spacecraft continued its orbital tour of Jupiter and made observations deep inside the inner magnetosphere as Cassini approached in the upstream solar wind and then in turn was out in the solar wind.

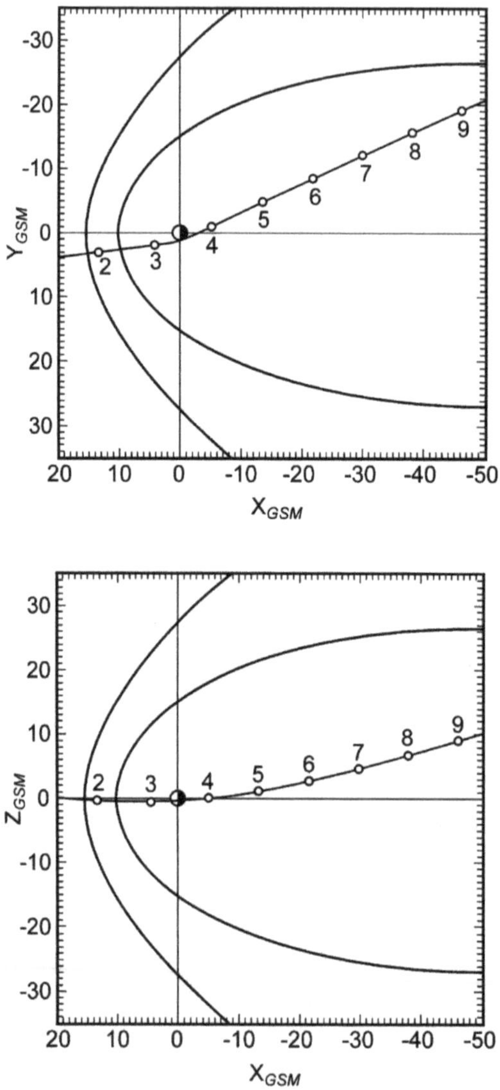

Figure 19. Spacecraft trajectory during the Earth swingby in both the (X, Y) and (X, Z) GSM planes.

A crucial aspect of these studies is that they will serve as a focal point for illuminating comparative studies of the Jovian and Saturnian magnetospheres. This data is largely unexplored to date since there is no science support in place for Cassini. However, a recent issue of the science journal Nature highlighted some of the science findings from this opportunity. Magnetic field data from the magnetometer onboard Cassini contributed to the findings of three of these papers (as described below). The two spacecraft conjunction at Jupiter in combination with remote sensing observations of the planet by the Earth orbiting Hubble Space Telescope and Chandra Telescope have allowed the effects of the solar wind on changes within the magnetosphere to be monitored. The magnetosphere of Jupiter and its hectometric radio emissions and aurorae are seen to respond to interplanetary shock waves within the solar wind (Gurnett *et al.*, 2002). The two spacecraft, separated by some 9 million kilometres, encountered the boundary of the magnetosphere while it was contracting in response to an increase in solar wind pressure (Kurth *et al.*, 2002). The presence of both spacecraft proved critical in enabling scientists to prove that the magnetosphere is indeed controlled by the solar wind. In addition a pulsating X-ray hotspot was observed near Jupiter's north pole (Gladstone *et al.*, 2002). This hotspot pulsates every 45 minutes, a periodicity previously observed near Jupiter in radio and energetic electron bursts. It does not appear to be influenced by any changes in the solar wind but by some as yet unknown factor inside the magnetosphere.

The three main topics which the magnetometer data set will be used to address are:

Solar-wind Jupiter magnetospheric interaction

We know that the dynamics of the rotationally driven Jovian magnetosphere are very different from that of the Earth's solar wind driven one, where at Jupiter the fast rotation imposed by the planet is communicated throughout the magnetosphere by the magnetic field. The important question is how do changes in the solar wind effect the internal magnetic activity within the magnetosphere?

Dynamics of the outer Jovian environment

How do the outer boundaries of the Jovian environment (the bow shock and magnetopause), respond to changes in the interplanetary medium and how quickly can these changes occur? How are these changes then fed into the outer Jovian environment? Two point measurements of these boundaries made by Cassini and Galileo during this period will allow some resolution of the dynamics of the system as well as the shape of the boundaries which is thought to be much less rigid than originally modelled. Some initial analysis of numerous occasions of mirror mode activity within the Cassini magnetosheath (André *et al.*, 2002) shows that the Cassini Jupiter flyby allows tracking of the mirror instability along flow lines down to the distant jovian magnetosheath, as yet unexplored by previous missions. The most distant observations of mirror mode activity were measured hundreds of radii downstream in the post-dusk magnetosheath where either isolated or very long sequence of magnetic bi-stated dropouts appeared.

Solar wind structure

Dual spacecraft measurements have not been made before of the interplanetary medium in the vicinity of Jupiter. Observations of similar but evolving solar wind structures at the two spacecraft have been observed separated by radial distance of some 750 R_J. Understanding the similarities and differences between such structures will allow us to gain a better understanding of the evolution of structure within the solar wind at 5 AU.

Acknowledgements

The instrument is the result of the work and co-operation of a large multi-national team. We would like to thank members of the team at Ultra Electronics, UK where the FGM was put together and the following team members where their contributions are as listed (*denotes this person was formerly at the relevant institute):

	Institute	Responsibility
M. Barlow	*Imperial	Former Instrument Manager
T. Beek	Imperial	Power System Design/Fabrication, Subchassis No. 1 Assembly
M. W. Dunlop	*Imperial	Science Planning and Science Operations
M. Malhotra	*Imperial	Design, Development and testing
D. Shaw	*Imperial	Ground Operations Software (SW)
P. Slootweg	Imperial	Operations Engineer
R. White	*Imperial	Operations Engineer, Ground Operations SW
M. Fong	JPL	V/SHM Engineer
R. Marquedant	JPL	V/SHM Principal Engineer
L. Wigglesworth	JPL	V/SHM Engineer
J. Wolf	JPL	Software Engineer
D. E. Huddleston	*UCLA	Calibration Data analysis
R. C. Snare	UCLA	Instrument Test and Spacecraft Integration Support
S. Szalai	KFKI	Ground Support Equipment (EGSE)
G. Musmann	TUB	Data Processing Unit (DPU) Manager, FGM Calibration, Magnetic Cleanliness
H. Hartje	TUB	Instrument Calibration
M. Rahm	TUB	Instrument Calibration
I. Richter	TUB	Instrument Calibration

For the tremendous effort made to ensure a magnetically clean spacecraft we would like to thank P. Narvaez of JPL and K. Mehlem of ESA as well as the Cassini project management who supported this effort. We had a number of engineers assigned at JPL to help with our interface to the project team there and we would like to express our thanks to the most recent of these B. J. Cheney and C. Elliot for their dedication and assistance. We would like to thank J. Feynman at JPL for her dedication and efforts during the years before launch as Instrument Scientist and M. Burton from JPL for her important and very supportive role as Discipline Scientist for the instrument following launch. J. Wolf and M. Burton also prepared figures included in this paper.

The contributions to this investigation have been supported by PPARC in the U.K., DARA in Germany and by NASA in the U.S. The work by KHG is finacially supported by the German Ministerium fuer Bildung und Wissenschaft and the Germann Zentrum fuer Luft- und Raumfahrt under contract 50OH99014.

References

Acuna, M. H., Connerney, J. E. P., and Ness, N. F.: 1983, *J. Geophys. Res.* **88**, 8771.
André, N., Erdos, G., and Dougherty, M. K.: 2002, *Geophys. Res. Letters*, in press.
Balogh, A., *et al.*: 1992, *Astron. Astrophys. Suppl. Ser.* **92**, 221.
Behannon, K. W., Connerney, J. E. P., and Ness, N. F.: 1981, *Nature* **292**, 753.
Belcher, J. W.: 1973, *J. Geophys. Res.* **78**, 6480.
Ben-Jaffel, L., Leers, V., and Sandel, B.: 1995, *Science* **269**, 951.
Bhardwaj, A. and Gladstone, G. R.: 1982, *Rev. Geophys.* **38**, 295.
Blanc, M., *et al.*: 2002, *Space Sci. Rev.*, in print.
Bunce, E. J. and Cowley, S. W. H.: 2002, *Ann. Geophysica*, submitted.
Carbary, J. F. and Krimigis, S. M.: 1982, *Geophys. Res. Lett.* **9**, 1073.
Caudal, G: 1986, *J. Geophys. Res.* **91**, 4201.
Connerney, J. E. P., Ness, N. F., and Acuna, M. H.: 1982, *Nature* **298**, 44.
Connerney, J. E. P., Acuna, M. H., and Ness, N. F.: 1983, *J. Geophys. Res.* **88**, 8779.
Connerney, J. E. P.: 1993, *J. Geophys. Res.* **98**, 18659.
Cramm, R., Glassmeier, K.-H., Stellmacher, M., and Othmer, C.: 1998, *J. Geophys. Res.* **103**, 11951.
Davis, L. Jr. and Smith, E. J.: 1968, *Trans. AGU* **49**, 257.
Davis, L. and Smith, E. J.: 1986, *J. Geophys. Res.* **91**, 1373.
Davis, L. and Smith, E. J.: 1990, *J. Geophys. Res.* **95**, 15257.
Dougherty, M. K., Southwood, D. J., Balogh, A., and Smith, E. J.: 1993, *Planet. Space Sci.* **41**, 291.
Dougherty, M. K. and Southwood, D. J.: 1993, *Adv. Space Res.* **13**, 10305.
Dougherty, M. K., Southwood, D. J., and Lachin, A.: 1997, *Adv. Space Res.* **20**, 215.
Dougherty, M. K., Dunlop, M. W., Prange, R., and Rego, D.: 1998, *Planet. Space Sci.* **46**, 531.
Elphic, R. C. and Russell, C. T.: 1978, *Geophys. Res. Lett.* **5**, 211.
Espinosa, S. A. and Dougherty, M. K.: 2000, *Geophys. Res. Lett.* **27**, 2785
Eviatar, A., Siscoe, G. L., Scudder, J. D., Sittler, E. C., and Sullivan, J. D.: 1982, *J. Geophys. Res.* **87**, 8091.
Giampieri, G. and Dougherty, M. K.: 2002, *Geophys. Res. Lett.*, submitted.
Gladstone, G. R., Waite, J. H., Grodent, Jr, D., Lewis, W. S., Crary, F. J., Elsner, R. S., Weisskopf, M. C., Majeed, T., Jahn, J.-M., Bhardwaj, A., Clarke, J. T., Young, D. T., Dougherty, M. K., Espinosa, S. A., and Cravens, T. E.: 2002, *Nature* **415**, 1000.
Goertz, C.: 1983, *Geophys. Res. Lett.* **10**, 455.
Grun, E., Goertz, C. K., Morfill, G. E., and Havnes, O.: 1992, *Icarus* **99**, 191.
Gurnett, D. A., Kurth, W. S., Hospodarsky, G. B., Persoon, A. M., Zarka, P., Lecacheux, A., Bolton, S. J., Desch, M. D., Farrell, W. M., Kaiser, M. L., Ladreiter, H.-P., Rucker, H. O., Galopeau, P., Louarn, P., Young, D. T., Pryor, W. R., and Dougherty, M. K.: 2002, *Nature* **415**, 985.
Haynes, P. L., Balogh, A., Dougherty, M. K., Southwood, D. J., Fazakerley, A., and Smith, E. J.: 1994, *Geophys.Res. Lett.* **21**, 405.
Hedgecock, P.C: 1975, *Space Sci. Instr.* **1**, 83.
Huddleston, D.E., Strangeway, R. J., Warnecke, J., Russell, C. T., Kivelson, M. G., and Bagenal, F.: 1997, *Geophys. Res. Lett.* **24**, 2143.
Huddleston, D. E., Strangeway, R. J., Warnecke, J., Russell, C. T., and Kivelson, M. G.: 1998, *J. Geophys. Res.* **103**, 19877.
Huddleston, D. E., Strangeway, R. J., Blanco-Cano, X., Russell, C. T., Kivelson, M. G., and Khurana, K. K.: 1999, *J. Geophys. Res.* **104**, 17479.

Kabin, K., Israelevich, P. L., Ershkovich, A., Neubauer, F. M., Gombosi, T. I., de Zeeuw, D. L., and Powell, K. G.: 2000, *J. Geophys. Res.* **105**, 10761.

Kaiser, M. L., Desch, M. D., Kurth, W. S., Lecacheux, A., Genova, F., Pedersen, B. M., and Evans, D. R.: 1984, *Saturn*, T. Gehrels and M. S. Matthews, eds. (Univ. Arizona Press), 378.

Keller, C. N., Cravens, T. E., and Jan, L.: 1994, *J. Geophys. Res.* **99**, 6511.

Kellock, S., Austin, P., Balogh, A., Gerlach, B., Marquedant, R., Musmann, G., Smith, E., Southwood, D., and Szalai, S.: 1996, *Proc. SPIE*, Denver, Colorado, **2803**, 141.

Khan H., Cowley, S. W. H., Kolesnikova, E., Lester, M., Brittnacher, M. J., Hughes, T. J., Kurth, W. S., McComas, D. J., Newitt, L., Owen, C. J., Reeves, G. D., Singer, H. J., Smith, C. W., Southwood, D. J., and Watermann, J. F.: 2001, *J. Geophys. Res.* **106**, 30141.

Khurana, K. K: 1997, *J. Geophys, Res.* **102**, 11295.

Kivelson, M. G., Khurana, K. K., Walker, R. J., Russell, C. T., Linker, J. A., Southwood, D. J., and Polanskey, C.: 1996a, *Science*, **273**, 337.

Kivelson, M. G., Khurana, K. K., Walker, R. J., Warnecke, J., Russell, C. T., Linker, J. A., Southwood, D. J., and Polanskey, C.: 1996b, *Science* **274**, 396.

Kivelson, M. G., Khurana, K. K., Walker, R. J., Warnecke, J., Russell, C. T., Coroniti, F. V., Polanskey, C., Southwood, D. J. and Schubert, G.: 1996c, *Nature*, **384** 537.

Kivelson, M. G., Khurana, K. K., Coroniti, F. V., Joy, S., Russell, C. T., Walker, R. J., Warnecke, J., Bennett, L., and Polanskey, C.: 1997, *Geophys. Res. Lett* **24**, 2155.

Kivelson, M. G., Khurana, K. K., Russell, C. T., Volwerk, M., Walker, R. J., and Zimmer, C.: 2000, *Science*, **289**, 1340.

Kivelson, M. G., Bagenal, F., Neubauer, F. M., Paranicas, C., and Saur, J.: 2002,*Jupiter*, F. Bagenal, ed., (Cambridge Univ. Press), in press.

Kurth, W. S., Gurnett, D. A., Hospodarsky, G. B., Farrell, W. M., Roux, A., Dougherty, M. K., Joy, S. P., Kivelson, M. G., Walker, R. J., Crary, F. J., and Alexander, C. J.: 2002, *Nature* **415**, 991.

Langel, R., Berbert, J., Jennings, T., Horner, R.: 1981, NASA/GSFC TM 82160.

Leamon, R. L., Dougherty, M. K., Southwood, D. J., and Haynes, P. L.: 1995, *J. Geophys. Res.* **100**, 1829.

Lepping, R. P., Desch, M. D., Klein, L. W., Sittler, E. C., Sullivan, J. D., Kurth W. S., and Behannon, K. W.: 1983, *J. Geophys. Res.* **88**, 8801.

Maurice, S. and Engle, I.: 1995, *J. Geophys. Res.* **100**, 17143.

McNutt, R. L., Jr.:1983, *Adv. Space Res.* **3**, 55.

Narvaez, P; 2003, *Space Sci.Rev.* this volume.

Neubauer, F. M., D. A. Gurnett, J. D. Scudder and R. E. Hartle: 1984, in Saturn, eds T. Gehrels and M. S. Matthews (Univ. Arizona Press), 760.

Neubauer, F. M.: 1999, *J. Geophys. Res.* **104**, 28671.

Richardson, J. D.: 1986, *J. Geophys. Res.* **91**, 1381.

Southwood, D. J., Dougherty, M. K., Leamon, R. J., and Haynes, P. L.: 1995, *Adv. Space Res.* **16**, 4177.

Southwood, D. J., Dougherty, M. K., Balogh, A., Cowley, S. W. H., Smith, E. J., Tsurutani, B. T.,Russell, C. T., Siscoe, G. L., Erdos, G., Glassmeier, K.-H., Gleim, F., and Neubauer, F. M.: 2001, *J. Geophys. Res.*, 30109.

Smith, E. J., Davis, L., Jones, Jr., D. E., Coleman, P. J., Colburn, Jr., D. S., Dyal, P., and Sonnet, C. P.: 1980a, *Science* **207**, 407.

Smith, E. J., Davis, L., Jones, Jr., D. E., Coleman, P. J., Colburn, Jr., D. S., Dyal, P., and Sonnet, C. P.: 1980b, *J. Geophys. Res.* **85**, 5655.

Smith, E. J., Dougherty, M. K., Russell, C. T., and Southwood, D. J.: 2001, *J. Geophys. Res.* 30129.

Stevenson, D. J: 1980, *Science* **208**, 746.

Thompson, P., Dougherty, M. K., Southwood, D. J.: 1995, *Planet. Space Sci.* **43**, 625.

Tsurutani, B. T., Smith, E. J., Anderson, R. R., Ogilvie, K. W., Scudder, J. D., Baker, D. N., and Bame, S. J.: 1982, *J. Geophys. Res.* **87**, 6060.

Tsurutani, B. T., Southwood, D. J., Smith, E. J., and Balogh, A.:1993, *J. Geophys. Res.* **98**, 21203.

Tsurutani, B. T., Southwood, D. J., Dougherty, M. K., Smith, E. J., Burton, M. E., Glassmeier, K.-H., Neubauer, F. M., Arballo, J. K., Galvan, C., Zhou, X.-Y., and Chao, J. K.: 2001, *J. Geophys. Res.*, 30223.

Tsyganenko, N. A: 1989, *Planet. Space Sci.* **37**, 5.

Tsyganenko, N. A. and Peredo, M.: 1994, *J. Geophys. Res.* **99**, 199.

Tsyganenko, N. A., Le, G., Russell, C. T., and Iyemori, T.: 1999, *J. Geophys. Res.* **104**, 10275.

THE MAGNETOSTATIC CLEANLINESS PROGRAM FOR THE CASSINI SPACECRAFT

P. NARVAEZ

Jet Propulsion Laboratory, 4800 Oak Grove Drive, Pasadena, CA, 91109, U.S.A.
(E-mail: pablo.narvaez@jpl.nasa.gov)

(Received 12 May 1998; Accepted in final form 26 November 2002)

Abstract. The Cassini spacecraft, launched in October 1997 and expected to reach Saturn in 2004, carries two magnetometer experiments on a 10-m boom, one at the mid-section of the boom and the other situated at the end of the boom. In order to gather valid scientific magnetic field data and avoid electromagnetic interference, the spacecraft had to comply with stringent magnetostatic cleanliness requirements. This paper describes the results of the Cassini magnetics cleanliness program that achieved the goal of minimizing the magnetic field interference with Cassini's DC magnetic field science instruments.

Keywords: spacecraft magnetic cleanliness, spacecraft magnetic tests

1. Introduction

One of several Cassini science instruments that will be performing measurements of Saturn will be a magnetometer science instrument. The instrument consists of two sensors; the fluxgate magnetometer (FGM) sensor and the vector/scalar helium magnetometer (V/SHM) sensor. Both of these sensors are located along a 11-m long magnetometer boom (Figure 1), with the FGM located 5.814 m from the center of the spacecraft and the V/SHM located at the tip of the boom 11.015 m from the center of the spacecraft. Although a magnetometer boom of such length would seem to provide ample reduction of spacecraft-related magnetic sources that fall off as the inverse-cube of their distances, the complex nature of the spacecraft clearly indicated that the potential for magnetic contamination of the FGM and V/SHM magnetic field data was a distinct possibility. Early in the project phase, several spacecraft engineering subsystems and science instruments were identified as potential magnetic interference sources. The survey revealed the spacecraft's various subsystems would include stepper motors, filter wheel motors, reaction wheel motors, traveling wave tube amplifiers, multiple latch valves, RF isolators/circulators, latch relays, etc. Preliminary calculations showed the permanent magnets contained in these subsystems could collectively impact the magnetometer science experiments. To minimize the interference caused by the magnetic fields yielded by these sources, a magnetics control program was implemented for the Cassini Project.

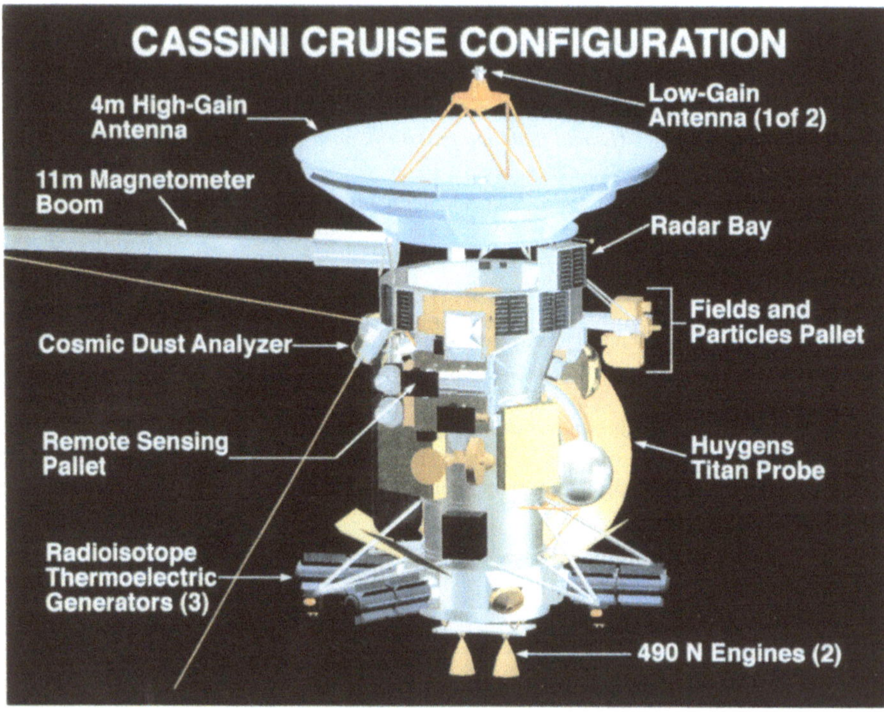

Figure 1. Cassini spacecraft and locations of magnetometers.

2. Magnetics Control Program

Spacecraft magnetics control programs are typically established to ensure that the spacecraft does not cause interference to sensitive science instruments. Since the goal of achieving a maximum allowable spacecraft magnetic field of 0.2 nanotesla (nT) at the V/SHM location was clearly being compromised due to multiple magnetic sources, a magnetic cleanliness program was also established for the Cassini Project to ensure that the spacecraft's magnetic fields from its various sources were controlled to less than 0.2 nT at the V/SHM location. In order to achieve these goals, it was necessary to impose design requirements on all Cassini subsystems hardware and to test and evaluate these subsystems to verify compliance with their respective design requirements.

A magnetics control review board (MCRB) was established to oversee the magnetics cleanliness progress. The board consisted of magnetometer science team members, including the principal investigators and co-investigators, members of the Electromagnetics Compatibility Group at the Jet Propulsion Laboratory and a member of the European Space Agency.

Early in the project, a magnetics cleanliness handbook was written and distributed to all Cassini hardware designers. In this document, subsystem design requirements, practices and guidelines on how to minimize magnetic fields within an instrument design were provided. With a requirement of 0.2 nT at the V/SHM location as a basis, all subsystems and assemblies were given an allocated dipole moment requirement. Engineering subsystems and science instruments were allocated a dipole moment of 25 Gauss-cm^3, which is equivalent to 5 nT when measured at a distance of 1 m. Since most subsystems consisted of several assemblies, each assembly was therefore allocated a dipole moment of 12.5 Gauss-cm^3. Components larger than typical subsystems were provided with higher magnetic dipole allocations. These included Huygens Probe and the propulsion module system (PMS).

3. Early Magnetic Interference Detection

Most magnetic cleanliness problems in past programs were due to subsystems that were already designed before the development of the spacecraft design. These subsystems were typically designed without much regard to magnetic cleanliness due to the absence of magnetic requirements. To avoid such recurrence on the Cassini project, the MCRB's first priority was to sensitize subsystem designers early in the design process on the need to maintain a magnetically clean hardware design. All Cassini subsystem cognizant engineers were informed of their respective magnetics requirements contained in the Magnetics Control Plan Handbook. To ensure that everyone understood their responsibilities within the context of the magnetics cleanliness program, a project-wide magnetics workshop was organized. All spacecraft hardware designers met with members of the magnetics control review team for a full day. At this meeting, participants were made aware of potential magnetic cleanliness problems and the many methods available to solve such problems. Discussed at this workshop were tutorials on when and how to shield, implement magnetic compensation, produce wiring layouts with the goal of minimizing loop areas, contain motor magnetic flux, modify a design by replacing magnetic materials with non-magnetic materials, etc. The designers were encouraged to minimize the use of soft magnetic materials such as invar and kovar, as well as hard magnetic materials such as ferromagnetic components, particularly permanent magnets. A set of guidelines was established that assisted hardware designers in developing the best design which produced the least amount of magnetic fields. The workshop also recommended that all hardware cognizant engineers verify the adequacy of their magnetic control design as early as possible by having the design assessed by knowledgeable MCRB members. The workshop heavily emphasized the importance of assessing problems as early as possible to allow for flexibility in the available solutions.

4. Cassini Subsystem Modifications

Based on the efforts of the MCRB team, many hardware designers adjusted their respective design to provide maximum cancellation of magnetic effects. For example, radio frequency subsystem (RFS) TWTAs were packaged inside their housing in a side by side manner, but with their respective magnetic field polarities in opposite directions, so as to completely cause cancellation of the fields. In another effort, the power/pyro distribution subsystem (PPS) designer arranged all magnetic latch relays in pairs, but with their respective magnetic poles opposite to each other to prove self-cancellation. In the case where there were an odd number of relays, a small compensation magnet was used to neutralize the odd relay. Another example of self-compensation was reduction of fields at the V/SHM from the radioisotope thermoelectric generators (RTGs). Four RTGs were magnetically measured, and a theoretical model was produced which provided an optimum compensation at the V/SHM of only 4 pT with the selected arrangement of clocking angles.

Many instruments contained redesigns of their respective wiring layout with the goal of minimizing loop areas. For example, the PPS shunt radiator's original design generated several hundred nanoTesla at 1 m. The radiator's design was modified, and all loop areas were eliminated by the implementation of backwiring. Since many instruments required scanning or sweeping along a field of view, all employed motors in their design. Hence, to minimize flux leakage from motors, instrument designers enclosed their motors inside high permeability shielding materials such as Metglass or mu-metal. The reaction wheel assemblies (RWA) were also enclosed in double-nested high permeability magnetic shields.

In several cases, subsystems substituted magnetic material with non-magnetic materials. For example, the radio plasma wave subsystem (RPWS) contained antenna mechanisms constructed of mainly non-magnetic materials, substituting magnetic sources with titanium, composites, and plastic gears. The antennas, which were originally made of steel, were fabricated out of non-magnetic beryllium copper. The cosmic dust analyzer (CDA) contained turntable plastic ball-bearings which replaced steel ball-bearings.

In instances where subsystems could not provide self-compensation or change highly magnetic materials to non-magnetic materials, magnetic compensation was employed. The most significant magnetic compensation occurred in the latch valves. The PMS latch valves had the highest dipole moments of any component on the spacecraft (720 Gauss-cm^3). Each latch valve was measured and magnetically compensated with magnets having the same dipole moment, but placed opposite to the moment of each respective latch valve. In order to avoid operational interference, the compensation occurred at a safe distance of 15 cm from each latch valve. Many other subsystems were compensated in a similar manner, including the radio frequency instrument subsystem (RFIS), which had a highly magnetic Ka-Band traveling wave tube amplifier that was compensated. In many cases, demagnetization of the subsystems provided the best means to reduce their respective residual magnetic fields. A typical demagnetization process involved submitting the subsystem to a slowly decaying and alternating magnetic field of 50 Gauss.

TABLE I

Cassini subsystems modified for magnetic cleanliness.

Cassini subsystem	Magnetic reduction method	Resulting magnetic dipole (Gauss-cm^3)
Bus interface unit (BIU)	Current loop reduction	5
Shunt radiator	Current loop reduction	5
Reaction wheel assembly (RWA), electronics (four total)	Mu-metal shielding	16
Imaging science subsystem, narrow angle camera, ISS NAC	Metglass shielding	17
Radio frequency subsystem (Bay 6)	Compensation magnets	19
Propulsion subsystem latch valves	Compensation magnets	20
Radio frequency subsystem (Bay 5)	TWT self-compensation metglass shielding	23
Radio frequency electronics (RFES)	Compensation magnets Mu-metal shielding	24
Power & pyro subsystem (PPS)	Compensation magnets current loop reduction self-compensation	28
Command data subsystem (CDS)	Mu-metal shielding	32
Imaging science subsystem, wide angle camera (ISS WAC)	Metglass shielding	34
Radar	Compensation magnets	39
Engine gimbal actuators (EGA) (four total)	Compensation magnets	42
Reaction wheel assembly (RWA), mechanisms (four total)	Double-nested Mu-metal shielding	43
Composite infrared spectrometer (CIRS)	Metglass shielding current loop compensation	50
Cosmic dust analyzer (CDA)	Mu-metal shielding	55
Ka-band radio frequency instrument subsystem (RFIS)	Compensation magnets	119
Visual & infrared mapping spectrometer (VIMS)	Compensation magnets	305
Radioisotope thermoelectric generators (RTGs)	Self-compensation	390

All Cassini subsystem designers contributed towards designing a magnetically clean spacecraft. There were several designers, however, who very significantly reduced their respective magnetic interferences by implementing a number of magnetic reduction methods and options presented to them in the magnetics workshop. These are listed in Table I.

5. Test and Verifcation

Each Cassini subsystem and assembly was tested to verify its respective magnetic cleanliness prior to final installation on the spacecraft. Magnetic field measurements were conducted inside a three-axes Helmholtz coil system used to buck-out the Earth's magnetic field and therefore provide a zero-field environment (Figure 2).

Magnetic testing of Cassini subsystems was typically performed early in the development process to allow for design modifications with the goal of meeting the magnetic cleanliness requirements. In testing, the subsystems were energized in worst-case modes to determine the magnetic field due to current flow.

Typical magnetic tests consisted of first measuring the subsystem inside a zero-magnetic field environment created by a Helmholtz coil, then demagnetizing the subsystem and, as a final step, measuring its residual magnetic field. From the measured magnetic field values, the subsystem's *X*, *Y* and *Z* dipole moment components were extracted from the data. The total spacecraft residual magnetic fields were determined at the V/SHM and FGM locations by analysis using data from the subsystem measurements in order to verify that the Cassini spacecraft's magnetic condition satisfied the magnetic requirement at the sensor locations. This piece-meal analytical approach was necessary because the large size of the fully integrated spacecraft prevented its measurement inside the Helmholtz coil system.

Figure 2. Three-axis Helmholtz coil used for demagnetizing and magnetics testing of Cassini subsystems.

6. Results of Magnetic Cleanliness Tests

The magnetic test results obtained from the measurement of each Cassini subsystem provided the magnetic model of the overall spacecraft magnetic field at the magnetometer sensor locations. The spacecraft's magnetic field at the V/SHM and FGM locations was calculated by extrapolating and summing the resulting magnetic fields from each subsystem's dipolar moment components. The extrapolation considered the relative orientation of the subsystem and the location of the center of mass of the subsystem in terms of spacecraft coordinates. Based on the dipole model, the overall Cassini spacecraft's total magnetic field at the V/SHM is shown in Table II. The table contains the final *X*, *Y* and *Z*-axis magnetic field results at the V/SHM location. Results also include the magnetic field from the Huygens Probe, which was measured separately by ESA, and the magnetic field contributions from the three RTGs:

From the above results, it is noted that the requirement of 0.2 nT at the V/SHM is satisfied, even with the Huygens Probe attached to the spacecraft.

After the Huygens Probe is released, the magnetic field at the V/SHM remains approximately the same, as indicated in Table III.

A plot of the spacecraft's magnetic field about 360 degrees in the spacecraft *X*-*Y* plane shows in Figure 3 a slight depression along the magnetometer boom at the V/SHM location and a high peak between the spacecraft's $+X$ and $-Y$ quadrants (see Mehlem, 1978). The fields are due to the compensated latch valves and TWTAs that lie in this space.

TABLE II
Overall spacecraft field at V/SHM sensor location.

Field@V/SHMx (nT)	Field@V/SHMy (nT)	Field@V/SHMz (nT)	Total (rss) (nT)
Hsuygens probe −0.048	Huygens probe 0.003	Huygens probe −0.016	Huygens Probe 0.051
Spacecraft 0.032	Spacecraft 0.092	Spacecraft 0.019	Spacecraft 0.099
Total(x) −0.016	Total(y) 0.095	Total(z) 0.003	Total(rss) 0.096
RTGs Compensation at V/SHM		0.0041 nT	
Overall Total		0.1001 nT	
Maximum Allowed		0.200 nT	

TABLE III
Overall spacecraft field at V/SHM sensor location

Field V/SHMx	Field V/SHMy	Field V/SHMz	Total (rss)
Spacecraft 0.032 nT	Spacecraft 0.092 nT	Spacecraft 0.019 nT	Spacecraft 0.099 nT
RTGs compensation at V/SHM		0.0041 nT	
Overall total		0.1031 nT	
Maximum allowed		0.200 nT	

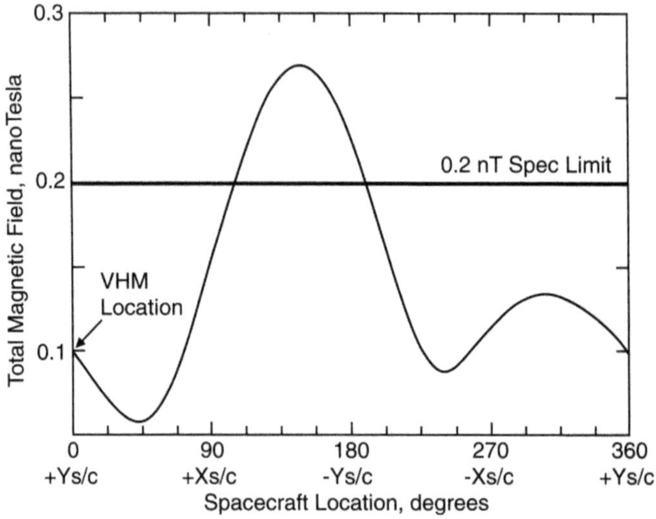

Figure 3. Cassini magnetic field at 11.015 m (after Huygens probe deployment).

Figure 3 also shows there is at least a margin of 0.1 nT between the spacecraft's science requirement of 0.2 nT and the expected spacecraft magnetic field at the V/SHM location. Note that the magnetometer boom (+*Y*s/c) is only approximately 45 degrees from the overall spacecraft minimum magnetic field.

A similar plot shown in Figure 4, but now in polar coordinates, indicates that a spacecraft composed of multiple dipole moments nearly resembles a dipolar field at far distances:

The log/linear plot in Figure 5 demonstrates that the spacecraft's field along the magnetometer boom falls off as the inverse cube of the distance. As the magnetometer boom is deployed, the V/SHM will see the spacecraft magnetic field decay gradually as the inverse cube of the distance.

7. Conclusion

The overall effort instituted in the magnetics control program included the implementation of subsystem magnetic field reduction and measurement and spacecraft level modeling. A total of over 100 magnetic field measurements were performed at the subsystem level, with 56 included in modeling the overall spacecraft magnetic field at the V/SHM sensor location. This work was carried out at the JPL magnetics test laboratory located in Pasadena, California. In addition, testing, data analysis and optimum self-compensation analysis of the three RTGs were carried out at EG& G, Miamisburg, Ohio, and at ESA-ESTEC, Noordwijk, The Netherlands (see Mehlem and Narvaez, 1999).

As noted in Table II, the overall goal of achieving 0.2 nT at the end of the boom was obtained with ample (a factor of 2) margin, a clear sign that the Cassini

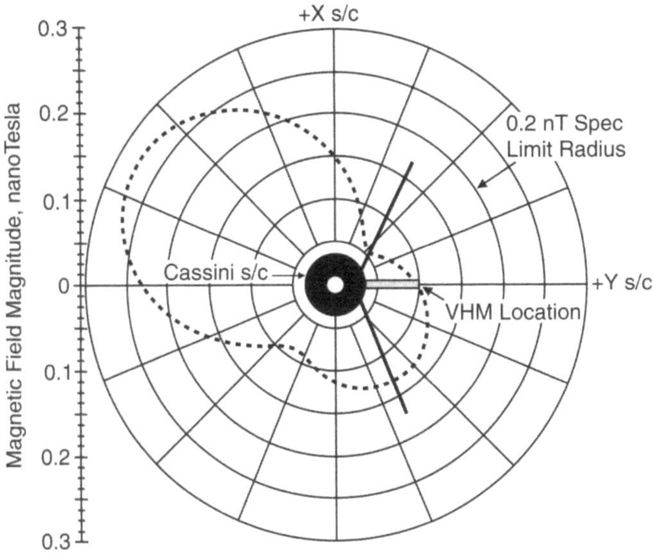

Figure 4. Cassini S/C magnetic field after Huygens probe deployment at 11.015 m (V/SHM location).

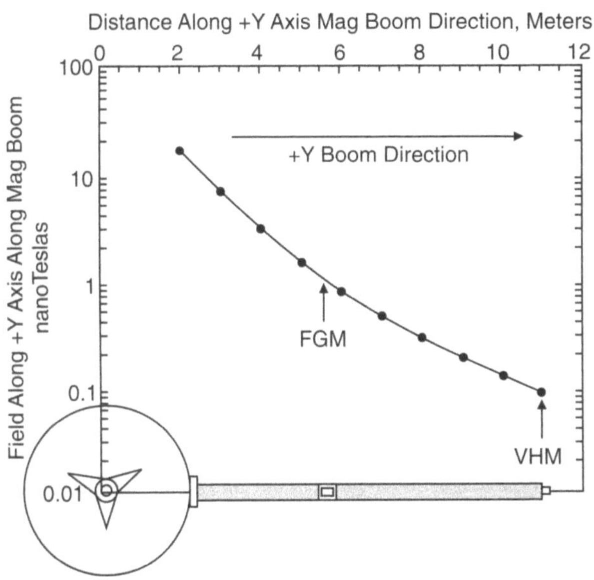

Figure 5. Cassini magnetic field along mag boom direction.

control program was very successful. Each Cassini spacecraft subsystem provided contributions which were necessary to achieve the goals set out by the magnetics control review board team. The resulting efforts have paid off for the magnetometer team. In the absence of a magnetic cleanliness program, the interfering magnetic

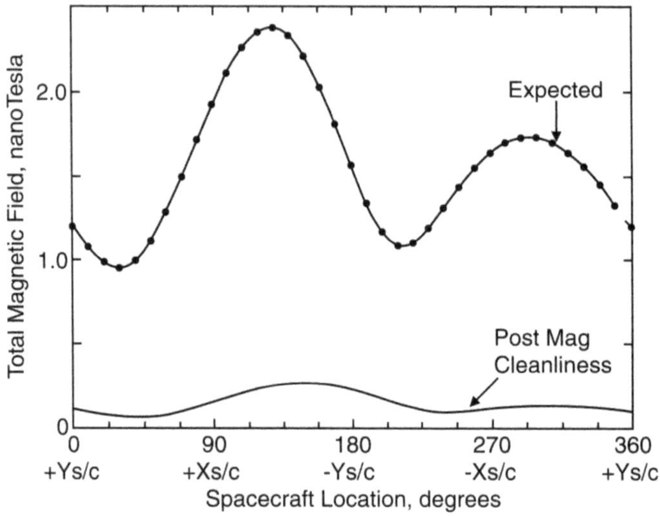

Figure 6. Cassini magnetic field at 11.015 m, post-magnetic cleanliness vs. expected.

fields would have been quite high and quite variable, resulting in ambiguous and corrupted data. Figure 6 provides a comparison between the Cassini magnetic field resulting from the magnetics cleanliness program and the fields expected for the spacecraft if no magnetics control program had been implemented.

Acknowledgments

The research described in this paper was carried out by the Jet Propulsion Laboratory, California Institute of Technology, under a contract with the National Aeronautics and Space Administration.

References

Mehlem, K.: 1978, *Multiple magnetic dipole modeling and field prediction of satellites,* IEEE Trans on Magnetics, September 1978.

Mehlem, K. and Narvaez, P.: 1999, *Magnetostatic Cleanliness of the Radioisotope Thermoelectric Generators (RTGs) of Cassini,* IEEE EMC Symposium, Seattle, WA, pp. 899–904.

THE CASSINI RADIO AND PLASMA WAVE INVESTIGATION

D. A. GURNETT,[1,*] W. S. KURTH,[1] D. L. KIRCHNER,[1] G. B. HOSPODARSKY,[1] T. F. AVERKAMP,[1] P. ZARKA,[2] A. LECACHEUX,[2] R. MANNING,[2,†] A. ROUX,[3] P. CANU,[3] N. CORNILLEAU-WEHRLIN,[3] P. GALOPEAU,[3] A. MEYER,[3] R. BOSTRÖM,[4] G. GUSTAFSSON,[4] J.-E. WAHLUND,[4] L. ÅHLEN,[4] H. O. RUCKER,[5] H. P. LADREITER,[5] W. MACHER,[5] L. J. C. WOOLLISCROFT,[6,†] H. ALLEYNE,[6] M. L. KAISER,[7] M. D. DESCH,[7] W. M. FARRELL,[7] C. C. HARVEY,[8] P. LOUARN,[8] P. J. KELLOGG,[9] K. GOETZ[9] and A. PEDERSEN[10]

[1]*Department of Physics and Astronomy, University of Iowa, Iowa City, IA 52242, U.S.A.*
[2]*Observatoire de Paris, Meudon, France*
[3]*CETP/UVSQ, Velizy, France*
[4]*Swedish Institute of Space Physics, Uppsala, Sweden*
[5]*Austrian Academy of Sciences, Space Research Institute, Graz, Austria*
[6]*Department of Automatic Control and Systems Engineering, University of Sheffield, Sheffield, U.K.*
[7]*NASA/Goddard Space Flight Center, Greenbelt, MD 20771, U.S.A.*
[8]*CESR/CNRS, Toulouse, France*
[9]*Department of Physics and Astronomy, University of Minnesota, Minneapolis, MN 55455, U.S.A.*
[10]*Department of Physics, University of Oslo, Blindern, Oslo, Norway*
*(*Author for correspondence: E-mail address: donald-gurnett@uiowa.edu)*

(Received 11 June 1998; Accepted in final form 5 September 2002)

Abstract. The Cassini radio and plasma wave investigation is designed to study radio emissions, plasma waves, thermal plasma, and dust in the vicinity of Saturn. Three nearly orthogonal electric field antennas are used to detect electric fields over a frequency range from 1 Hz to 16 MHz, and three orthogonal search coil magnetic antennas are used to detect magnetic fields over a frequency range from 1 Hz to 12 kHz. A Langmuir probe is used to measure the electron density and temperature. Signals from the electric and magnetic antennas are processed by five receiver systems: a high frequency receiver that covers the frequency range from 3.5 kHz to 16 MHz, a medium frequency receiver that covers the frequency range from 24 Hz to 12 kHz, a low frequency receiver that covers the frequency range from 1 Hz to 26 Hz, a five-channel waveform receiver that covers the frequency range from 1 Hz to 2.5 kHz in two bands, 1 Hz to 26 Hz and 3 Hz to 2.5 kHz, and a wideband receiver that has two frequency bands, 60 Hz to 10.5 kHz and 800 Hz to 75 kHz. In addition, a sounder transmitter can be used to stimulate plasma resonances over a frequency range from 3.6 kHz to 115.2 kHz. Fluxes of micron-sized dust particles can be counted and approximate masses of the dust particles can be determined using the same techniques as Voyager. Compared to Voyagers 1 and 2, which are the only spacecraft that have made radio and plasma wave measurements in the vicinity of Saturn, the Cassini radio and plasma wave instrument has several new capabilities. These include (1) greatly improved sensitivity and dynamic range, (2) the ability to perform direction-finding measurements of remotely generated radio emissions and wave normal measurements of plasma waves, (3) both active and passive measurements of plasma resonances in order to give precise measurements of the local electron density, and (4) Langmuir probe measurements of the local electron density and temperature. With these new capabilities, it will be possible to perform a broad range of studies of radio emissions, wave-particle interactions, thermal plasmas and dust in the vicinity of Saturn.

[†]Deceased

Space Science Reviews **114:** 395–463, 2004.

1. Introduction

This paper describes the Cassini radio and plasma wave science investigation, hereafter referred to by the acronym RPWS. The basic objectives of the RPWS investigation are to study radio emissions, plasma waves, thermal plasma, and dust in the vicinity of Saturn. The Voyager 1 and 2 spacecraft, which made the first radio and plasma wave measurements at Saturn, clearly showed that many complex types of wave phenomena exist in Saturn's magnetosphere (Warwick *et al.*, 1981; Gurnett *et al.*, 1981; Warwick *et al.*, 1982; Scarf *et al.*, 1982). For a description of the radio and plasma wave instrumentation on Voyager see Warwick *et al.* (1977) and Scarf and Gurnett (1977), and for an overview of radio and plasma wave observations in the vicinity of Saturn see Kaiser *et al.* (1984a) and Scarf *et al.* (1984). The Cassini RPWS instrument is considerably more complex and has greatly improved capabilities compared to the Voyager radio and plasma wave instruments. Voyager used two orthogonal 10-m electric antennas for electric field measurements. The radio instrument used these antennas as two independent monopoles (for both intensity and limited polarization measurements) and the plasma wave instrument used them as a single-axis dipole. In contrast, the Cassini RPWS uses three nearly orthogonal 10-m electric antennas for electric field measurements, three orthogonal search coil magnetometers for magnetic field measurements, and a Langmuir probe for electron density and temperature measurements. Electric field measurements are made over a frequency range from 1 Hz to 16 MHz, magnetic field measurements are made over a frequency range from 1 Hz to 12 kHz, and electron density fluctuation measurements are made over a frequency range from 1 Hz to 2.5 kHz. Signals from the electric and magnetic antennas are processed by five on-board receivers, all considerably more advanced than those flown on Voyager. These include a high frequency digital spectrum analyzer to perform high frequency (3.5 kHz to 16 MHz) spectral, polarization, and direction-finding measurements, a medium frequency analog receiver to perform medium-frequency (24 Hz to 12 kHz) spectral measurements, a low frequency (1 Hz to 26 Hz) digital receiver to perform low frequency spectral measurements, a five-channel waveform receiver to perform wave normal and fine-scale plasma structure measurements, and a digital wideband receiver to provide high-resolution frequency-time spectrograms. In addition, a sounder transmitter can be used to excite local plasma resonances, and a Langmuir probe is used to provide measurements of the electron density and temperature. As with Voyager (Gurnett *et al.*, 1983; Aubier *et al.*, 1983), the RPWS will be able to measure the flux of dust and make estimates of the mass of the dust particles from the signature of impacts with the spacecraft.

Since the Cassini spacecraft is to be the first orbiter of Saturn, this spacecraft will provide much better spatial and temporal coverage than was possible with Voyagers 1 and 2. Specifically, the orbit of Cassini is expected to provide equatorial coverage of all local times for radial distances ranging from about 3.1 R_S

(Saturn radii) to more than 60 R_S. This spatial coverage will provide numerous passes through the bow shock, the magnetopause, the outer magnetosphere, and the inner magnetosphere. During a latter phase of the mission, the inclination of the orbit will be increased to about 75°, thereby providing coverage of the magnetosphere at high latitudes in the vicinity of the auroral field lines. In addition to the magnetospheric coverage, the spacecraft will provide more than 44 close flybys of Saturn's large moon Titan, and close flybys of Saturn's small icy satellites (Mimas, Enceladus, Tethys, Dione, and Rhea). A prime objective of these flybys will be to study the plasma environment of the moons and their interaction with the magnetosphere of Saturn. The flybys of Titan are of particular interest, since Titan is known to have a dense atmosphere (Hunten *et al.*, 1984) and acts as a major source of plasma in the outer regions of the magnetosphere (Neubauer *et al.*, 1984). The small icy moons, Enceladus and Dione in particular, are suspected to be a source of plasma in the inner magnetosphere (Scarf *et al.*, 1984). For an overview of the scientific objectives of the Cassini mission see Matson *et al.* (this issue). For a review of the magnetospheric objectives for the mission, see Blanc *et al.* (2002).

The RPWS investigation is a multi-institutional effort involving 22 investigators from 6 countries and 10 institutions. A list of the investigators, their affiliation and their responsibilities is given in Table I. The investigation is led by Donald A. Gurnett of the University of Iowa, who is the Principal Investigator. The Deputy Principal Investigator is William S. Kurth. During the hardware phase the Project Manager was Richard L. Huff, and the Principal Engineer was Donald L. Kirchner, all at the University of Iowa. Four institutions produced hardware for the project: the University of Iowa, the Observatoire de Paris, the Centre d'etudes des Environnements Terrestre et Planetaires (Universite Versailles Saint Quentin), and the Swedish Institute of Space Physics. The University of Sheffield developed various parts of the flight software, and the Space Research Institute of the Austrian Academy of Sciences carried out radiation pattern studies of the electric antennas. The scientific objectives of the investigation, a description of the instrument, and a discussion of the in-flight operations and data analysis are given in the following sections, which are organized as follows. Section 2 discusses the scientific objectives, Section 3 describes the instrumentation requirements needed to achieve the scientific objectives, Section 4 describes the instrumentation, Section 5 describes the calibrations and performance, and Section 6 describes the in-flight operations, data analysis and data archiving.

2. Scientific Objectives

The scientific objectives of the Cassini RPWS investigation can be broken down into five topics: radio emissions, plasma waves, lightning, thermal plasma, and dust, each of which is discussed below.

TABLE I

RPWS investigators.

RPWS investigator	Institution (Country)	Main responsibility
D. A. Gurnett (P.I.) and W. S. Kurth (Deputy P.I.)	University of Iowa (USA)	Data processing unit. Medium frequency receiver. Wideband receiver; Waveform receiver. Electric antenna motor driver. Electric antenna procurement. Data analysis.
A. Lecacheux and P. Zarka	Observatoire de Paris (France)	High frequency receiver. Sounder; Power supply. Electric preamplifiers. Electric antenna procurement. Data analysis.
P. Canu, N. Cornilleau-Wehrlin, P. Galopeau and A. Roux	CETP/UVSQ (France)	Magnetic antennas. Data analysis.
R. Boström, G. Gustafsson and J.-E. Wahlund	Swedish Inst. of Physics (Sweden)	Langmuir probe. Langmuir probe preamplifier. Langmuir probe digital processor. Data analysis.
H. P. Ladreiter and H. O. Rucker	Austrian Academy of Sciences (Austria)	Antenna modeling. Direction-finding calibration. Data analysis.
H. Alleyne and L. J. C. Woolliscroft*	University of Sheffield (UK)	Data compression software. Data analysis.
M. D. Desch, W. M. Farrell and M. L. Kaiser	NASA/Goddard Space Flight Center (USA)	Data analysis.
C. C. Harvey and P. Louarn	CESR/CNRS (France)	Data analysis.
K. Goetz and P. J. Kellogg	University of Minnesota (USA)	Data analysis.
A. Pedersen	University of Oslo (Norway)	Data analysis.

*Deceased

2.1. Radio Emissions

It is now known that the magnetospheres of five magnetized planets, Earth, Jupiter, Saturn, Uranus, and Neptune, are sources of non-thermal radio emissions (Kaiser, 1989; Zarka, 1998). In addition, one of Jupiter's moons, Ganymede, is also known to produce non-thermal radio emissions (Gurnett *et al.*, 1996; Kurth *et al.*, 1997). These magnetospheric radio emissions are important not only because of the basic plasma processes involved, but also because they provide a valuable remote sensing tool. Since the beaming pattern of the radio emission source is in all cases strongly controlled by the large-scale magnetic field of the planet, the modulation of the radio emission intensity as the planet rotates provides a direct link with the deep interior of the planet. At the present time the only method of accurately determining the rotation rate of the interior of the outer planets is by monitoring the rotational modulation of magnetospheric radio emissions.

Although various radio emission mechanisms have been identified in planetary magnetospheres, one mechanism stands out above all others in terms of radiated power: the cyclotron maser mechanism (Wu and Lee, 1979). Cyclotron maser radiation is produced by a coherent electromagnetic plasma instability and is generated at frequencies very close to the electron cyclotron frequency, which is the frequency of an electron's cyclotron motion in a magnetic field. The electron cyclotron frequency is given by $f_{ce} = 28\text{B}$ Hz, where B is the magnetic field strength in nT (Stix, 1992). Since the coupling in the cyclotron maser mechanism is with the electrons, the radiation is primarily right-hand polarized with respect to the magnetic field (i.e., the electric field vector rotates in the same sense as the cyclotron motion of an electron). At Earth, cyclotron maser radiation is closely associated with discrete auroral arcs (Gurnett, 1974) and is often called auroral kilometric radiation (AKR). The free energy source responsible for the cyclotron maser radiation was originally believed to be a loss-cone anisotropy in the auroral electron distribution (Wu and Lee, 1979; Le Queau *et al.*, 1984). However, more recent measurements suggest that the free energy source is either due to a ring distribution (Louarn *et al.*, 1990) formed by electrons that are electrostatically trapped by the magnetic mirror force of the planetary magnetic field and the parallel electric field responsible for the auroral electron acceleration, or by a shell distribution (Ergun *et al.*, 2000) formed by the acceleration due to the parallel electric field. For a discussion of the cyclotron maser instability and its application to planetary radio emission, see Zarka (1998). In addition to the cyclotron maser radiation, several other considerably weaker radio emission mechanisms have also been identified in planetary magnetospheres. These include synchrotron radiation (Carr *et al.*, 1983), mode conversion from electrostatic upper hybrid waves (Jones, 1976), and mode conversion from Langmuir waves (Cairns and Melrose, 1985). For a review of these and various other radio emission mechanisms, see Goldstein and Goertz (1983) and Zarka (1998).

Most of our present knowledge of radio emissions from Saturn's magnetosphere comes from Voyagers 1 and 2 (Warwick *et al.*, 1981, 1982; Gurnett *et al.*, 1981; Scarf *et al.*, 1982). The Voyager measurements show that the radio emission spectrum of Saturn extends over a very broad frequency range, from about 3 kHz to 1.2 MHz, and has a broad peak extending from about 100–400 kHz. Since the peak intensity occurs at kilometer wavelengths, this radiation is called Saturn kilometric radiation (SKR). The intensity and shape of the SKR spectrum is shown in Figure 1, and is very similar to terrestrial AKR. The total radiated power ranges from about 10^8–10^{10} Watts, and the radiation is generated near the local electron cyclotron frequency primarily in the right-hand polarized extraordinary mode. Because of the characteristic frequency and polarization, it is now widely believed that SKR is generated by the cyclotron maser mechanism (Galopeau *et al.*, 1989). Studies of the radiation pattern (Kaiser *et al.*, 1981; Kaiser and Desch, 1982; Lecacheux and Genova, 1983) indicate that the radiation is generated over the dayside auroral

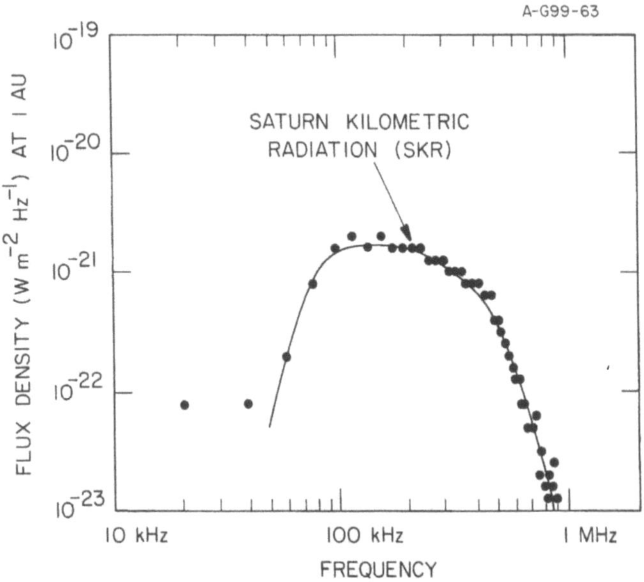

Figure 1. A typical spectrum of Saturn kilometric radiation, which is an intense radio emission that occurs in a frequency range from about 100 to 800 kHz. This radio emission is believed to be generated at low altitudes in the dayside auroral regions by the cyclotron maser mechanism. The intensities have been normalized to a radial distance of 1 AU (astronomical unit).

zone on magnetic field lines that pass through or near the dayside polar cusps. The inferred source field lines projected down to the cloud tops of Saturn are shown in Figure 2 (from Galopeau *et al.*, 1995). These projected source locations are in good agreement with the location of Saturn's aurora from the Hubble Space Telescope (Trauger *et al.*, 1998). Our present knowledge of the rotational period of Saturn (10 h 39.4 min) comes almost entirely from the rotational modulation of SKR (Desch and Kaiser, 1981). One surprising and still unexplained feature is that a strong rotational modulation occurs even though the magnetic dipole axis of Saturn is aligned within 1° of the rotational axis. Most likely the magnetic field of Saturn has significant deviations from a dipole field near the planet (Galopeau *et al.*, 1991; Galopeau and Zarka, 1992). Also, there is evidence of long-term variations in the SKR period that may ultimately limit the accuracy with which the rotational period of Saturn can be determined (Lecacheux *et al.*, 1997; Galopeau and Lecacheux, 2000; Zarka and Cecconi, 2002).

In addition to the rotational modulation, the SKR intensity also displays considerable variability on time scales ranging from seconds to months. Studies by Desch (1982) show that the SKR intensity is strongly influenced by solar wind pressure variations. Higher SKR intensities are associated with higher solar wind pressures. Kurth *et al.* (1981) have also shown that the orbital position of Saturn's large icy moon Dione may play a role in modulating the SKR intensity, possibly suggesting an Alfvén wave-like interaction with the magnetosphere of Saturn, similar to

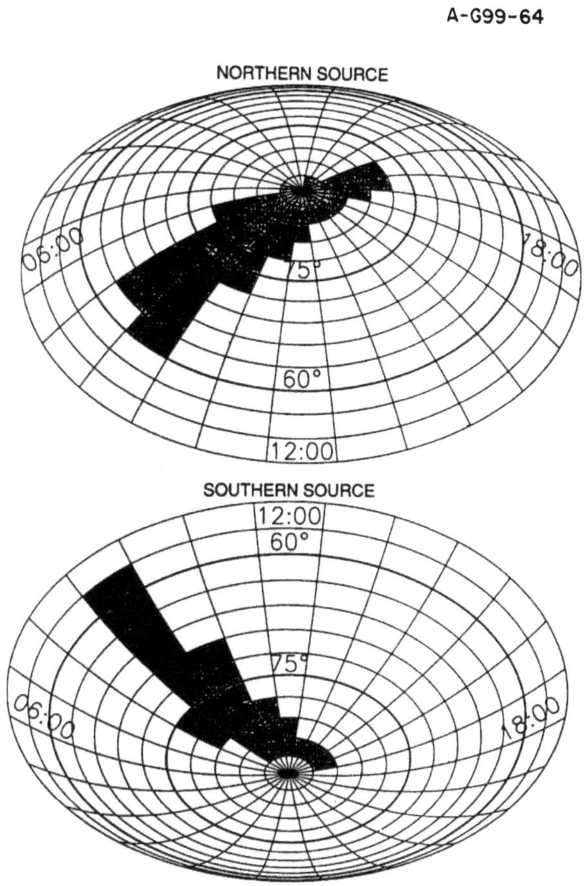

Figure 2. SKR source locations projected down Saturn's magnetic field lines to the cloud tops (from Galopeau *et al.*, 1995). These source positions are in good agreement with the location of Saturn's aurora, as determined from ultraviolet images taken by the Hubble Space Telescope (Trauger *et al.*, 1998).

the interaction of Jupiter's moon Io with the magnetosphere of Jupiter (Goldreich and Lynden-Bell, 1969). For a discussion of the Alfvén wave interaction at Io, see Neubauer (1980). It has long been suspected that gaseous emissions from Dione and various other icy moons of Saturn, such as Enceladus and Tethys, are significant sources of plasma in the inner regions of the magnetosphere. Therefore, these moons could very well induce an Alfvén wave interaction similar to the interaction that occurs at Io. Evidence also exists that the occurrence of SKR is linked with the occurrence of spokes in the rings (Mendis *et al.*, 1984). Zarka (1998) has also shown that the SKR has considerable fine structure on time scales of a fraction of a second or less, possibly similar to the fine structure observed in terrestrial AKR (Gurnett *et al.*, 1979a). For a more detailed review of the Voyager SKR observations and their interpretation, see Kaiser *et al.* (1984a) and Zarka (1998).

In addition to the SKR, numerous other much weaker radio emissions were observed by Voyagers 1 and 2 at somewhat lower frequencies, ranging from a few kHz to several tens of kHz (Gurnett *et al.*, 1981). These emissions have very narrow bandwidths and are believed to be produced by mode conversion from electrostatic upper hybrid waves near half-integral harmonics of the electron cyclotron frequency, similar to various types of narrowband emissions observed in the magnetospheres of Earth and Jupiter (Kurth and Gurnett, 1991). The basic condition for the generation of these narrowband emissions is that the upper hybrid resonance frequency, f_{UHR}, must be near a half-integral harmonic of the electron cyclotron frequency, $(n + 1/2)f_{\mathrm{ce}}$. The upper hybrid resonance frequency is given by $f_{\mathrm{UHR}} = \sqrt{f_{\mathrm{ce}}^2 + f_{\mathrm{pe}}^2}$, where $f_{\mathrm{pe}} = 8980\sqrt{n_{\mathrm{e}}}$ Hz is the electron plasma frequency and n_e is the electron density in cm^{-3} (Stix, 1992). The frequencies of some of these emissions appear to correspond closely with half-integral harmonics of the electron cyclotron frequency near the orbits of the moons Tethys, Dione and Rhea. This relationship suggests that these moons are somehow directly involved in the generation of radio emissions, possibly by the ejection of gas that is subsequently ionized and converted to an unstable velocity distribution by an interaction with the co-rotating magnetospheric plasma. At even lower frequencies, the Voyager data suggest that electromagnetic radiation is trapped in the magnetospheric cavity at frequencies below the solar wind electron plasma frequency, very similar to continuum radiation trapped in the magnetospheres of Earth and Jupiter (Kurth, 1992). Since the low frequency cutoff of the trapped continuum radiation is usually at the local electron plasma frequency (Gurnett and Shaw, 1973; Scarf *et al.*, 1979), this cutoff could potentially provide a very accurate measurement of the local electron density, which is a basic plasma parameter of the magnetospheric plasma.

It is clear that our knowledge of radio emissions in the magnetosphere of Saturn is incomplete. Much more observing time and spatial coverage in the vicinity of Saturn are needed to fully understand the modulation effects imposed by planetary rotation, solar wind pressure variations, and interactions with Saturn's moons and rings. Direction-finding and polarization measurements are needed to constrain the generation mechanisms and to compare the source locations with other in-situ and remote sensing observations. High-time resolution measurements are needed to study the fine structure of SKR. The Cassini mission with its four years of in-orbit observations and its advanced radio wave instrumentation provides an ideal opportunity to carry out these studies. With the Cassini RPWS we expect to achieve the following specific results relative to radio emissions from Saturn's magnetosphere.

- Improve our knowledge of the rotational modulation of Saturn's radio sources, and hence of Saturn's rotation rate.
- Determine the location of the SKR source as a function of frequency, and investigate the mechanisms involved in generating the radiation.
- Obtain a quantitative evaluation of the anomalies in Saturn's magnetic field by performing direction-finding measurements of the SKR source.

- Establish if gaseous ejections from the moons Rhea, Dione, and Tethys are responsible for the low frequency narrow-band radio emissions.
- Determine if SKR is controlled by Dione's orbital position.
- Establish the nature of the solar wind-magnetosphere interaction by using SKR as a remote indicator of magnetospheric processes.
- Investigate the relationship between SKR and the occurrence of spokes and other time dependent phenomena in the rings.
- Study the fine structure in the SKR spectrum, and compare with the fine structure of terrestrial and Jovian radio emissions in order to understand the origin of this fine structure.

A recent review of some of the planetary radio astronomy issues to be addressed by Cassini at Saturn is given by Kurth and Zarka (2001).

2.2. Plasma Waves

Since plasma waves can transfer energy and momentum between particles in a plasma, plasma waves are now recognized as being of fundamental importance for understanding the dynamics of space plasmas. For example, in planetary magnetospheres, it has been known for many years (Kennel and Petschek, 1966) that wave-particle interactions play a fundamental role in controlling the loss of radiation belt particles. Charged particles trapped in a planetary magnetic field are lost if their trajectory intersects the surface or atmosphere of a planet or moon. This loss process produces a cone-shaped hole in the particle velocity distribution function called the loss cone. In the absence of plasma waves, magnetically trapped particles would bounce back and forth along the magnetic field with an essentially infinite lifetime. However, the velocity space anisotropy caused by the loss cone leads to the growth of plasma waves. In the case of electrons, the waves generated are primarily electron-cyclotron and whistler-mode waves, and in the case of ions, ion-cyclotron waves. Once generated, the waves interact resonantly with trapped particles, causing small perturbations in their trajectories relative to the magnetic field. From very general principles, it can be shown that these perturbations cause the radiation belt particles to diffuse toward the loss cone, ultimately causing the particles to be lost when they strike the planet's atmosphere (Kennel and Petschek, 1966). This loss process is the dominant mechanism that limits the lifetime of particles trapped in a planetary radiation belt. The resulting energy flux striking the atmosphere can sometimes be substantial, causing aurora and numerous other effects. In addition to radiation belt dynamics, wave-particle interactions are involved in many other processes. These include the heating of plasma at the bow shock and magnetopause, the thermalization of plasma escaping from moons and rings, and the acceleration of plasma in regions of strong field-aligned currents.

As with radio emissions, our present knowledge of plasma waves in the magnetosphere of Saturn comes almost entirely from Voyager 1 and 2 measurements

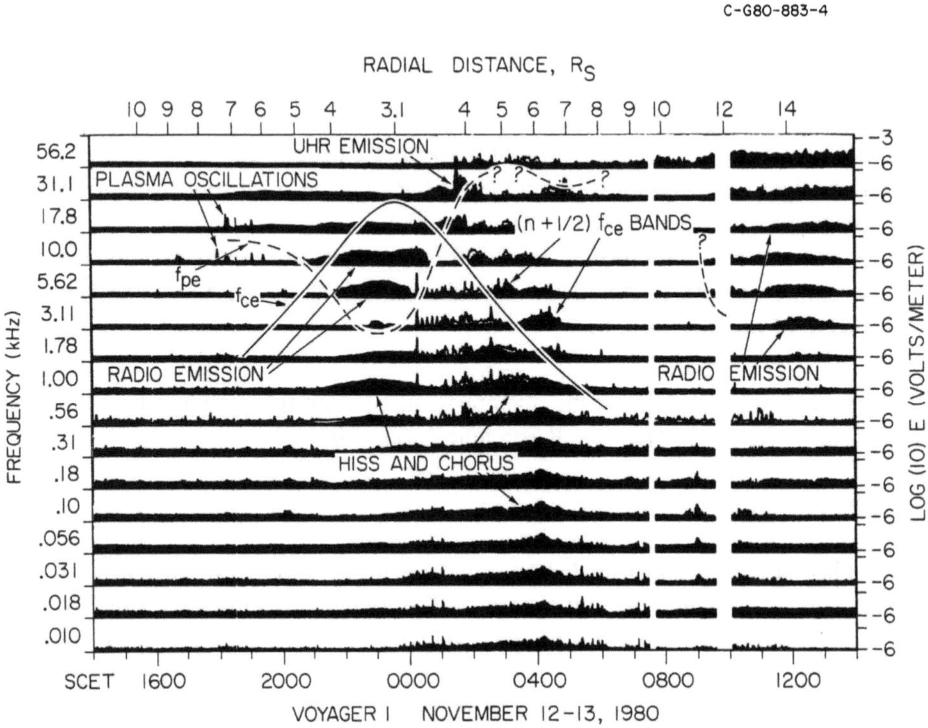

Figure 3. A 16-channel plot showing the plasma wave electric field intensities detected by the Voyager 1 plasma wave instrument during the November 12–13, 1980 flyby of Saturn (from Gurnett *et al.*, 1981). The highest plasma intensities were observed in the inner region of the magnetosphere, inside of about 10 R_S (Saturn radii).

(Gurnett *et al.*, 1981; Scarf *et al.*, 1982; Scarf *et al.*, 1984). The Voyager 1 and 2 data show that a wide variety of plasma wave phenomena are present in the magnetosphere of Saturn, and near Saturn's large moon Titan. Figure 3 gives an overview of the plasma wave electric field intensities observed during the Voyager 1 flyby of Saturn. The intensities are shown in 16 logarithmically spaced frequency channels with center frequencies extending from 10 Hz to 56.2 kHz. Spacecraft event time (SCET) in hours and minutes is shown at the bottom of the plot and the radial distance from Saturn is shown at the top of the plot in Saturn radii ($1R_s = 60{,}330$ km). As can be seen, the plasma wave phenomena occurring in the inner region of the magnetosphere are extremely complex. The plasma wave modes that can be easily identified include electron plasma oscillations, upper hybrid resonance (UHR) emissions, $(n + 1/2) f_{ce}$ electron-cyclotron waves, and whistler-mode chorus and hiss. The UHR emissions occur at the upper hybrid resonance frequency, f_{UHR}, and the electron-cyclotron waves occur near half integral harmonics of the electron cyclotron frequency, $(n + 1/2) f_{ce}$. Of these, the electron-cyclotron waves and the whistler-mode emissions are of particular importance because they control the pitch-angle scattering and loss of energetic electrons from the radiation belt.

The relatively smooth intense emissions labeled "radio emission" are the narrowband electromagnetic emissions described in the preceding section. These radio emissions are believed to be produced by mode conversion from the electrostatic UHR waves (Gurnett *et al.*, 1981). Several other types of waves are also observed that are not shown in Figure 3 . These include electrostatic waves associated with the bow shock, a broad region of low frequency MHD turbulence between the magnetopause and the rigidly co-rotating inner region of the magnetosphere, and broadband electrostatic noise in the distant magnetotail. For a detailed review of these and other types of plasma waves observed in the magnetosphere of Saturn, see Scarf *et al.* (1984).

During the Voyager 1 flyby of Saturn the spacecraft made a very close flyby of Saturn's large moon Titan, which orbits Saturn at a radial distance of about 20.2 R_S. At this large radial distance, which is near the dayside magnetopause, Titan is sometimes in the magnetosphere of Saturn and sometimes in the solar wind, depending on the solar wind pressure. During the Voyager 1 flyby, the solar wind pressure was such that Titan was inside the magnetosphere. Since the magnetosphere co-rotates with Saturn, Titan was exposed to the co-rotational plasma, which was flowing past Titan at a speed of about 120 km/s (Neubauer *et al.*, 1984). As indicated by the sketch at the top of Figure 4, the spacecraft flew almost directly through the geometric wake of Titan at a radial distance of about 2.6 Titan radii (R_T). The interaction of Titan with the rapidly flowing magnetospheric plasma produced a number of very interesting plasma wave effects. The electric field intensities from the 16-channel plasma wave spectrum analyzer are shown in Figure 4 . Two regions of intense low frequency electric field noise are evident, the first near the entry into the wake region and the second, of somewhat lower frequency, near the exit from the wake region. In the upper frequency channels, a weak narrowband emission can be seen at a frequency that has been identified as an UHR emission. Since the electron cyclotron frequency is small compared to the electron plasma frequency, the upper hybrid resonance frequency, $f_{\mathrm{UHR}} = \sqrt{f_{\mathrm{ce}}^2 + f_{\mathrm{pe}}^2}$ is essentially at the electron plasma frequency. The electron density, n_e, can then be computed directly from the upper hybrid resonance frequency, and is shown by the scale on the left-hand side of the plot. As can be seen the electron density has two well-defined peaks, the first at the entry into the wake and the second at the exit from the wake. These density peaks are believed to be caused by ions escaping from Titan's ionosphere and carried downstream by the co-rotational plasma flow. The low frequency electric field noise, which occurs in the same region, is believed to be caused by a two-stream interaction between the escaping exospheric ions and the rapidly flowing magnetospheric plasma. It is believed that the waves responsible for the low frequency electric field noise play a crucial role in thermalizing and assimilating the ions from Titan's ionosphere into the magnetospheric plasma.

Although the Voyager 1 and 2 observations provide an initial overview of plasma waves in the vicinity of Saturn, much still remains to be learned. Plasma waves in the inner magnetosphere are very complex and difficult to analyze. Since very little

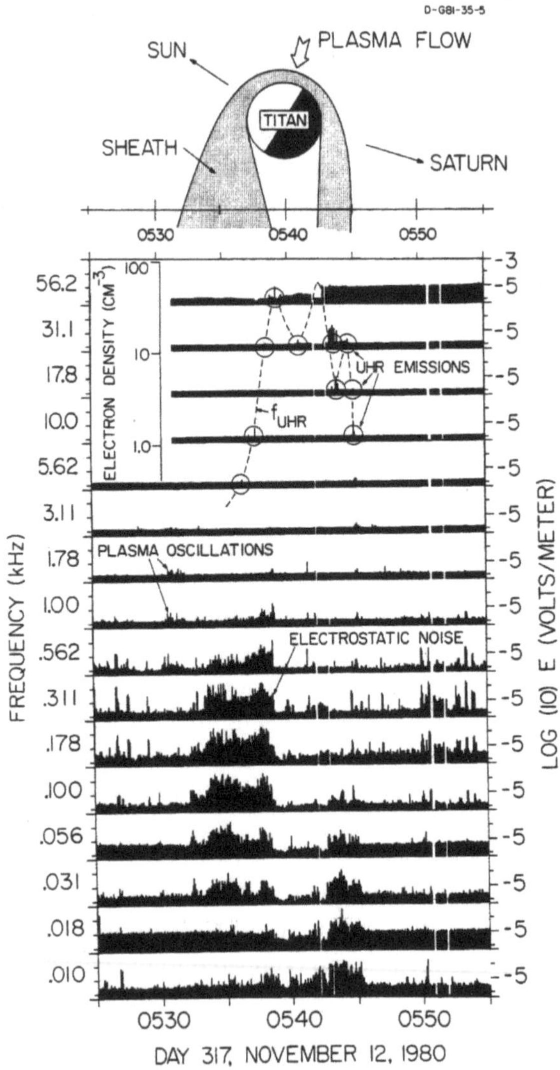

Figure 4. A 16-channel plot showing the plasma wave electric field intensities detected by the Voyager 1 plasma wave instrument during the November 12, 1980 flyby of Titan (from Gurnett *et al.*, 1981). The narrowband emission indicated by the dashed line labeled "UHR emissions" gives the local electron density, which is indicated by the scale on the left. The electron density is strongly enhanced in the region immediately downstream of Titan, indicating that plasma is being carried away from Titan by an interaction with the rapidly rotating magnetosphere of Saturn. Strong, low frequency electrostatic noise is also evident in this same region.

high-rate waveform data were obtained at Saturn, the limited frequency and time resolution of the Voyager low-rate data (as in Figures 3 and 4) makes it very difficult to distinguish various plasma wave modes. Since no wave magnetic field or density fluctuation measurements were made with the Voyager plasma wave instrument, it is also often very difficult to distinguish electrostatic waves from electromagnetic waves, and therefore to identify the nature of the generation process. For example, it is not known whether the low frequency electric field noise in Titan's wake region (see Figure 4) is due to electrostatic or electromagnetic waves. Since the Cassini RPWS instrument provides magnetic field measurements, the electrostatic/electromagnetic character of these and other emissions can be determined. The Cassini instrument also has considerably better frequency and time resolution than the Voyager plasma wave instrument. These improvements in the instrumentation, together with the greatly improved spatial coverage provided by the Cassini mission will provide a major advance in our knowledge and understanding of plasma waves in the Saturnian system. With the Cassini RPWS we expect to achieve the following specific results relative to plasma waves in the Saturnian system.

- Establish the spectrum and types of plasma waves associated with gaseous emissions from Titan, the rings, and the icy satellites.
- Determine the role of plasma waves in the interaction of Saturn's magnetospheric plasma (and the solar wind) with the ionosphere of Titan.
- Establish the spectrum and types of plasma waves that exist in the radiation belt of Saturn.
- Determine the wave-particle interactions responsible for the loss of radiation belt particles.
- Establish the spectrum and types of waves that exist in the magnetotail and polar regions of Saturn's magnetosphere.
- Determine if waves driven by field-aligned currents along the auroral field lines play a significant role in the auroral charged particle acceleration.
- Determine the electron density in the magnetosphere of Saturn, near the icy moons, and in the ionosphere of Titan.

2.3. Lightning

Lightning is a fundamental process in Earth's atmosphere and is thought to occur at five other planets: Venus (Ksanfomality *et al.*, 1983; Russell, 1991) (although this is controversial (Gurnett *et al.*, 2001)), Jupiter (Gurnett *et al.*, 1979b; Kurth *et al.*, 1985), Saturn (Warwick *et al.*, 1981; Kaiser *et al.*, 1983; Zarka and Pedersen, 1983), Uranus (Zarka and Pedersen, 1986), and Neptune (Gurnett *et al.*, 1990; Kaiser *et al.*, 1991). A comparison of the characteristics of lightning at these planets is given by Zarka *et al.* (1995). Lightning is usually associated with convective storms, although in rare cases it is also known to be caused by volcanic activity. At Earth, lightning discharges occur when the electric field caused by charge separation in a convective storm exceeds the breakdown field of the atmosphere (Uman, 1987). Because of the high temperatures generated along the discharge path, lightning induces chemical reactions that otherwise would not occur in planetary atmospheres. For example, in Jupiter's atmosphere the dissociation of CH_4, NH_3, H_2S, and H_2O

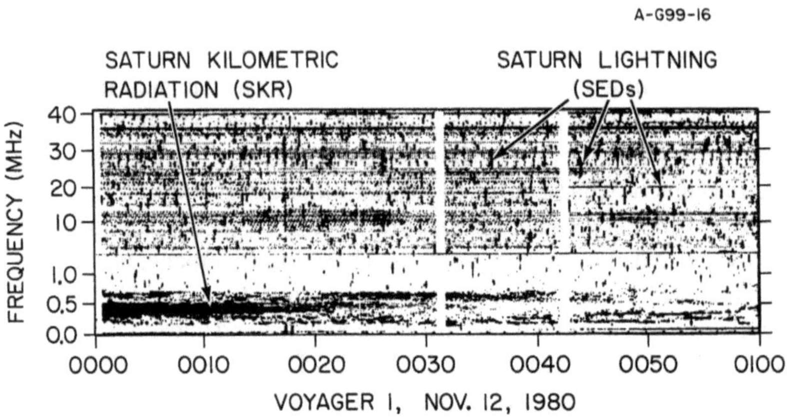

Figure 5. A frequency-time spectrogram of the electric field intensities detected by the planetary radio astronomy (PRA) instrument during the Voyager 1 flyby of Saturn (illustration adapted from Kaiser *et al.*, 1984b). The impulsive bursts scattered randomly around the spectrogram are believed to be due to lightning in Saturn's atmosphere and are called Saturn Electrostatic Discharges (SEDs). A persistent level of Saturn kilometric radiation (SKR) can also be seen near the bottom of the spectrogram, from about 100 to 800 kHz.

is believed to be primarily caused by the high temperatures associated with lightning discharges (Prinn and Owen, 1976). Similar processes are believed to occur in Saturn's atmosphere (Prinn *et al.*, 1984). When organic molecules are present in the atmospheric gas, it has also been demonstrated that electrical discharges can generate complex amino acids that are the building blocks of living organisms (Miller, 1955).

Lightning was first discovered in the atmosphere of Saturn by Voyager 1. During the Voyager 1 flyby, very intense impulsive bursts of radio noise, called Saturn electrostatic discharges (SEDs), were detected by the Planetary Radio Astronomy experiment (Warwick *et al.*, 1981). A frequency-time spectrogram showing a large number of SEDs is shown in Figure 5. At first the origin of these bursts was not clear. However, later analyses by Burns *et al.* (1983) and Kaiser *et al.* (1983) provided strong evidence that the bursts originated from lightning in the equatorial region of Saturn's atmosphere. The rate of occurrence of SEDs showed a persistent periodicity ranging from about 10.0 to 10.16 h, substantially less than the 10.66-h periodicity of the SKR, which as described earlier gives the rotational period of Saturn's deep interior. The shorter period of the SED events is believed to be due to strong prograde equatorial winds that carry convective storms eastward relative to the rotation of Saturn's interior. Typical SED bursts have a duration of 30–400 ms, and involve the radiation of 10^6–10^9 Joules of energy at radio frequencies. The frequency range of these bursts extends from as low as 20 kHz to as high as 40 MHz (Zarka and Pedersen, 1983), and the spectrum has a well-defined low frequency cutoff. The low frequency cutoff varies from about 6 to 15 MHz for storms located on the dayside of Saturn, to less than 300 kHz for storms located on the nightside of Saturn. The low frequency cutoff is believed to be caused in the ionosphere of Saturn by the well-known cutoff of free-space electromagnetic radiation at the local electron plasma frequency. This cutoff has been used to measure the electron density in Saturn's ionosphere (Kaiser *et al.*, 1984b). Typical peak electron densities vary from about 4×10^5 cm^{-3} on the dayside of Saturn to less than 10^3 on the nightside of Saturn. Under some circumstances absorption is also believed to play a role in the low frequency cutoff of the lightning signal as it passes through the ionosphere (Zarka, 1985a). For a review of SEDs and the effect of the ionosphere on the received lightning spectrum, see Zarka (1985b). Although wideband waveform measurements were obtained at frequencies below about 10 kHz by both the Voyager 1 and 2 plasma wave instruments, no whistlers were detected from lightning, most likely because the spacecraft did not pass through magnetic field lines that connected with the low-latitude lightning source.

One of the outstanding results from the Voyager 1 flyby of Saturn was the discovery of organic molecules in the atmosphere of Titan (Hanel *et al.*, 1981). These organic molecules are believed to be produced primarily by photosynthesis due to solar ultraviolet radiation. However, since lightning could be important for synthesizing complex organic molecules (Sagan and Khare, 1971), a careful search for lightning in Titan's atmosphere was carried out during the Voyager 1 flyby of Titan (Desch and Kaiser, 1990). No evidence of lightning was found. However, because of the poor sensitivity of the Voyager radio instrument at high frequencies, and the lack of complete line-of-sight coverage of the entire atmosphere, it is still believed that weak or very localized lightning discharges could exist in Titan's atmosphere (Lammer *et al.*, 2001).

Because of the long observing time in the vicinity of Saturn, and the numerous close flybys of Titan, the Cassini mission provides an ideal opportunity to improve our knowledge of lightning in the atmosphere of Saturn, and to definitely establish whether lightning exists at Titan. The RPWS instrument is expected to have a sensitivity that is about a factor of one hundred better than Voyager for detecting impulsive lightning-like signals at high frequencies (above the ionospheric cutoff), and a sensitivity comparable to Voyager for detecting whistler-mode waves produced by lightning. In addition the RPWS has the capability of performing direction-finding measurements, a capability that did not exist with Voyager. Direction-finding measurements can be used, for example, to determine the location of lightning storms in the atmosphere of Saturn and to track these storms as they evolve. With the Cassini RPWS we expect to achieve the following specific results relative to lightning.

- Establish the long-term morphology and temporal variability of lightning in the atmosphere of Saturn.
- Determine the spatial and temporal variation of the electron density in Saturn's ionosphere from the low frequency cutoff and absorption of lightning signals.
- Carry out a definitive search for lightning in Titan's atmosphere during the numerous close flybys of Titan.

- Perform high-resolution studies of the waveform and spectrum of lightning in the atmosphere of Saturn, and compare with terrestrial lightning.

2.4. Thermal Plasma

The electron number density and temperature of the thermal plasma are among the most important parameters of a plasma. Together with the magnetic field they control the dispersion and damping of waves and are key quantities in many other plasma processes. Although these parameters are important, they are also difficult to measure, particularly with direct particle detection techniques. In fact, one of the major problems encountered in interpreting the Voyager observations was the uncertainty in the plasma density (Scarf *et al.*, 1984; Neubauer *et al.*, 1984). Two effects contribute to these difficulties. First, photoelectron emission from the spacecraft body produces a local non-thermal plasma that makes it difficult to detect the ambient plasma, particularly at low energies below a few tens of electron volts. Second, charges that accumulate on the spacecraft body produce strong local electric fields that prevent certain classes of particles from reaching the detector (electrons if the spacecraft is charged to a negative potential and ions if the spacecraft is charged to a positive potential). Although these effects can be minimized by various techniques, they are almost always a major concern, particularly when the plasma is hot and tenuous.

Plasma wave measurements of various characteristic plasma resonances provide a powerful and very accurate technique for determining certain basic plasma parameters, such as the electron density and temperature. Since the wavelengths of the waves responsible for the plasma resonances are usually much larger than the dimensions of the spacecraft, the resonance frequencies are usually completely unaffected by the presence of the spacecraft. Probably the most important characteristic resonance frequency of a plasma is the electron plasma frequency, f_{pe}, which as described earlier is directly proportional to the square root of the electron density. The upper hybrid resonance, f_{UHR}, is another very important resonance that can be used to determine the electron density. If the electron cyclotron frequency, f_{ce}, is known from magnetic field measurements, as it usually is, then it is easy to show that the electron density can be determined from the upper hybrid resonance frequency by using the equation, $n_e = [(f_{pe}^2 - f_{ce}^2)/8980]\ \text{cm}^{-3}$, where the frequencies are in Hz.

Two plasma wave techniques can be used to determine the characteristic resonance frequencies of a plasma: passive and active. Passive measurements rely on thermal fluctuations or plasma instabilities to excite the characteristic resonance frequencies in the plasma. A good example of the passive technique is illustrated in Figure 6 , which shows a frequency-time spectrogram of the electric field intensities observed by the Galileo plasma wave instrument during the close flyby of Earth that occurred on December 8, 1992. The narrow line labeled “thermal plasma line” is caused by thermal excitation of electrostatic oscillations at the electron plasma

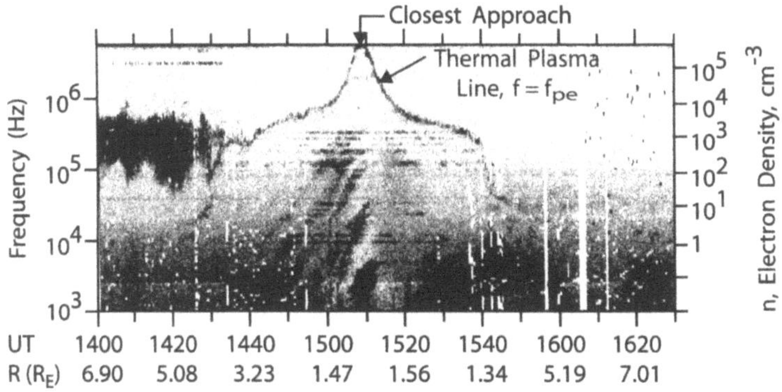

Figure 6. A frequency-time spectrogram of the electric field intensities detected by the Galileo plasma wave instrument during the close flyby of Earth that occurred on December 8, 1992. The narrow emission line labeled "thermal emission" is caused by thermal excitation of electrostatic oscillations at the electron plasma frequency. This emission line can be used to determine the local electron density, which is shown by the scale of the right.

frequency (also called Langmuir waves). Since the electron plasma frequency is directly proportional to the square root of the electron density, the oscillation frequency can be converted directly to electron density. A corresponding electron density scale is shown on the right-hand side of the plot. The sharp peak centered on closest approach is due to the passage through Earth's ionosphere. Although not shown, the electron temperature also can be computed from the spectral shape of the emission line. For a discussion of the techniques used to compute the electron temperature, see Meyer-Vernet and Perche (1989). The main requirement for reliable measurements with the passive technique is that the length of the electric antenna be substantially greater than the Debye length, λ_D, which is given by $\lambda_D = 6.9\sqrt{T_e/n_e}$ cm, where T_e is the electron temperature in °K, and n_e is the electron density in cm^{-3} (Stix, 1992). Otherwise, the thermal plasma line becomes very diffuse and poorly defined. Since the Debye length increases as the electron temperature increases and the density decreases, the technique works best in cold dense plasmas, such as planetary ionospheres. During the Cassini mission the primary application of the passive resonance technique is expected to be the ionosphere of Titan. As the spacecraft passes through the ionosphere of Titan we expect to obtain electric field spectrograms comparable to Figure 6 , from which the electron density can be directly determined to an accuracy of a few percent.

The main disadvantage of the passive resonance technique is that the thermally excited plasma line is very weak, typically only a few dB above the receiver noise level, and becomes very diffuse and difficult to measure when the Debye length is large. Therefore, under some circumstances it is not possible to measure the electron density using passive techniques. For example, in Figure 6 the thermal plasma line becomes very difficult to detect in the hot tenuous magnetospheric plasma before about 1425 UT, and after about 1540 UT. To assure that we can measure the electron density under all circumstances, the RPWS also includes an active method of exciting plasma resonances. Active excitation of plasma resonances has been performed for many years in Earth's magnetosphere, and is usually called plasma sounding (Lockwood, 1963; Benson, 1970; Etcheto and Bloch, 1978; Harvey *et al.*, 1979; Decreau *et al.*, 1987). Plasma sounding consists of transmitting a short pulse at a fixed frequency, and then "listening" for a resonant response from the plasma. After each pulse the transmitter frequency is increased by a small amount, eventually scanning the entire frequency range of interest. When a resonance is encountered, the received signal is strongly enhanced. The duration of the resonant response depends on a number of factors, including thermal (Landau) damping of the excited wave and convection of the wave packet away from the spacecraft. Resonant effects are typically observed at the electron plasma frequency, f_{pe}, the upper hybrid resonance frequency, f_{UHR}, the harmonics of the electron cyclotron frequency, nf_{ce}, and a series of resonances called the q-resonances, f_q, that occur between the electron cyclotron harmonics. The resonance at the electron plasma frequency can provide very accurate measurements of the electron density, and if the q-resonances are sufficiently separated from the electron cyclotron harmonics these resonances can provide measurements of the electron temperature (Etcheto *et al.*, 1983).

In addition to the plasma wave method of studying the thermal plasma, the RPWS instrument also includes a Langmuir probe for measuring the electron density and temperature. The Langmuir probe technique consists of applying a bias voltage to the probe, a sphere in our case, and measuring the current to the probe as a function of the bias voltage (Mott-Smith and Langmuir, 1926). For a positive bias voltage it can be shown that the current is directly proportional to the electron number density. For negative bias voltages the current depends mainly on the thermal motions of the electrons. Only electrons with enough kinetic energy to overcome the electrostatic potential barrier imposed by the bias voltage contribute to the current. In this negatively biased regime it can be shown that the electron temperature is inversely proportional to the slope of the logarithm of the current plotted as a function of the bias voltage. Thus, the Langmuir probe can also be used to determine the electron temperature, a quantity that is difficult to measure using wave techniques. One problem with the Langmuir probe technique is that spacecraft charging effects cause an offset in the voltage scale relative to the potential in the plasma. However, this offset can be determined from the shape of the current-voltage characteristic of the probe. Photoelectrons also complicate the interpretation of Langmuir probe data. For positive bias voltages of more than a few volts, all of the photoelectrons are attracted back to the probe. Photoelectrons from the probe then have no effect on the current-voltage characteristics of the probe. However, photoelectrons from the spacecraft are now attracted to the probe, and these give a lower limit on the electron number density that can be measured. For a 1 eV plasma at Saturn, this lower limit is about 1 cm^{-3}. Somewhat lower densities are reachable in eclipse conditions. The spacecraft body and its surrounding plasma sheath also affects the current-voltage characteristics of the probe. To avoid disturbances from the spacecraft body the Debye shielding length, λ_D, should be significantly smaller than the distance from the probe to the spacecraft. For the probe geometry used on Cassini and a plasma temperature of 1 eV, this restriction places a lower limit on the electron densities that can be determined accurately of about 25 cm^{-3}. Lower densities are measurable, but the interpretation will be more difficult. Since the electron densities in the ionosphere of Titan are expected to be much greater than 25 cm^{-3}, the Langmuir probe should be able to provide accurate measurements of the electron density and temperature in the ionosphere of Titan. If the plasma density is sufficiently high, it is also expected that the Langmuir probe can provide useful measurements in the vicinity of the icy moons and in the inner regions of the magnetosphere.

With the RPWS we expect to achieve the following specific results relative to the density and temperature of the thermal plasma.

- Determine the spatial and temporal distribution of the electron density and temperature in Titan's ionosphere.
- Characterize the escape of thermal plasma from Titan's ionosphere in the downstream wake region.
- Constrain and, when possible, measure the electron density and temperature in other regions of Saturn's magnetosphere.

2.5. Dust

The Saturn system is known to contain a large number of very small dust particles, with sizes on the order of a few microns (Smith *et al.*, 1981; Esposito *et al.*, 1984). Imaging measurements with Voyagers 1 and 2 show that the micron-sized particles are heavily concentrated near the ring plane at radial distances near and inside the G ring. These particles are most likely produced by micro-meteoroid impacts on the rings and icy moons. Such very small particles are important for several reasons. First, since they evaporate rapidly, these particles are an important source of magnetospheric plasma. Second, because of their large numbers they also act as a significant absorber of energetic radiation belt particles, thereby contributing to the loss of particles from Saturn's radiation belt. Third, since the dust particles are maintained in a charged state by photoelectron emission, electrostatic and electromagnetic forces can lead to a variety of non-Keplerian effects (Mendis *et al.*, 1984), including acceleration to high energies. Goertz and Morfill (1983) have suggested that the interaction of charged dust particles with the magnetosphere of Saturn could be a significant source of field-aligned currents in the inner regions of Saturn's magnetosphere.

During the Voyager 2 flyby of Saturn, the plasma wave and radio astronomy instruments detected impacts of micron-sized particles on the spacecraft (Scarf *et al.*, 1982; Warwick *et al.*, 1982). The signal detected by these instruments is caused by an impact ionization effect. When a small particle strikes a solid surface at

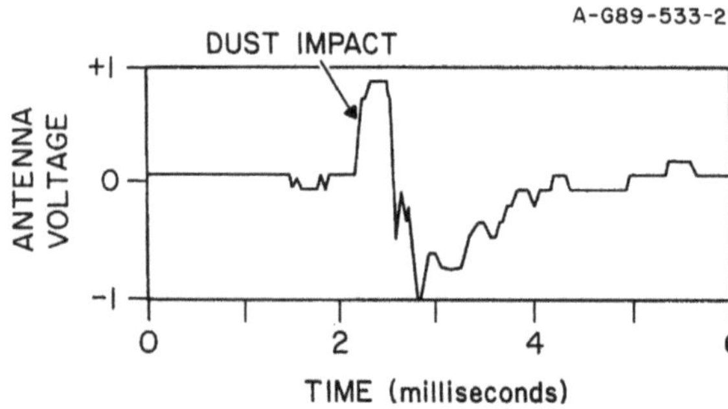

Figure 7. An example of the voltage pulse detected on the Voyager 2 electric field antenna due to a dust impact during the Saturn ring plane crossing on August 26, 1981. These pulses are believed to be due to micron-size particles striking the spacecraft body. Impact rates as high as 600 s^{-1} were observed at the center of the ring plane.

a high velocity (greater than a few km/s), the particle and part of the surface material are vaporized and heated to a very high temperature. The high temperature, $\sim 10^5$ K, causes a substantial amount of the material to be ionized (Hornung and Drapatz, 1981). As the resulting cloud of ionized gas expands over the electric antenna, some of the charge is collected by the antenna, thereby producing a voltage pulse in the receiver. An example of such a voltage pulse is illustrated in Figure 7, which shows the voltage waveform produced by a dust impact during the Voyager 2 ring plane crossing at Saturn (Gurnett *et al.*, 1983). Laboratory measurements show that the charge released is directly proportional to the mass of the particle (Grün, 1981). Since the resulting voltage on the antenna is directly proportional to the charge collected, the pulse amplitude is believed to be directly proportional to the mass of the impacting particle. During the Voyager 2 crossing of Saturn's ring plane, which occurred at 2.88 R_S, the plasma wave instrument detected impact rates as high as several hundred per second. Impact rates of at least 0.1 per second were observed at all radial distances out to at least 35 R_S (Gurnett *et al.*, 1983). Similar dust impact effects were observed by the Voyager 2 plasma wave instrument at Uranus and Neptune (Gurnett *et al.*, 1987, 1991). At sufficiently high impact rates the individual waveforms cause a quasi-continuous spectrum of voltage noise on the antenna, similar to "shot noise" in a vacuum tube. Aubier *et al.* (1983), Meyer-Vernet *et al.* (1986, 1996), Pedersen *et al.* (1991) and others have shown that this shot noise spectrum can be modeled to give information on the number density and mass distribution of the impacting particles. Similar impact ionization effects were also observed and analyzed during the Vega 1 and 2 flybys of Halley (Trotignon *et al.*, 1987; Oberc and Parzydlo, 1992).

With the RPWS we can study the distribution of dust particles in the Saturnian system by using dust impact detection techniques very similar to those used with the Voyager plasma wave and radio astronomy instruments. Several very significant improvements have been made in the RPWS that enhance our ability to detect and analyze dust impacts. On Voyager the waveform measurements were made by taking the potential difference between the two antenna elements, so that the antenna was operated as a dipole. Because of the dipole response, it was very difficult to estimate the charge released from the impact, and therefore, to estimate the mass of the impacting particle. On Cassini, we plan to directly measure the voltage on the antenna relative to the spacecraft body, similar to the technique that was used by the Voyager radio astronomy investigators. With this system, the charge released by the impact can be determined directly from the amplitude of the voltage pulse on the antenna, thereby giving a signal that is proportional to the mass of the impacting dust particle. On Voyager, the impact waveforms were often saturated because the waveform had only 4-bit resolution. This limitation made it impossible to obtain a meaningful mass estimate from the amplitude of the voltage pulse. The Cassini RPWS uses an automatic gain control and a waveform system with a larger dynamic range (8 bit). With the improved dynamic range it should be possible to routinely obtain mass distributions from pulse height analyses of the observed impact waveforms. An on-board microprocessor algorithm will be used to continuously identify and analyze dust impacts throughout the mission. This method of detecting and analyzing dust particles is expected to complement the measurements from the cosmic dust analyzer (CDA) (see Grün *et al.*, this issue). Although the CDA has a much better capability to analyze the properties of the particles detected, RPWS is expected to be able to detect much lower fluxes since the detecting area for the RPWS is much greater than the aperature area of the CDA. With the RPWS measurements we expect to achieve the following specific results relative to dust impacts.

- Determine the spatial distribution of micron-sized dust particles through out the Saturnian system.
- Measure the mass distribution of the impacting particles from pulse height analyses of the impact waveforms.
- Determine the possible role of charged dust particles as a source of field-aligned currents.

3. Required Instrument Characteristics

To achieve the scientific objectives of the Cassini mission, the instrumentation must have certain design and performance characteristics. This section gives the rationale for the RPWS instrument design and describes the performance required to achieve the scientific objectives described in the previous section. The topics discussed include: (1) field sensors, (2) frequency ranges, (3) frequency and time resolutions, and (4) sensitivities and dynamic ranges.

3.1. Field Sensors

A basic question that arises in the design of all radio and plasma wave instruments is what types of antennas should be used, electric or magnetic, and in each case, how many field components should be detected. For measurements of radio waves at frequencies well above the local characteristic frequencies of the plasma (i.e., above the electron cyclotron frequency and the electron plasma frequency), the propagation is essentially unaffected by the local plasma. At these frequencies, it is not necessary to measure both the electric and magnetic fields, since they have a constant known ratio, $E = cB$. In principle, it does not matter whether the electric field or the magnetic field is measured. However, for various reasons, both electrical and mechanical, it turns out that an electric antenna can operate at much higher frequencies and with a much greater sensitivity than a magnetic antenna of comparable size and weight (Gurnett, 1998). Therefore, for high frequency radio measurements, such as for SKR and SEDs, an electric dipole antenna is preferred. For plasma waves, which occur at lower frequencies, near or below either the electron cyclotron frequency or the electron plasma frequency, the situation is more complicated. Plasma waves can be either electrostatic, with no magnetic field, or electromagnetic, with both an electric field and a magnetic field. Although an electric antenna can detect both types of waves, the only way to conclusively distinguish an electrostatic wave from an electromagnetic wave is to use both an electric antenna and a magnetic antenna. If a wave can be detected with both an electric antenna and a magnetic antenna, then it is an electromagnetic wave. If it can be detected with an electric antenna, but not with a magnetic antenna, then it is an electrostatic wave. A major shortcoming of the Voyager plasma wave instrument, which used only an electric antenna, was the inability to distinguish electrostatic waves from electromagnetic waves. Therefore, for plasma wave measurements (i.e., at frequencies below either the electron cyclotron frequency or the electron plasma frequency) both electric and magnetic antennas should be used.

Next we consider the number of components to be measured. The Voyager electric field antenna consisted of two cylindrical elements mounted in a V configuration, see Scarf and Gurnett (1977) and Warwick *et al.* (1977). For the plasma wave instrument the two elements were used as a dipole, and for the radio astronomy instrument they were used as two orthogonal monopoles. Although the V configuration provided a limited capability to perform polarization measurements, no capability existed for performing direction-finding measurements. Since the RPWS scientific objectives require accurate determinations of source positions, as well as polarization measurements, it is essential that the RPWS be able to perform both direction finding (i.e., wave normal, $\hat{\mathbf{k}}$) and polarization measurements of high frequency radio signals (i.e., $f > f_{ce}$ and $f > f_{pe}$). Since at high frequencies the electric field is always perpendicular to the wave normal ($\mathbf{E} \cdot \hat{\mathbf{k}} = 0$), the wave normal direction can be determined by measuring the plane of rotation of the electric field (provided the wave is not linearly polarized). Therefore, the RPWS must include full three-axis electric field measurements. The easiest way to achieve this is to use three orthogonal electric monopoles. In the low frequency plasma wave part of the spectrum ($f < f_{pe}$ or $f < f_{ce}$), it is also important to carry out wave normal measurements. Unfortunately, when the anisotropic effects of the plasma are considered, the wave normal of an electromagnetic wave cannot be determined from electric field measurements. From Poisson's equation, $\nabla \cdot \mathbf{E} = \rho$, one can see that because the charge density, ρ, in a plasma is in general not zero, the electric field is no longer perpendicular to the wave normal, $\mathbf{E} \cdot \hat{\mathbf{k}} \neq \mathbf{0}$. Fortunately, Maxwell's equation $\nabla \cdot \mathbf{B} = \mathbf{0}$ always implies that $\mathbf{B} \cdot \hat{\mathbf{k}} = \mathbf{0}$. Therefore, to make wave normal measurements in the low frequency plasma wave part of the spectral three-axis magnetic field measurements are required. In addition, to assure that electrostatic waves can be detected and to resolve the ambiguity in the direction of propagation (i.e., along $\hat{\mathbf{k}}$ or $-\hat{\mathbf{k}}$), electric field measurements are also required, although these do not necessarily have to be three-axis measurements. In addition to electric field measurements, it is also useful to have the capability to measure electron density fluctuations in the plasma wave frequency range. This capability is easily achieved by biasing the electric antenna element with a known fixed bias current (as though it were a Langmuir probe). With this bias condition, it can be shown that the voltage variations on the element are proportional to the fractional electron density variation, $\delta n_e/n_e$. For electrostatic waves simultaneous measurements of both the electric field and the electron density variation can provide information on the wavelength of the wave (Kelley and Mozer, 1972; Wahlund *et al.*, 1998).

3.2. Frequency Ranges

To decide on the frequency ranges required for the electric and magnetic field measurements, we must consider the characteristic frequencies of the radio and plasma wave phenomena to be studied in the vicinity of Saturn. The frequency ranges and spectrums of all radio and plasma wave phenomena known or predicted to occur in the vicinity of Saturn are summarized in Figures 8–10. Figure 8 shows a model of the electron plasma frequency, f_{pe}, and electron cyclotron frequency, f_{ce}, as a function of radial distance near the equatorial plane on the dayside of Saturn; and Figures 9 and 10 show representative electric and magnetic field spectrums at a radial distance of 10 R_S. First, we consider the upper limit of the frequency range. Two high frequency limits must be considered, one for the electric field and the other for the magnetic field. The high frequency limit for the electric field is determined by the highest radio emission frequency of interest, and the high frequency limit for the magnetic field is determined by the highest plasma wave frequency of interest. As can be seen in Figure 8, lightning from Saturn's atmosphere (SEDs) has the highest frequencies. From the Voyager observations it is known that the SED spectrum extends up to at least 40 MHz. Since the shape of the high frequency part of the SED spectrum is already known (Zarka and Pedersen, 1983), we do not plan to make electric field measurements as high as 40 MHz. Our primary objectives relative to SEDs are to monitor the long-term occurrence of lightning and to study variations in the low frequency cutoff imposed by Saturn's ionosphere. For these purposes it is only necessary to measure the SED spectrum to frequencies slightly greater than the maximum ionospheric cutoff frequency. Since the maximum plasma frequency

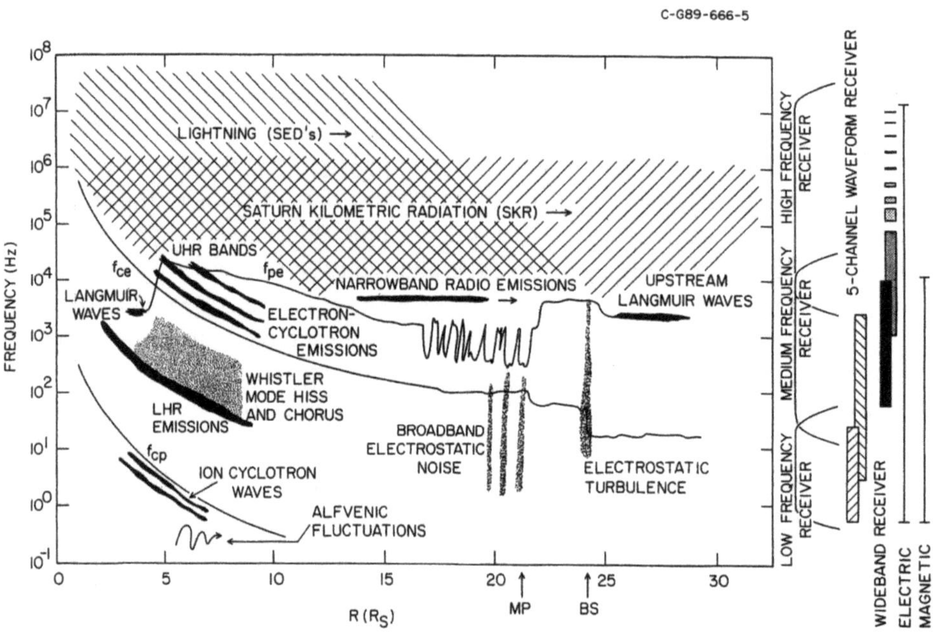

Figure 8. A summary of all the radio and plasma wave phenomena known and predicted to occur in the vicinity of Saturn displayed as a function of frequency and radial distance from Saturn. The bow shock (BS) and magnetopause (MP) positions, indicated by the arrows at the bottom of the plot, are representative of the dayside subsolar region.

of the dayside ionosphere is typically about 5 MHz, we have selected 16 MHz as the upper frequency limit for the electric field measurements. Since the ionospheric cutoff frequency varies as $f_{\text{cutoff}} = f_{\text{pe}}/\cos\theta$, where θ is the angle of incidence, this upper frequency limit allows the detection of lightning over a large range of incidence angles, approximately $0 \leq \theta \leq 70°$.

For the upper frequency limit of the magnetic field measurements we must consider the highest plasma wave frequency that is likely to be encountered in Saturn's magnetosphere. As discussed earlier, magnetic field measurements have two main purposes: (1) to distinguish electrostatic waves from electromagnetic waves, and (2) to determine the wave normal direction of electromagnetic waves. During the baseline tour, the periapsis radial distance is expected to be in the range from about 4 to 6 R_S, and the apoapsis is in the range from about 20 to 130 R_S. As can be seen from Figure 8 , the whistler mode, which is the highest frequency electromagnetic plasma wave mode that is likely to be of interest, is generally in the range from 3 to 10 kHz, with a maximum of about 12 kHz. Based on this magnetic field model we have selected 12 kHz as the upper limit for the magnetic field measurements. This allows magnetic field measurements with a simple tri-axial search coil magnetometer, which can be easily designed to respond to frequencies up to 12 kHz. In selecting this upper limit we realize that it will not be possible to

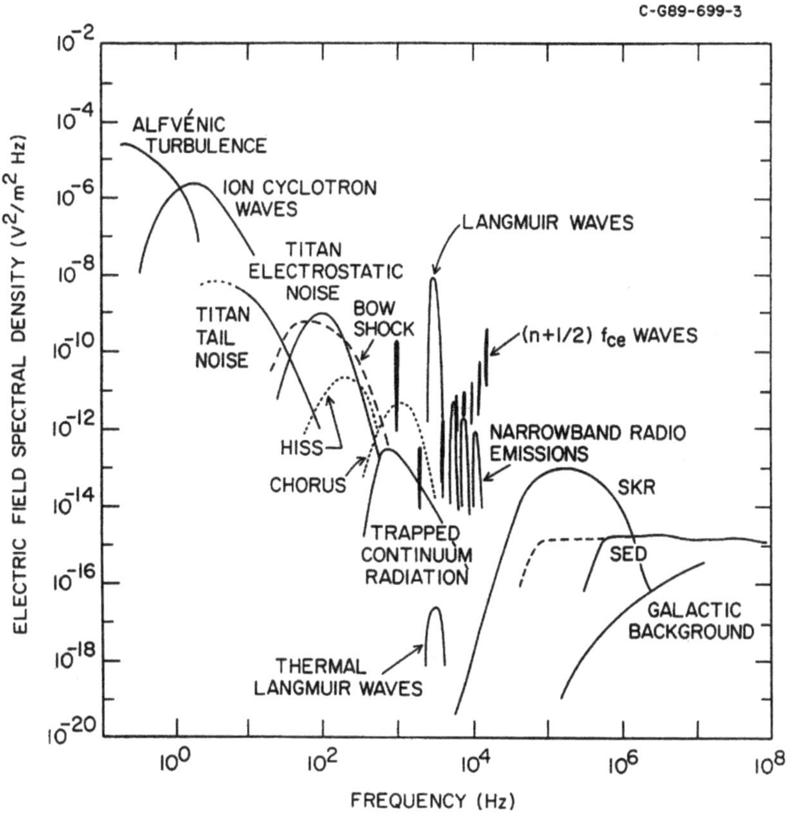

Figure 9. Typical electric field spectral densities of various radio emissions and plasma waves that are known to occur in the magnetosphere of Saturn. To a first approximation the SKR and SED intensities vary inversely as the square of the radial distance from Saturn, and are shown at a representative radial distance of 10 R_S.

detect the magnetic field of certain plasma wave modes that occur at frequencies above 12 kHz. For example, during the Titan flybys UHR emissions are expected to extend up to frequencies as high as several hundred kHz, and during Saturn orbit insertion (SOI), which is at about 1.3 R_S, electrostatic $(n+1/2)f_c$ electron cyclotron waves could extend up to frequencies as high as several hundred kHz. However, in all of these cases we feel confident that it will be easy to identify these electrostatic modes. For example, at Titan the UHR emissions will be at frequencies of ten to several hundred kHz, well above the electron cyclotron frequency, which is the highest frequency for the whistler mode. The situation is more complicated near SOI, where whistler-mode emissions could occur at frequencies as high as 500 kHz. To provide magnetic field measurements at such high frequencies would require the use of a loop antenna (Gurnett, 1998). Although such antennas have been flown in the past, they are large and very difficult to accommodate on the spacecraft, so a decision was made that high frequency magnetic field measurements were not

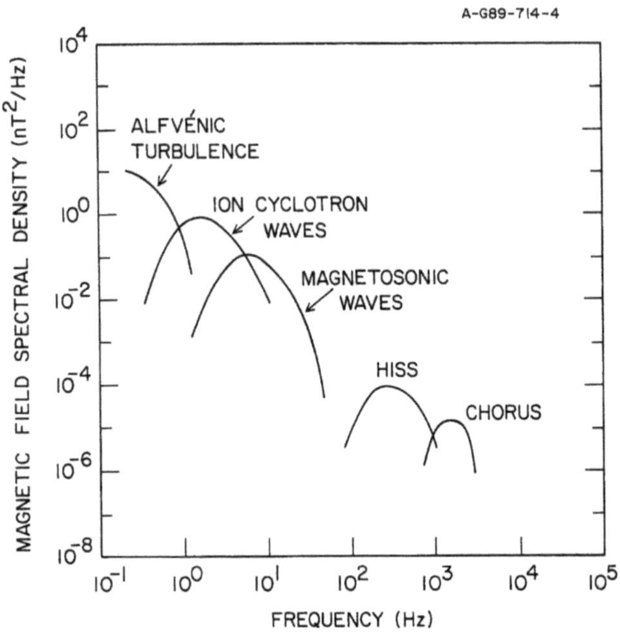

Figure 10. Typical magnetic field spectral densities of various radio emissions and plasma waves that are known to occur in the magnetosphere of Saturn.

justified to achieve this limited objective, given the additional resources that would be required.

Next, we consider the low frequency limit of the plasma wave spectrum. Since scientifically interesting plasma wave phenomena exist down to essentially zero frequency, the low frequency cutoff of the RPWS electric and magnetic field sensors is determined almost entirely by technical considerations. For example, conventional static field magnetometers tend to have better sensitivities than search coil magnetometers at frequencies below about 1 Hz. Since a tri-axial static field magnetometer (MAG) is included on Cassini (Southwood *et al.*, this issue), there is no reason to extend the frequency range of the RPWS magnetic antenna below about 1 Hz. Therefore, we have adopted 1 Hz as the low frequency cutoff of the RPWS magnetic field measurements. For the electric field antennas the situation is more complicated. The frequency response of the electronics for the electric field measurements could easily be extended below 1 Hz. However, because of sheath effects around the spacecraft body, such low frequency, quasi-static, electric field measurements require very long antennas, typically with lengths at least ten times the maximum dimension of the spacecraft body. Since such long antennas (~100 m, tip-to-tip) could not be accommodated on Cassini due to spacecraft dynamics considerations, we have arbitrarily defined the low frequency cutoff of the RPWS electric field measurements to be the same as for the magnetic field measurements (i.e., 1 Hz).

3.3. Frequency and Time Resolutions

It is well known that the frequency and temporal structure of Saturnian radio emissions and plasma waves vary over an extremely large range (Scarf *et al.*, 1984; Zarka, 1998). Some types of waves, such as continuum radiation and whistler-mode hiss, have smooth continuous spectrums that can be resolved with very modest frequency and time resolution. On the other hand, certain other types of waves, such as lightning-generated whistlers, whistler-mode chorus, and cyclotron maser radiation, have extremely complicated frequency-time structures. These structures often extend down to frequency and time resolutions on the order of $\Delta f \, \Delta t \sim 1$. Furthermore, in some cases it is necessary to stress high frequency resolution (i.e., small Δf), such as in the analysis of plasma resonances, whereas in other cases it is necessary to stress high-time resolution (i.e., small Δt), such as in the analysis of lightning and dust impacts. Resolving these conflicting demands is one of the main challenges that must be faced in designing a radio and plasma wave investigation.

What little is known about the fine structure of radio emissions and plasma waves at Saturn comes almost entirely from the Voyager measurements. The Voyager plasma wave instrument had a 16-channel spectrum analyzer spanning the frequency range from 10 Hz to 56 kHz, and a wideband waveform receiver that covered the frequency range from 50 Hz to 10 kHz. The time resolution of the 16-channel measurements was 4 s. During selected intervals, the wideband waveform provided high-resolution 48-s "snapshots" of the electric field waveform with a sample rate of 28,800 samples/s. However, the number of wideband frames that could be transmitted was severely limited by data rate considerations. The Voyager planetary radio astronomy instrument, which made measurements from 20 kHz to 40 MHz, had somewhat better frequency resolution ($\Delta f / f \sim$ 1–5%), but relatively poor time resolution (~6 s/sweep). Although the radio astronomy instrument also included a high-rate mode that allowed rapid (millisecond) sampling of a selected channel, again the amount of high-rate data collected was severely limited.

Since it is highly likely that radio emissions and plasma waves in Saturn's magnetosphere have fine-scale structures comparable to those observed in the Earth's magnetosphere, it is important that the RPWS instrument be designed with sufficient frequency and time resolution to resolve these structures. In particular, the RPWS should provide a substantial improvement relative to the Voyager radio and plasma wave instruments. Our basic approach to achieving this goal is to make measurements on two widely different frequency and time scales: nearly continuous low-rate spectral measurements with a frequency resolution on the order of a few percent and a time resolution on the order of a few seconds, and short-duration high-rate wideband waveform measurements with frequency and time resolutions approaching the limit, $\Delta f \Delta t \sim 1$. The low-rate survey measurements will allow us to produce frequency-time spectrograms comparable to Figure 6 (from the Galileo plasma wave instrument). To provide flexibility, the frequency and time resolution of the low-rate measurements must be controlled by a reprogrammable microprocessor, so that

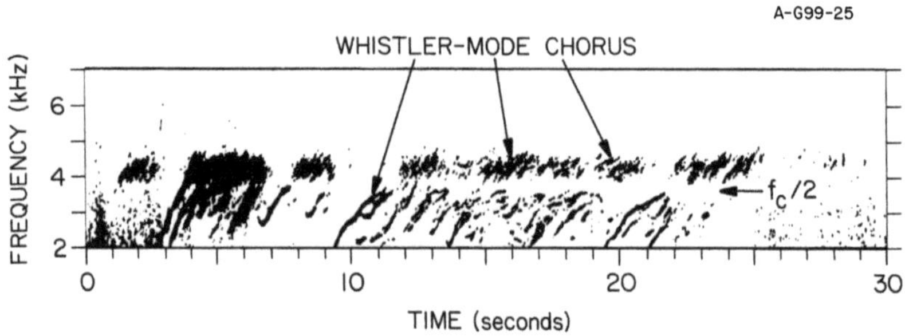

Figure 11. A high-resolution frequency-time spectrogram of whistler-mode chorus observed in Earth's magnetosphere. These emissions are produced by a cyclotron resonance interactions with energetic radiation belt electrons. The fine structure is caused by nonlinear resonant particle trapping effects. Note the prominent gap at one-half the electron cyclotron frequency, $f_c/2$.

they can be changed to accommodate unexpected results. The high-rate wideband waveform measurements will allow us to produce frequency-time spectrograms comparable to Figure 11, which shows a spectrogram of whistler-mode chorus in Earth's magnetosphere. Since the entire waveform is transmitted to the ground, these measurements have the advantage that the frequency and time resolution of the spectral processing can be adjusted during the ground processing to provide the optimum resolution for the phenomena being investigated, the only limit being that $\Delta f \Delta t \gtrsim 1$. To provide high-resolution measurements of high frequency radio emissions, the waveform receiver must also have a frequency conversion mode of operation that can provide waveform measurements in selected frequency bands at high frequencies. This mode of operation will allow us to determine whether SKR has fine structure comparable to terrestrial AKR. A frequency-time spectrogram showing the fine structure of terrestrial AKR using this type of frequency conversion mode of operation is shown in Figure 12. As can be seen, terrestrial AKR has fine structure extending down to bandwidths of only a few hundred Hz and time scales of tens of milliseconds, or less. Such high-resolution measurements of SKR will provide fundamental constraints on the mechanism by which these radio emissions are generated.

3.4. Sensitivities and Dynamic Ranges

The RPWS instrument must have sufficient sensitivity to detect the weakest signals of interest in the vicinity of Saturn, and still have adequate dynamic range to respond to the strongest signals without saturating. Figures 9 and 10 show the range of electric and magnetic field strengths that must be measured for various phenomena in the vicinity of Saturn. For some phenomena, such as the SKR and SED events, the intensities vary considerably. In these cases, the spectrums were selected from periods of relatively high intensity at a radial distance of 10 R_S. As

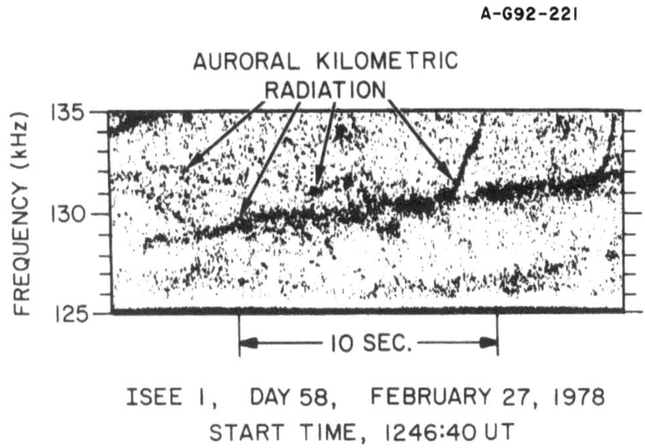

Figure 12. A high-resolution frequency-time spectrogram of terrestrial auroral kilometric radiation (AKR) showing the extremely complicated fine structure of this radiation. The fine structure is most likely caused by nonlinear resonant particle trapping effects, similar to chorus. The Cassini RPWS should be able to determine whether SKR has similar fine structure.

can be seen, the intensities vary over a wide range. For electric fields the spectral densities that must be measured range from a minimum of about 10^{-18} V^2 m^{-2} Hz^{-1} at 10^6 Hz, to a maximum of about 10^{-6} $V^2m^{-2}Hz^{-1}$ at 1 Hz, a total range of 120 dB. For the magnetic field the spectral densities that must be measured range from a minimum of about 10^{-7} nT^2 Hz^{-1} at 10^3 Hz to a maximum of about 10^1 nT^2 Hz^{-1} at 1 Hz, a total range of 80 dB. Since it is very difficult to achieve a total dynamic range of 120 dB, special techniques must be used to accommodate this very large range of signal strengths. For example, since the intensities tend to increase toward lower frequencies, the gain near the front end of the receiving system must be decreased at low frequencies in order to avoid saturation when strong low frequency signals are present. Also, since digital waveforms typically cannot accommodate such large dynamic ranges, automatic gain control systems must be used for all waveform measurements. These and a variety of other techniques must be used in order to assure that the instrument can perform reliable measurements over the large range of field strengths illustrated in Figures 9 and 10 .

4. Instrument Description

The RPWS instrumentation consists of three electric antennas, three magnetic antennas, a Langmuir probe and its associated electronics, and five specialized receivers designed to address the scientific objectives and performance requirements discussed in the two previous sections. In this section we describe each of these elements of the instrument.

4.1. Block Diagram

A simplified block diagram of the RPWS instrument is shown in Figure 13. Three monopole electric field antennas, labeled E_u, E_v, and E_w, are used to provide electric field signals to the various receivers. The orientations of these three antennas relative to the x, y, and z axes of the spacecraft are shown in Figure 14. By electronically taking the difference between the voltages on the E_u and E_v monopoles, these two antennas can be used as a dipole, E_x, aligned along the x-axis of the spacecraft. The E_u and E_v antennas also can be used to sound the local plasma by transmitting short pulses. In an alternate mode of operation, they can be biased and used as Langmuir probes to measure the phase velocity of density structures in the plasma. The tri-axial search coil magnetic antennas, labeled B_x, B_y, and B_z in Figure 13, are used to detect three orthogonal magnetic components of electromagnetic waves. The search coil axes are aligned along the x, y, and z axes of the spacecraft. The spherical Langmuir probe, shown at the bottom of the block diagram, is used for electron density and temperature measurements. Both the electric antennas and the Langmuir probe can be used to detect dust impacts.

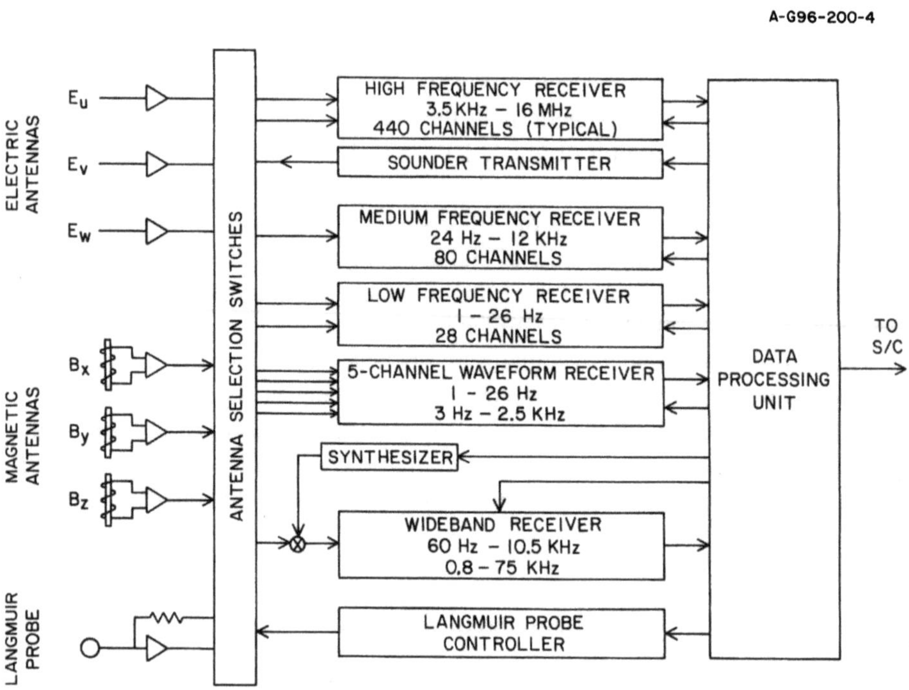

Figure 13. A functional block diagram of the RPWS instrument. The seven sensors, three electric, three magnetic, and the Langmuir probe are shown on the left side. In the middle, the several receivers which the instrument uses to analyze the signals are shown. The data management and control functions, as well as the interface with the spacecraft are shown as the data processing unit on the right side.

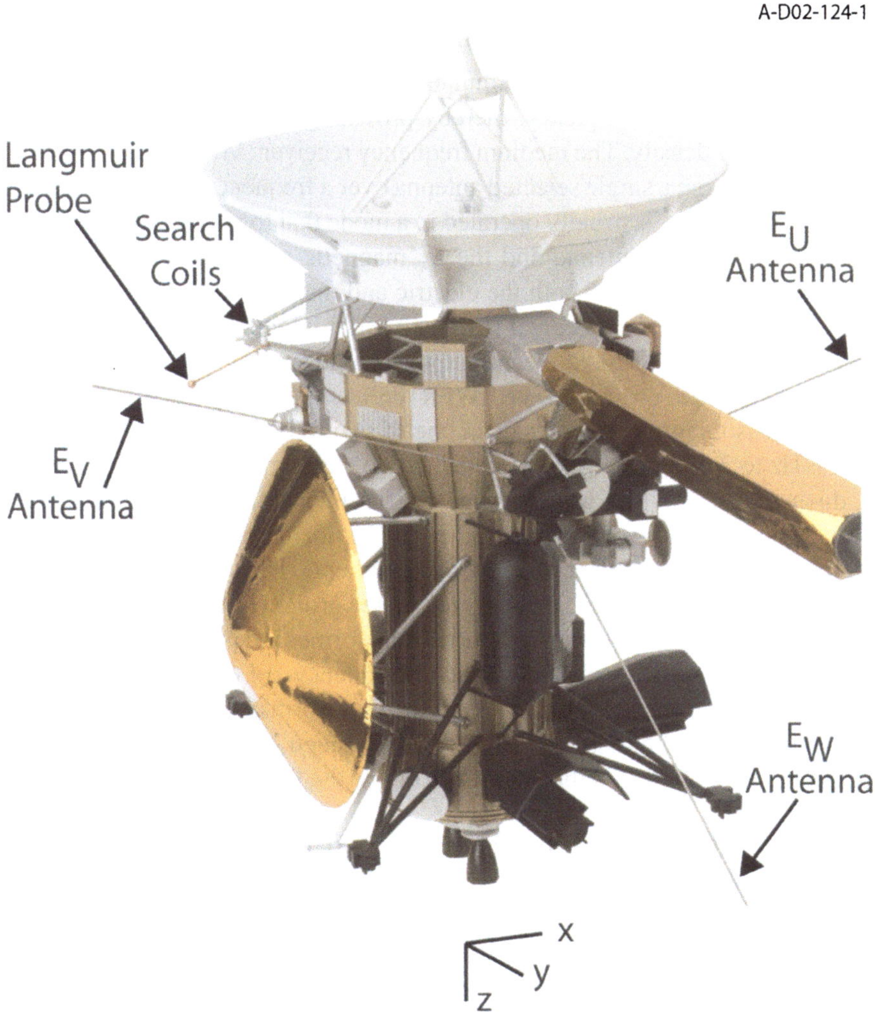

Figure 14. The Cassini spacecraft showing the locations of the RPWS sensors and their relationship with other structures on the spacecraft.

Next, we consider the function of the five receivers shown in the middle column of the block diagram in Figure 13 . These receivers are connected to the antennas described above by a network of switches in the block labeled "antenna selection switches," the details of which are not shown. The high frequency receiver (HFR) provides simultaneous auto- and cross-correlation measurements from two selected antennas over a frequency range from 3.5 kHz and 16 MHz. By switching the two inputs of this receiver between the three monopole electric antennas, this receiver can provide direction-of-arrival measurements, plus a full determination of the four Stokes parameters. The high frequency receiver includes a processor that performs all of its digital signal processing, including data compression. The high frequency receiver also includes a sounder transmitter that can be used to transmit short square wave pulses from 3.6 to 115.2 kHz. When used in conjunction with the high frequency receiver, the sounder can stimulate resonances in the plasma, most notably at the electron plasma frequency, thereby providing a direct measurement of the electron number density. The medium frequency receiver (MFR) provides intensity measurements from a single selected antenna over a frequency range from 24 Hz to 12 kHz. This receiver is usually operated in a mode that toggles every 32 s between the E_{x} electric dipole antenna and the B_{x} magnetic search coil, thereby providing spectral information for both the electric and magnetic components of plasma waves. The low frequency receiver (LFR) provides intensity measurements from 1 to 26 Hz, typically from the E_{x} electric dipole antenna and the B_x magnetic antenna. The five-channel waveform receiver (WFR) collects simultaneous waveforms from up to five sensors for short intervals in one of two frequency bands, either 1 to 26 Hz, or 3 Hz to 2.5 kHz. When connected to two electric and three magnetic antennas, this receiver provides wave normal measurements of electromagnetic plasma waves. The wideband receiver is designed to provide nearly continuous wideband waveform measurements over a bandwidth of either 60 Hz to 10.5 kHz, or 800 Hz to 75 kHz. These waveforms can be analyzed on the ground in either the temporal domain, or in the frequency domain (Fourier transformed) to provide high-resolution frequency-time spectrograms. In a special frequency-conversion mode of operation, the high frequency receiver can provide waveforms to the wideband receiver in a 25-kHz bandwidth that is tuneable to any frequency between 125 kHz and 16 MHz. Table II summarizes the characteristics of the five basic receivers described above.

The Langmuir probe controller shown in the block diagram is used to sweep the bias voltage of the probe over a range from –32 to +32 V in order to obtain the current-voltage characteristics of the probe, and thereby the electron density and temperature. The controller can also set the bias voltage on the E_{u} and E_{v} monopoles over a range from –10 to +10 V in order to operate them in a current collection mode for $\delta n_{\mathrm{e}}/n_{\mathrm{e}}$ measurements.

The RPWS data processing unit shown on the right-hand side of the block diagram consists of three processors. The first processor, called the low-rate processor, controls all instrument functions, collects data from the high frequency receiver, the medium frequency receiver, the low frequency receiver, and the Langmuir probe, and carries out all communications with the spacecraft command and data system (CDS). The second processor, called the high-rate processor, handles data from the wideband and five-channel waveform receivers and passes the data along to the low-rate processor for transmission to the CDS. The third processor, called the data compression processor, is primarily used for data compression, but can also perform specialized operations such as on-board dust detection by using waveforms from the wideband receiver.

As shown in Figure 14, the RPWS hardware is mounted in several locations around the Cassini spacecraft. The main electronics package, which includes the

TABLE II
Summary of RPWS receiver characteristics.

Receiver section	Measurements	Spectral range	Spectral resolution	Temporal resolution
High frequency receiver (HFR)	$E_1 = E_u$, E_v or E_x $E_2 = E_w$; $\|E_1\|^2$, $\|E_2\|^2$, Re($E_1 \bullet E_2$*), Im($E_1 \bullet E_2$*)	3.5–318 kHz 0.125–16.125 MHz	$\Delta f/f = 5$, 10, 20% (log); $\Delta f = 3.1$ kHz to $n \times 25$ kHz (linear)	0.1–10 s/spectrum
Medium frequency receiver (MFR)	One of: E_x, E_u, E_v, E_w, B_x, B_z	24–180 Hz; 180 Hz–1.5 kHz; 1.5–12 kHz	$\Delta f/f \simeq 13\%$; $\Delta f/f \simeq 7\%$; $\Delta f/f \simeq 7\%$	16 s/spectrum; 16 s/spectrum; 8 s/spectrum
Five-channel waveform receiver (WFR)	Five of: E_x, E_u, E_v, E_w, B_x, B_y, B_z, LP	~1–26 Hz; 3 Hz–2.5 kHz	0.2 Hz, typical; 3.5 Hz typical	1 data set/5 min, typical, 1 data set/16 s, max
Low frequency receiver (LFR)	Two of: E_x, E_u, E_v, E_w, B_x, B_y, B_z, LP	~1 - 26 Hz	$\Delta f/f = 12\%$	$1E + 1B$ spectrum/16s, typical
Wideband receiver (WBR)	One of: E_x, E_u, E_v, E_w, B_x, LP	60 Hz–10.5 kHz; 0.8–75 kHz	13.6 Hz, typical; 109 Hz, typical	125ms/spectrum, typical

medium frequency receiver and the digital processing unit, resides in bay 4 of the main spacecraft bus. The electric antenna deployment mechanisms and the high frequency receiver are mounted on a bracket on the $+y$ side of the spacecraft, just below the base of the magnetometer boom. The magnetic search coil assembly is supported by a short boom that is attached to the high gain antenna support structure on the $-x$ side of the spacecraft. The Langmuir probe is mounted on the end of an 0.8-m hinged boom that extends outward from the magnetic search coil assembly.

4.2. Electric Antennas

The three electric monopole antennas and their deployment mechanisms were provided by Orbital Sciences Corporation. The antenna elements consist of conducting cylinders, each 10 m long and 2.86 cm in diameter. The elements are made of beryllium-copper, silver plated on the exterior surface, and painted black on the interior for thermal control. Approximately 12% of the surface area is perforated with small holes to allow sunlight to heat the shaded side of the element in order to reduce thermal bending. The elements themselves are formed by two opposing semi-cylindrical strips with interlocking tabs. For launch, the elements are flattened

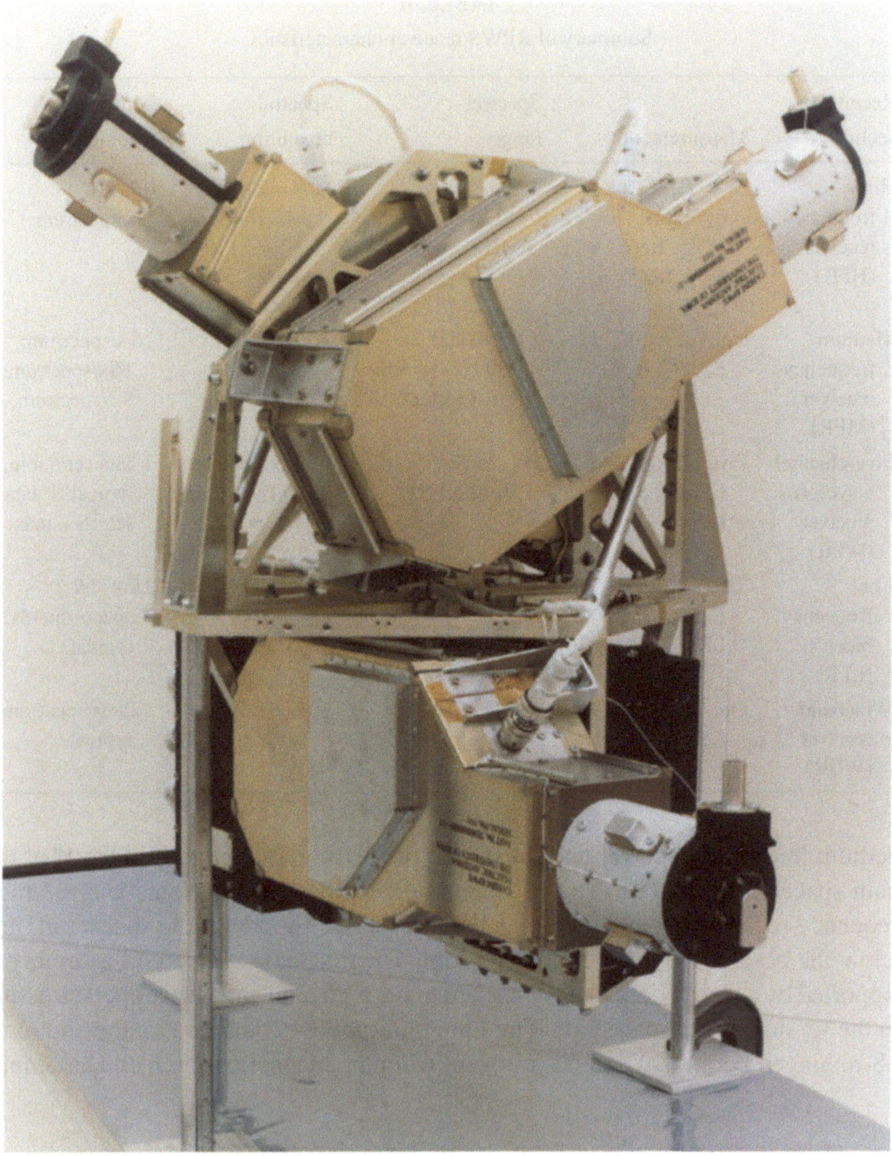

Figure 15. A photograph of the RPWS electric antenna assembly. The three electric monopole deployment mechanisms are most obvious in this view, however, both the motor control electronics and the high frequency receiver are also part of this assembly.

and rolled onto a spool. During deployment after launch, a motor-driven deployment mechanism feeds the element through a forming channel that expands the strips into a cylindrical tube. A photograph of the three electric antenna deployers, the high frequency receiver, and the associated mounting bracket is shown in Figure 15.

TABLE III

Comparison of physical and electrical orientations of the electric antennas.

	Physical orientation*		Electrical orientation (Rheometry)	
Antenna†	θ (deg)	ϕ (deg)	θ (deg)	ϕ (deg)
E_u	107.5	24.8	107.9	16.5
E_v	107.5	155.2	107.3	162.7
E_w	37	90	31.4	91.2

*The angle θ is the polar angle measured with respect to the spacecraft $+z$ axis, ($\cos\theta = z/\sqrt{x^2+y^2+z^2}$), (and the angle ϕ is azimuth angle measured with respect to the $+x$ axis ($\tan\phi = x/y$).

†In some previous papers (Ladreiter *et al.*, 1995; Rucker *et al.*, 1996) and various engineering documents, E_u, E_v, and E_w antennas are labeled E_{x+}, E_{x-}, and E_z. The u, v, w, notations are used in this paper to avoid confusion with the spacecraft x, y, and z axes.

The physical orientations of the three elements with respect to the x, y, and z axes of the spacecraft are provided in Table III. The E_u and E_v elements are extended symmetrically at angles of 60° with respect to the spacecraft y-z plane. The plane containing these two elements is rotated 37° with respect to the spacecraft x-y plane. These two elements can be used to provide a dipole antenna (E_x) with a tip-to-tip length of 18.52 m parallel to the spacecraft x-axis. The E_w antenna is extended perpendicular to the plane formed by the E_u and E_v elements (i.e., at an angle of 37° from the spacecraft z-axis). Several considerations led to this antenna configuration. First, the three monopoles provide the nearly orthogonal tri-axial configuration required for direction finding. Second, as previously described, the E_u and E_v elements can be operated as a dipole (E_x,), which minimizes common-mode coupling in order to provide the lowest possible level of spacecraft-generated interference. The E_u and E_v elements are also symmetric with respect to the magnetometer boom, which minimizes the effect this boom has on the dipole antenna pattern. Third, the w-axis element is rotated somewhat away from the spacecraft in order to reduce electrical coupling and interference from electrical equipment in the lower bay of the spacecraft. Fourth, the E_u and E_v elements are rotated away from the spacecraft x-y plane so that they do not interfere with the field of view of the stellar reference unit. This orientation also provides clearance for release of the Huygens probe.

The three electric antennas were deployed to their full 10-m length during a 30-h activity on October 25, 1997, ten days after launch. Full deployment was confirmed by both a potentiometer reading and a limit switch indication, the latter being the most reliable indication of full deployment. A brief set of measurements performed subsequent to the deployments verified that each antenna was properly connected to the instrument and that there were no shorts to the spacecraft structure. There are no plans to retract or otherwise change the length of the elements for the duration of the mission.

Due to the complex shape of the Cassini spacecraft, the effective electrical axes of the monopole antennas differ significantly from their physical orientations. Electrical measurements using a 1:30 scale model of the spacecraft in a tank of electrolytic fluid (rheometry) were performed by Rucker *et al.* (1996). These measurements show angular offsets of the electrical axes by as much as 5 to 6 degrees for the E_w antenna, and 7 to 8 degrees for the E_u and E_v antennas (see Table III). Ladreiter *et al.* (1995) have developed an analysis technique that can be used to perform in-flight determinations of the electrical axes by using a point source with known polarization characteristics. Jupiter provides the best source for this purpose and in-flight calibrations of the electrical axes of the antennas were carried out during the Jupiter flyby, which took place in December 2000 and January 2001. Preliminary results of this calibration are given by Vogl *et al.* (2001). Although this calibration occurred with the Huygens probe attached to the spacecraft, the rheometry results show that the electrical axes shift by only a small amount (less than one degree) when the probe is released.

4.3. Magnetic Antennas

The tri-axial search coil magnetic antennas were provided by CETP in Velizy, France. These antennas are mounted on a short, fixed boom just under the high gain antenna on the $-x$ side of the spacecraft, as shown in Figure 14 . The three axes of the search coils are aligned with the x, y, and z axes of the spacecraft. Each search coil utilizes 10,000 turns of 0.07 mm wire wound around a μ-metal core 25 cm long with a cross-sectional area of 15 mm^2. The sensors have a usable frequency range of 1 Hz to 20 kHz. A flux feedback winding is used in each search coil to flatten the frequency response over the range from 30 Hz to 18 kHz. The transfer function in this frequency range is approximately 150 mV/nT. Preamplifiers are mounted at the base of the mounting boom. When the spacecraft $-z$ axis is pointed toward the Sun, as it is whenever the spacecraft is in the inner solar system, the high gain antenna acts as a sun shade for the search coils. Two radioisotope heating units are used to provide an internal heat source for thermal control. A thermal blanket system surrounds the assembly to minimize the radiated heat loss. A photograph of the search coil magnetic antenna assembly is shown in Figure 16.

4.4. Langmuir Probe

The Langmuir probe consists of a 5-cm diameter titanium sphere mounted on the end of a 0.8-m hinged boom that folds outward from one leg of the boom that holds the search coil magnetic antenna assembly. A preamplifier is mounted near the base of the search coil boom. To minimize the influence of photoelectrons,

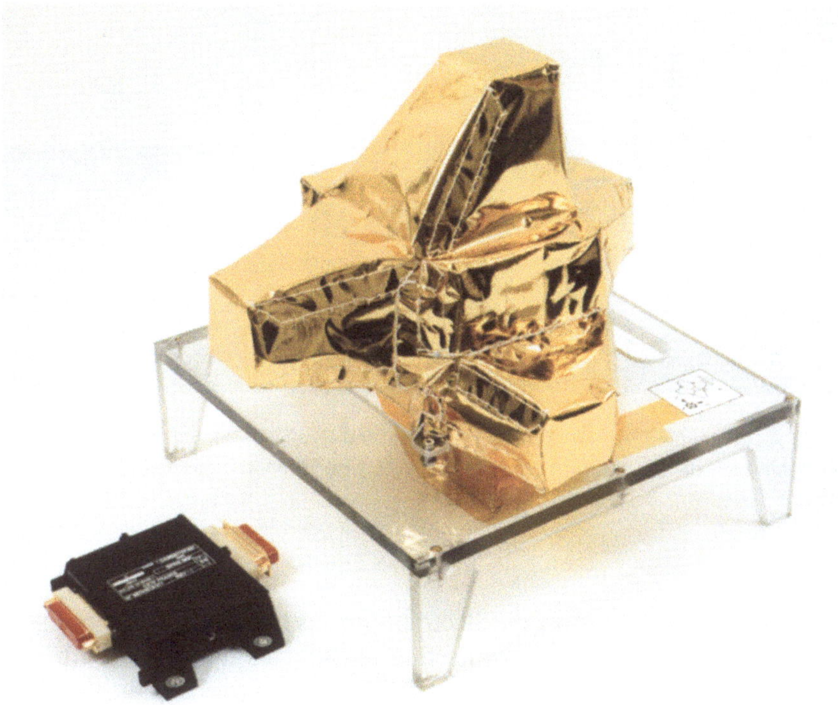

Figure 16. A photograph of the tri-axial magnetic search coil assembly. Thermal blankets are shown as flown.

the outermost 10.9-cm segment of the boom is made as thin as possible (diameter = 6.35 mm) and has an outer guard shield that is held at the same voltage as the probe sphere. In its deployed configuration, the probe is approximately 1.5 m from the nearest spacecraft surface. When the $-z$-axis of the spacecraft is sun-pointed, as it is in the inner solar system, the probe is in the sunlight. The surface of the probe is coated with titanium nitride by baking at high temperature in a nitrogen atmosphere (Whalström *et al.*, 1992). This treatment provides a durable surface with a constant work function so that the potential of the probe surface is uniform. (If the potential varies over the surface, the current-voltage characteristic is adversely affected.) This coating effectively eliminates hysteresis effects, which can occur for other surface coatings. A photograph of the Langmuir probe is shown in Figure 17.

The boom holding the Langmuir probe was deployed on 15 October 1997 as part of the launch sequence. Deployment was verified by noting that the probe emitted the expected photoelectron current when biased to a negative voltage (the spacecraft orientation was such that the probe was in sunlight when deployed).

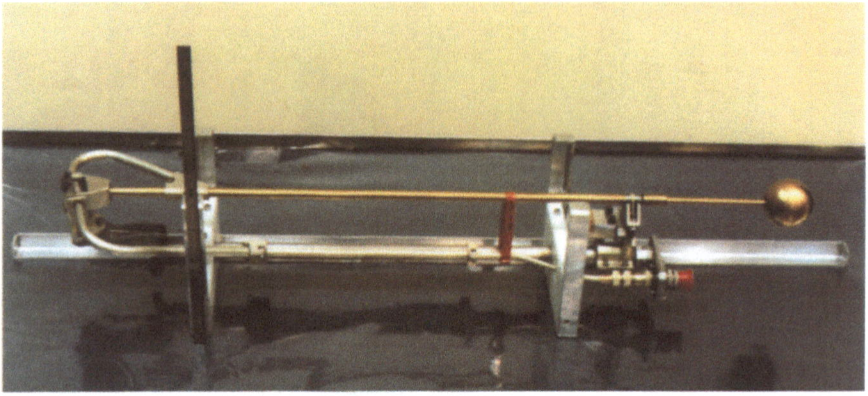

Figure 17. A photograph of the Langmuir probe and boom assembly in its stowed configuration on one of the struts of the tripod that also supports the magnetic search coil assembly.

As mentioned earlier, two of the electric antennas (E_u, E_v) can also be operated as Langmuir probes in a current-collecting mode. This mode is used primarily for studying plasma density variations associated with waves and turbulence. The current measurement range is 1 nA to 100 μA for positive bias voltages, and 100 pA to 100 μA for negative bias voltages.

4.5. High Frequency Receiver

The high frequency receiver consists of two sets of four analog receivers followed by a digital signal processing unit. The design is based on a similar receiver flown on the Wind spacecraft (Bougeret *et al.*, 1995). One set of receivers is connected to the E_w monopole. The other set of receivers can be connected either to the E_x dipole or toggled between the E_u and E_v monopoles. Three of the four receivers have fixed-frequency filters (A, B, and C) that together cover a frequency range from 3.5 to 319 kHz. Each filter covers 2.2 octaves in frequency (i.e., a factor of 4.5). Digital spectral analysis is performed within each of the three filter bands to provide 8, 16, or 32 logarithmically spaced frequency channels, yielding spectral resolutions of 20, 10, or 5%, respectively. The fourth (HF) receiver is tunable over a frequency range from 100 kHz to 16.1 MHz and has a bandwidth of 25 kHz. This receiver can be tuned in two different modes, HF1 and HF2. HF1 provides measurements over a frequency range from 100 kHz to 4.125 MHz in 25-kHz steps. HF2 provides measurements over a frequency range from 125 kHz to 16.125 MHz in 50-kHz steps. Within the 25 kHz passbands of HF1 or HF2, it is possible to have either 1, 2, 4, or 8 linearly spaced channels, thereby giving frequency resolutions of 25, 12.5, 6.25, and 3.125 kHz. The high frequency receiver has an automatic

TABLE IV

Characteristics of the high frequency receiver.

Band	Frequency range	Frequency steps	Frequency resolution	Integration times
A	3.5–16 kHz	8, 16, or 32 (logarithmic)	20, 10, or 5%	0.125, 0.25, 0.5, or 1 s
B	16–71 kHz	8, 16, or 32 (logarithmic)	20, 10, or 5%	0.125, 0.25, 0.5, or 1 s
C	71–319 kHz	8, 16, or 32 (logarithmic)	20, 10, or 5%	0.125, 0.25, 0.5, or 1 s
HF1	125 kHz–4.125 MHz	n × 25 kHz; 1, 2, 4, or 8 linear channels within 25-kHz band	3.125, 6.25, 12.5, 25, or n × 25 kHz	20, 40, 80, or 160 ms
HF2	125 kHz–16.125 MHz	n × 50 kHz; 1, 2, 4, or 8 linear channels within 25-kHz band	3.125, 6.25, 12.5, 25, or n × 50 kHz	10, 20, 40, or 80 ms

gain control (AGC) that provides a dynamic range of about 90 dB. In addition, a switchable attenuator provides an additional 26 dB of dynamic range. At any specific gain setting of the AGC, the digital spectrum analyzer has a dynamic range of about 30 dB. Table IV provides a summary of the basic characteristics of the high frequency receiver.

Using the two sets of receivers, complex auto- and cross-correlation measurements can be performed between the E_x dipole and E_w monopole antennas, thereby giving the amplitude and relative phase of signals detected by the two antennas. This mode of operation allows for polarization and direction-finding measurements of purely circularly polarized radio emissions such as cyclotron maser radiation. In an alternate mode of operation, the first of the two sets of receivers can be toggled rapidly between the E_u and E_v monopoles, with the second set on the E_w antenna, in order to provide direction-finding measurements.

For low frequency space-borne radio measurements, the wavelength λ is usually much larger than the length L of the antenna. Under these conditions ($\lambda >> L$) angular resolution is virtually nonexistent. Nevertheless, the direction to the center of the source can be obtained by comparing the amplitudes and relative phases of signals from various combinations of antennas. On a rotating spacecraft, direction-finding measurements can be performed with only two antennas. This technique was used by the Ulysses spacecraft to derive the location of Jovian low frequency radio sources with an accuracy of approximately 1–2° (Reiner *et al.*, 1993; Ladreiter *et al.*, 1994). The main limitation of the rotating antenna technique is that the radio emission characteristics must remain essentially constant during one spacecraft rotation, which usually takes 10 s or more. This is a serious limitation for auroral radio emissions, which often have rapid intensity variations, sometimes on time scales of a fraction of a second or less. Such rapid variations explain why the technique could be applied to only limited portions of the Jovian radio spectrum (Ladreiter *et al.*, 1994). Since Cassini is three-axis stabilized the rotating antenna technique obviously cannot be used. Ladreiter *et al.* (1995) have shown that direction-finding and full polarization information (i.e., four Stokes parameters) can be obtained with three nearly orthogonal antennas. Direction-finding and polarization measurements are strongly interdependent parameters and, ideally, measurements should be made simultaneously on all three antennas. However, since the high frequency receiver can only process signals from two antennas at a time, direction-finding and polarization measurements are made using consecutive correlation measurements with two different pairs of antennas, (E_u, E_w) and (E_v, E_w). The time between these measurements is typically 45 to 325 ms (above 125 kHz) or 250 to 2000 ms (below 125 kHz). Because of the very short time interval between successive measurements, this technique is expected to be relatively insensitive to short-term intensity variations.

The high frequency receiver utilizes digital filters that have an out-of-band rejection greater than 45 dB. They are preceded by an anti-aliasing filter and an automatic gain control (AGC). The AGC normalizes the signal amplitude to a level that provides optimal performance for the digital signal processor. The first receiver has an input, which for simplicity we call E_1, that is selectable from either E_x, E_u, or E_v. The second receiver has an input, which we call E_2, from the E_w antenna. The digital processor provides two auto-correlations, $|E_1|^2$ and $|E_2|^2$, and two cross-correlation $\mathrm{Re}(E_1 \cdot E_2^*)$ and $\mathrm{Im}(E_1 \cdot E_2^*)$, for each frequency channel. In the normal mode of operation, a series of auto- and cross-correlation measurements is averaged in order to reduce fluctuations to the digital quantization level (1 bit = 0.375 dB). The exact number of spectrums integrated is controlled by command and depends on the detailed choices of frequency and bandwidth. The intrinsic noise level of the A, B, and C bands is 4.9×10^{-17} V^2 Hz^{-1} for the E_1 receiver and 2.5×10^{-17} V^2 Hz^{-1} for the E_2 receiver. For HF1 and HF2, these noise levels are typically double the A, B, and C noise levels due to mixer noise.

The high frequency receiver also has the capability of providing a frequency-converted signal to the wideband receiver. In the frequency conversion mode of operation, the high frequency receiver translates high frequency signals downward into a 50–75-kHz passband, which is then sent to the wideband receiver. This passband can be shifted to any frequency between 4.125 and 16 MHz, except those centered at multiples of 50 kHz, and to any frequency between 125 kHz and 4.125 MHz.

The high frequency receiver also includes a capability to measure amplitudes in a selected frequency channel with millisecond temporal resolution. In this mode of operation any frequency between 125 kHz and 16 MHz can be selected. This capability allows very short duration signals such as lightning to be detected and can be used for envelope sampling of bursty emissions.

TABLE V
Sounder transmission characteristics.

Band	Frequency range (kHz)	1/Pulse width (Hz)
1	3.6–7.2	200
2	7.2–14.4	400
3	14.4–28.8	800
4	28.8–57.6	1600
5	57.6–115.2	3200

Since the high frequency receiver is under the control of its own processor, virtually all of the receiver parameters can be selected by command, making for an extremely flexible instrument. This flexibility allows for regular surveys of the radio frequency spectrum of Saturn at low data rates (typically about 450 bps), as well as specialized studies of short duration phenomena at very high spectral resolution (the maximum number of frequency channels across the frequency range of the receiver is approximately 3300) as well as direction-finding and full polarization measurements. To minimize the data rate for a given mode, the processor also includes both lossless (Meander code) and lossy (Rice code) compression algorithms.

4.6. Sounder Transmitter

The high frequency receiver also includes a sounder transmitter. The transmitter design is based on the Ulysses/URAP instrument (Stone *et al.*, 1992). In the sounder mode of operation, a short pulse consisting of a 26-V peak-to-peak square wave is transmitted on the E_x dipole antenna, and the receiver then "listens" for a resonance response from the plasma. After each pulse the transmitter frequency is increased by a small increment. This process is continued across the entire frequency range of interest, thereby producing a spectrum of the plasma resonances. When a resonance is encountered, the received signal is strongly enhanced for a short period of time after the pulse has been transmitted. The sounder can measure electron densities from 0.2 to 164 cm^{-3} ($f_{pe} \approx 3.6$ to 115.2 kHz), and is expected to be active for a few seconds out of every 5 to 10 min.

The sounder can transmit in five bands over a frequency range from 3.6 to 115.2 kHz. The five bands are described in Table V. Each band is divided into 18 channels for a total of 90 frequencies. The default bandwidths are given in Table V. The upper and lower frequency limits of the frequency sweep can be selected by command.

Two modes of operation for the sounder are currently planned, designated PAA and AAA. Each of these two modes involves taking three measurements at the selected frequency. In the PAA (Passive Active) mode, a "passive" measurement is taken before the pulse is transmitted, and then two "active" measurements are

TABLE VI

Medium frequency receiver characteristics.

Band	Frequency range	$\Delta f/f$ (%)	Effective bandwidth (Hz)	Sweep time (s)
1	24–180 Hz	13	5.6	16
2	180 –1500 Hz	7	19.4	16
3	1.5–12 kHz	7	139	8

taken; the first at a time T_0 after the pulse, and the second at a time T_1 after the first measurement. In the AAA mode, three "active" measurements are taken sequentially at time intervals of T_0, T_1, and T_2 after transmission of the pulse. The parameters T_0, T_1, and T_2 can be selected by command.

4.7. Medium Frequency Receiver

The medium frequency receiver (MFR) system is based on similar receivers flown on the ISEE, Galileo, and Polar spacecraft (Gurnett *et al.*, 1978, 1992, 1995). The purpose of the medium frequency receiver is to provide continuous spectral measurements over a frequency range from 24 Hz to 12 kHz, with moderate frequency and temporal resolution and a relatively low data rate. This receiver system has the capability of processing signals from the B_x, B_z, E_u, E_v, E_w and E_x, antennas. In a typical mode of operation, the receiver toggles between the E_x dipole antenna and the B_x magnetic antenna in order to provide alternating electric and magnetic spectrums.

The medium frequency receiver consists of three frequency bands designated 1, 2, and 3, each covering a 3 octave frequency range. Band 1 is divided into 16 logarithmically spaced frequency channels ($\Delta f/f \simeq 13$ %) and bands 2 and 3 each have 32 frequency channels ($\Delta f/f \simeq 7$ %). The effective bandwidths for the three bands are given in Table VI. The frequencies are selectable using a two-stage frequency conversion scheme. Each band of the medium frequency receiver uses a mixer to convert the selected frequency to a fixed intermediate frequency (IF) filter. A second mixer then down-converts the IF signal to a baseband filter. The baseband signal is logarithmically compressed, rectified, and summed to provide a 0–5 volt DC output, and then converted to an 8-bit binary number by an analog-to-digital converter. The dynamic range of the medium frequency receiver, from the lowest signal that can be detected to the saturation level, is about 110 dB.

The upper frequency limits of the three medium frequency receiver frequency bands are determined by three bandpass filters at the front end of the receiver. The lower frequency limits are imposed by a combination of the bandpass filters and the low frequency limit of the front end (IF) mixer. The filter frequencies are scaled upward by a factor of eight for each successive higher band, thereby providing frequency coverage over nine octaves. The low-pass filters provide 55 dB of rejection for out-of-band signals.

TABLE VII
Waveform receiver input selections.

Channel	Input selection
1	E_x dipole, E_u*
2	E_w, E_v*
3	B_x, Langmuir probe*
4	B_y
5	B_z

*Operating in Langmuir probe (current collecting) mode.

4.8. Five-Channel Waveform Receiver

The five-channel waveform receiver provides simultaneous waveforms from up to five separate sensors in passbands of either 1 to 26 Hz, or 3 Hz to 2.5 kHz. The purpose of this receiver is to provide high-resolution spectral measurements and to determine the polarization and wave normal of low frequency plasma waves. When connected to the Langmuir probe (including the E_u and E_v monopoles operating in the current collecting mode), the waveform receiver can also provide $\delta n_e/n_e$ waveforms. The waveforms are sampled with 12-bit resolution once every 10 ms for the 1–26 Hz bands and once every 140 μs for the 26 Hz and 2.5 kHz bands. Simultaneous waveform samples from up to five receiver channels are stored in a buffer memory until they are read out via the science telemetry. The waveforms are processed on the ground to produce the auto- and cross-correlations that are needed to compute the wave normal and polarization.

The waveform receiver consists of five parallel analog input channels. The five inputs can be connected to various sensors. Table VII summarizes the sensor selections for each of the five channels. Signals from the five sensors are routed directly to a gain select stage that has commandable gains of 0, 10, 20, or 30 dB for each receiver channel. Channels 1 and 2 have independent gain settings, but channels 3–5 (which are usually connected to the B_x, B_y, and B_z magnetic sensors) share the same gain control lines. The programmable gain amplifiers are similar to those used in the wideband receiver. The amplifier gains are controlled by the high rate processor and can be operated in either a fixed gain mode or an automatic ranging mode. The output of the programmable gain amplifiers goes to a 26-Hz low-pass filter and a 3-Hz to 2.5-kHz bandpass filter. The desired analysis passband is selected by spacecraft command and the waveform outputs are sampled simultaneously by sample-and-hold circuits and converted to digital signals by a 12-bit analog-to-digital converter. The five-channel waveform receiver can also be commanded to measure one, two, three or four channels. These special modes allow greater resolution for special observations. For example, the single-channel waveform receiver mode using the 3-Hz to 2.5-kHz passband effectively allows a third wideband receiver channel in addition to the 60-Hz to 10.5-kHz and 0.8–75-kHz channels that are included in the wideband receiver. Also, the waveform receiver provides signals to the digital processing unit in order to generate the passband for the low frequency receiver function (see Section 4.9).

Since the primary purpose of the five-channel waveform receiver is to provide the amplitude and relative phase of the five measured field components, each receiver channel must have known amplitude and phase responses. To achieve the required performance, all five low-pass passive filters have been carefully matched and are phase-stable to within a one degree between 1 Hz and 2.5 kHz.

4.9. Low Frequency Receiver

In order to have continuous spectral information in the frequency range from 1–26 Hz, waveforms from the five-channel waveform receiver are Fourier transformed by the data compression processor in the digital processing unit. The three-channel mode (E_x, E_w, B_x) of the five-channel waveform receiver is used, although typically only two of these channels, E_x and B_x, are analyzed, thereby providing simultaneous electric and magnetic spectrums. Processor speed limits the temporal resolution between successive spectrums to about 16 s. The Fourier transform processing utilizes a 512-point waveform to produce a 256-frequency component spectrum with 0.2-Hz resolution. To minimize the data volume and to produce a spectrum with resolution similar to that produced by the medium frequency receiver, the linear components of the spectrum are binned to form a logarithmically spaced spectrum with 32 frequencies covering a frequency range from 0.2 to 26 Hz.

4.10. Wideband Receiver

The wideband receiver is similar to the wideband receivers previously used on the Voyager, Galileo, Polar, and Cluster spacecraft, and provides high-resolution electric and magnetic field waveform measurements in passbands of either 60 Hz to 10.5 kHz, or 0.8–75 kHz. The wideband receiver also serves as the front end for the on-board dust detection function. The data compression processor (described below) is used to search the waveforms for the signature of dust impacts, based on those observed by the Voyager plasma wave instrument (see Figure 7).

The wideband receiver processes signals from a single selected sensor (either E_u, E_v, E_x, E_w, B_x, or Langmuir probe). To provide the capability for obtaining waveforms at higher frequencies, the wideband receiver input can be connected to the frequency conversion output from the high frequency receiver (see Section 4.5).

The instantaneous dynamic range of the wideband receiver is 48 dB. Because of the expected large dynamic range of the input signals, a set of discrete gain amplifiers and an automatic gain control are used to amplify the signal to the proper level in steps of 10 dB over a range of 0–70 dB. This system provides a total dynamic range of over 100 dB for the wideband receiver. The feedback loop in the automatic gain control has a time constant of 0.1 s. However, gain updates are only made prior to a waveform capture. Since waveform captures typically occur once per multiple of 125 ms, the effective gain update can be much slower than 0.1 s. The output from the discrete gain amplifiers goes to the two bandpass filters (60 Hz to 10.5 kHz or 0.8–75 kHz). The output of the selected bandpass filter is sent to an 8-bit analog-to-digital converter. The sampling rate is 27,777 samples/s for the 10-kHz channel and 222,222 samples/s for the 75-kHz channel.

4.11. Data Processing Unit

The RPWS data processing unit uses three microprocessors, the low-rate processor (LRP), the high-rate processor (HRP), and the data compression processor (DCP), to perform the tasks necessary for the instrument operation. These processors are responsible for collecting the science and housekeeping data, data formatting and data compression, and provide the interface to the spacecraft command and data system.

The tasks and responsibilities of the three processors are as follows. The low-rate processor is considered the primary processor since it has the sole interface with the Cassini spacecraft via the bus interface unit (BIU). It also provides the interface to the high frequency receiver, the interface and control for the medium frequency receiver and the analog housekeeping analog-to-digital converter, and the control for the antenna motor subsystem. The high-rate processor is used to control the waveform and wideband receivers, and the Langmuir probe instrument. The high-rate processor contains a data compression chip for compression of waveform signals. The data compression processor contains a dedicated math processor and is used only for data processing tasks, such as data compression. This processor performs the on-board dust-detection and also the Fourier transforms for the low frequency receiver.

The essential elements of each of the three processors are an 80C85 microprocessor, two or more random access memory chips, and one or two read-only memory chips. The 80C85 runs with a 3-MHz clock and is used in the expanded configuration which uses an address latch to give a full 16-bit address bus and a bi-directional buffer to expand the data bus. Each processor can communicate with the other two processors via the interprocessor communication bus. The 80C85 has a memory address bus width of 16 bits giving a total addressing capability of 2^{16} or 64 kilobytes (kB). The memory in each of the processors consists of both a read-write memory and a read-only memory.

The low-rate and high-rate processors each have more than 64 kB of random access memory available. The low-rate processor contains 64 kB of read/write memory with 16 kB write-protected and 4 kB of read-only memory. The bus interface unit appears to the processor as an additional 16 kB of random access memory.

TABLE VIII
Summary of RPWS physical characteristics.

Element	Mass (kg)	Power (peak W)	Volume (cm^3)	Location
Main electronics	5.59	5.09	$41.7 \times 17.8 \times 16.8$	Main Bus, Bay 4
Antenna bracket assembly	29.77	11.14	$61 \times 67 \times 67$	Upper shell structure of Bay 4
Magnetic search coils	1.05	0	$30 \times 30 \times 30$	1-m boom attached to high gain antenna support structure
Magnetic search coil preamps	0.28	0.1	$12.6 \times 3.2 \times 8.8$	Base of search coil boom
Langmuir probe	0.85	0	$74.3 \times 10 \times 11.5$	0.8-m boom on search coil boom
Langmuir probe preamp	0.135	0.05	$8.7 \times 9 \times 4.8$	Base of search coil boom
Total	37.68	16.38		

The high-rate processor has 96 kB of random access memory. The excess memory is handled by switching memory chips under program and hardware control.

All communication with the spacecraft is over a redundant 1-MHz bi-directional serial data bus conforming to MIL STD 1553. The interface is implemented via a bus interface unit.

4.12. Mass, Power, and Telemetry

The RPWS has a total mass of 37.68 kg and requires a power of 16.38 W in its fully operational science modes. This power value does not include a momentary ~3 W increase when the bus interface unit is active. By project policy, this bus interface power is allocated to the command and data system and not to the science instrument. Mass and power breakdowns for the various elements of the instrument are provided in Table VIII.

Because of its many different receivers, operating modes, and because of the use of several different data compression schemes, the RPWS has a highly variable science telemetry rate. In all of the defined telemetry modes that include RPWS science telemetry, there are four data pickup rates, or rates at which the RPWS can send data to the command and data system. These rates are 30.464, 60.928, 182.784, and 365.568 kbps. However, because of limitations on the Cassini solid state recorder data volume as well as limits on the total data volume that can be telemetered to the ground in any given day, the RPWS can utilize such high data rates only occasionally and for relatively short periods of time. Therefore, a number of different observing modes have been defined that use minimal data rates for basic survey information and utilize the higher data rates for special observations, usually involving the wideband or five-channel waveform receivers. In the usual case where the actual science telemetry rate is less than the current telemetry mode's data pickup rates, RPWS outputs "zero-length" packets that are discarded by the command and data system to reduce the net data production rate.

The basic low-rate survey mode we have defined generates approximately 1 kbps before compression. Based on limited in-flight experience to date, the actual data rate in this mode after compression is about 700 bps. However, since most of the compression techniques used are data content dependent, it is not clear that this compressed data rate is representative of what will be encountered at Saturn. In addition to the continuous low-rate observations, the RPWS survey includes occasional wideband receiver samples (typically about one minute every couple hours) that are essential to the interpretation of the lower resolution data. On average, these high rate samples add approximately 600 bps to the RPWS survey rate. Typical observation modes that require higher data rates include the 75-kHz wideband receiver observations which generate close to 360 kbps before compression, the maximum duty cycle mode of the five-channel waveform receiver in the 2.5-kHz bandpass which generates approximately 130 kbps before compression, the low-duty cycle mode of the 10-kHz wideband receiver measurements which generates approximately 30 kbps before compression, and the high-resolution temporal and/or spectral modes of the high frequency receiver which generate data rates up to about 4 kbps.

Because of the uncertainty in the compression factors and data volume allocations imposed due to the limited downlink and onboard storage, the RPWS has the ability to monitor its actual data volume and compare it to a model. In the case where the actual data production is exceeding the model, various steps can be taken to decrease the data rate. For example, the rate at which waveform samples are acquired by the five-channel waveform receiver can be slowed down. This approach is effective when no wideband data are being acquired. During wideband receiver operations, the effective data rate can be reduced by decreasing the duty cycle of the wideband data. For example, instead of acquiring a 2048-sample data set once every 125 ms, a data set could be acquired every 250 ms.

5. Calibrations and Performance

An extensive series of amplitude calibrations, frequency responses, phase calibrations, and instrument performance checks were carried out on the RPWS prior to launch, both before and after integration on the spacecraft. These tests and calibrations were performed at room temperature (25 °C), –20 °C, and 40 °C. While there are calibration signals available in the instrument for in-flight calibration purposes, these are mainly used to check for drifts due to aging or radiation exposure. The primary calibration information to derive physical units (spectral density, etc.) is derived from the prelaunch tests.

5.1. Calibration Procedure

5.1.1. *Electric Antennas*

Each of the RPWS electric antennas is connected to an amplifier located in the high frequency receiver. In the long wavelength regime where the wavelength is much greater than the tip-to-tip length of the antenna, L, the potential between the two elements of an electric dipole is given by $\Delta V = EL_{\mathrm{eff}}$, where L_{eff} is a quantity called the effective length (Gurnett, 1998). This relation ignores the presence of any electrical load. For a dipole consisting of collinear elements, to a very good approximation L_{eff} is one-half the tip-to-tip length of the antenna. For a V configuration, L_{eff} is the distance between the geometric centers of the two elements. Taking into account the 120° included angle between the two elements of the E_x dipole and the finite distance between the roots of the two elements, the effective length of the Cassini E_x dipole is $L_{\mathrm{eff}} = 9.26$ m. Next we must consider the effect of an electrical load on the antenna. At frequencies above a few tens of Hz, where the antenna impedance is primarily capacitive, the ratio of the output voltage to the input voltage is given by a simple capacitive divider,

$$\frac{V_{\mathrm{out}}}{EL_{\mathrm{eff}}} = \frac{C_{\mathrm{A}}}{C_{\mathrm{A}} + C_{\mathrm{L}}}. \tag{1}$$

In the above equation C_{A} and C_{L} are the antenna and load capacities, respectively. For a cylindrical antenna, to a good approximation, the antenna capacity is given by

$$C_{\mathrm{A}} = \frac{2\pi\varepsilon_0(L/2)}{[\ell n(L/2a) - 1]} \tag{2}$$

where $L/2$ is the length of the element, a is the radius of the element and ε_0 is the permittivity of free space. For $L/2 = 10$ m and $a = 0.014$ m, $C_A = 100.2$ pF.

TABLE IX

Measured base capacitances for the electric antennas.

	Base capacitance (pF)		
Contributor	E_u	E_v	E_w
Mechanism	50.4 ± 0.2	50.8 ± 0.2	50.7 ± 0.2
Coax cable	34.7 ± 0.3	34.3 ± 0.4	37.4 ± 0.4
HFR input	33.8	28.0	17.6
Total	118.9	113.1 ± 0.6	105.7 ± 0.6

The load capacity can be considered to be the sum of two parts: (1) the base capacity, which consists of the internal capacities of all circuits and mechanical structures connected to the base of the antenna; and (2) the capacity between the antenna and the spacecraft structure. Since the electric antenna cannot be fully extended to the in-flight configuration while the spacecraft is on the ground, there is no way to directly measure the capacitance between the antenna and the spacecraft structure. Thus, the total base capacitances listed in Table IX only give lower limits to the load capacities. Another, and perhaps better way to estimate the load capacities is to determine the half-wavelength resonance frequency of the antenna. This resonance frequency is shifted downward by a small amount due to the load capacity. During the instrument checkout in December 1998 and January 1999, a sharp peak was observed in the E_x dipole noise level at 8.5 MHz (see the discussion of noise levels in Section 5.2). This peak is due to the half-wave resonance. An antenna modeling program was then used with an accurate representation of the spacecraft geometry to determine the load capacitance required to place the resonant frequency at 8.5 MHz. The resulting value was $C_L = 150$ pF. This is our best current estimate of the load capacity for the E_x antenna. Using the capacitive divider relation above with $C_A = 100.2$ pF and $C_L = 150$ pF gives a loading loss of 7.9 dB for the E_x antenna.

Due to the complex shape of the Cassini spacecraft that acts as the ground plane for these antennas, the electrical axes of the antennas are expected to be rotated significantly from their physical orientations (Rucker *et al.*, 1996). Radio signals from Jupiter were used as a calibration source during the Cassini flyby of Jupiter in late 2000 and early 2001 to determine the electrical axes of the antennas. The orientations of the antenna axes can be determined by inverting the direction-finding software, using the direction to Jupiter as the source direction. Preliminary results from this calibration are given by Vogl *et al.* (2001). Additional antenna calibrations are scheduled on approach to Saturn and while in orbit after the Huygens probe is released.

Electric field calibrations of the receivers were performed by applying signals with known amplitudes at the preamplifier inputs and relating the input signal strength with the resulting telemetry value. As described below, the telemetry values from the various receivers are related to the input signal strength either via a set of look-up tables or through an analytical function that fits ground calibration data.

5.1.2. *Magnetic Antennas*

The sensitivity, frequency response, phase response, and noise levels of the magnetic search coil antennas were calibrated at a low-noise magnetic field observatory near Chambon La Foret, France. The calibrations were performed using a Helmholtz coil driven by a known AC current source to produce a magnetic field with a known magnitude and phase. The thermal blankets were installed in order to account for any conductivity effects that the blanket may have on the sensor response. The resulting data relate input field strength to the voltage at the magnetic preamplifier output. Additional calibrations were then used to relate the voltage input to the various receivers to an output telemetry value. Combining these two steps provides an overall end-to-end calibration of the magnetic field measurements. This calibration was verified through a series of tests performed with the search coil antennas connected to the RPWS instrument after integration on the spacecraft. An end-to-end check of the phase response was also carried out after integration on the spacecraft. Noise levels were measured on the spacecraft by placing the search coils in a μ-metal chamber, which shields the sensors from external noise sources.

5.1.3. *High Frequency Receiver*

The calibration of the high frequency receiver was performed at Meudon, France. Sine wave and white noise sources with known spectral properties were applied to the input of the electric antenna amplifiers and the corresponding telemetry values were recorded. The calibration data were then fit to analytical models for each of the receiver bands. From these models, the telemetry values can be converted to physical units. A series of tests was then performed after integration of the instrument on the spacecraft, to verify the high frequency receiver calibrations.

5.1.4. *Medium Frequency Receiver*

The medium frequency receiver uses compressors with a piecewise-linear approximation to a logarithmic amplitude response. Over the range of amplitudes, the response of each compressor consists of a series of five distinct linear segments that deviate slightly from a true logarithmic response. Because the compressors for each of the three bands have different amplitude sensitivity characteristics, the amplitude response of each compressor must be measured separately. The amplitude response of a typical logarithmic compressor is shown in Figure 18 for channel 8 of Band 1 of the medium-frequency receiver. This curve was produced by applying a signal to the input of the electric antenna preamplifier at the center frequency of the filter (52.82 Hz). The amplitude was stepped in 2 dB increments to cover the complete amplitude range of the receiver. Similar amplitude responses were measured for Band 2 (channel 16) and Band 3 (channel 10).

Since all the frequency channels of a given medium frequency receiver band utilize the same logarithmic compressors, it is only necessary to measure the amplitude response at one frequency channel for each band. The amplitude responses for the remaining channels can be obtained by calibrating the gain through the entire system as a function of frequency at a fixed amplitude. This calibration is called a channel-to-channel gain test. By combining the amplitude response with the channel-to-channel gain calibration, a complete set of calibrated look-up tables can be produced that convert the output telemetry values to signal strengths at the preamplifier inputs.

The frequency response of the three filters used in the medium frequency receiver was also determined by sweeping a sine wave signal over frequency while monitoring the output of a fixed channel. Figure 19 illustrates responses of the single sideband filter for Band 1 and the double sideband filters for Bands 2 and 3 of the medium frequency receiver, as well as the center frequencies for each of the frequency steps in the receiver at the top of the figure.

5.1.5. *Low Frequency Receiver*

The low frequency receiver employs onboard fast Fourier transform processing of one or two channels of the five-channel waveform receiver to produce spectrums over a frequency range from 1 to 26 Hz. The calibration of the low frequency receiver involved both amplitude and frequency calibrations. The amplitude calibration is similar to the medium frequency receiver amplitude calibration. A signal was applied to the input of the electric antenna amplifier at the center frequency of every fourth Fourier frequency component. The amplitude was then stepped in 2 dB increments to cover the complete amplitude range of the receiver. Figure 20 shows the amplitude response of a typical low frequency receiver channel (channel 16). These tests were repeated for the various gain states, and at –20 °C and +40

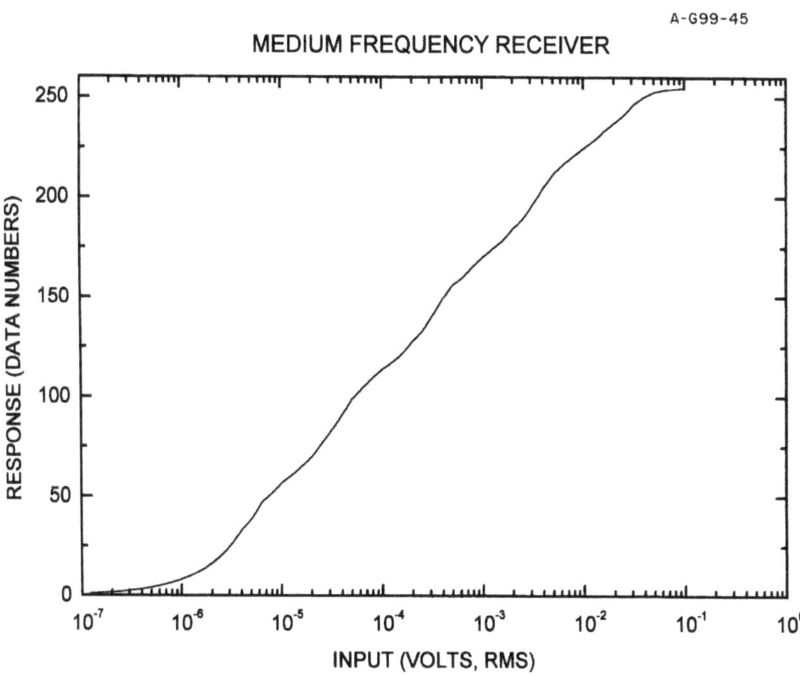

Figure 18. A typical medium frequency receiver calibration curve relating the input signal strength to the output of the log compressor. This particular curve is for the middle channel (channel 8) of the lowest frequency band (Band 1) of the receiver.

°C. The flattening of the curve at high input signal strength is due to saturation and the flattening at low input amplitudes is due to the preamplifier noise level.

The sensitivity of the low frequency receiver varies from channel to channel across the band. These frequency variations affect the calibration. A channel-channel gain test, similar to that performed on the medium frequency receiver, was conducted for each of the gain states of the low frequency receiver. This test was performed by applying an input signal of fixed amplitude to the input of the electric preamplifiers, and sweeping the signal from the lowest frequency channel of the low frequency receiver to the highest frequency channel. A combination of the amplitude response and the channel-channel gain measurements is then used to complete the calibration of the low frequency receiver for each gain state. Figure 21 shows the frequency response of the 32 quasi-logarithmically spaced channels. As discussed earlier, these channels are generated on board by binning (summing) Fourier components in such a way so as to create approximately logarithmically spaced channels.

5.1.6. *Wideband and Five-Channel Waveform Receivers*

The response of the wideband and five-channel waveform receivers was determined by applying signals of known frequency and amplitude to the electric preamplifiers, and determining the gain factors required to convert the telemetry values

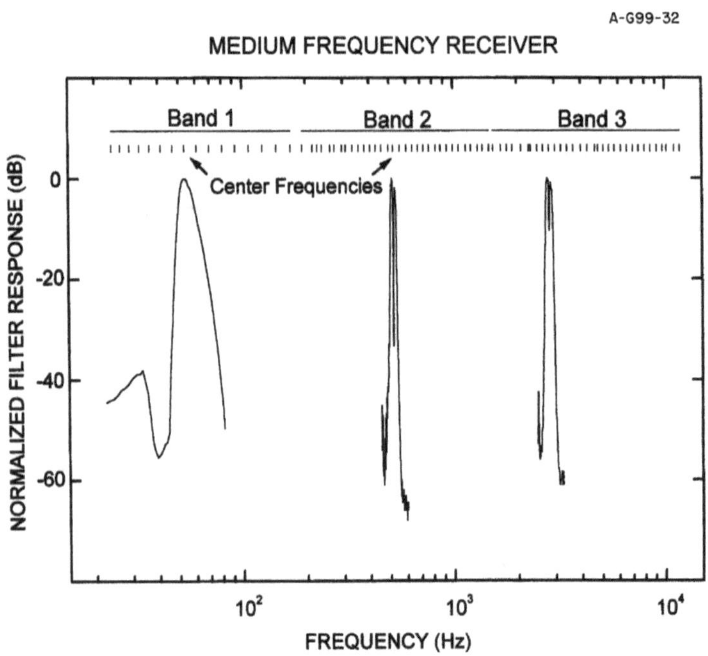

Figure 19. Frequency response curves and center frequency locations for the medium frequency receiver.

into physical units. These gain factors provide calibrations for both the time-series waveform and the spectrum produced by a Fourier transform of the waveform. The amplitude response of the wideband and five-channel waveform receivers was determined for each gain state, and for every filter mode. The frequency and phase response of the wideband receiver and the five-channel waveform receiver was determined by applying input signals of fixed amplitude to the input of the individual receivers, and sweeping the signal across the frequency band of the receiver. Figure 22 shows the frequency response of the two wideband receiver bands (10 and 75 kHz), and Figure 23 shows the frequency response of the 1–26-Hz and 3-Hz to 2.5-kHz bands of the five-channel waveform receiver. End-to-end calibration checks were performed by repeating the frequency response test and by applying an input signal of white noise with known spectral properties to the input of the electric and magnetic preamplifiers.

5.2. Performance

Cassini was successfully launched on 15 October 1997. At the time of this publication, the RPWS has been operated extensively during the cruise to Saturn: (1) for a period of about 30 h on 25–26 October 1997 in conjunction with the electric antenna deployment, (2) for about 2 h during the first Venus flyby on 26 April 1998 in a very limited mode designed to search for lightning, (3) for several days during

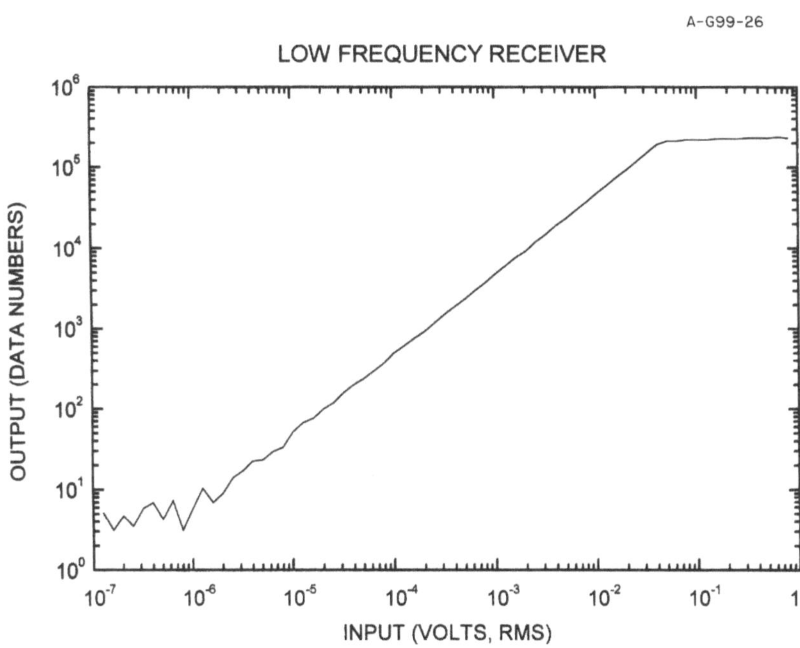

Figure 20. A typical low frequency receiver calibration curve relating the input signal strength to the output of the log compressor. This particular curve is for the 16th channel of the receiver.

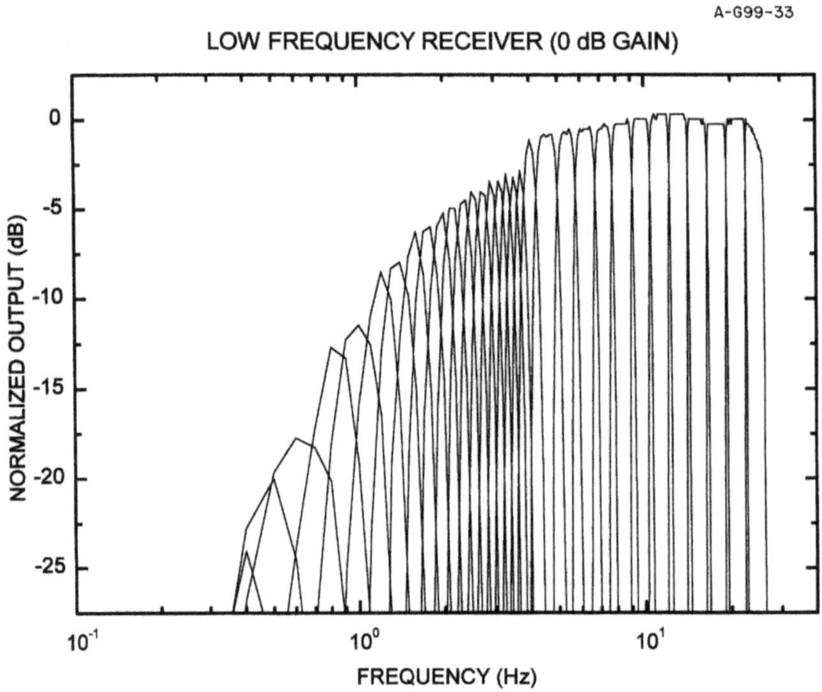

Figure 21. Frequency response curves for the quasi-logarithmically-spaced channels constructed on board from Fourier transform components for the low frequency receiver.

the period from 30 December 1998 to 19 January 1999 for instrument checkout, (4) for a few-hour period on 24 June 1999 during the second Venus encounter, (5) for about a month during the Earth flyby from 13 August to 14 September 1999, and (6) more-or-less continuously since February 2000 including a long period in the vicinity of Jupiter. The deployment data set was limited to modes designed to ensure that the electric antennas were properly deployed and provide some indication of in-flight performance. The first Venus flyby observations were of such a limited nature (28 channels between 80 kHz and 16 MHz) that they are not very useful for characterizing the in-flight operation of the instrument. The most complete data in hand are those from the instrument checkout and the Earth and Jupiter flybys. These data include a complete survey of noise levels obtained prior to other instruments being turned on as well as a limited quiet test which provided a first opportunity to observe any interference that might be generated by the spacecraft and other instruments. More importantly they provide extensive opportunities to study plasma waves in the solar wind and magnetospheric environments as well as a variety of solar and planetary radio emissions.

Figure 24 shows the noise levels from December 30, 1998 as observed on the E_x dipole, and Figure 25 shows the corresponding noise level on the B_x antenna.

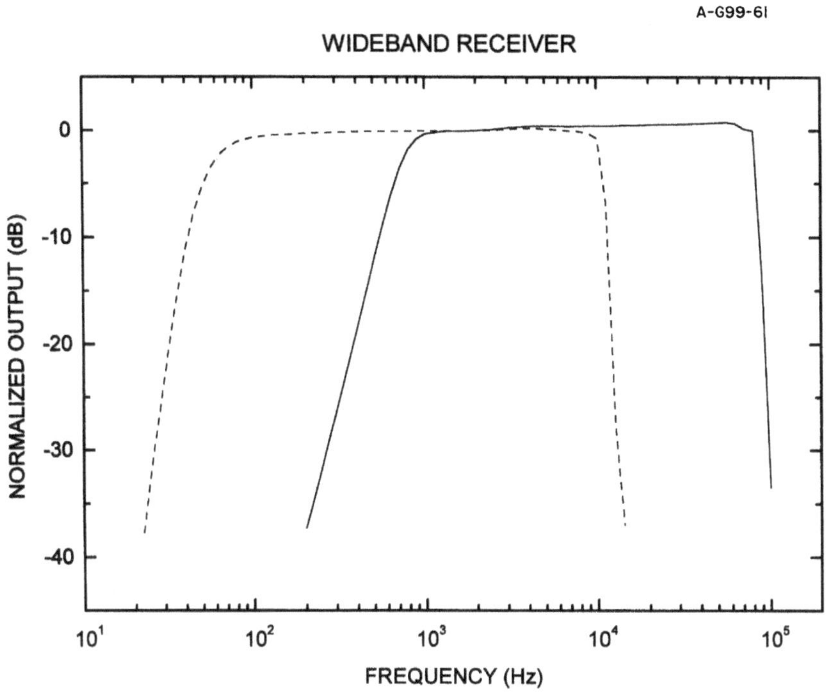

Figure 22. Frequency response curves for the 10- and 75-kHz wideband receiver filters.

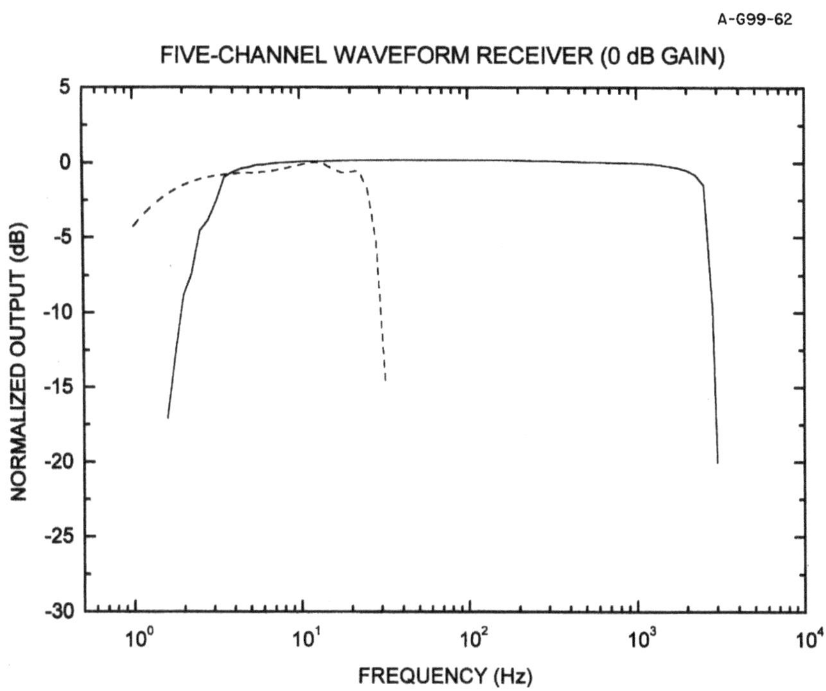

Figure 23. Frequency response curves for the 26-Hz and 2.5-kHz waveform receiver filters.

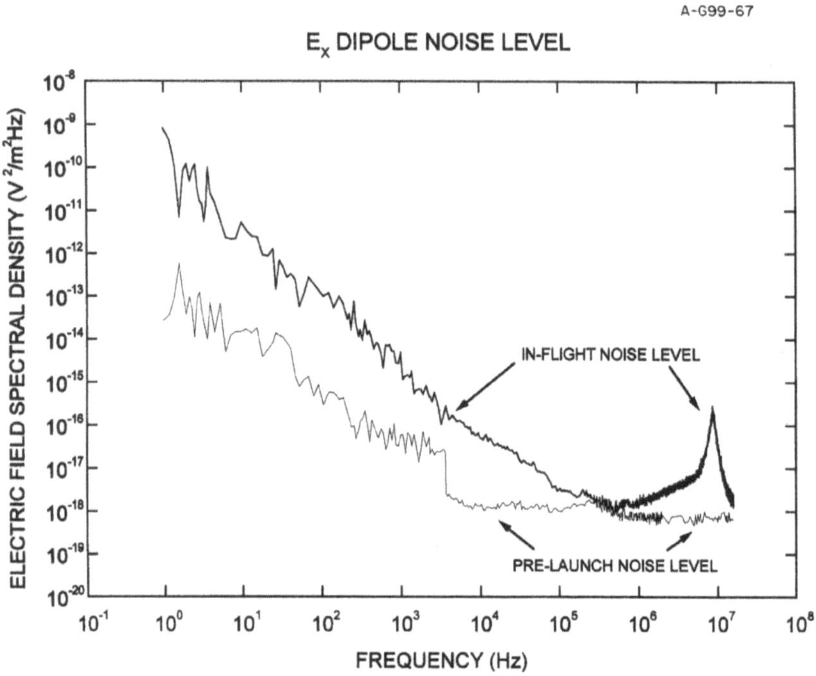

Figure 24. Plots of the pre-launch and in-flight noise levels for the E_x dipole antenna. The enhanced in-flight noise levels at low-frequencies (below about 10^5 Hz) is due to thermal plasma noise. The enhanced in-flight noise levels at high frequencies (above about 10^6 Hz) is due to galactic radio noise. The sharp peak in the noise level at about 8.5 MHz is due to the half-wavelength antenna resonance.

For comparison, the pre-launch noise levels are also shown. The in-flight magnetic noise levels are very similar to the pre-launch noise levels. The in-flight electric field noise levels appear to be significantly higher than the pre-launch noise levels, particularly at low frequencies. These elevated in-flight noise levels are mainly due to thermal plasma noise, which is a function of electron density and temperature and characteristics of the antenna such as its ratio of length to radius (Meyer-Vernet and Perche, 1989). At Saturn the thermal plasma noise levels are expected to be significantly lower because of the lower electron densities in the vicinity of Saturn. The enhanced in-flight noise levels at frequencies above about 1 MHz are due to the galactic background (Brown, 1973). The peak at 8.5 MHz is due to the half-wave resonance of the antenna. Comparisons with Ulysses and Wind observations indicate the sensitivity of the RPWS instrument compares favorably with the radio investigations on those spacecraft, even though they both employ significantly longer antennas than Cassini.

The performance of the RPWS during the cruise phase of the mission has been excellent. A number of examples of cruise data serve to illustrate this. Figure 26 shows an overview of the radio and plasma wave spectrum observed during the Earth flyby, which occurred on August 18, 1999 (Kurth *et al.*, 2001a). This spectrogram

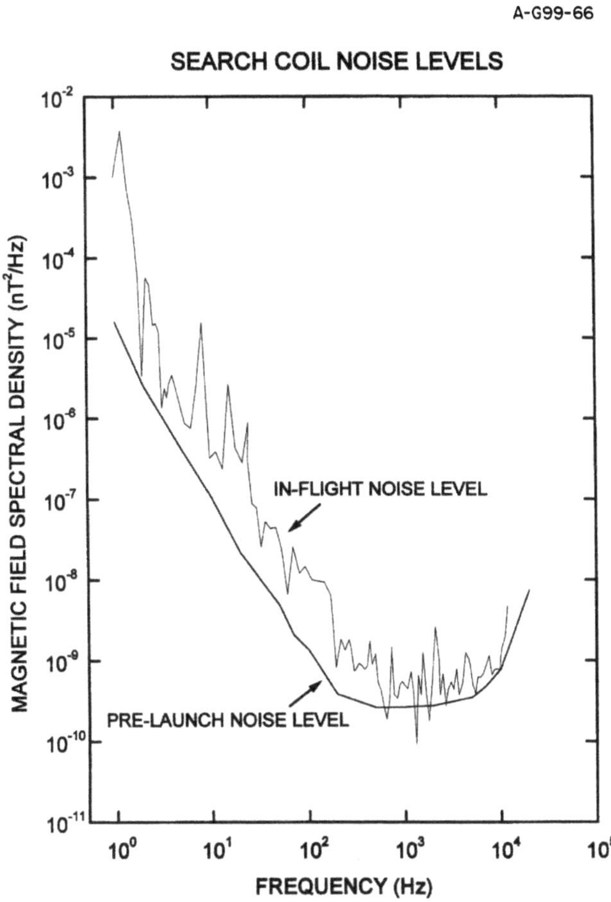

Figure 25. Plots of the pre-launch and in-flight noise levels of the search coil magnetic sensor. The slightly enhanced in-flight noise levels below about 100 Hz are believed to be due to a combination of spacecraft-generated interference (the sharp spikes) and solar wind plasma waves.

shows solar, terrestrial, and man-made radio emissions at the highest frequencies as well as the significant boundaries of the terrestrial magnetosphere and many of the plasma wave modes normally found within it. During the Earth flyby the instrument was configured to obtain intensive observations in the 2.5-kHz waveform receiver mode and subsequent analyses of chorus and other terrestrial plasma wave phenomena have demonstrated the use of this receiver for wave normal and Poynting flux analyses (Hospodarsky *et al.*, 2001). Also during the Earth flyby, both the sounder and Langmuir probe were successfully operated in the magnetospheric conditions for which they were designed. As shown in Kurth *et al.* (2001a) both portions of the instrument functioned well and returned excellent electron density information.

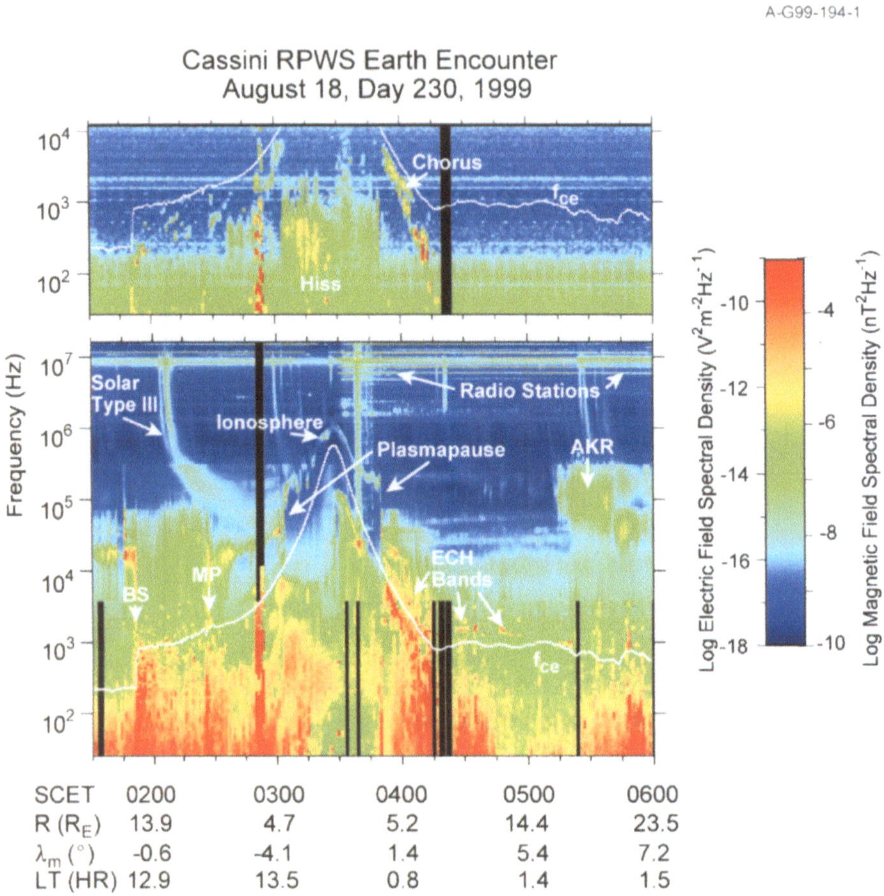

Figure 26. An overview of radio and plasma wave phenomena observed during the Earth flyby (Kurth *et al.*, 2001a). The upper panel shows magnetic fields and the lower panel the electric fields. Noted in the illustration are various boundaries of the magnetosphere (BS: bow shock and MP: magnetopause) as well as the primary emissions identifiable at the resolution of these spectrograms. The burst of noise at about 0250–0255 is due to a high duty-cycle sounder test and does not represent a natural emission. Normal sounder operations will occur on a much lower duty cycle and should not be as disruptive.

The high frequency receiver's capabilities have also been demonstrated in flight. Figure 27 shows examples of many of the various types of Jovian radio emissions as observed during the Jupiter flyby in late 2000 (Gurnett *et al.*, 2002; see also Lecacheux, 2001). The sensitivity and noise characteristics of the high frequency receiver are evident in this display. Pending completion of the analysis of the Jupiter antenna calibration, it appears that the polarization and direction-finding capabilities of this receiver will also be shown to be as designed (Lecacheux, 2001). The high and medium frequency receivers were used in new studies of the influence of the solar wind on the Jovian magnetosphere (Gurnett *et al.*, 2002; Kurth *et al.*, 2002).

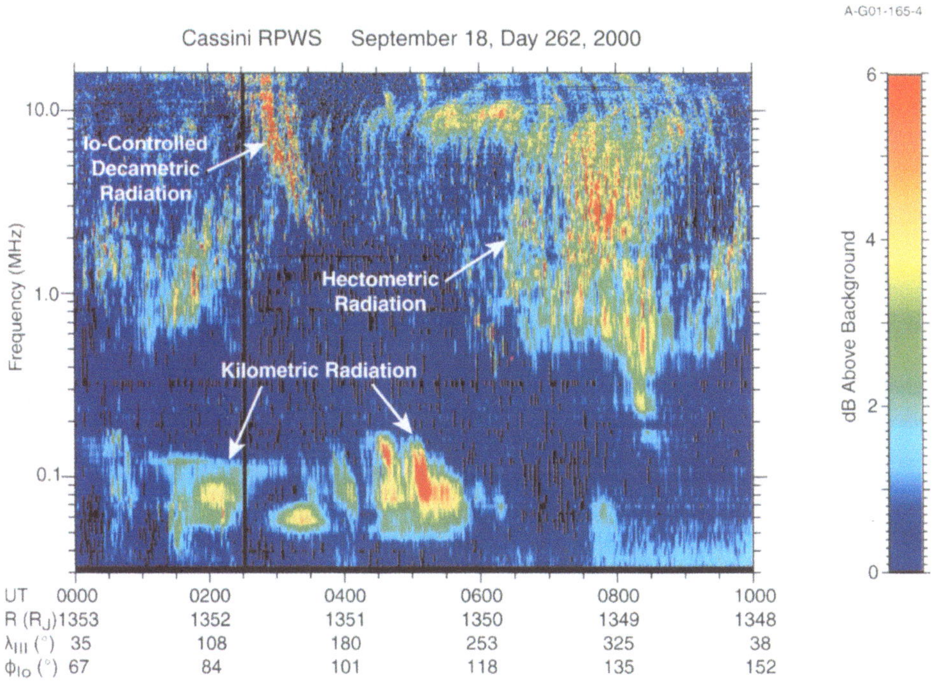

Figure 27. Observations of many of the various types of Jovian radio emissions (Gurnett *et al.*, 2002). This spectrogram shows examples of decametric, hectometric, and kilometric radiation during one Jovian rotation. Although the spacecraft is two-thirds of an AU from Jupiter at the time, there is considerable signal-to-noise ratio for all of the emissions and clear evidence of the arc-like structures which comprise the decametric and hectometric emissions.

The Jupiter flyby also provided a unique opportunity to study the fine temporal and spectral detail of a number of Jovian radio emissions for the first time using the capabilities of both the wideband receiver and the high frequency receiver (Kurth *et al.*, 2001b; Lecacheux *et al.*, 2001; Kaiser *et al.*, 2001).

Figure 28 displays the first confirmed detection of Saturn in the RPWS data while Cassini was at a distance of greater than 2.5 AU from Saturn. A high-density region in the solar wind was detected at 1 AU and ballistically propagated radially to determine when the region would pass by Jupiter and Saturn. Within a day or so of the predicted time, low frequency Jovian radio emissions were clearly observed by Cassini. Subsequently, and again within a day or so of the ballistic propagation, evidence of the arrival of the high density region at Saturn was marked by the detection of the radio emissions shown in Figure 28 by Cassini. Given this expected response of SKR in response to the arrival of a high density solar wind at Saturn and the similarity in spectrum and temporal variability to SKR observed by Voyager, we are confident that these emissions are, indeed Saturn kilometric radiation. A number of other examples of SKR are also present in the RPWS

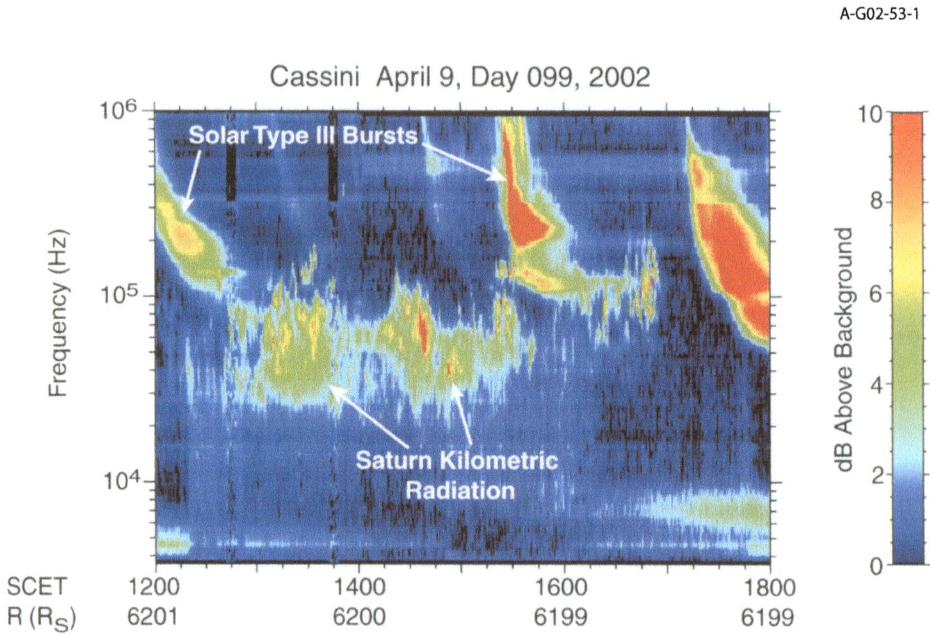

Figure 28. One of the first confirmed detections of Saturn kilometric radiation made while Cassini was still 2.5 AU from Saturn. The Saturnian emissions extend from about 30 to 200 kHz in this example and last for several hours. This radio activity is thought to be the result of a high-density region in the solar wind interacting with Saturn's magnetosphere.

measurements from early 2002, onward. Hence, it is clear that Cassini's study of the Saturnian magnetosphere and its response to the solar wind input is well under way.

6. In-Flight Operations, Data Processing, and Archiving

As can be seen from the preceding description, the RPWS instrument is quite complex and has a number of different operating modes. This complexity has a strong impact on the in-flight operation of the instrument and on the analysis and archiving of the data. In this section we provide an overview of the instrument operations, data analysis, and data archiving.

6.1. In-Flight Operations

The different types of receivers used to cover the spectral and temporal range covered by the RPWS does not lend itself to a monolithic, synchronous mode of operation. Nevertheless, to reduce the magnitude of the in-flight operations to an acceptable level requires that many of the measurements be scheduled in a systematic way. The approach is to attempt to acquire survey information in the form of uniform spectral and temporal information at a low enough data rate, ~1 kbps, to ensure that such coverage is available for the entire Saturnian tour and for a large portion of the cruise and approach to Saturn. The survey data set will support spatial mapping, statistical studies, and studies of dynamical effects in the magnetosphere and their possible correlation with solar wind variations. In addition to the survey information, special observations will be added (sometimes at greatly increased data rates) at specific locations or times to provide enhanced information required by several of the science objectives. The special observations may include full polarization and direction-finding capability or high spectral or temporal resolution observations by the high frequency receiver, wideband measurements at one of the possible bandwidths, acquisition of $\delta n_e/n_e$ measurements, or intensive wave-normal analysis afforded by acquiring five-channel waveforms on an accelerated schedule. While minimizing the number of different modes in which the instrument is operated both simplifies operations and yields a more manageable data set, flexibility (for example in the duty cycle of wideband measurements) increases the likelihood that enhanced measurements can be integrated successfully with the resource requirements of the other instruments. One of the resources which will be most limited on Cassini is the overall data volume; RPWS requires large data volumes for some of its measurements.

To simplify operations and to be consistent with Cassini Project policies, virtually all RPWS observational modes are defined by internal sequencing constructs termed instrument expanded blocks. In the RPWS, an instrument expanded block is, in essence, a small subroutine that sets up each of the receivers as required for the desired mode. In some cases, the instrument expanded block cycles the instrument through a small number of modes having a relatively short cycle period with each step in the cycle lasting several seconds to several minutes. Each observation mode, then, is initiated by a trigger command. In most cases, the instrument will continue operating in the triggered mode until another trigger command is received which initiates the next desired mode. Early in the tour, it is anticipated that the set of instrument internal blocks will change with each background sequence load (approximately once per month). However, it is possible and desirable that the required set of instrument expanded blocks will stabilize with flight experience and each sequence later in the tour will only require a new set of trigger commands using the established "library" of modes.

The RPWS team has provisionally defined a small number of observation modes to date and these have been exercised in a very limited way in flight. Here, we illustrate a few of these modes.

Survey Mode: The survey mode is designed to acquire basic information on a continuous basis from all of the RPWS receivers with low to moderate temporal and spectral resolution at a low data rate in order to support many of the scientific objectives of the investigation, most notably those dealing with the configuration of the magnetosphere and its dynamics. The survey data also provide context for more intensive measurements at specific times and locations such as satellite flybys. This mode acquires one electric field and one magnetic field spectrum every

16 and 32 s in the low and medium frequency receivers, respectively, covering the range from 1 Hz to 12 kHz and collects 180 channels between 3.5 kHz and 16 MHz with the high frequency receiver once per 16 s. Approximately once per two and one-half minutes, a set of five-channel waveforms is acquired. The waveform alternate between the low (26 Hz) and high (2.5 kHz) band so that waveforms in both bands are acquired approximately once every five minutes. Waveform samples are acquired from the E_w antenna once every 5 s and analyzed onboard for signatures of dust impacts based on algorithms used to identify impacts in the Voyager plasma wave waveform data. The sounder is operated for a few seconds every five or ten minutes and the Langmuir probe is swept every five minutes, 40 s and density measurements at fixed bias are acquired every 16 s in between sweeps. The data rate for this mode is about 700 bps, after compression. In addition to these low rate measurements, a cyclic (repeating) activity will run with a periodicity of about two hours that will acquire a $\sim$1 min set of wideband data each cycle. Beyond about 15 R_S the wideband data will be in the 10-kHz mode. Inside of this distance, brief samples will be acquired in both the 10- and 75-kHz modes. The long-term average data rate of these wideband samples is approximately 600 bps, after compression.

Wideband Mode: This mode adds to the Survey Mode the acquisition of intensive wideband data in either the 10- or 75-kHz bandwidth for extended periods of time (several minutes to a few hours) with a duty cycle that is commensurate with the data volume allocated for the measurement. A high duty cycle mode would acquire 2048-sample waveforms every 125 ms with an uncompressed data rate of about 131 kbps. This mode would be used, for example, to study the detailed spectral and temporal character of whistler-mode chorus. The duty cycle is decreased by acquiring 1 of every n possible 2048-sample data sets. Duty cycling employed in this manner retains the spectral resolution afforded by the 2048 contiguous samples, but sacrifices the temporal resolution between the data sets in order to minimize the data rate. One specific use of this duty-cycled wideband mode is to acquire waveforms from the E_w antenna at ring plane crossings to determine the thickness and scale height of the E ring. The elevated impact rates during ring plane crossings will not likely be handled well with the onboard dust detection algorithm, hence, acquiring the waveforms for analysis on the ground is required.

Other Modes: Other, more specific modes will be developed during cruise which will be used for special observations. For example, during campaigns designed to study the interactions of the rings and icy satellites with the magnetosphere, a modified version of the survey mode that increases the rate of waveform captures will enable a more intensive examination of waves associated with ion pickup. During traversals through the high latitude, low altitude regions later in the tour when it is possible that auroral field lines would be crossed, wideband and/or waveform data will be acquired on an enhanced schedule that will permit the identification of electrostatic solitary structures. The high frequency receiver will also be reconfigured to support observations near the source of Saturn kilometric radiation by increasing spectral or temporal resolution or by focusing on the direction-finding capabilities of the receiver.

6.2. Data Processing

The RPWS team is large and spans several institutions. Also, since hardware is provided by a number of different institutions, the expertise for different portions of the instrument resides at different institutions. This situation requires a distributed data analysis approach. However, the desire to make data and analysis results available to all team members and the need to ensure that the results of a given analysis are uniform, no matter where the analysis takes place, present a particular challenge. Our approach to this challenge involves a number of elements.

First, the basic data products will be produced in a centralized fashion at The University of Iowa. This is primarily dictated by the fact that the project supplies a single data port per principal investigator and this port is located at the principal investigator's home institution. All raw telemetry packets and ancillary data such as ephemeris and attitude information will be collected at The University of Iowa. The raw telemetry packets will be edited and decompressed (where necessary) and stored in a small number of file types whose format is agreed to by the team. It is anticipated that these same file types will be directly archivable with the planetary data system (PDS) with the appropriate labels and documentation. The resulting files will be stored both at The University of Iowa and CETP in France. The two databases ensure easy access to both U.S. and European co-investigators and also provide a level of security against loss.

Second, the institution responsible for providing specific hardware (see Table I) will provide algorithms and look-up tables as required to convert the telemetry values into physical units. These calibration algorithms and look-up tables will be maintained in the RPWS calibration documentation developed under the direction of the principal investigator. Hence, any changes in calibrations will be propagated to the team in a well-controlled manner.

Third, the cognizant institutions will also develop analysis tools and distribute these to other team members, as requested. The most generic of these tools, i.e., those providing general analysis and display capabilities of interest to many team members, will be accessible through a web-based system located at The University of Iowa (and perhaps mirrored at a location in Europe). Such a system will make the tools available to the widest set of investigators without requiring that the software be ported to numerous computing platforms. More specific analysis tools, needed for highly specialized analyses, will likely reside only in the institutions where they were developed. Results from these more specialized tools will be available to all team members and will be archived with the planetary data system as high-level data sets.

6.3. Archiving

The RPWS team intends to provide a complete archive to the PDS, consistent with the Cassini Project and PDS requirements. The University of Iowa is the outer planets subnode of the planetary plasma interactions node of the PDS, which should ensure the timely archive of data with compliant formats. Currently, several different types of archive data sets are envisioned. The basis of the RPWS archive will be a full-resolution, calibrated data set (CODMAC Level 3) with documentation. High volume data sets including wideband and waveform receiver data will be archived in uncalibrated, edited telemetry format (CODMAC Level 2) with calibration tables and procedures. This approach is dictated by the already large data volumes of these data sets.

A browse data set will also be archived. This data set will be a fully calibrated set of amplitudes as a function of frequency and time but at somewhat lower resolution than that ultimately achievable by the instrument. This data set will enable quick and easy access to the RPWS data in order to search for events or to support high level surveys of the data.

It is also likely that a number of highly specialized data sets will be archived after significant detailed analyses. Examples of such data sets include the results of the Langmuir probe measurements and radio direction-finding results. Plans for these highly specialized archives will evolve over time.

Acknowledgements

The authors would like to express their thanks to R. Barrie, R. Brechwald, A. Cooper, M. DeBower, R. Huff, R. Johnson, S. Kutcher, M. Mitchell, B. T. Pham, J. Phillips, R. Randall, S. Remington, W. Robison, W. Schintler, J. Schwartz, P. Sheyko, D. Tomash, and E. Williams at the University of Iowa, for their role in designing, assembling, and testing the instrument; to P. Fédou, N. Monge, D. Carriere, J. M. Boulben, R. Knoll, J. P. Mengue, C. Gueriau, L. Belmon, G. Nicol, A. Rapin, and L. Friel at the Observatoire de Paris, for their role in designing, assembling, and testing the high frequency receiver; to P. Chauveau and S. Louis at CETP, for their role in designing, assembling, testing of the search coil magnetic sensors, and to V. Bouzid for her role in software development; to H. Gunnarsson, B. Holback, S.-E. Jansson, and H. Thomas at the Swedish Institute of Space Physics, for their role in designing, assembling, and testing the Langmuir probe; to I. Willis at the University of Sheffield, for his role in the software development; to R. Kramer of Orbital Sciences Corporation for his role in constructing the electric antennas; to W. Fawcett, K. Breitenbach, M. Lane, R. Poynter, P. Liewer, and S. J. Bolton at the Jet Propulsion Laboratory for their role in coordinating the experiment integration and testing at the Jet Propulsion Laboratory. Special thanks are also given to R. Spehalski, R. Draper, T. Gavin, and E. Miner at the Jet Propulsion Laboratory for their help in solving many of the spacecraft interface problems associated with a complex instrument of this type. The U.S. funding for the instrument development, assembly, and testing was provided by NASA through contract 959515 through the Jet Propulsion Laboratory. The post-launch U.S. mission operations and data analysis phase of this investigation is supported by NASA through contract 961152, also through the Jet Propulsion Laboratory. The Observatoire de Paris contributions were supported by the Centre National d'etudes Spatiales. The CETP contributions are supported by Centre National d'etudes Spatiales under contract 204.

References

Aubier, M. G., Meyer-Vernet, N., and Pedersen, B. M.: 1983, *Geophys. Res. Lett.* **10**, 5–8.

Benson, R. F.: 1970, in: Thomas, J. O., and Landmark, B. J. (eds.), *Plasma Waves in Space and Laboratory, 2,* Edinburgh University Press, Edinburgh, pp. 25–54.

Blanc, M., Bolton, S., Bradley, J., Burton, M., Cravens, T. E., Dandouras, I., Dougherty, M. K., Festou, M. C., Feynman, J., Johnson, R. E., Gombosi, T. G., Kurth, W. S., Liewer, P. C., Mauk, B. H., Maurice, S., Mitchell, D., Neubauer, F. M., Richardson, J. D., Shemansky, D. E., Sittler, E. C., Tsurutani, B. T., Zarka, P., Esposito, L. W., Gruen, E., Gurnett, D. A., Kliore, A. J., Krimigis, S. M., Southwood, D., Waite, J. H., and Young, D. T.: 2002, *Space Sci. Rev.*, **104**, 253–346.

Bougeret, J.-L., Kaiser, M. L., Kellogg, P. J., Manning, R., Goetz, K., Monson, S.J., Monge, N., Friel, L., Meetre, C.A., Perche, C., Sitruk, L., and Hoang, S.: 1995, *Space Sci. Rev.* **71**, 231.

Brown, L. W.: 1973, *Astrophys. J.* **180**, 359–370.

Burns, J. A., Showalter, M. R., Cuzzi, J. N., and Durisen, R. H.: 1983, *Icarus* **54**, 280–295.

Cairns, I. H., and Melrose, D. B.: 1985, *J. Geophys. Res.* **90**, 6637–6640.

Carr, T. D., Desch, M. D., and Alexander, J. K.: 1983, in: Dessler, A. J. (ed.), *Physics of the Jovian Magnetosphere*, Cambridge University Press, Cambridge, pp. 226–284.

Decreau, P. M., Hamelin, E. M., Massif, R., DeFeraudy, H., Pawela, E., Perraut, S., Pottelette, R., and Bahnsen, A.: 1987, *Ann. Geophys.* **5A**, 181–185.

Desch, M. D.: 1982, *J. Geophys. Res.* **87**, 4549–4554.

Desch, M. D., and Kaiser, M. L.: 1981, *Geophys. Res. Lett.* **8**, 253–256.

Desch, M. D., and Kaiser, M. L.: 1990, *Nature* **343**, 442–444.

Ergun, R. E., Carlson, C. W., McFadden, J. P., Delory, G. T., Strangeway, R. J., and Pritchett, P. L.: 2000, *Astrophys. J.* **538**, 456–466.

Esposito, L. W., Cuzzi, J. N., Holberg, J. B., Marouf, E. A., Tyler, G. L., and Porco, C. C.: 1984, in: Gehrels, T. and Matthews, M. S. (eds.), *Saturn*, University of Arizona Press, Tucson, pp. 463–545.

Etcheto, J., and Bloch, J. J.: 1978, *Space Sci. Rev.* **22**, 597–610.

Etcheto, J., Belmont, G., Canu, P., and Trotignon, J. G.: 1983, *Eur. Space Agency Spec. Publ.* **SP-195**, 39–46.

Galopeau, P., and Zarka, P. 1992, Reply to the comment by J. E. P. Connerney and M. D. Desch, *J. Geophys. Res.* **97**, 12,291–12,297.

Galopeau, P. H. M., and Lecacheux, A.: 2000, *J. Geophys. Res.* **105**, 13089–13101.

Galopeau, P., Zarka, P., and Le Queau, D.: 1989, *J. Geophys. Res.* **94**, 8739–8756.

Galopeau, P., Ortega-Molina, A., and Zarka, P.: 1991, *J. Geophys. Res.* **96**, 14,129–14,140.

Galopeau, P., Zarka, P, and Le Queau, D.: 1995, *J. Geophys. Res.* **100**, 26,397–26,410.

Goertz, C. K., and Morfill, G. E.: 1983, *Icarus* **53**, 219–229.

Goldreich, P., and Lynden-Bell, D.: 1969, *Astrophys. J.* **156**, 59–78.

Goldstein, M. L., and Goertz, C. K.: 1983, in: Dessler, A. J. (ed.), *Physics of the Jovian Magnetosphere,* Cambridge University Press, pp. 317–352.
Grün, E.: 1981, *Eur. Space Agency Spec. Publ.* **SP-155**, 81–90.
Grün *et al.*, this issue.
Gurnett, D. A.: 1974, *J. Geophys. Res.* **79**, 4227–4238.
Gurnett, D. A.: 1998, in: Pfaff, R. F., Borovsky, J. E., and Young, D. T. (eds.), *Measurement Techniques in Space Plasmas: Fields, Geophys. Monograph 103,* American Geophysics Union, Washington, DC, pp. 121–136.
Gurnett, D. A., and Shaw, R. R.: 1973, *J. Geophys. Res.*, **78**, 8136–8149.
Gurnett, D. A., Scarf, F. L., Fredricks, R. W., and Smith, E. J.: 1978, *IEEE Trans. Geosci. Electron.* **GE-16**, 225–230.
Gurnett, D. A., Anderson, R. R., Scarf, F. L., Fredricks, R. W., and Smith, E. J.: 1979a, *Space Sci. Rev.* **23**, 103–122.
Gurnett, D. A., Shaw, R. R., Anderson, R. R., and Kurth, W. S.: 1979b, *Geophys. Res. Lett.* **6**, 511–514.
Gurnett, D. A., Kurth, W. S., and Scarf, F. L.: 1981, *Science* **212**, 235–239.
Gurnett, D. A., Grün, E., Gallagher, D., Kurth, W. S., and Scarf, F. L.: 1983, *Icarus* **53**, 236–254.
Gurnett, D. A., Kurth, W. S., Scarf, F. L., Burns, J. A., Cuzzi, J. N., and Grün, E.: 1987, *J. Geophys. Res.* **92**, 14,959–14,968.
Gurnett, D. A., Kurth, W. S., Cairns, I. H., and Granroth, L. J.: 1990, *J. Geophys. Res.* **95**, 20,967–20,976.
Gurnett, D. A., Kurth, W. S., Granroth, L. J., and Allendorf, S. C.: 1991, *J. Geophys. Res.* **96**, 19,177–19,186.
Gurnett, D. A., Kurth, W. S., Shaw, R. R., Roux, A., Gendrin, R., Kennel, C. F., Scarf, F. L., and Shawhan, S. D.: 1992, *Space Sci. Rev.* **60**, 341–355.
Gurnett, D. A., Persoon, A. M., Randall, R. F., Odem, D. L., Remington, S. L., Averkamp, T. F., DeBower, M. M., Hospodarsky, G. B., Huff, R. L., Kirchner, D. L., Mitchell, M. A., Pham, B. T., Phillips, J. R., Schintler, W. J., Sheyko, P., and Tomash, D. R.: 1995, *Space Sci. Rev.* **71**, 597–622.
Gurnett, D. A., Kurth, W. S., Roux, A., Bolton, S. J., and Kennel, C. F.: 1996, *Nature* **384**, 535–537.
Gurnett, D. A., Zarka, P., Manning, R., Kurth, W. S., Hospodarsky, G. B., Averkamp, T. F., Kaiser, M. L., and Farrell, W. M.: 2001, *Nature* **409**, 313–315.
Gurnett, D. A., Kurth, W. S., Hospodarsky, G. B., Persoon, A. M., Zarka, P., Lecacheux, A., Bolton, S. J., Desch, M. D., Farrell, W. M., Kaiser, M. L., Ladreiter, H.-P., Rucker, H. O., Galopeau, P., Louarn, P., Young, D. T., Pryor, W. R., and Dougherty, M. K.: 2002, *Nature* **415**, 985–987.
Hanel, R. A., Conrath, B. J., Flasar, F. M., Kunde, V. G., Maguire, W., Pearl. J. C., Pirraglia, J. A., Samuelson, R., Hearth, L., Allison, M., Cruikshank, D. P., Gautier, D., Gierasch, P., Horn, L., Koppany, R., and Ponnamperuna, C.: 1981, *Science* **212**, 192–200.
Harvey, C. C., Etcheto, J., and Mangeney, A.: 1979, *Space Sci. Rev.* **23**, 39–58.
Hornung, D. H., and Draptz, S.: 1981, *Eur. Space Agency Spec. Publ.* **SP-155**, 23.
Hospodarsky, G. B., Averkamp, T. F., Kurth, W. S., Gurnett, D. A., Dougherty, M., Inan, U., and Wood, T.: 2001, *J. Geophys. Res.* **106**, 30,253–30, 269.
Hunten, D. M., Tomasko, M.G., Flasar, F. M., Samuelson, R. E., Strobel, D. F., and Stevenson, D. J.: 1984, in: Gehrels, T., and Matthews, M. S. (eds.), *Saturn*, University of Arizona Press, Tucson, pp. 671–759.
Jones, D.: 1976, *Nature* **260**, 686–689.
Kaiser, M. L.: 1989, in: Tsurutani, B. T., and Oya, H. (eds.), *Plasma Waves and Instabilities at Comets and Magnetospheres*, AGU Monograph **53**, American Geophysics Union, Washington, DC, pp. 221–237.
Kaiser, M. L., and Desch, M. D.: 1982, *J. Geophys. Res.* **87**, 4555–4559.
Kaiser, M. L., Desch, M. D., and Lecacheux, A.: 1981, *Nature* **292**, 731–733.
Kaiser, M. L., Connerney, J. E. P., and Desch, M. D.: 1983, *Nature* **303**, 50–53.

Kaiser, M. L., Desch, M. D., Kurth, W. S., Lecacheux, A., Genova, F., Pedersen, B. M., and Evans, D. R.: 1984a, in: Gehrels, T. and Matthews, M. S. (eds.), *Saturn*, University of Arizona Press, Tucson, pp. 378–415.

Kaiser, M. L., Desch, M. D., and Connerney, J. E. P.: 1984b, *J. Geophys. Res.* **89**, 2371–2376.

Kaiser, M. L., Zarka, P., Desch, M. D., and Farrell, W. M.: 1991, *J. Geophys. Res.* **96**, 19,043–19,047.

Kaiser, M. L., Farrell, W. M., Desch, M. D., Hospodarsky, G. B., Kurth, W. S., and Gurnett, D. A.: 2001, in: Rucker, H. O., Kaiser, M. L., and Leblanc, Y. (eds.), *Planetary Radio Emissions V*, Austrian Academy of Sciences Press, Vienna, pp. 41–48.

Kelley, M. C., and Mozer, F. S.: 1972, *J. Geophys. Res.* **77**, 6900–6903.

Kennel, C. F., and Petschek, H. E.: 1966, *J. Geophys. Res.* **71**, 1–28.

Ksanfomality, L. V., Scarf, F. L., and Taylor, W. W. L.: 1983, in: Hunten, D. M., Colin, L., Donahue, T. M., and Moroz, V. I. (eds.), *Venus*, University of Arizona Press, Tucson, pp. 565–603.

Kurth, W. S.: 1992, in: Rucker, H. O., Bauer, S. J., and Kaiser, M. L. (eds.), *Planetary Radio Emissions III*, Austrian Academy of Science, pp. 329–350.

Kurth, W. S., and Gurnett, D. A.: 1991, *J. Geophys. Res.* **96**, 18,977–18,991.

Kurth, W. S., and Zarka, P.: 2001, in: Rucker, H. O., Kaiser, M. L., and Leblanc, Y. (eds.), *Planetary Radio Emissions V*, Austrian Academy of Sciences Press, Vienna, pp. 247–259.

Kurth, W. S., Gurnett, D. A., and Scarf, F. L.: 1981, *Nature* **292**, 742–745.

Kurth, W. S., Strayer, B. D., Gurnett, D. A., and Scarf, F. L.: 1985, *Icarus* **61**, 497–507.

Kurth, W. S., Gurnett, D. A., Roux, A., and Bolton, S. J.: 1997, *Geophys. Res. Lett.* **24**, 2167–2170.

Kurth, W. S., Hospodarsky, G. B., Gurnett, D. A., Kaiser, M. L., Wahlund, J.-E., Roux, A., Canu, P., Zarka, P., and Tokarev, Y.: 2001a, *J. Geophys. Res.* **106**, 30,239–30,252.

Kurth, W. S., Hospodarsky, G. B., Gurnett, D. A., Lecacheux, A., Zarka, P., Desch, M. D., Kaiser, M. L., and Farrell, W. M.: 2001b, in: Rucker, H. O., Kaiser, M. L., and Leblanc, Y. (eds.), *Planetary Radio Emissions V*, Austrian Academy of Sciences Press, Vienna, pp. 15–28.

Kurth, W. S., Gurnett, D. A., Hospodarsky, G. B., Farrell, W. M., Roux, A., Dougherty, M. K., Joy, S. P., Kivelson, M. G., Walker, R. J., Crary, F. J., and Alexander, C. J.: 2002, *Nature* **415**, 991–994.

Ladreiter, H. P., Zarka, P., and Lecacheux, A.: 1994, *Planet. Space Sci.* **42**, 919–931.

Ladreiter, H. P., Zarka, P., Lecacheux, A., Macher, W., Rucker, H. O., Manning, R., Gurnett, D. A., and Kurth, W. S.: 1995, *Radio Sci.* **30**, 1699–1712.

Lammer, H., Tokano, T., Fischer, G., Stumptner, W., Molina-Cuberos, G. J., Schwingenschuh, K., Rucker, H. O.: 2001, *Planet. Space Sci.* **49**, 561–574.

Lecacheux, A.: 2001, in: Rucker, H. O., Kaiser, M. L., and Leblanc, Y. (eds.), *Planetary Radio Emissions V*, Austrian Academy of Sciences Press, Vienna, pp. 1–13.

Lecacheux, A., and Genova, F.: 1983, *J. Geophys. Res.* **88**, 8993–8998.

Lecacheux, A., Galopeau, P., and Aubier, M.: 1997, in: Rucker, H. O., Bauer, S. J., and Lecacheux, A. (eds.), *Planetary Radio Emissions IV*, Austrian Academy of Science Press, Vienna, pp. 313–325.

Lecacheux, A., Kurth, W. S., and Manning, R.: 2001, in: Rucker, H. O., Kaiser, M. L., and Leblanc, Y. (eds.), *Planetary Radio Emissions V*, Austrian Academy of Sciences Press, Vienna, pp. 29–39.

Le Queau, D., Pellat, R., and Roux, A.: 1984, *Phys. Fluids* **27**, 247–265.

Lockwood, G. E. K.: 1963, *Can. J. Phys.* **41**,190–194.

Louarn, P., Roux, A., De Feraudy, H., Le Queau, D., Andre, M., and Matson, L.: 1990, *J. Geophys. Res.* **95**, 5983–5995.

Matson *et al.*, this issue.

Mendis, D. A., Hill, J. R., Ip, W.-H., Goertz, C. K., and Grün, E.: 1984, in: Gehrels, T. and Matthews, M. S. (eds.), *Saturn*, University of Arizona Press, Tucson, pp. 546–589.

Meyer-Vernet, N. and Perche, C.: 1989, *J. Geophys. Res.* **94**, 2405–2415.

Meyer-Vernet, N., Aubier, M. G., and Pedersen, B. M.: 1986, *Geophys. Res. Lett.* **13**, 617–620.

Meyer-Vernet, N., Lecacheux, A., and Pedersen, B. M.: 1996, *Icarus* **123**, 113–128.

Miller, S. L.: 1955, *J. Amer. Chem. Soc.* **77**, 2351–2361.

Mott-Smith, H. M., and Langmuir, I.: 1926, *Phys. Rev.* **28**, 727–762.

Neubauer, F. M.: 1980, *J. Geophys. Res.* **85**, 1171–1178.

Neubauer, F. M., Gurnett, D. A., Scudder, J. D., and Hartle, R. E.: 1984, in: Gehrels, T., and Matthews, M. S. (eds.), *Saturn*, University of Arizona Press, Tucson, pp. 760–787.

Oberc, P., and Parzydlo, W.: 1992, *Icarus* **98**, 195–206.

Pedersen, B. M., Meyer-Vernet, N., Aubier, M. G., and Zarka, P.: 1991, *J. Geophys. Res.* **96**, 19,187–19,196.

Prinn, R. G., and Owen, T.: 1976, in *Jupiter*, Gehrels, T. (ed.), University of Arizona Press, Tucson, 319–371.

Prinn, R. G., Larson, H. P., Caldwell, J. J., and Gautier D.: 1984, in: Gehrels, T., and Matthews, M. S. (eds.), *Saturn*, University of Arizona Press, Tucson, pp. 88–149.

Reiner, M. J., Fainberg, J., and Stone, R. G.: 1993, *Geophys. Res. Lett.* **20**, 321–324.

Rucker, H. O., Macher, W., Manning, R., and Ladreiter, H. P.: 1996, *Radio Science* **31**, 1299–1311.

Russell, C. T.: 1991, *Space Science Rev.* **55**, 317–356.

Sagan, C., and Khare, B. N.: 1971, *Astrophys. J.* **168**, 563–569.

Scarf, F. L., and Gurnett, D. A.: 1977, *Space Sci. Rev.* **21**, 289–308.

Scarf, F. L., Gurnett, D. A., and Kurth, W. S.: 1979, *Science* **204**, 991–995.

Scarf, F. L., Gurnett, D. A., Kurth, W. S., and Poynter, R. L.: 1982, *Science,* **215,** 587–594.

Scarf, F. L., Frank, L. A., Gurnett, D. A., Lanzerotti, L. J., Lazarus, A., and Sittler, Jr., E. C.: 1984, in: Gehrels, T. and Matthews, M. S. (eds.), *Saturn*, University of Arizona Press, Tucson, pp. 318–353.

Smith, B. A., *et al.*: 1981, *Science* **212,** 163–190.

Southwood, D., *et al.*, this issue.

Stix, T.: 1992, *Waves in Plasmas, American Institute of Physics*, New York, 6.

Stone, R. G., Bougeret, J. L., Caldwell, J., Canu, P., DeConchy, Y., Cornilleau-Wehrlin, N., Desch, M. D., Fainberg, J., Goetz, K., Goldstein, M. L., Harvey, C. C., Hoang, S., Howard, R., Kaiser, M. L., Kellogg, P. J., Klein, B., Knoll, R., Lecacheux, A., Lengyel-Frey, D., MacDowall, R. J., Manning, R., Meetre, C. A., Meyer, A., Monge, N., Monson, S., Nicol, G., Reiner, M. J., Steinberg, J.-L., Torres, E., deVilledary, C., Wouters, F., and Zarka, P.: 1992, *Astron. Astrophys. Supp. Ser.* **92**, 291–316.

Trauger, J. T., Clarke, J. T., Ballester, G. E., Evans, R. W., Burrows, C. J., Crisp, D., Gallagher, J. S., Griffiths, R. E., Hester, J. J., Hoessel, J. G., Holtzman, J. A., Krist, J. E., Mould, J. R., Sahai, R., Scowen, P. A., Stapelfeldt, K. R., Watson, A. M.: 1998, *J. Geophys. Res.* **103**, 20,237–20,244.

Trotignon, J. G., Beghin, C., Grard, R., Pedersen, A., Formisano, V., Mogilevsky, M., and Mikhailov, Y.: 1987, *Astron. Astrophys.* **187**, 83–88.

Uman, M. A.: 1987, *The Lightning Discharge,* Academic Press, N. York.

Vogl, D. F., Ladreiter, H. P., Zarka, P., Rucker, H. O., Macher, W., Kurth, W. S., Gurnett, D. A., and Fischer, G.: 2001, in: Rucker, H. O., Kaiser, M. L., and Leblanc, Y. (eds.), *Planetary Radio Emissions V*, Austrian Academy of Sciences Press, Vienna, pp. 357–365.

Wahlund, J.-E., Eriksson, A. I., Holback, B., Boehm, M. H., Bonnell, J., Kintner, P. M., Seyler, C. E., Clemmons, J. H., Eliason, J. L., Knudsen, D. J., Norqvist, P., and Zanetti, L. J.: 1998, *J. Geophys. Res.* **103**, 4343–4375.

Warwick, J. W., Pearce, J. B., Peltzer, R. G., and Riddle, A. C.: 1977, *Space Sci. Rev.* **21**, 309–327.

Warwick, J. W., Pearce, J. B., Evans, D. R., Carr, T. D., Schauble, J. J., Alexander, J. K., Kaiser, M. L., Desch, M. D., Pedersen, B. M., Lecacheux, A., Daigne, G., Boischot, A., and Barrow, C. W.: 1981, *Science* **212**, 239–243.

Warwick, J. W., Pearce, J. B., Evans, D. R., Romig, J. H., Alexander, J. K., Desch, M. D., Kaiser, M. L., Aubier, M., Leblanc, Y., Lecacheux, A., and Pedersen, B. M.: 1982, *Science* **215**, 582–687.

Whalström, M. K., Johansson, E., Veszelei, E., Bennich, P., Olsson, M., and Hogmark, S.: 1992, *Thin Solid Films* **220**, 315–320.

Wu, C. S., and Lee, L. C.: 1979, *Astrophys. J.* **230**, 621–626.
Zarka, P.: 1985a, *Icarus* **61**, 508–520.
Zarka, P.: 1985b, in: Rucker, H. O. and Bauer, S. J. (eds.), *Planetary Radio Emissions,* Austrian Academy of Science, Vienna, pp. 237–270.
Zarka. P.: 1998, *J. Geophys. Res.* **103**, 20,159–20,164.
Zarka, P., and Cecconi, B.: 2002, *J. Geophys. Res.*, in preparation.
Zarka, P., and Pedersen, B. M.: 1983, *J. Geophys. Res.*, **88**, 9007–9018.
Zarka, P., and Pedersen, B. M.: 1986, *Nature* **323**, 606–608.
Zarka, P., Pedersen, B. M., Lecacheux, A., Kaiser, M. L., Desch, M. D., Farrell, W. M., and Kurth, W. S.: 1995, in *Neptune and Triton,* Cruikshank, D. P. (ed.), University of Arizona Press, 341–387.

THE CASSINI COSMIC DUST ANALYZER

R. SRAMA[1,*], T. J. AHRENS[3], N. ALTOBELLI[1], S. AUER[4], J. G. BRADLEY[2], M. BURTON[2], V. V. DIKAREV[1,21], T. ECONOMOU[5], H. FECHTIG[1], M. GÖRLICH[11], M. GRANDE[6], A. GRAPS[1], E. GRÜN[1,20], O. HAVNES[7], S. HELFERT[19], M. HORANYI[17], E. IGENBERGS[8], E. K. JESSBERGER[9], T. V. JOHNSON[2], S. KEMPF[1], A.V. KRIVOV[18], H. KRÜGER[1], A. MOCKER-AHLREEP[1], G. MORAGAS-KLOSTERMEYER[1], P. LAMY[10], M. LANDGRAF[22], D. LINKERT[1], G. LINKERT[1], F. LURA[11], J. A. M. MCDONNELL[12], D. MÖHLMANN[11], G. E. MORFILL[13], M. MÜLLER[12], M. ROY[2], G. SCHÄFER[1], G. SCHLOTZHAUER[11], G. H. SCHWEHM[14], F. SPAHN[18], M. STÜBIG[1], J. SVESTKA[15], V. TSCHERNJAWSKI[11], A. J. TUZZOLINO[5], R. WÄSCH[11] and H. A. ZOOK[16]

[1]*Max-Planck-Institut für Kernphysik, Postf. 103980, 69029 Heidelberg, Germany*
[2]*Jet Propulsion Laboratory, 4800 Oak Grove Drive, Pasadena, CA 91103, U.S.A.*
[3]*Seismological Laboratory, CALTECH, Pasadena, CA, U.S.A.*
[4]*Post Office Box 421, Basye, VA 22810, U.S.A.*
[5]*Enrico Fermi Institute, University of Chicago, Chicago, IL 60637-1433, U.S.A.*
[6]*Rutherford Appleton Laboratory, Chilton, Didcot, Oxon, OX11 0QX, U.K.*
[7]*Auroral Observatory, University of Tromso, 9000 Tromso, Norway*
[8]*Fachgebiet Raumfahrttechnik, TU München, Boltzmannstrasse 15, 85748 Garching, Germany*
[9]*University Münster, Schloßplatz 2, 48149 Münster, Germany*
[10]*Laboratoire d'Astronomie Spatiale, B.P. 8, 13376 Marseille Cedex 12, France*
[11]*DLR Berlin, Rutherfordstrasse 2, 12489 Berlin, Germany*
[12]*Planetary and Space Science Research Institut, Open University, Walton Hall, Milton Keynes MK7 6AA, U.K.*
[13]*Max-Planck-Institut für Physik und Astronomie, Postf. 1603, 85740 Garching, Germany*
[14]*ESA-ESTEC, Planetary and Space Science Division, P.O. Box 299, 2200 AG Noordwijk, The Netherlands*
[15]*Prague Observatory, Petrin 205, 11846 Prague 1, C.R.*
[16]*NASA Johnson Space Center, SN3, Houston, TX 77058, U.S.A.*
[17]*Laboratory for Atmospheric and Space Physics, 1234 Innovation Drive, Boulder, CO 80303-7814, U.S.A.*
[18]*AG nonlinear dynamics, University of Potsdam, Am neuen Palais 19, 14469 Potsdam, Germany*
[19]*Lehrstuhl Praktische Informatik, University of Mannheim, A5, 68131 Mannheim, Germany*
[20]*HIGP, University of Hawaii, 1680 East West Road, Honolulu, HI 96822, U.S.A.*
[21]*Astronomy Institute of St. Petersburg State University, Russia*
[22]*ESA/ESOC, Robert-Bosch-Straße 5, 64293 Darmstadt, Germany*
*(*Author for correspondence: E-mail address: ralf.srama@mpi-hd.mpg.de)*

(Received 12 March 1998; Accepted in final form 7 January 2002)

Abstract. The Cassini-Huygens Cosmic Dust Analyzer (CDA) is intended to provide direct observations of dust grains with masses between 10^{-19} and 10^{-9} kg in interplanetary space and in the jovian and saturnian systems, to investigate their physical, chemical and dynamical properties as functions of the distances to the Sun, to Jupiter and to Saturn and its satellites and rings, to study their interaction with the saturnian rings, satellites and magnetosphere. Chemical composition of interplanetary meteoroids will be compared with asteroidal and cometary dust, as well as with Saturn dust, ejecta

Space Science Reviews **114:** 465–518, 2004.

from rings and satellites. Ring and satellites phenomena which might be effects of meteoroid impacts will be compared with the interplanetary dust environment. Electrical charges of particulate matter in the magnetosphere and its consequences will be studied, e.g. the effects of the ambient plasma and the magnetic field on the trajectories of dust particles as well as fragmentation of particles due to electrostatic disruption.

The investigation will be performed with an instrument that measures the mass, composition, electric charge, speed, and flight direction of individual dust particles. It is a highly reliable and versatile instrument with a mass sensitivity 10^6 times higher than that of the Pioneer 10 and 11 dust detectors which measured dust in the saturnian system. The Cosmic Dust Analyzer has significant inheritance from former space instrumentation developed for the VEGA, Giotto, Galileo, and Ulysses missions. It will reliably measure impacts from as low as 1 impact per month up to 10^4 impacts per second. The instrument weighs 17 kg and consumes 12 W, the integrated time-of-flight mass spectrometer has a mass resolution of up to 50. The nominal data transmission rate is 524 bits/s and varies between 50 and 4192 bps.

Keywords: CDA, Cassini, dust sensor, E-ring, interplanetary dust

1. Introduction

The Cassini-Huygens mission provides the opportunity for a thorough investigation of the interplanetary dust complex and the saturnian dust environment. The scientific objectives were stated in our proposal dated 1989: "The overall objective of the proposed investigation is the exploration of the physical, chemical and dynamical properties of small dust particles in interplanetary space and in the saturnian environment. The parameters to be studied include mass, composition, electric charge, speed and flight direction of individual particles. The impact rate, mass distribution, average composition, angular distribution, and charge will be determined with respect to heliocentric and saturnian distances, to the distance from rings and satellites and to magnetospheric coordinates."

Cassini-Huygens is the first spacecraft which will orbit Saturn. The current experimental results are based on former Voyager and Pioneer flyby measurements and give only snapshots of the complex dust environment of Saturn. Current simulations of dusty rings in the saturnian system are still based on the result of the former Voyager missions. Now, the dust experiment on Cassini-Huygens promises much better results in quantity and quality. The long measurement time of over 3 years around Saturn will allow extensive studies of the ring details. Especially the results of the Galileo dust detector in the jovian system lead to ideas and predictions of dusty phenomena such as dust atmospheres around small moons (e.g. Krüger *et al.*, 2000; Thiessenhusen *et al.*, 2000), dust streams (Horanyi, 2000) and halo orbits (Howard and Horanyi, 2001) in the saturnian system. Table I shows a summary of the former dust detectors on interplanetary spacecrafts and Table II summarizes the scientific goals.

TABLE I

Mass sensitivities and measurement ranges of different interplanetary dust detectors.

Spacecraft	Mass threshold (kg)	Dynamic range	Sensitive area (m^2)	Reference
Pioneer 8/9	2×10^{-16}	10^2	0.010	Berg and Richardson (1968)
Pioneer 10	2×10^{-12}	—	0.26	Humes *et al.* (1974)
Pioneer 11	1×10^{-11}	—	0.26 (0.57)	Humes (1980)
HEOS 2	2×10^{-19}	10^4	0.010	Hoffmann *et al.* (1975)
Helios 1 and 2	9×10^{-18}	10^4	0.012	Dietzel *et al.* (1973)
Ulysses	2×10^{-18}	10^6	0.10	Grün *et al.* (1983)
Galileo	2×10^{-18}	10^6	0.10	Grün *et al.* (1992)
Cassini	5×10^{-19}	10^6	0.10	This work

The mass thresholds refer to 20 km/s impact speed. The Pioneer 10 and 11 detectors are threshold detectors.

2. Scientific Background

2.1. Study of Dust in the Outer Solar System

The manifestations of dust in the solar system have long been known. Comets and their dusty tails impressed mankind in living memory. Huygens recognized the ring around Saturn in 1655 and a few years later the zodiacal light phenomenon was correctly explained by Cassini. Dust in this context means particulate matter which does not manifest itself as isolated bodies but from some distance can be recognized as an ensemble of indistinguishable particles. It is known today that all these phenomena are closely related. The Cosmic Dust Analyzer (CDA) investigation will significantly enhance our knowledge on many aspects of this complex.

New comets have been observed to emit dust as distant as 15 AU from the sun. However, most of this dust after a short visit will leave the solar system on hyperbolic orbits (Mukai *et al.*, 1989). Also, short period comets have been observed to inject dust far from the Sun. An example is comet Schwassmann-Wachmann 1 which is beyond Jupiter in a nearly circular orbit ($a = 6.38, e = 0.13$) and undergoes sudden changes in brightness of as much as six magnitudes (Jewitt, 1989). Observations of 2060 Chiron indicate that this "asteroid" shows cometary activity at distances of 12–17 AU from the Sun (Bus *et al.*, 1989). At the distance of the asteroid belt several bands of thermal dust emission parallel to the ecliptic were detected by the IRAS satellite (Low *et al.*, 1984) and by COBE (Reach *et al.*, 1995).

Interstellar grains were detected by the Ulysses dust detector in our planetary system which allows the study of interstellar matter by in-situ measurements outside of 1 AU. Although the Cassini dust detector is as sensitive and reliable as the

TABLE II

Scientific objectives of the cosmic dust analyzer.

Cruise science	
	Extend studies of interplanetary dust to the orbit of Saturn.
	Sample the chemical composition of dust in interplanetary space and across the asteroid belt.
	Determine the flux of interstellar particles during solar maximum conditions.
	Search for dust streams originating from Saturn.
Jupiter flyby	
	Investigate the dynamics of the Io dust streams as discovered by Ulysses and Galileo. Characterize their direction, size-mass-distribution and correlation with the jovian and interplanetary magnetic field.
	Investigate the dust stream fluxes caused by the jovian system with respect to the Jupiter distance.
	Analyze dust stream particles at a different epoch from Galileo.
	Characterize the elemental composition of dust stream particles.
Rings	
	Map size distribution of ring material, search for ring particles beyond the known E-ring.
	Analyze the chemical composition of ring particles.
	Study dynamical processes (erosional and electromagnetic) responsible for the E-ring structure, study interactions between the E-ring and Saturn's magnetosphere, search for electromagnetic resonances.
	Determine dust and meteoroid distribution both in the vicinity of the rings and in interplanetary space.
Icy satellites	
	Define the role of meteoroid impacts as mechanism of surface modifications.
	Obtain information on the chemical composition of satellites from the analysis of gravitationally bound ejecta particles in the vicinity of the satellites (within Hill spheres).
	Investigate interactions with the ring system and determine the importance of the various satellites as a source for ring particles.
Magnetosphere of Saturn	
	Determine the role that dust plays as source and sink of charged particles in the magnetosphere.
	Search for electromagnetically dominated dust (small particles) and for dust streams.

Ulysses and Galileo detector, the Cassini trajectory does not allow the monitoring of interstellar dust for many years (the interstellar grains are shielded by the Sun, Landgraf *et al.*, 1999). Just in 1999 and in the late tour (after 2007) the measurement of interstellar grains is possible in order to support the recent results (Landgraf *et al.*, 2000; Grün and Landgraf, 2000). Recently, Altobelli et al. (2003) presented the discovery of ISD.

In-situ observations of dust out to nearly 20 AU were provided by the "beer can" detectors on board the Pioneer 10 and 11 spacecraft (Humes *et al.*, 1974; Humes, 1980; Dikarev and Grün, 2002). These highly reliable but relatively insensitive impact detectors recorded impacts of a total of 182 dust particles during their voyages outside the Earth's orbit. Fifteen and four of these impacts were recorded in the vicinity of Jupiter and Saturn, respectively. While the dust concentration detected between Jupiter and Saturn is mainly due to the cometary components, the dust outside Saturn's orbit is dominated by grains originating from the Edgeworth-Kuiper belt (Landgraf *et al.*, 2002). Surprisingly, the plasma wave experiment on board the Voyager 2 spacecraft picked up charge signals from expanding plasma clouds generated by dust impacts onto the spacecraft during its passage through Saturn's ring plane (Gurnett *et al.*, 1983).

Until recently most of our knowledge about Saturn's ring system was obtained through Earth based astronomical observations. Images taken by Voyagers' TV cameras showed that the rings are far more complex than one had ever imagined before. However, a detailed look at this complexity by the Cassini-Huygens mission may provide the key to a realistic view of planetary ring physics. In order to study Saturn's rings one has to understand the environment in which they are embedded. An important aspect of this environment is the interplanetary and interstellar dust. Gurnett *et al.* (1997) used the Voyager PWS experiment to detect particles in the interplanetary space even beyond Saturn.

2.2. Interplanetary Dust

A great deal already is known about the interplanetary meteoritic complex from laboratory and ground based studies on stratospheric dust collections (e.g., Brownlee, 1985; Jessberger *et al.*, 2001), on photographic meteors, and on radar detected meteors (Baggeley, 1999), as well as results from zodiacal light measurements (Leinert and Grün, 1990). A great deal more has been learned from measurements taken from in-situ experiments flown on spacecraft (McDonnell, 1978). From these measurements it has been shown (Whipple, 1967; Grün *et al.*, 1985) that the meteoritic complex in the inner solar system is self destructive with a time scale of the order of 10^5 years. Therefore most dust grains in this complex have been derived, relatively recently, from larger bodies. The Cosmic Dust Analyzer will have the excellent capability to provide fundamentally new information as well as to solve a number of enigmas that have arisen as the result of previous space meteoroid investigations.

The dust experiments on the solar orbiting Pioneer 8 and 9 spacecraft measured a relatively large flux of micron to sub-micron sized dust grains that were deduced by Berg and Grün (1973) to be in hyperbolic orbits, that are leaving the solar system to become interstellar grains. Zook and Berg (1975) called these particles beta meteoroids and deduced that they were probably primarily produced as debris resulting from mutual collisions between larger meteoroids that were sunward from the Pioneer 8 and 9 sensors. Whipple (1975) found that most beta meteoroids, because of their directional characteristics, must have derived from outside of 0.5 AU. Further analyses (Zook, 1975) showed that the flux of beta meteoroids appeared to be increasing with increasing heliocentric distance near, and just outside of, 1 AU.

Almost simultaneously, however, McDonnell *et al.* (1975) were able to explain the same data by assuming that the production rate of beta meteoroids varied instead with heliocentric longitude. But uncertainty remains. First, zodiacal light data show that the heliocentric radial variation of meteoritic particles (or, more precisely, their cross-sectional area per unit volume) varies as $r^{-\alpha}$ (r = heliocentric distance) where $\alpha = -1.3$ inside of 1 AU (Leinert *et al.*, 1981) and $\alpha = -1.5$, or even more negative, outside of 1 AU (Hanner *et al.*, 1976). Why should the radial distribution of zodiacal particles change character right at, or near, 1 AU? Second, to add to the mystery, Jackson and Zook (1989) found, through numerical modelling, that many dust grains ejected from asteroids in the main belt would be expected to be trapped into heliocentric orbital period resonances with the Earth, and would be concentrated around and external to 1 AU. The Cosmic Dust Analyzer, with its large area (10 times that of the Pioneer 8 and 9 sensors), its high sensitivity, and with the ability to vary its pointing direction, is well suited to solve the mystery.

Meteoroid penetrations through the 25 μm thick stainless steel meteoroid penetration sensor on the Pioneer 10 spacecraft gave a spatial density of meteoroids that decreased from 1 to about 1.8 AU from the Sun, and then remained constant, or even increased with increasing heliocentric distance, out to nearly 20 AU (Humes *et al.*, 1974; Humes, 1980). This was a very enigmatic result for two reasons: first, the zodiacal light sensor on the same spacecraft gave a spatial density of meteoroids that decreased from 1 to 3.3 AU, except for an approximate 30% additional asteroid belt contribution between 2.3 and 3.3 AU, and with a zero spatial density after 3.3 AU. More insight into this puzzle was obtained when sensors on the IRAS satellite detected bands of thermal infra-red emission parallel to the ecliptic (Low *et al.*, 1984). Sykes and Greenberg (1986) interpreted these asteroidal bands as due to collisions within the Eos, Koronis, and Themis families in the main asteroid belt. This leads one to ask whether main belt asteroid collisions could be giving rise to mostly coarse grained material that is providing most of the scattering cross-section observed by zodiacal light sensors, while comets are contributing most of the small particles detected by the penetration sensors. The CDA instrument is able to sense both large and small grains, as well as to observe any enhanced production of beta meteoroids in the asteroid belt. Its directionality measurements will permit discrimination between cometary and asteroidal sources.

The second reason that the Pioneer 10 penetration data present an enigma, is due to the lack of spatial density fall-off with increasing heliocentric distance beyond 2 AU. Poynting-Robertson drag (e.g., see Burns *et al.*, 1979) would be expected to set up a meteoroid spatial density population that increases with decreasing heliocentric distance inside the source region of meteoroids, with a zero spatial density outside that region. This marked contradiction of theoretical expectations compared to the actual penetration data led Zook (1980) to suggest that meteoroids made of water ice were penetrating the Pioneer 10 sensor at far distances from the Sun, but were evaporating at close distances from the sun, with few, or none, inside 2 AU. Humes (1980) found that the meteoroid penetration data obtained between

3.5 and 5 AU from the 50 μm thick penetration sensors on the Pioneer 11 satellite could not be explained as due to penetrations by meteoroids in largely prograde heliocentric orbits. The model that gave him the best fit to that data was a model that assumed that meteoroids were in highly eccentric orbits and that they were approximately randomly inclined to the ecliptic plane. Both asteroids and short period comets are nearly all in prograde orbits about the Sun. So the puzzle is, what could be the source of meteoroids that are in such highly inclined and highly eccentric orbits? Are they related to long period comets? The CDA sensor can provide more information about the true orbital and compositional characteristics of this family of particles, and lead us to their source.

2.3. The Jovian System

In 1973 when the Pioneer 10 spacecraft flew by Jupiter micron sized dust particles were detected within the jovian system for the first time (Humes *et al.*, 1974). Almost 20 years later the jovian system was recognized as a source of intermittent streams of sub-micron sized dust particles when the Ulysses spacecraft flew by the planet (Grün *et al.*, 1993). Similar streams were later detected within 2 AU from Jupiter during Galileo's approach to the planet (Grün *et al.*, 1996).

It was immediately recognized (Horanyi *et al.*, 1993; Hamilton and Burns, 1993) that Jupiter's magnetosphere would eject sub-micron sized dust particles if they existed in the magnetosphere. At two places small dust in abundance had been observed by Voyager's cameras: (1) the jovian ring at $1.8R_J$ (Jupiter radius, $R_J =$ 71,492 km) and its weak extension out to $3R_J$ (Showalter *et al.*, 1995), and (2) Io's volcanic plumes, that reach heights of about 300 km above Io's surface. Both phenomena have been suggested as the source of the dust streams.

Electromagnetic interaction of the particles making up the dust streams was evident both in the Ulysses and Galileo data when both spacecraft were outside the jovian magnetosphere: the arrival direction showed significant correlations with the ambient interplanetary magnetic field (Grün *et al.*, 1993, 1996). Zook *et al.* (1996) demonstrated that only particles in the 10 nm size range can couple strongly enough to the interplanetary magnetic field to show the effects observed by Ulysses. The corresponding impact speeds were deduced to be in excess of 200 km/s.

Within the jovian magnetosphere Galileo has detected streams of submicrometer sized dust particles which must originate in the inner jovian system within several R_J from Jupiter (Grün *et al.*, 1996, 1997). These streams show highly variable impact rates correlated with Jupiter's rotation period, implying that the particles strongly interact with the planet's magnetic field (Graps *et al.*, 2000; Horanyi, 2000). In addition to these sub-micron sized dust stream particles Galileo has identified two more types of dust: a concentration of small dust impacts at the times of Ganymede and Europa closest approach, and big micron sized dust particles concentrated in the inner jovian system between the Galilean satellites (Krüger *et al.*, 2000; Thiessenhusen *et al.*, 2000; Krivov *et al.*, 2002).

Cassinis closest approach to Jupiter was on December 30, 2000 at a distance of $137 R_J$, whereas at this time Galileo was 10 times closer to the planet. This opportunity allowed a combined measurement of the jovian dust streams simultaneously with two spacecraft. Dust trajectories exist which intersect both the Galileo and the Cassini-Huygens orbit. The idea of this unique measurement was to identify peaks in the dust fluxes of both instruments which provides a direct measure to determine the time-of-flight of the grains between about 14 and $137 R_J$. The current analysis shows that the dust magnetosphere interaction is more complex than originally thought and further modelling is necessary. The record of impact mass spectra of dust stream particles were achieved by CDA and the data are still in analysis. However, such tiny projectiles show only weak signatures in the mass spectra such that the spectra are dominated by peaks of the target material.

2.4. Chemical Composition of Cosmic Dust and its Parents

Previous in-situ sampling and remote chemical analyses of solar system objects has been, to date, limited to the moon, Venus, Mars, comet Halley and, to a very limited degree, Phobos. Sampling and analysis of ejecta during close fly-bys of saturnian satellites and rings will make a whole new class of objects available to in-situ chemical analysis. Comparison of these icy objects with other solar system bodies will shed light on their mutual interrelation.

It is common to compare the isotopic, chemical and molecular composition of solar system bodies with the composition of meteorites, since these are the only extraterrestrial objects which can be studied in great detail in the laboratory (Anders and Grevesse, 1989). Such studies have resulted in a wealth of information. The genetic links between different classes of meteorites and their constituents set stringent boundary conditions on early solar system materials and processes.

The other extraterrestrial objects, which can be studied in the laboratory – albeit less completely than meteorites – are interplanetary dust particles (IDPs). Some IDPs closely resemble chondritic meteorites in their chemical composition. Others differ from any known meteorite class (Brownlee, 1985; Bradley, 1988; Jessberger *et al.*, 1992). Since their sources and the interrelations of individual IDPs are unknown, the interpretation of this observation is difficult (Sutton and Flynn, 1988; Arndt *et al.*, 1996).

A third and even less direct, but nevertheless extremely exciting source of information derives from in-situ measurements near comets such as the Halley experiments (Kissel *et al.*, 1986; Jessberger, 1999).

Overall, the presence of three major classes of material is to be expected (Jessberger *et al.*, 1988): Ice, CHON, and silicates. The higher the variability is from grain to grain of the relative proportions of ice, CHON, and silicate components, the more probable will be the identification of the constituents. Or, likewise, the smaller the grains are, the higher are the chances to encounter mono-component grains which may ease identification. Therefore chemical composition will be optimized for the smallest particles. The chemical composition of dust grains does not only characterize its own identity but also carries memories of its parents. Everywhere in the solar system dust is a short lived phenomenon which requires permanent replenishment. Dynamic effects dissipate dust away from its sources until it is swept up by larger bodies or lost to interstellar space. Cassini on its path to Saturn will cross several source regions of interplanetary dust.

Dust grains that are ascertained to be asteroidal in origin will be compositionally compared with ground-based spectral studies of asteroids. Both asteroid albedo and compositional type are found to vary with distance and this effect will simplify comparison of Cassini results with ground-based data. Compositions of clearly identifiable cometary grains may be obtained, especially beyond the asteroid belt. In some cases individual grains will be related to unique comets; grain compositions will then be compared with ground-based studies of those comets. The future NASA and ESA missions to comets which may even bring back samples and the dust-dedicated STARDUST mission will significantly contribute to solve this question.

2.5. Saturn's Rings

The diversity of Saturn's ring structure, from the diffuse and voluminous E-ring to the extremely fine structure of the main rings, the complex variability in many rings, examplified by the F-ring with its kinks, braids and multiple strands, is one of the most challenging problems of planetary physics today. We expect Cassini-Huygens to lead to answers to most of the fundamental questions related to the formation, structure, dynamics, evolution and lifetime of rings. The observations provided by the Cosmic Dust Analyzer, will surely be one of the cornerstones on which to build a complete understanding of ring systems. Measurements by the CDA will provide the spatial distribution of dust and dust properties such as mass, composition, charge and trajectory in the parts of the ring system through which the Cassini Orbiter will pass. The Cassini-Huygens investigation of dust, charged particles and fields, together with photometer, spectrometer and camera data, will surely provide a wealth of information on dusty plasma properties which will be invaluable in our effort to understand such diverse subjects as planetary ring systems, cosmogony, the physics of colloidal plasmas and dynamics of single dust particles.

In-situ measurements of dust in the saturnian system will be done repeatedly throughout the whole E-ring and beyond. During Saturn Orbit Insertion (SOI) there will be a one-time passage, when the spacecraft will move above the main rings at a height of 10,000 to 30,000 km and reach a periapsis distance of $1.3 R_s$ from Saturn. Dust which would have been detected above the rings during SOI would have given testimony of impacts onto the main rings. Correlation with simultaneous spoke observations (Grün *et al.*, 1983) would allow us to test the hypothesis that spokes are triggered by impacts of large meteoroids onto the B-ring (Goertz and Morfill, 1983). The meteoroid flux is influential for planetary ring structure and evolution both because of its erosional effect on ring material and since it adds a substantial amount of mass to the rings. It is possible that the rings are impacted by their own mass of meteoritic material over the age of the solar system. It is also likely that this bombardment and associated ballistic and electromagnetic transport of the impact ejecta can substantially redistribute angular momentum within rings (Goertz and Morfill, 1988) and affects the compositon (Cuzzi and Estrada, 1998). However, the spacecraft pointing and operational constraints did not allow CDA perform measurements driving SOI.

The most complete measurements of dust will be in the E-ring region and further out. Little or nothing is known about the dust outside the E-ring and Cassini will reveal if additional faint rings are present there. The mapping of the incident dust flux and size distribution will be an important task here. Questions which will be answered by the CDA and Cassini are how far inside Enceladus the ring material extends and also a determination of its outer boundary position and sharpness.

It is most often assumed that Enceladus is the major source of E-ring material since the ring brightness has a profound maximum around, or near to, its orbit (Haff *et al.*, 1983). The true role of this and other moons as sources or sinks for ring material will be deduced from a mapping of dust chemical composition, size and mass distributions throughout the ring. Passages closer than 1000 km to Enceladus (Spahn *et al.*, 1999) and Dione, with the possibility that the CDA will detect newly injected dust, and their mass and velocity distribution, should lead to a much better understanding of dust injection processes (Krivov *et al.*, 2001).

An answer to whether the dust in the E-ring is a source for neutral gas and plasma (Morfill *et al.*, 1990) or whether dust condenses out of neutral gas injected by the moons (Johnson *et al.*, 1989) will emerge from a comparison of the radial dependence of dust size and composition with neutral gas and plasma conditions.

The evolution of the E-ring is thought to consist mainly of an outward drift of dust which is being sputtered. The proposed drift mechanisms are gyrophase drift (e.g., Burns and Schaffer, 1989) and plasma drag (Morfill *et al.*, 1990). The magnitude and relative importance of these two effects is poorly known. However, plasma drag tends to circularize particle orbits (Dikarev, 1999), whereas the gyrophase drift often leads to elliptic orbits with high eccentricity. Mapping of individual dust particle trajectories as a function of position, dust size and charge, compared with plasma density, composition and energy distribution will tell us what role the different transport effects play.

The large thickness of the E-ring, possibly in excess of 10,000 km (Hamilton, 1994) is also a major puzzle. The expected evolution of a ring dominated by gravitational effects where collisions occur is a collapse towards the ring plane. Is the large thickness of the E-ring possible because its dust particles have short lifetimes and are constantly replenished? Can the great thickness be related to the dust injection process, the initial dust orbit inclination being conserved as the dust particles drift radialy outwards, or are there other processes acting to increase the ring thickness beyond this injection spread? Other questions, regarding physical processes will be answered by the Cassini experiment: How important is the electrostatic support of the E-ring and other rings (Havnes and Morfill, 1984). This effect is crucially dependent on the dust mass and charge together with the plasma condition. The CDA and plasma experiments should give the parameters necessary to determine the role of this mechanism. Can electromagnetic resonance effects on single dust particles (Schaffer and Burns, 1987) also be important for Saturn's E-ring? A possibility is that oscillations in an electrostatically supported dust ring can be in resonance with corotating magnetospheric sector features (Melandso and Havnes, 1991) and that this can lead to increases in the ring thickness. Intriguing major resonances are within the E-ring and at the exact location of the G-ring. This opens up the possibility for vertical oscillations in the E and G-ring with periods of the order of a day.

We expect that important checks on the dust optical properties will result from the combined results of Cassini photometers, spectrometers, cameras and the dust analyzer. Remote sensing by measurements of light scattering at dust is an important tool, however, it suffers from inherent uncertainties related to size, structure (refractive index) and shape of dust. A comparison of scattering properties with in-situ measurements by the CDA will likely lead to more confidence in the results from other parts of the ring system which will not be made directly accessible to the Cassini orbiter.

2.6. Charges on Interplanetary and Saturnian Dust Particles

There exist a variety of mechanisms by which cosmic dust particles can be electrically charged: the capture of ambient electrons and ions, secondary electron emission by energetic electron and ion impacts, photoemission due to short-wavelength electromagnetic radiation, field emission of electrons, triboelectric effects, and field evaporation of ions. Interplanetary and interstellar dust particles acquire a positive charge in the solar wind and can be strongly influenced by the Lorentz force as they pass through planetary magnetospheres. There, the charge on the particles changes rapidly when they pass through different plasma environments (Colwell *et al.*, 1998).

In the case of interplanetary dust particles outside the earth's orbit the only relevant charging processes are interactions with solar wind electrons and ions and photoemission by solar UV radiation. Depending upon conditions in the solar wind (low and high speed) the proton number density varies from 12 to 4 cm^{-3} at 1 AU, the bulk speed of the wind varies from 330 to 700 km/s, and the mean thermal energy increases from 3 to 20 eV for protons and decreases from 11 to 9 eV for electrons (Morfill *et al.*, 1986). The flux of photoelectrons from a metal surface at 1 AU was estimated by Wyatt (1969) to be equal to 2.5×10^{10} cm^{-2} s^{-1}. The photoelectron flux from silicate and graphite surfaces can be up to one order of magnitude lower; the same is expected for the flux from icy surfaces. From these numbers it follows that charging of interplanetary dust particles is dominated by photoemission, which leads generally to positively charged particles with surface potentials of several volts. The value of the potential depends on the photoemission yield of the particle material, e.g. silicate particles attain potentials of 2.5–5 V which are practically constant in the size interval 0.1–10 μm (Lamy *et al.*, 1985). Potentials of particles from conducting materials will be a few volts higher. In case of very small particles with dimensions small compared to the wavelength of light, the photoemission yield can be enhanced by a factor of 2–3 which results in higher surface potentials.

The potential of dust particles is practically independent of distance from the Sun because both solar UV flux and fluxes of solar wind particles decrease with the second power of distance from the sun (Tiersch and Notni, 1989). Non-zero electric charge of dust particles influences their dynamics in the interplanetary medium, see e.g. review of Morfill *et al.* (1986) and references therein. In case of saturnian particles the equilibrium charge will be determined again by their interaction with the ambient plasma and UV radiation. Parameters of plasma in the Saturnian environment cover a very broad range: densities from 10^{-2} cm^{-3} to several 100 cm^{-3} and energies of electrons from $\sim$1 eV to more than 1 keV, see e.g. Grün *et al.* (1984) and Horanyi. In regions where the plasma flux is low (e.g. above Saturn's A- and B-rings) the situation will be similar to interplanetary space and the potential of particles will be positive and equal to several volts.

In dense plasma regions where the electron flux is dominant, the sign and value of the potential will be determined by the energy of the electrons. At electron energies above about 10 eV secondary electron emission becomes important, which results in a reduction of the negative potential. At energies above a few hundred eV the secondary electron emission yield becomes >1 which causes a change of the potential from negative to positive. Generally, the yield reaches a maximum value $\delta_{\rm m}$ at energies $E_{\rm m}$ between 300 and 2000 eV. At higher energies the secondary yield becomes again <1. Furthermore, the yield is material dependent. The maximum yield for metals and semiconductors is of the order of unity and for insulators it is equal to 2–30. According to measurements of Hashimov and Tarakanov (1982) with water ice at temperatures of 90–95 K, the yield attains a maximum value of about 2 at an energy of the primary electrons of 900 eV. The yield is <1 for energies below 60 eV.

As a consequence of nonlinear charging equations two identical dust particles in the same environment but with different histories can have charges of opposite sign. In addition, small changes in the environmental conditions can cause large and rapid changes of the charge state of dust particles.

The process of secondary electron emission is very sensitive to the particle size and the physical properties of the dust particles. At electron energies >1 keV the range of electrons in a "compact" dust particle is >0.1 μm and, therefore, electrons can penetrate through small particles and moreover cause secondary electron emission from the exit side. One must also take into account the fact that the yield increases with increasing angle of incidence of primary electrons by up to an order of magnitude compared to plane surfaces (Draine and Salpeter, 1979). On the other hand, if the particle is in form of an aggregate with a rough surface, the yield will be lower.

If a small particle is hit by electrons with high energies, the electrons are penetrating through the particle, and the negative potential is low or may become even positive due to secondary electron emission from the exit side.

The total electric charge of dust grains is limited by either field evaporation, or field emission. These processes become dominant at field strengths $>3\times10^{10}$ V/m (positive charge) and $>10^9$ V/m (negative charge), respectively. Electrostatic repulsive forces can destroy particles (electrostatic fragmentation) at much lower field strength if they become higher than the maximum tensile strength of the material. The latter process is particularly important in case of fluffy particles which might have tensile strengths of about 10^3 Pa. For such a fluffy particle with a radius of 1 μm electrostatic fragmentation can occur already at surface potentials of as low as 10 V.

Electric charges on Saturnian dust particles strongly influence their dynamics, see e.g. reviews of Grün *et al.* (1984) and Mendis *et al.* (1984). If distances among dust particles are smaller than the Debye length in the ambient plasma, their mutual interactions become important which results in lowering the charges of individual particles, see e.g. Havnes (1987) and references therein.

Morfill *et al.* (1990) estimated the surface potential of dust grains for plasma conditions in the Saturnian magnetosphere. The results predict surface potentials in a range of -1200 to $+1.5$ V for Saturn distances from 4 to $9R_s$ (1 keV electrons and δ_m from 0 to 1).

2.7. Interplanetary Cumulative Dust Fluxes

2.7.1. *The Staubach Model*

Interplanetary dust fluxes were estimated (Table 3) on the basis of the Grün *et al.* (1985) fluxes at masses $m = 10^{-12}$, 10^{-9}, and 10^{-6} g. These fluxes were doubled in an effort to match what was observed with Pioneer 10 (Humes, 1980) due to the spacecraft motion. The $1/R$ fall-off with R (for $R > 3$ AU) for $m = 10^{-15}$ g, and the $1/R$ falls off with $R(R > 1$ AU) for $m = 10^{-12}$ g, are pure assumptions with little or no data to support them. The $m = 10^{-9}$ g flux follows that of the Pioneer 10 flux (i.e. it is constant beyond 2 AU). The $1/R^2$ flux fall off for $m = 10^{-6}$ g was assumed to try to approximately match zodiacal light data. At Saturn a twofold increase of the interplanetary meteoroid flux was assumed due to gravitational enhancement.

The following fluxes and their variation with radial distance went into the creation of Table 3:

1. At $m = 10^{-15}$ g, flux $= 1.1 \times 10^{-3}$ m^{-2} s^{-1} at 1 AU and is constant to 3 AU, then falls off as $3/R$ for $R > 3$ AU.

TABLE III

Interplanetary dust flux and fluence.

dr (AU)	dt (d)	Flux	Fluence, $m = 10^{-15}$ g	Flux	Fluence, $m = 10^{-12}$ g	Flux	Fluence, $m = 10^{-9}$ g	Flux	Fluence, $m = 10^{-6}$ g
0.9–2	611	95	58,000	4.0	2460	0.40	240	0.006	4
2–3	423	95	40,000	2.4	1020	0.26	109	0.001	1
3–4	146	81	12,000	1.7	250	0.26	38	0.0007	0
4–5	224	63	14,000	1.3	300	0.26	58	—	
5–6	208	52	11,000	1.1	230	0.26	54	—	
6–7	229	44	10,000	0.93	210	0.26	59	—	
7–8	261	38	9900	0.80	210	0.26	68	—	
8–9.1	327	34	11,000	0.70	230	0.26	85	—	
9.1	1490	68	100,000	1.4	2100	0.52	770	—	
Total	2429		265,900		7010		1481		5

The columns are heliocentric distance intervals in AU, time in days spent in the corresponding distance interval, fluxes onto a flat plate sensor which is mounted to a spinning platform in number/(m^2 day) and fluences in number/m^2 per distance interval.

2. At $m = 10^{-12}$ g, flux $= 7 \times 10^{-5}$ m^{-2} s^{-1} at 1 AU and falls off as $1/R$ for $R > 1$ AU.
3. At $m = 10^{-9}$ g, flux $= 6 \times 10^{-6}$ m^{-2} s^{-1} at 1 AU and falls to 3×10^{-6} m^{-2} s^{-1} at 2 AU and is then constant for $R > 2$ AU.
4. At $m = 10^{-6}$ g, flux $= 1 \times 10^{-7}$ m^{-2} s^{-1} at 1 AU and falls off as $1/R^2$ for $R > 1$ AU.

The CDA instrument consists of two subsystems, the dust analyzer (DA) and the high rate detector (HRD). Therefore the numbers above have to be multiplied by the sensitive area F of the different sensors ($F_{\mathrm{DA}} = 0.1$ m^2, $F_{\mathrm{HRD}} = 0.006$ m^2) and by a factor Ω/π which takes into account the effective solid angle of the respective sensor ($\Omega_{\mathrm{DA}} = 0.6$ sr, $\Omega_{\mathrm{HRD}} = 3.0$ sr). About 100 particles of 10^{-12} g or larger and several thousand smaller particles are expected to be recorded by the dust analyzer and the high rate detector.

2.8. Saturnian Dust Fluxes

The Saturnian dust fluxes (Table 4) have been calculated from the physical models provided. The maximum cumulative (mass greater or equal to m) flux in the ring plane during one ring plane crossing have been calculated.

The three models have been constructed in order to get a conservative estimate of the spacecraft hazard due to Saturnian dust particles. This is certainly true for model 3 for which it is assumed that all the mass is in particles of a

TABLE IV

Saturnian dust fluxes: maximum flux in the ring plane (in number/m^2 s) during one ring plane crossing for $m = 10^{-12}$ g (model A), $m = 10^{-9}$ g (model B) and $m = 10^{-6}$ g (model C).

R	Model 1, $m = 10^{-12}$ g (A), flux (1/m^2s)	Model 2, $m = 10^{-9}$ g (B), flux (1/m^2s)	Model 3, $m = 10^{-6}$ g (C), flux (1/m^2s)
2.9	1.8×10^3	2.7	0.45
4.0	8.3×10^2	1.2	0.21
6.0	22	3.2×10^{-2}	5.4×10^{-4}
9.0	4.9×10^{-4}	7.3×10^{-7}	1.2×10^{-7}

single mass (in our case 10^{-6} g). The values given for 10^{-6} g are upper limits. Model B relies mostly on energetic particle data from Pioneer 11 and both Voyager spacecraft. Therefore this model describes best the intermediate sized particles (approximately 10^{-9} g) for which energetic particle effects are effective. Model A is only based on ground based and Voyager optical data of the E-ring. This model describes best the micron sized particles which are most effective light scatterers. No evidence has been acquired so far for sub-micron sized particles. However, there are good reasons to believe that sub-micron sized particles exist in abundance.

Contrary to the interplanetary dust flux the Saturnian dust flux is expected to be highly anisotropic. Therefore, depending on whether this flux is within the field-of-view of the sensor, dust particles are recorded or not. The highest fluxes are expected during ring plane crossings in the inner Saturnian system. The value of 1.8×10^3 m^{-2} s^{-1} given for $m = 10^{-12}$ g is considered a low value for the flux of the smallest detectable particles and therefore the instrument will be able to record fluxes temporarily up to two orders of magnitude higher.

3. Instrument Description

3.1. Experimental Approach

The stated objectives of this investigation require a versatile instrument consisting of several components which are optimized individually for different tasks. Therefore the detection of dust particle impacts is accomplished by two different methods: (1) a high rate detector subsystem, using two separate polyvinylidene fluoride (PVDF) sensors, for the determination of high impact rates during Saturnian ring plane crossings and (2) a Dust Analyzer (DA) using impact ionization. The DA measures the electric charge carried by dust particles, the impact direction, the impact speed, mass and chemical composition, whereas the high rate detector is

TABLE V

The Cassini-Huygens Cosmic Dust Analyzer team in 2001.

Principal investigator	Ralf Srama
Investigation senior scientist	Eberhard Grün
Science planning lead/OTL	Sascha Kempf
Deputy operations technical lead	Georg Moragas-Klostermeyer
Co-investigators	Thomas J. Ahrens
	Siegfried Auer
	Hugo Fechtig
	Manuel Grande
	Ove Havnes
	Mihaly Horanyi
	Eduard Igenbergs
	Torrence V. Johnson
	Elmar K. Jessberger
	Sascha Kempf
	Harald Krüger
	Philippe Lamy
	Franz Lura
	J. Anthony M. McDonnell
	Dietrich Möhlmann
	Ernst Pernicka
	Gerhard H. Schwehm
	Frank Spahn
	Jiri Svestka
	Anthony J. Tuzzolino
	Richard Wäsch
	Herbert A. Zook
Electronics chief engineer	Dietmar Linkert
Software engineer	Stefan Helfert
Laboratory technician	Gerhard Schäfer

capable of determining particle mass for particles with a known speed. The DA itself consists of three subsystems, the charge detector (entrance grids, EG), the impact ionization detector (IID) and the chemical analyzer (CA) (Bradley *et al.*, 1996; Srama, 1997). The Chemical Analyzer was developed by the University of Kent, Canterbury, U.K., under the leadership of J.A.M. McDonnell (now at Open University). Table 5 shows the Cosmic Dust Analyzer team in 2001 and Figure 1 shows the flight model of the Cosmic Dust Analyzer. Figure 2 shows a cut through the instrument with the subsystems labeled.

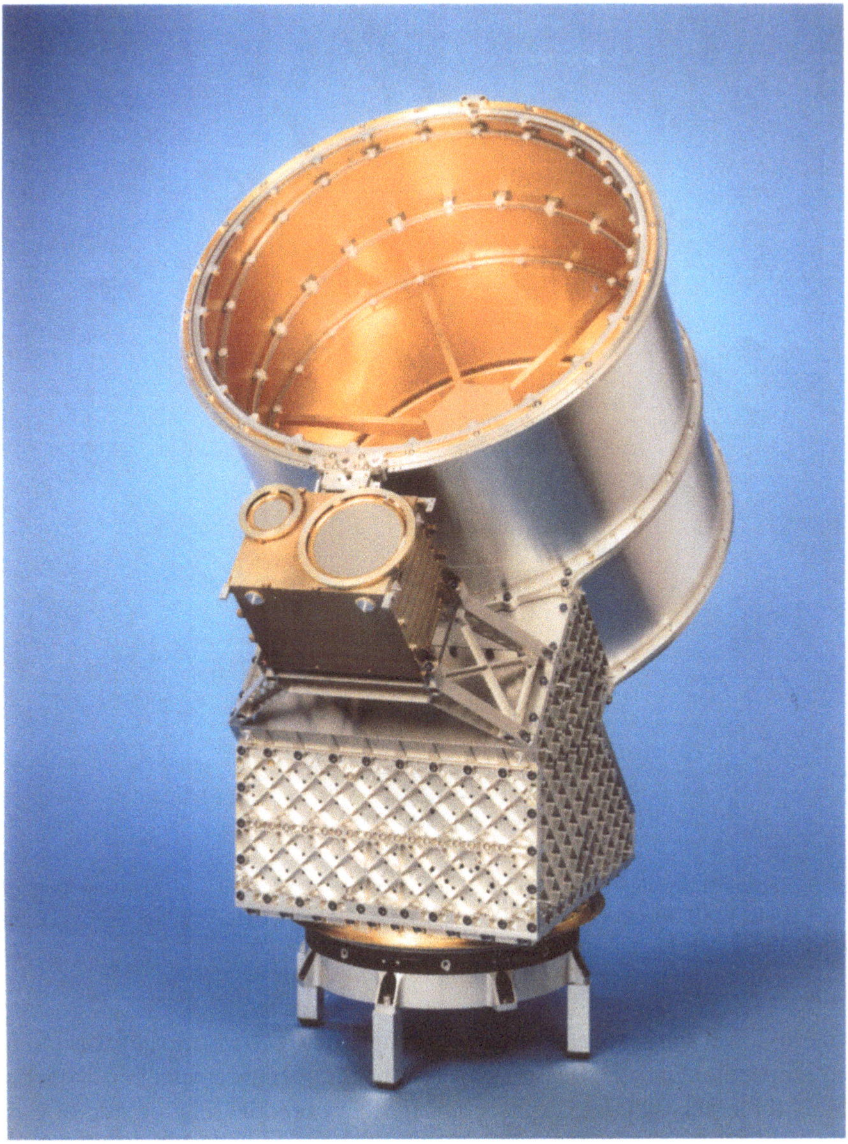

Figure 1. The Cosmic Dust Analyzer.

3.2. Angular Sensitivity, Sensor Pointing and Field-of-Views

The general purpose of this instrument is to cover the whole hemisphere with its field-of-view. On Galileo, this was achieved with the wide aperture of $\pm 70^\circ$ and a mounting of the instrument by 55° with respect to the Galileo spin axis. Originally the Cassini-Huygens design included a continuously rotating pointing platform for the fields and particles instruments which was canceled during a descoping process

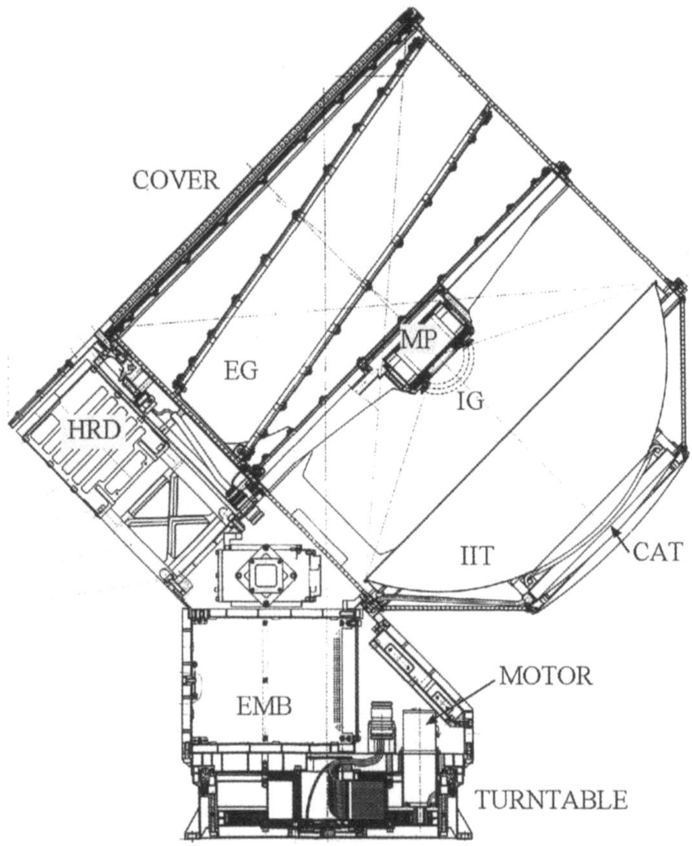

Figure 2. Technical drawing of the Cosmic Dust Analyzer. CAT, Chemical Analyzer Target; EG, Entrance Grids; EMB, Electronics Main Box; HRD, High Rate Detector; IG, Ion Grids; IIT, Impact Ionization Target; MP, Multiplier.

in order to lower the spacecraft costs. Although the CDA instrument was mounted nearly perpendicular to the Cassini spin axis, wide coverage cannot be obtained with a mainly three-axis stabilized spacecraft. Furthermore, the rotation rate of Cassini is restricted to the maximum value of 0.26 °/s and, during high activity periods, other instruments determine the orientation of the spacecraft. All these constraints lead to a redesign of the instrument and a turntable was added at the interface to the spacecraft.

The mounting vector of the turntable points 15° below the spacecraft x–y plane (Figure 3). Furthermore, this vector points 30° away from the $+y$-axis towards -x. The coordinates of the articulation axis with respect to the spacecraft x–y–z coordinate system are $(-0.483; 0.837; 0.259)$. The Dust Analyzer detectors (IID, CAT and HRD) are mounted at 45° with respect to the articulation axis. The boresight vector of the field-of-view has the coordinates $(-0.250; 0.433; 0.866)$ in the launch position (0° position, downwards to $+z$). The turntable enables the instrument to rotate

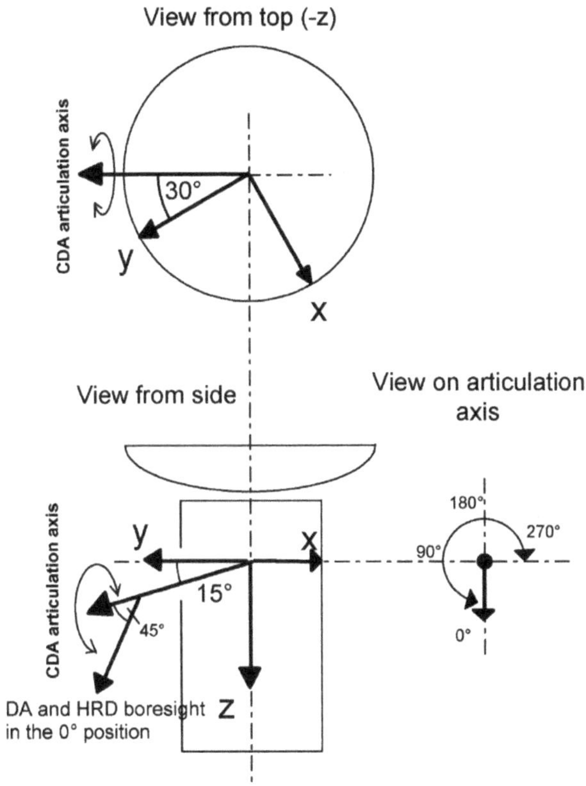

Figure 3. Mounting geometry of the CDA onboard Cassini. The CDA articulation axis is mounted 30° away from the $+y$-axis and points 15° downwards towards $+z$. CDA (DA and HRD) is mounted by 45° with respect to the articulation axis. The CDA boresight is shown for its 0° position (lower left). CDA can articulate by 270° (lower right).

by 270°. The cable wrap drum inside the turntable does not allow a full revolution. The "lower right" quarter of the full circle cannot be reached by the instrument. The spacecraft coordinate system is such that the x–y plane is perpendicular to the spacecraft spin axis z. The $+z$ direction points to the main engine, whereas the Huygens probe points towards $-x$. Besides the high gain antenna (which points towards $-z$), Cassini-Huygens has two low gain antennas (LGA). LGA 1 points towards the $-z$ direction whereas LGA 2 points towards the Huygens probe ($-x$-axis). During the inner cruise, the three-axis stabilized spacecraft has an orientation such that the high gain antenna points towards the Sun and the selected low gain antenna points towards the earth as precisely as possible. Figure 4 shows the field-of-view of the CAT for a variety of articulation angles in the spacecraft coordinate system.

A computer simulation program was used to calculate the geometric detection probability of the sensors. Under the condition of an isotropic flux of particles with an incidence angle θ the sensitive area of the Impact Ionization Target and the Chemical Analyzer Target were calculated. The result is shown in Figure 5.

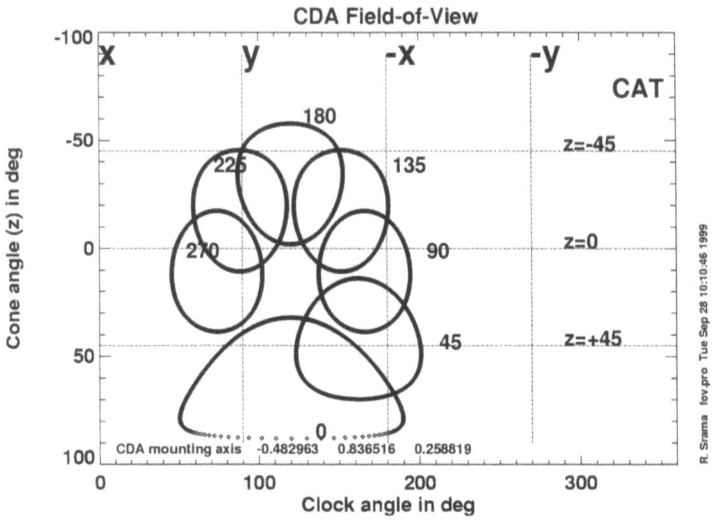

Figure 4. The field-of-view of the chemical analyzer target (CAT) for different articulation angles (0, 45, 90, 135, 180, 225 and 270) in the spacecraft coordinate system. The spacecraft axes are labled. The z-asymmetry is caused by the mounting of CDA 15° below the x–y plane. The field-of-view of the CAT is ±28°. The field-of-view of the IID is ±45° (not shown).

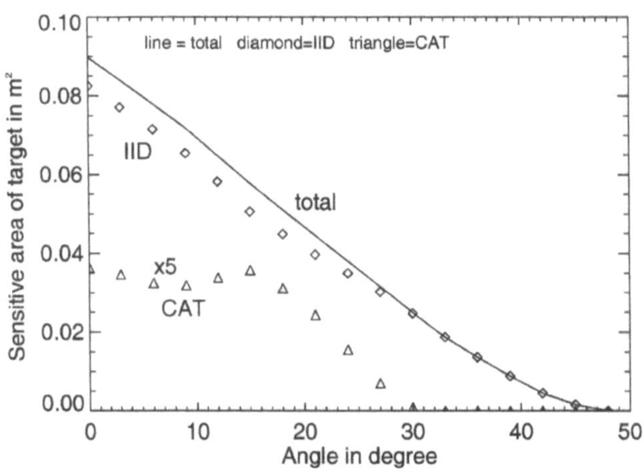

Figure 5. Sensitive area of the Impact Ionization Target (IID, diamonds) and the Chemical Analyzer Target (CAT, triangles). The total field-of-view is shown by the black line. The field-of-view of the CAT is much more constrained. The curve contains also the obscuration by the multiplier mounting structure. The transmission of the entrance grids of 95% for each grid is taken into account.

The calculations clearly show the decrease of sensitivity for increasing incidence angles. The chemical analyzer target does not detect any impacts with incidence angles larger than 28° and the impact ionization target has a limit of 45°. These limits are due to the shielding by the side walls of the detector cone. The calculations

TABLE VI

Overview of the apertures of the CDA subsystem in comparison with the Galileo dust detector subsystem.

Subsystem	Aperture in ±degree	Solid angle in sr
Impact Ionization Detector	45	0.59
Chemical Analyzer Target	28	0.47
High Rate Detector	88	ca. 3
Galileo dust detector subsystem	70	1.45

The aperture of the CDA cone is smaller than the aperture of the former Galileo detector because of a longer cylindrical housing.

have taken the obscuration of the multiplier into account. Therefore, the sensitivity of the Chemical Analyzer Target is not maximal for normal incidence ($\theta = 0\,^\circ$). The IID curve is based on a $\cos^4$ function, whereas the CAT sensitivity can be described by a $\cos^3$ function for angles larger than $15\,^\circ$. Caused by the axial symmetry, the solid angle interval is $d\Omega = 2\pi \sin\theta \; d\theta$. The relative sensitivity $I(\theta)$ is therefore given by

$$I(q)\,dW = 2p\sin q\;dq\,A(q)/A\,(q = 0)$$

An integration of this function leads to the effective solid angle interval covered by the detector which is 0.473 sr for the chemical analyzer and 0.591 sr for the impact ionization detector (Table VI).

3.3. The Dust Analyzer (DA)

3.3.1. *General Description*

Figure 2 shows a cut through the CDA. The instrument consists of the sensor housing with its entrance grids, impact targets, the high rate detector, the electronics box and the turntable. The interior of the sensor housing was purged with dry nitrogen until launch in order to avoid any contamination of the sensitive multiplier and the rhodium target of the Chemical Analyzer. The cover avoids contaminations of the sensor targets until 3 weeks after launch. A redundant pyro device moves a lever which unlatches the cover, and preforced springs jettison the cover to a normal direction. All major parts were made of milled aluminum while a honeycomb structure provided the required stiffness for the cover and the cylindrical sensor housing. The pre-amplifier box is located directly above the main electronics and occupies a separate housing to keep the input cables as short as possible and to minimize any interference with the main electronics.

The turntable of the instrument allows a rotation of $270\,^\circ$. The turn limit is given by the capability of the integrated cable wrap drum and the mechanical end stops. A design with two layers of plastic balls (PEEK) and a bearing diameter of

TABLE VII

Measured temperatures of CDA during day 311 in 2001 (HRD switched on).

Pyro subsystem	−34 °C
Dust analyzer multiplier	−34 °C
Electronics main box	−7.5 °C
High rate detector	−17 °C
Spacecraft interface	+7 °C
Chemical analyzer target	−52 °C
Impact ionization target	−46 °C

The operating temperatures are −15 to +40 °C (CDA main electronics) and −25 to +40 °C (HRD), respectively. The targets, the multiplier and the pyro can operate at lower temperatures.

240 mm was selected and qualified. The torque necessary for the turn is provided by a Phytron ZSS32 stepper motor and a gear with a total gear ratio of approximately 1000:1. Special electronics were developed by Phytron to achieve very low power consumption and a maximum torque. The motor has a compensating pole configuration and a Mu-metal shielding to keep the stray magnetic fields as low as possible. The motor can be operated by four different motor currents between 150 and 300 mA and consumes between 2 and 5 W. The turn speed of the platform can be set and is normally in the range of 10 °/min.

The instrument thermal design incorporates a number of techniques for controlling the temperatures. The acceptable operating temperature ranges of the subsystems are listed in Table VII. The instrument is thermally isolated from the S/C by its mount and by multilayer thermal blankets covering the turntable, electronics boxes, the HRD and the cylindrical housing. The normal operating power of the instrument produces an acceptable overall temperature without supplementary heaters. When the instrument is switched off, the temperature is maintained by the S/C-controlled replacement heater attached on the top of the main electronics box. The instrument monitors five temperature locations internally when it is turned on, and the S/C provides monitoring of seven locations at all times. A special instrument-controlled heater is provided for periodic decontamination of the Chemical Analyzer Target through heating to about 90 °C (for at least 10 h). Because of the high depth to diameter ratio, the cylindrical housing aperture is a very effective radiator, and the interior required a special prepared gold coating to ensure a stable and low emissivity for this long duration mission.

3.3.2. *Technical Description and Measurement Principle*

The grid system (EG) at the front end allows measurement of the dust charge and velocity (Auer *et al.*, 2001). This configuration is based on a method first described

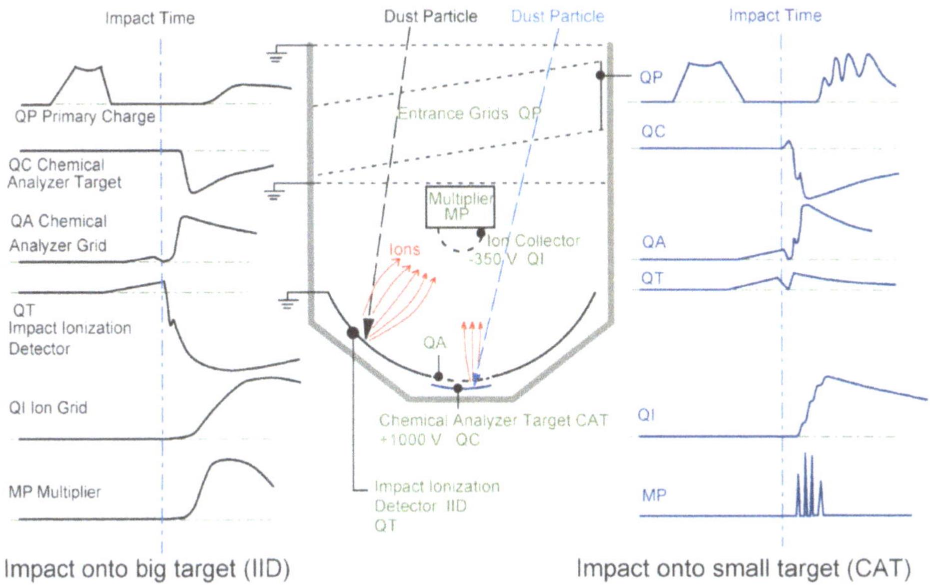

Figure 6. Hypervelocity impact signals of dust grains onto the big Impact Ionization Target (left) and onto the Chemical Analyzer Target (right). The chemical analyzer is a time-of-flight mass spectrometer and provides the elemental composition of the impacting dust grain.

in Auer (1975). The grids are made of stainless steel and each of them has a transmission of 95%. The innermost and outermost of the four grids are grounded, the other two grids are connected to a charge sensitive amplifier. A charged dust particle entering the sensor will induce a charge which corresponds directly to the charge of the particle. When the dust particle is far away from the sensor walls, all field lines are ending on the grids and the error in charge measurement is small. The output voltage of the amplifier will rise until the particle passes the second grid (Figures 6 and 7). As long as the particle is located between the second and third grid the output voltage remains more or less constant. As soon as the dust particle has passed the third grid, the voltage begins to fall until the fourth grid is passed. The distance of the fourth grid from the first grid, divided by the duration of the charge signal is equal to the particle's velocity component normal to these grids. Due to the inclination of 9° for the inner two grids, the path length between the grids depends on the angle of incidence, and allows a determination of the directionality of the incident particle in one plane. The choice of 9° is a compromise between angular resolution and tube length of the detector. The larger the angle the better the angular resolution, but the bigger and heavier the instrument. The detection of particle charges as low as 10^{-15} C has been achieved although the grid capacitance is high (approximately 200 pF) (Kempf 2004). The speed obtained by the entrance grid system (EG) is used to verify and calibrate the indirect determination of particle speed based on the rise times of the impact ionization signals.

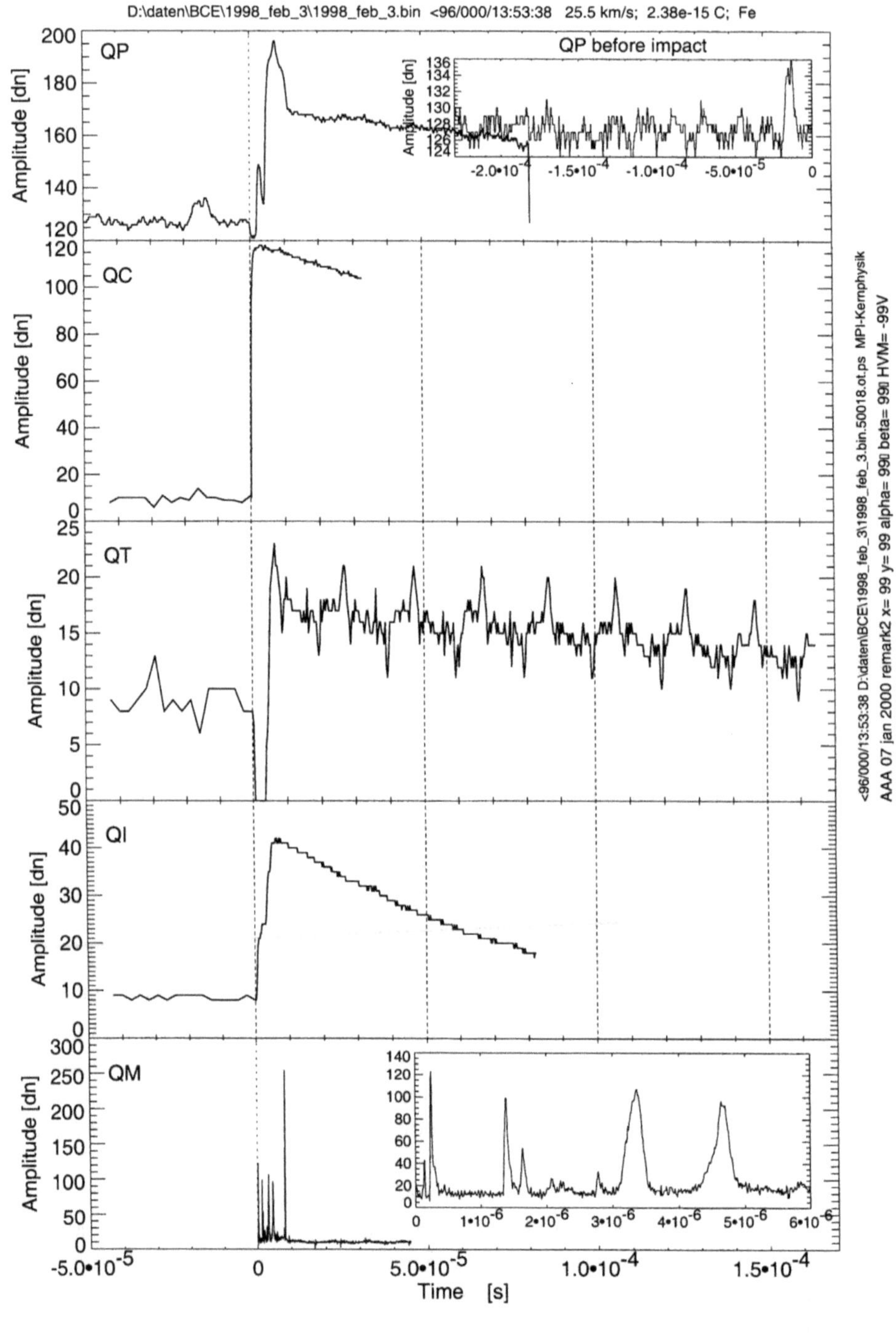

Figure 7. Unprocessed raw data of an 25 km/s particle impact on the Chemical Analyzer Target. The iron particle size was 150 nm. The peaks in the spectrum of the bottom diagram correspond to the ions of H, C, O, Fe and Rh. The data were recorded with the CDA flight spare unit at the Heidelberg dust accelerator facility.

A particle can impact either on the big gold plated Impact Ionization Target or the small rhodium Chemical Analyzer Target (CAT). In both cases the impact physics is the same. The impact produces particle and target fragments (ejecta), neutral atoms, ions and electrons (impact plasma). An electric field separates electrons (collected by the targets) and ions (collected by the ion grid). Charge sensitive amplifiers collect the charges at the various targets and grids. Amplifiers are connected at the Chemical Analyzer Target (QC), the Chemical Analyzer Grid (QA), the Impact Ionization Target (QT), the Ion Grid (QI), the Entrance Grids (QP), the multiplier anode (QMA) and the multiplier dynodes (DLA) (Figure 8). In order to increase the dynamic range, the amplifiers for QC, QT and QI are working with two measurement ranges.

The signals at the output of the electron multiplier must cover an exceptionally large dynamic range for two reasons. A wide dynamic range is required for measurement of a large range of ion abundances for any one impact, but, more importantly, a wide dynamic range is needed to make compositional measurements over the desired six orders of magnitude in range of particle masses impacting the system. Because of the random nature of the impact events and the short ion time-of-flight, it is clearly impossible to make real time gain changes for each event. Ordinary logarithmic amplifiers are not fast enough and do not have sufficient dynamic range for the time-of-flight measurements. An innovative solution to this problem has been created through the development of the dynode logarithmic amplifier. This system sums the linear signals from six different dynodes of the Johnston MM-1 multiplier in such a way that for large impacts the amplifiers for highest gain dynodes produce fixed (saturated) outputs that sum with an unsaturated low gain dynode signal. Thus it is a fast, low-noise, piece-wise linear approximation to true logarithmic performance. This special electronics was developed by the Rutherford Appleton Laboratory, U.K. (RAL). The main electronics was developed at MPIK whereas the mechanical design was done by G. Pahl (Munich). The mechanics was manufactured and the environmental tests for space qualification of CDA were carried out by DLR Berlin. The chemical analyzer is under the responsibility of Open University, U.K. (J.A.M. McDonnell).

All the outputs of the amplifiers are continuously compared with a channel-specific reference value (threshold), and if it is exceeded an event trigger is released. What happens now is that the sampling frequencies for the QC, QT, QI channels and the DLA are increased and the signals are digitized and stored in memory. The data processing by the 6 MHz MA31750 microprocessor system includes the calculation of signal rise times, amplitudes and integrals. A wavelet algorithm allows signal smoothing and a lossy compression. A lossless RICE compression algorithm can reduce the raw data by a factor of three. Approximately 1500 bytes are necessary for the lossless storage of one data frame. The data processing time limits the dead time of the instrument to 1 s. The calculated signal parameters are used for onboard data classification. Each event increases one of the 20 counter values. About half of the instrument memory is needed for the execution of the onboard software. The

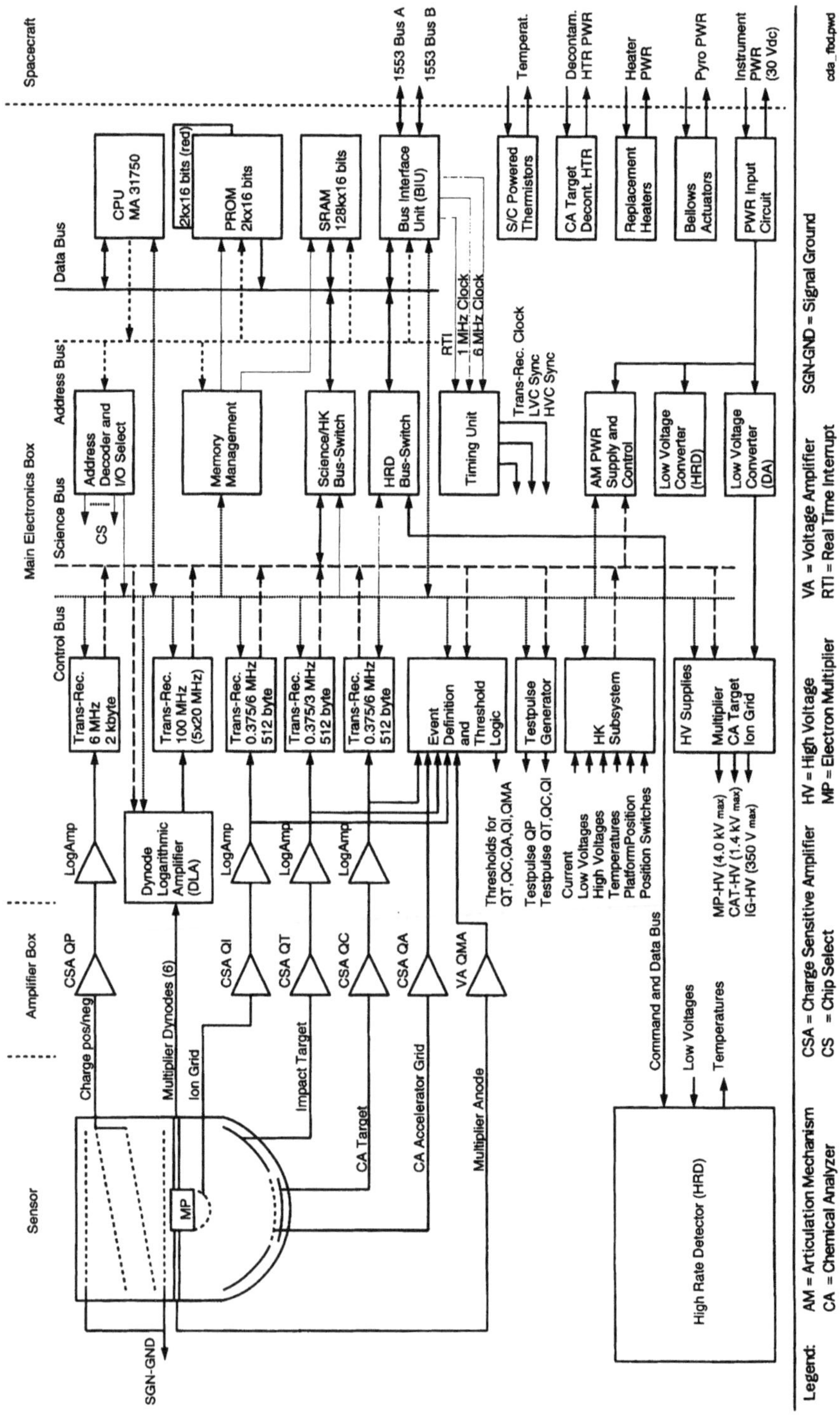

Figure 8. Functional block diagram of the cosmic dust analyzer.

remaining memory is used to store event data. The classification and priorization of detected events is a very complex procedure and is still under development. The onboard program was written in ADA using a TARTAN development system (KCS GmbH and University of Mannheim). Onboard data processing algorithms were developed by V. Tschernjawski (DLR Berlin, QP signal detection) and G. Schlotzhauer (DLR Berlin, wavelet compression).

3.3.3. *Instrument Data*

An electrically charged particle flying through the two inclined entrance grids at the front of the DA will induce charge signals on the grids (Figure 6). This induced charge is directly proportional to the charge of the particle and allows therefore a direct determination of its electric charge. The duration of the charge signal is equal to the particle's time-of-flight through EG and allows a determination of its speed. The inclined grid geometry leads to asymmetric signal shapes allowing the measurement of the particle direction in one plane. The particle can impact either on the outer big gold plated impact ionization detector IID (QT signal) or on the small inner chemical analyzer target, CAT (signal QC), which has a diameter of 16 cm. The gold target has a diameter of 41 cm.

The impact generates charged and uncharged fractures (ejecta), atoms, ions and electrons. The electrons of this plasma are collected by the target, and the ions are accelerated towards the inner grids (ion grids, signal QI) by an applied field of −350 V. Some of the ions fly through the grids producing a multiplier signal (MP).

The integrated chemical analyzer consists of the chemical analyzer target, the chemical analyzer grid (68% transmission) and the multiplier. The chemical analyzer grid is located 3 mm in front of the target and electrically grounded whereas the target is on a potential of +1000 V. The strong electric field between target and grid separates the impact charges very quickly and accelerates the ions towards the multiplier. The curved shape of the target and grid provides a better focusing of the ions onto the multiplier. This time-of-flight mass spectrometer has a flight path length of 230 mm and gives information about the elemental composition of the micrometeoroids (Ratcliff *et al.*, 1996). The chemical analyzer is a development of the University of Kent, Canterbury, U.K.

A complete overview of the signals measured and their measurement ranges are given in Table 8. The signals of the channels QP, QC, QT, QI and MP are digitized with 8-bit resolution and are stored in a raw data frame. Typical impact signals are shown in Figure 7. The sampling frequency is different for the various channels and the sampling frequency alters even within one channel. The channels QC, QT and QI are digitized continuously with 0.375 MHz until an event is detected by exceeding the specified thresholds. After this trigger the sampling frequencies are increased to 6 MHz (QC and QI) and 3 MHz (QT), respectively (Figure 8). The last 16 samples preceding the trigger are stored together with the following sampling points. This preserves the signal shape before the impact is detected by the electronics and allows a later reconstruction of the entire slope. Signals with slow rise times benefit

TABLE VIII

Signals of the Dust Analyzer monitored by the electronics.

Measurement channel	Measured quantity	Data aquisition frequency (MHz)	Signal amplitude resolution (bit)	Measurement range of charge and rise time	Particle parameter
QP Entrance grids (EG)	Induced charge Positive Negative	6	8	10^{-15} to 6×10^{-13} C 10^{-15} to 6×10^{-13} C Rise time = 0.25–80 μs	Speed: 2–40 km/s Trajectory in one plane Electric charge: 2×10^{-15} to 5×10^{-13} C
QT (IID) Impact ionization detector	Electrons (negative charge) generated upon impact	3	8	2×10^{-14} to 10^{-8} C Rise time = 0.5–150 μs	Mass: 1×10^{-15} to 1×10^{-9} g Speed: 2–40 km/s
QC (CAT) Chemical analyzer target	↑	6	8	1×10^{-14} to 10^{-8} C Rise time = 0.2–10 μs	↑
QA Chemical analyzer grid	Ions (positive charge) generated upon impact	—	—		(Just trigger/coincidence) —Not digitized
QI Ion grid	Ions (positive charge) generated upon impact	6	8	1×10^{-14} to 10^{-8} C Rise time = 0.2–90 μs	Mass Speed
MP Multiplier dynode signal	Ions (positive charge) generated upon impact	100	8	4×10^{-14} to 4×10^{-8} C (output of multiplier) Rise time = 10 ns	Chemical composition $m/\Delta m = 20$–50
QMA Multiplier anode signal	Ions (positive charge) generated upon impact	—	—	—	(Just as trigger) —Not digitized

The measurement ranges of particle properties are indicated. The channels QP, QT, QC, QI and MP are digitized for data analysis.

TABLE IX
Summary of digitized channels and their corresponding sampling depth.

Channel	Before impact: Sampling frequency (MHz)	Before impact: Number of samples	Before impact: Total record time before impact (μs)	After impact: Sampling frequency (MHz)	After impact: Number of samples	After impact: Total record time after impact (μs)	Total number of samples
QP	6	1362	227.0	6	510	85.0	1872
QC	0.375	16	42.66	6	196	32.66	212
QT	0.375	16	42.66	3	494	164.66	510
QI	0.375	16	42.66	6	494	82.33	510
MP (DLA)	—	0	0	100	640	6.4	1024
	—	—	—	10	384	38.4	

especially from this system since otherwise the first part of the signal would be missing and the determination of the rise time would be impossible. The sampling frequency of 375 kHz is a compromise between acceptable time resolution and low power consumption. An overview of the record parameters is given in Table IX.

An event causes an instrument dead time of 1 s to allow for complex data processing. The major processing steps are listed below.

(a) Integer wavelet transformation of raw data signals.
(b) Set small coefficients to zero (leads to a smooth curve after reverse transformation).
(c) Reverse transformation with most significant coefficients.
(d) Determination of peaks, rise times, amplitudes and times at 50% of the full amplitude.
(e) Classification of the event and increment the appropriate counter by 1.
(f) Further compression of the already wavelet transformed data by a RICE compression algorithm.
(g) Enable the event trigger and awaiting the next impact.

The informations from step (e) will be used to identify the impact location (big or small target) and to classify the event as good, poor or noise in order to increase the appropriate counter. The chosen integer reversible wavelet transformation provides an almost lossless tool to smooth and reduce the data without loosing the original signal shape and height.

Twenty counters provide a characterization in size, speed and impact location. Each counter has 16 bits and the counters are merged in four priority groups. The priorities will be used to give memory readout guidelines by command to the instrument. All these efforts have only the goal to compress the data. The highest compression level is given by the contents of the 20 counters. The next compression level are the extracted data of rise times and amplitudes. The lowest compression level are the wavelet treated sampling curves of an event. In order to adjust the memory readout data rate to the spacecraft storage capabilities and telemetry modes, a special data frame was developed which allows a readout of selected information. Therefore it will be possible to readout, e.g. only counter-data together with the multiplier raw data. It is clear that ancillary data such as impact time and spacecraft boresight information belong to each data frame. Further capabilities are to put science data in housekeeping data frames and vice versa. This will give the CDA instrument further flexibilities for spacecraft downlink capabilities of 40 bps and lower as they occured between launch and summer 2000.

3.3.4. *Calibration of the Dust Analyzer*

There are two components to the science calibration. The primary component is the measurement of particle impact characteristics on the flight and flight spare model instruments in ground based laboratory accelerators. The second part is the measurement of impact characteristics in the well-understood environment of space at 1 AU. The dust environment at 1 AU is very well characterized in flux and size distribution by orbiting instruments such as LDEF. Measurements in this environment allow the possibility of measuring many impacts at a relatively high rate that will provide a strong cross correlation to all earlier and future measurements.

The calibration principle of the impact signals is similar to the former Galileo and Ulysses experiments (Grün *et al.*, 1992). The yield of charge of the impact plasma is a function of the dust mass for a constant dust speed and the signal rise times are dependent on the impact speed. A good overview of the calibration of the Galileo dust detector is given in (Göller and Grün, 1989). Although this work provides the basis for the DA calibration much further work has to be done in order to understand the interference of the newly implemented small chemical analyzer target and the big impact ionization target. Furthermore the time-of-flight mass spectrometer is a completely new subsystem which has to be understood.

The particle parameters affecting the signals are the mass, speed, density, impact location on the target and the impact angle. Most of these parameters such as speed, mass and impact location can be varied easily at the dust accelerator facility in Heidelberg. However, based on the accelerator properties, there are some limitations with respect to the mass and density (material) of the selected dust particles (Stübig *et al.*, 2001). In order to extend the calibration range some tests have been performed at the University of Canterbury, the University of Munich, and at Caltech. The University of Canterbury (recently this group moved to Open University, U.K.) focuses on the investigations of mass spectra and their calibration (Goldsworthy *et al.*, 2002), whereas the University of Munich and Caltech are using different accelerator types enabling them to use bigger particles and to select almost any material as a dust source. In addition, a short pulse UV laser was used at CALTECH in order to achieve mass spectra of a variety of materials (Jyoti *et al.*, 1999). An overview of the particles used so far for calibration work at MPI-K is given in

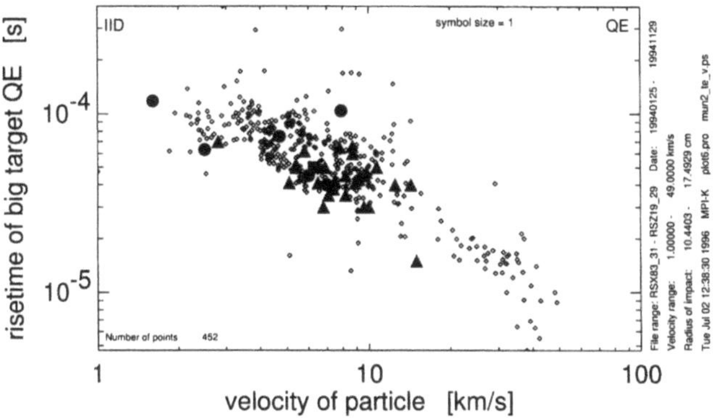

Figure 9. Mass and speed ranges of particles used for calibration and functional testing. The filled symbols represent measurements at the dust accelerator at the University of Munich.

Figure 9. For each single impact, signals of all channels (QP, QT, QC, QI and MP) of DA were recorded and analyzed in terms of their rise times, amplitudes and their relative timing. The experimental determined detection threshold T for an impacting particle (mass threshold) of DA can be described by T (kg) $= 3.04 \times 10^{-13} \times v^{-3.75}$ with the particle speed v in km/s. All particles with a speed of 10 km/s and masses above 5.9×10^{-17} kg should be detected by DA.

Before applying a function to calculate speed and mass of a dust particle, the impact location (CA or IID) have to determined. This will be mainly achieved by evaluating the ratio between the amplitude of the target signals QT and QC. Then, the particle speed is calculated by the rise time of the target and ion grid signals. Finally, the amplitude of these signals determine the mass of the dust particle. Note, however, that whenever a dust charge is detected, the particle speed can be derived directly from the duration of the charge signal. This feature allows one to calibrate the speed measurement that is based on signal rise times and to improve the accuracy of the particle mass.

In order to eliminate noise events, it is not sufficient to evaluate the signal amplitudes. The time differences between channels is relevant, too. Impacts on the IID show a time difference of 70 μs between the target and ion grid signals in the particle speed range between 5 and 11 km/s (the QI signal is later). This kind of information helps to separate real impacts from noise events. The detailed calibration results are reported elsewhere.

The time-of-flight mass spectrometer can be characterized by its mass resolution which is defined as mass of a peak (amu) divided by its mass equivalent of the peak half width (amu). The analysis of mass spectra obtained at different velocities and using a variety of particle materials has shown resolutions between 10 and 50. Furthermore, the peaks show non-symmetric shapes caused by the simple design of this spectrometer. The ions of the impact plasma have a broad direction and energy distribution which leads to different flight times from the target to the multiplier. The consequences are broad and asymmetric peaks, but an ion energy focusing design as for the dust instrument on Giotto or Stardust was not possible due to the geometric constraints inside the sensor.

An engineering calibration of the amplifier chain is performed by commanding a selected pulse size and shape to be applied to each of the inputs of the analog signal chains. The pulse output then triggers an event process and the output is recorded in exactly the same manner as a dust impact. By comparing the test pulse results over the duration of the mission, any changes in amplifier gain, noise or speed can be detected and corrected on the ground.

The Galileo instrument determines the impact speed with an accuracy within a factor of 1.6. Since the calculation of the particle mass is dependent on the speed, the mass determination was not better than within a factor of six. These numbers seem quite high but they are caused by the complex physics behind the impact process. Local impact pressures and plasma temperatures are dependent on the target and dust material, density, surface geometry, impact angle, particle speed and so on, but they determine the number of electrons and ions produced which will be detected by the amplifiers. Therefore it is expected that the CDA instrument cannot achieve much better accuracies than the Galileo and Ulysses dust experiments and the onboard improvements in electronics and software will not help too much in this aspect. But they will be useful in terms of evaluating the quality and probability of an impact, which means that noise events are eliminated more easily and the evaluated dust fluxes have very low uncertainties.

3.4. The High Rate Detector (HRD)

The overall objective of the HRD is to carry out quantitative measurements of particle flux and mass distribution throughout the Saturn ring system. The particle impact rate and particle mass distribution will be determined with respect to Saturnian distances, distance from the rings, and to magnetospheric coordinates. The particle mass range covered by the HRD (assuming a particle impact velocity of 15 km/s) ranges from 8×10^{-13} to 8×10^{-8} g for differential and cumulative flux measurements, and $>8 \times 10^{-8}$ g for cumulative flux measurements.

3.4.1. *General Description*

The HRD was designed, built and tested at the University of Chicago and measures differential and cumulative particle fluxes. The HRD has a high counting rate capability (up to 10^4 random impacts s^{-1} with <5% corrections) which will be particularly important during Saturn ring plane crossings, where fluxes are large enough to saturate the counting rate of DA (1 s^{-1}).

The HRD has significant inheritance from the University of Chicago Dust Counter and Mass Analyzer instrument (DUCMA) flown earlier on the Vega-1 and Vega-2 spacecraft to Comet Halley (Perkins *et al.*, 1985). The instrument employs

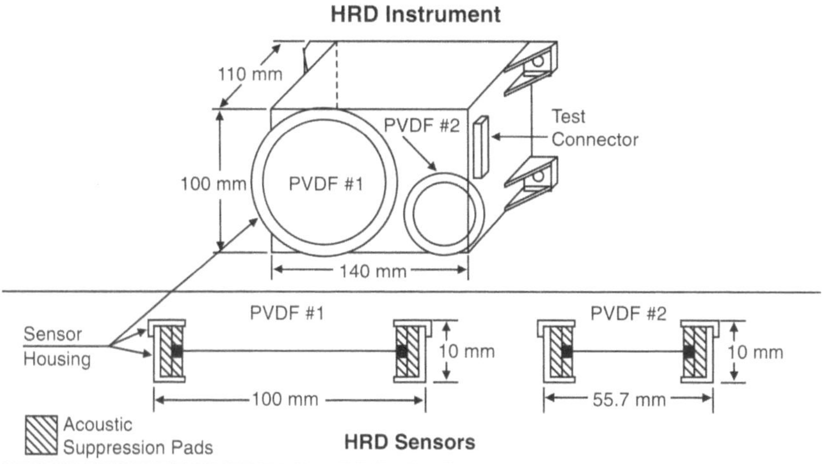

HRD Characteristics

Sensors: #1-50 cm^2, 28μm thick PVDF detector; #2-10 cm^2, 6 μm thick PVDF detector;
Particle velocity: >1 km/s; **Thresholds:** each sensor has four mass thresholds.
Particle mass (at 15 Km/s): differential and integral flux≈8 x 10^{-13}g to integral flux > 8x10^{-8}g; integral flux > 8x 10^{-8}g.
Discrete events: recording of impact time (1 second accuracy) and threshold firings for each impact.
Counting rates: up to 10^4 s^{-1} no corrections; 10^4 to 10^5 s^{-1}, known corrections.
Data Storage: microprocessor system with 60 k bytes data storage memory.
Flight modes:
a) **Normal (Cruise):** continuous recording of impact time, threshold firings and integral counts.
b) **Fast:** integral counts recorded and stored each Δt seconds during ring-plane crossings. Δt selectable at 0.1, 0.2,...,0.9, 1.0 seconds.
c) **Calibrate:** periodic calibration with in flight-calibrator.

Figure 10. Schematic of the HRD instrument and PVDF dust sensors.

the dust particle detection technique described by Simpson and Tuzzolino (1985) and consists of two polyvinylidene fluoride sensors with associated electronics (Figure 10). The sensors are mounted on the front of the HRD electronics box and the HRD detects individual particles impacting the PVDF sensors and provides continuous measurements of cumulative particle fluxes for particle masses greater than four mass thresholds for each of the two sensors (see Table 2).

The HRD is an independent instrument containing its own memory and processor. The only interface to the DA is via the power and data cables shown in Figure 11. HRD power is supplied by the DA main electronics and data transfer responds by latching the appropriate data into the HRD data output register. The latching of the data generates an interrupt to DA indicating that the data is ready to be read by DA and stored into DA memory. The HRD is rigidly mounted to the DA so that as the CDA turntable is rotated, the HRD scans different particle arrival directions. HRD properties can be described as follows.

Sensors: #1—50 cm^2, 28 μm thick PVDF detector; #2—10 cm^2, 6 μm thick PVDF detector

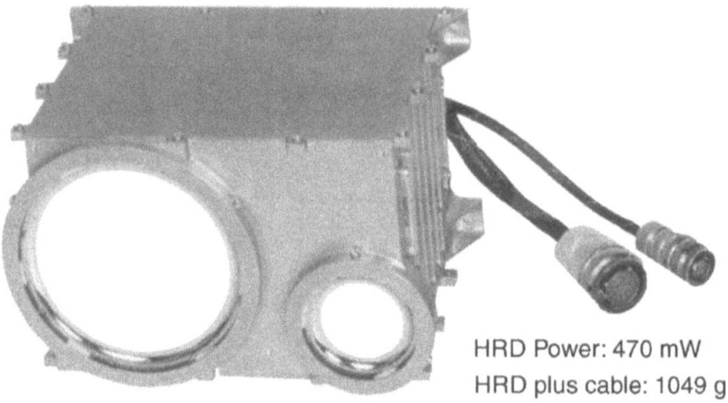

Figure 11. Photograph of the HRD flight instrument.

Particle velocity: >1 km/s, each sensor has four mass thresholds
Particle mass @ 15 km/s: differential and integral flux $\sim 8 \times 10^{-13}$ to 8×10^{-8} g, integral flux $>8 \times 10^{-8}$ g
Discrete events: recording of impact time (1 s accuracy) and threshold firings for each impact
Counting rates: up to 10^4 s^{-1} no corrections, 10^4 to 10^5 s^{-1} known corrections
Data storage: microprocessor system with 60 kbytes data storage memory *Operating modes:*

- NORMAL MODE (CRUISE): Continuous recording by the HRD of individual particle impact time, threshold firings for each impact, and integral counts. This operating mode will be sued for all interplanetary data collection. It may also be used during ring plane crossings up to particle impact rates of approximately few hundred impacts s^{-1}.
- FAST MODE (ENCOUNTER, RING PLANE): Integral counts are recorded and stored by the HRD each Δt seconds during ring plane crossings. The time interval Δt is selectable at 0.1, 0.2, ..., 0.9, 1.0 s. At the highest time resolution ($\Delta t = 0.1$ s) the spatial resolution for the counting rates will be $\sim$1 km.
- CALIBRATE MODE: Period electronic calibration of the HRD with the HRD inflight calibrator. This mode allows assessment of the electronic stability of the HRD throughout the mission.

3.4.1.1. PVDF Dust Sensors. The theory, fabrication and details of PVDF dust detector operation have been described in earlier reports (Simpson and Tuzzolino, 1985; Perkins *et al.*, 1985; Simpson *et al.*, 1989). A PVDF sensor (Figure 12) consists of a thin film of permanently polarized material. A hypervelocity dust particle impacting the sensor produces rapid local destruction of dipoles (crater or

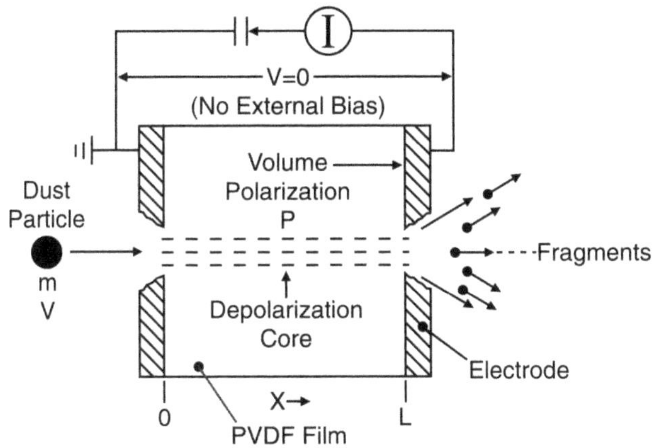

Figure 12. Schematic drawing of a polarized PVDF sample with conducting contact electrodes. An incident dust particle penetrates the sample resulting in complete depolarization along its track. The impact generates a fast current pulse I. The total charge carried by the current pulse is a function of particle mass and velocity.

penetration hole) which results in a large and fast current pulse at the input to the electronics (ns range). The output pulse (Figures 13–15) is sharp in time, with a maximum amplitude depending on impacting particle mass and velocity.

Since the depolarization induced current pulse is fast, the output pulse shape is determined by the choice of electronic time constants for the pre-amplifiers and shapers. Electronic time constants (amplifier shaping time constants, discriminator width) in the few microsecond range permit a high counting rate capability for the HRD sensor-electronics combination (10^4 random impacts s^{-1} with $<5\%$ corrections), as illustrated in Figure 16. The high counting rate capability of the HRD is of particular importance for Saturn ring plane crossings, where high dust particle fluxes are encountered.

3.4.1.2. Acoustic Signal Suppression. PVDF detectors have a secondary but minor mode of response due to their piezoelectric properties. Therefore, background acoustic disturbances of sufficient intensity could trigger the HRD PVDF detector thresholds, and these threshold firings could be mistakenly recorded as dust particle impacts. During the Cassini-Huygens mission, possible sources of acoustic disturbances include spacecraft gas jets, moving platforms, and large particle impacts on structures near the mounting positions of the HRD PVDF sensors.

To minimize the effects of the possible acoustic backgrounds, the HRD PVDF sensors are mounted in sound absorbing pads (Figures 10 and 17). This design was highly effective in suppressing the acoustic response from mechanical shocks (Perkins *et al.*, 1985).

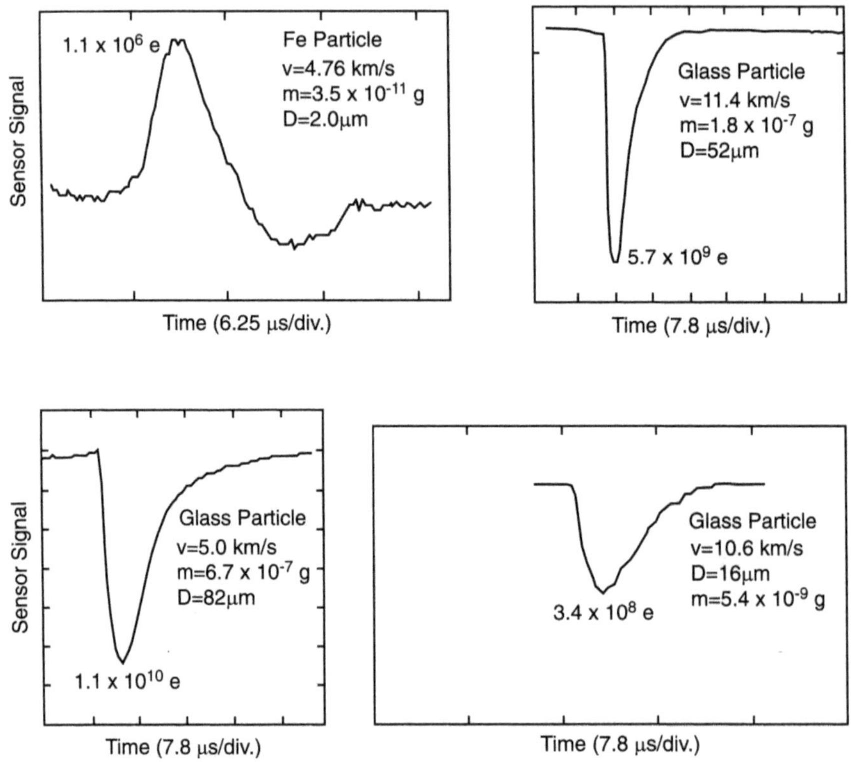

Figure 13. Examples of output signals from 28 μm thick PVDF dust sensors. Listed are particle velocity, mass and diameter. Output signal amplitudes are expressed in units of number of electron charges (e).

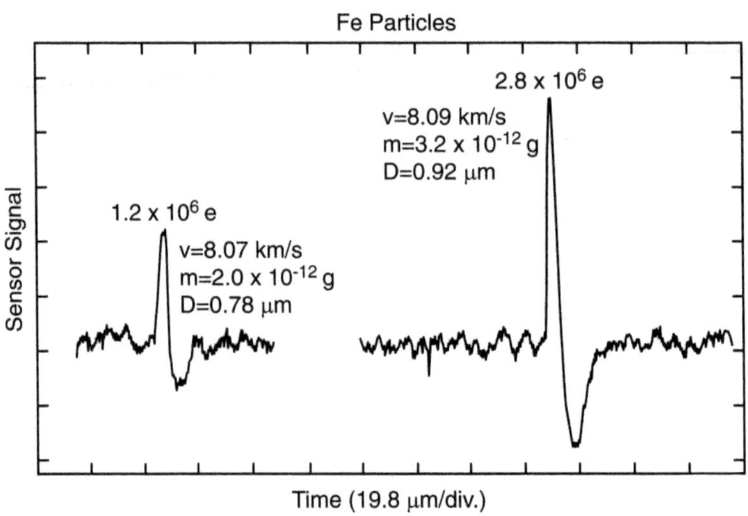

Figure 14. Output signals from 6 μm thick PVDF dust sensors. Listed are iron particle velocity, mass and diameter. Output signal amplitudes are expressed in units of number of electron charges.

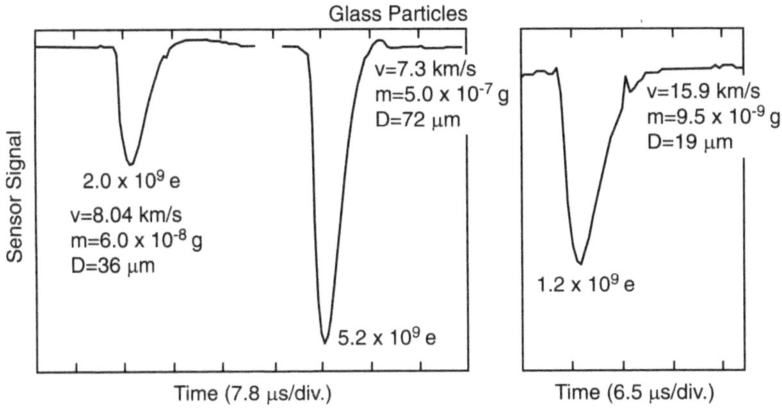

Figure 15. Output signals from 6 μm thick PVDF dust sensors. Listed are glass particle velocity, mass and diameter. Output signal amplitudes are expressed in units of number of electron charges.

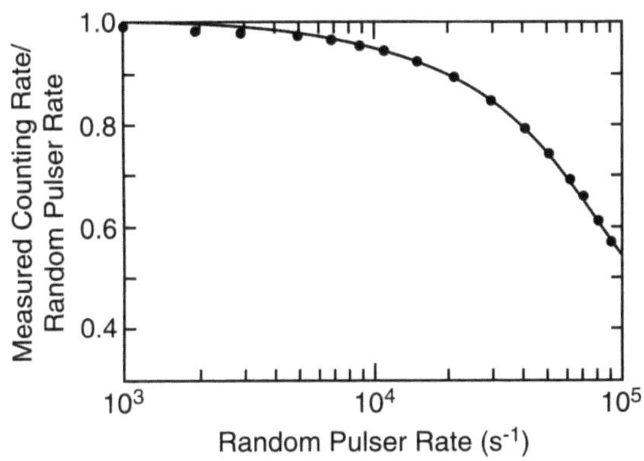

Figure 16. Illustration of counting rate capability of PVDF sensor/electronics system. Data taken random electronic pulser to simulate a random particle impact rate. Up to impact rates of 10^4 s^{-1}, counting losses are <5%.

3.4.2. *Thermal Aspects*

3.4.2.1. Thermal Control. Although our studies have shown that PVDF sensors may be operated at temperatures up to ~80 °C for long periods of time (weeks) with small (<5%) degradation in dust particle response, our HRD sensor mounting technique, discussed above, acoustically and thermally insulates the sensor from its surroundings. Under these conditions, the effective sensor emissivity ε and absorbtivity α are such that exposure of a PVDF sensor in space to direct solar illumination for a short period of time (~1 min) would result in sensor temperatures high enough to destroy the sensor.

Since there have been the possibility of direct exposure of the HRD sensors to solar illumination at a radial distance of 0.68 AU from the Sun during the Cassini-Huygens mission (Jaffe and Herrell, 1997), we studied different sensor coating

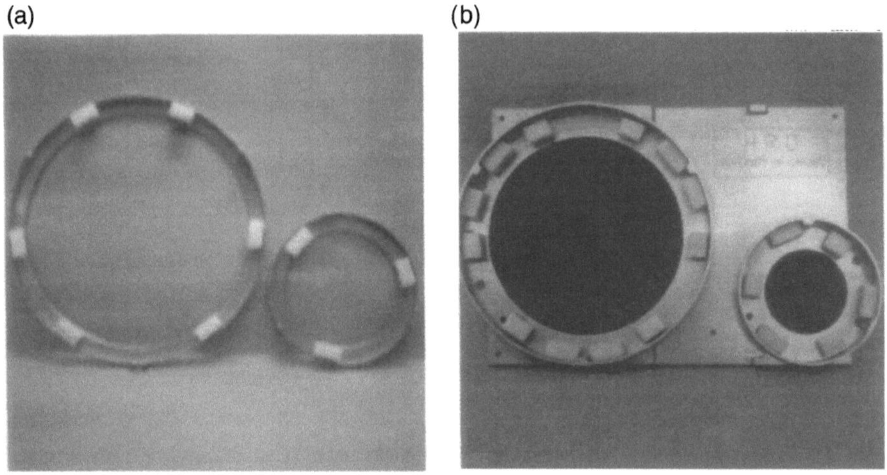

Figure 17. (a) Photograph of the HRD PVDF sensor covers containing sound absorbing pads attached to the inner portions of the covers. (b) Photograph of the front face of the HRD flight instrument with the flight sensors removed. Sound absorbing pads are attached to the inner portions of the sensor mounts. For each sensor, the central portion of the sensor mount is coated with a thick coating ($\sim$45 μm) of Chemglaze Z-306.

techniques which would restrict the sensor temperature to $\leq$80°C during solar exposure at 0.68 AU.

The basis for our studies is illustrated in Figure 18. Assuming the deposition of a coating material having emissivity ε_2 on both the back surface of the sensor and on the front of the HRD electronics box (Figure 17), sensor temperatures T, as a function of R, and HRD electronics box temperature may be calculated. Figure 18 shows calculated results for $\varepsilon_2 = 0.84$, where it is seen that the sensor temperature remains <80 °C at all radial distances $\geq$0.61 AU for all HRD electronics box temperatures $\leq$20 °C. The normal HRD operating temperature at distances >4 AU is between -20 and -10 °C.

One technique which leads to a value $\varepsilon_2 = 0.84$ consists of applying a (spray brush) Chemglaze Z-306 coating to one of the sensor surfaces and measured values of ε_2 versus Z-306 thickness are plotted in Figure 19. For HRD sensors, a Z-306 thickness of $\sim$40 μm has been used, with a value $\varepsilon_2 = 0.84$. Figure 20 shows the front and back surfaces of the HRD flight sensors.

PVDF sensors coated with Z-306 have been thermal cycled over the temperature range -200 to 115 °C and the coatings have shown excellent adherence to the PVDF surface and complete mechanical stability. Coated sensors have also successfully passed launch environments (random vibration up to 55 grms, shock acceleration with peak acceleration of 1750 g).

3.4.2.2. Temperature dependence of sensor dust particle output signal. During the Cassini-Huygens mission, the extreme temperature limits for the HRD sensors

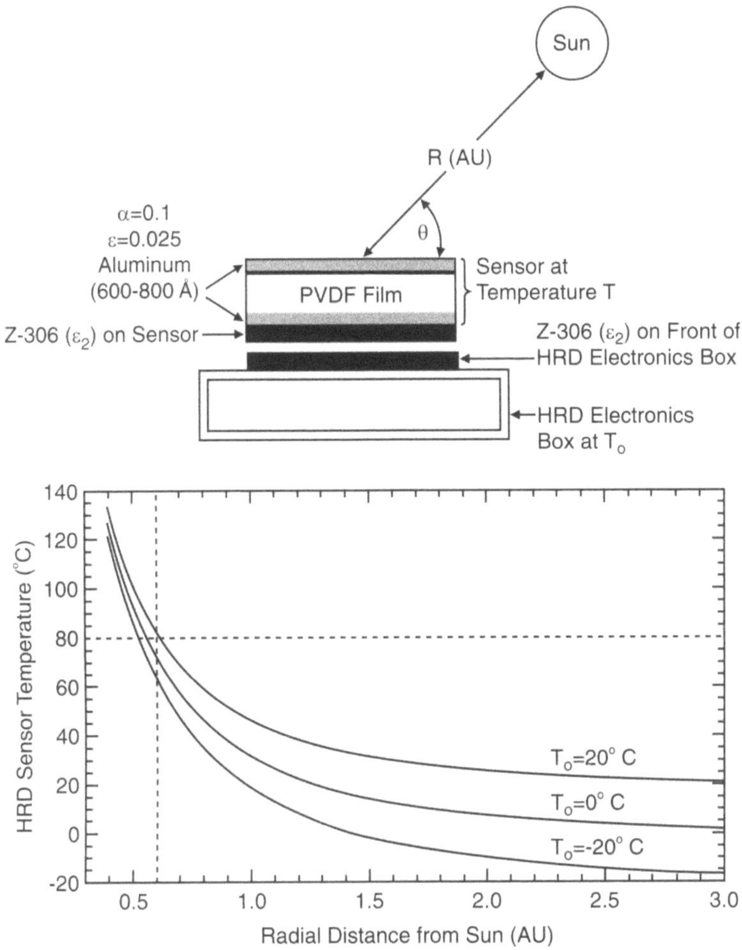

Figure 18. (a) Model assumed for HRD solar exposure at a radial distance R (AU) from the Sun. The sensor surface facing the Sun (front surface) has absorptivity $\alpha = 0.1$, and emissivity $\varepsilon = 0.025$. The back surface of the sensor is coated with Z-306, having emissivity ε_2 and is close (few mm) to a Z-306 coating having ε_2 which has been applied to the front of the HRD electronics box. The sensor is at temperature T and the HRD box is maintained at temperature T_0.

are predicted to range from -50 to $+80\,°\mathrm{C}$. For dust particles of fixed mass and velocity impacting a HRD sensor, two temperature dependent parameters determine the variation of output signals from the sensor-linear electronics combination as the temperature of the HRD changes:

- the variation in sensor capacitance with temperature, and;
- the variation of the dust particle depolarization induced charge signal with temperature.

These two contributions are discussed later.

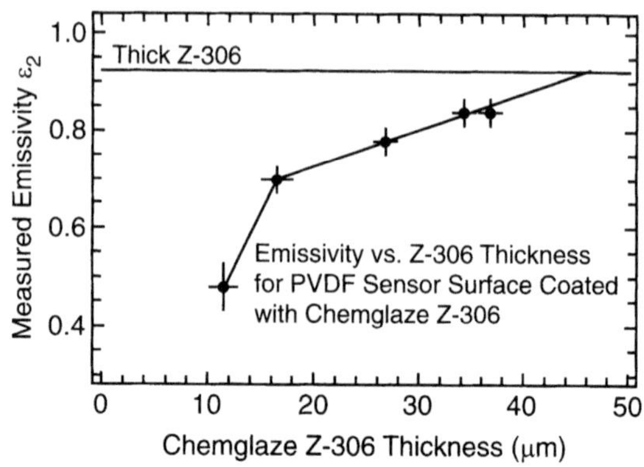

Figure 19. Measured emissivity ε_2 vs. Z-306 thickness for a PVDF sensor surface coated with Chemglaze Z-306.

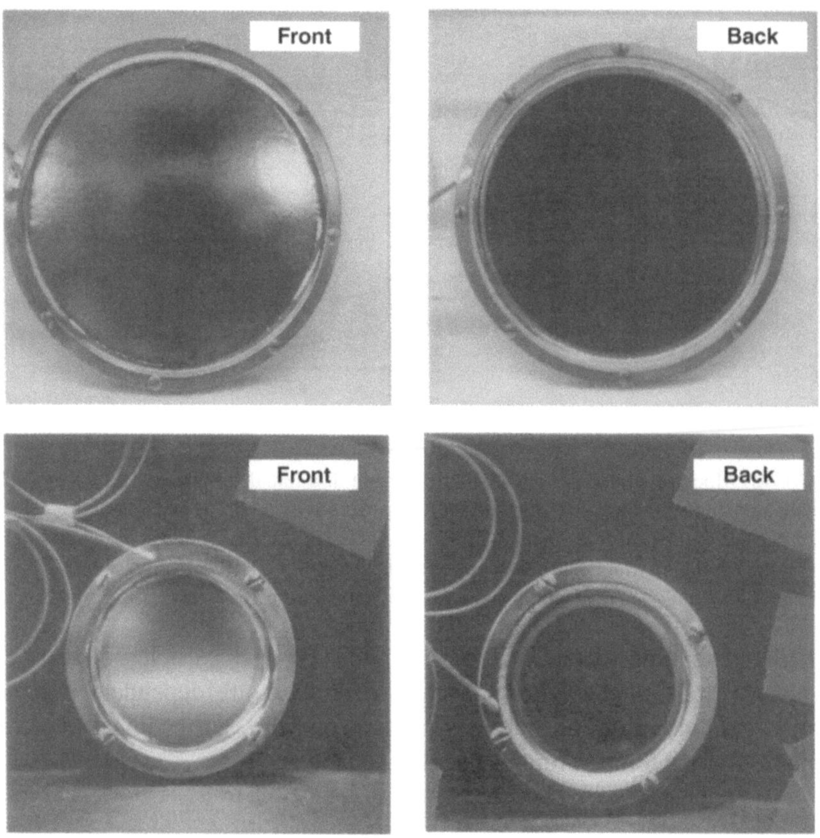

Figure 20. Upper panels: photograph of the front and back surfaces of HRD flight sensor #1 (50 cm^2, 28 μm thick). The back surface is coated with Z-306 ($\sim$40 μm thick). Lower panels: Photograph of front and back surfaces of HRD flight sensor #2 (100 cm^2, 6 μm thick). The back surface is coated with Z-306 ($\sim$40 μm thick).

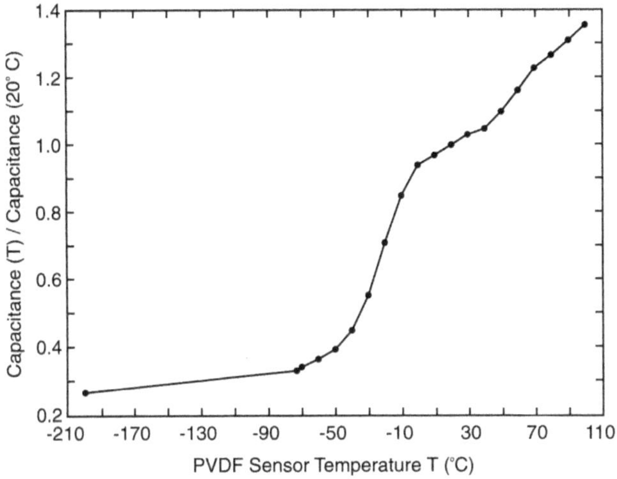

Figure 21. Measured PVDF sensor capacitance vs. sensor temperature. Capacitance values are normalized to the value at room temperature.

3.4.3. *Detector Capacitance and Amplifier Design*

The two HRD sensors have capacitances of 18 nF (50 cm^2, 28 μm thick) and 12 nF (10 cm^2, 6 μm thick) at room temperature. Figure 21 shows the measured dependence of PVDF sensor capacitance on temperature. Over the expected temperature range for the HRD sensors during the Cassini-Huygens mission (-50 to $+80$ °C), Figure 21 establishes the range of sensor capacitance which the pre-amplifier must accommodate. The final design of the HRD sensor pre-amplifier was such that over the temperature range -50 to $+80$ °C, variations in pre-amplifier pulse height (generated by an electronic pulser) were less than 2% resulting from variations in sensor capacitance.

The large HRD sensor capacitance also had to be considered in the design of the linear electronics with respect to the counting rate requirements for the Cassini-Huygens mission. The final design of the sensor linear electronics is using 2 μs time constants for the post amplifier pulse shaping and a discriminator width of 4 μs.

To determine an optimum choice for the HRD sensor's lowest mass threshold, the electronic noise (FWHM) for the combination HRD sensor, pre-amplifier (CSA) and shaping amplifier must be considered. If the lowest mass threshold is set too low, electronic noise counts will contribute to the total measured counts. If it is set too high, low-mass particles are lost.

The measured electronic noise (FWHM) for the HRD linear electronics versus capacitance at the CSA input is shown in Figure 22. For each HRD sensor (18 and 12 nF), the electronic noise is $\sim 3.5 \times 10^5$ e (FWHM) or $\sim 1.5 \times 10^5$ e (rms). By setting the lowest electronic threshold for each HRD sensor at a value $\sim 2.0 \times 10^6$ e (e.g., a factor 13.3 above the rms noise), electronic noise counts are completely excluded.

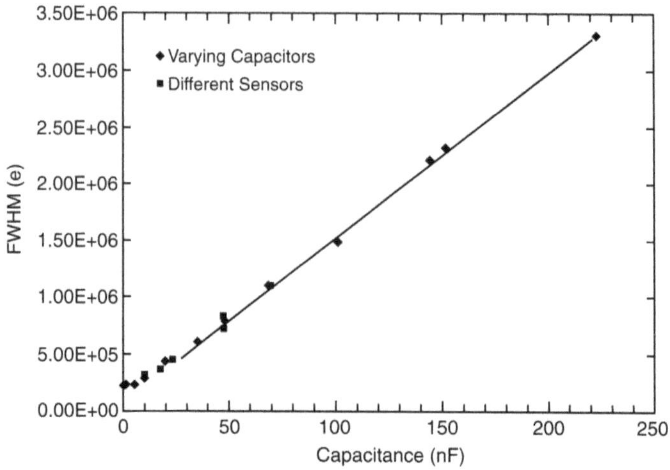

Figure 22. Measured electronic noise (FWHM) expressed in units of number of electrons (e) at the input to the HRD sensor charge sensitive pre-amplifier (CSA) vs. sensor capacitance. For the HRD sensors, the measured electronic noise is $\sim 3.5 \times 10^5$ e (FWHM).

3.4.4. *Detector Charge Signal*

Of the several PVDF material parameters which determine the magnitude of the PVDF detector depolarization charge signal resulting from an impacting dust particle (Simpson and Tuzzolino, 1985), the volume polarization magnitude, P, and its temperature dependence are the most important for determining the temperature dependence of the signal amplitude. Our laboratory measurements have shown that over the temperature range -50 to $+80\,^{\circ}$C, the charge signal amplitude will vary by less than 6% from the values measured at room temperature. Thus, we expected an overall possible variation in output signal amplitude resulting from both detector capacitance and depolarization charge signal temperature effects of less than 10% over the expected temperature range for the detector. This maximum 10% effect will contribute a negligible uncertainty for the particle mass thresholds.

3.4.5. *HRD Electronics, Digital Data and Commands*

Figure 23 shows a block diagram of the principal electronic functions of the HRD instrument and its interface to the CDA main electronics, and Figure 24 shows a diagram of the data flow from HRD to the CDA main electronics.

3.4.5.1. Linear (Analog) Electronics. The HRD linear electronics consists of charge sensitive pre-amplifiers (CSA), shaping amplifiers (SHAPER) and threshold discriminators. Each SHAPER in Figure 23 provides single integration – single differentiation RC shaping with a shaping time constant of 2 μs.

A particle impact on either sensor #1 or sensor #2 will result in output signals from the shapers (Figures 13–15) which may trigger the M1, M2, M3, M4 thresholds

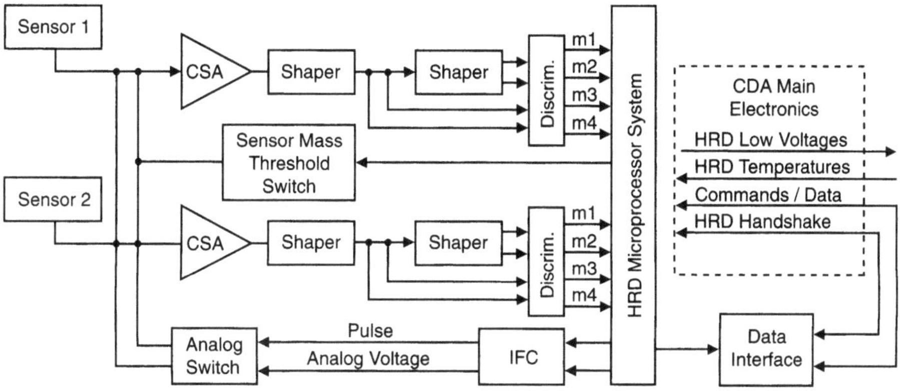

Figure 23. Block diagram of the principal electronic functions of the HRD and its interface to the CDA main electronics.

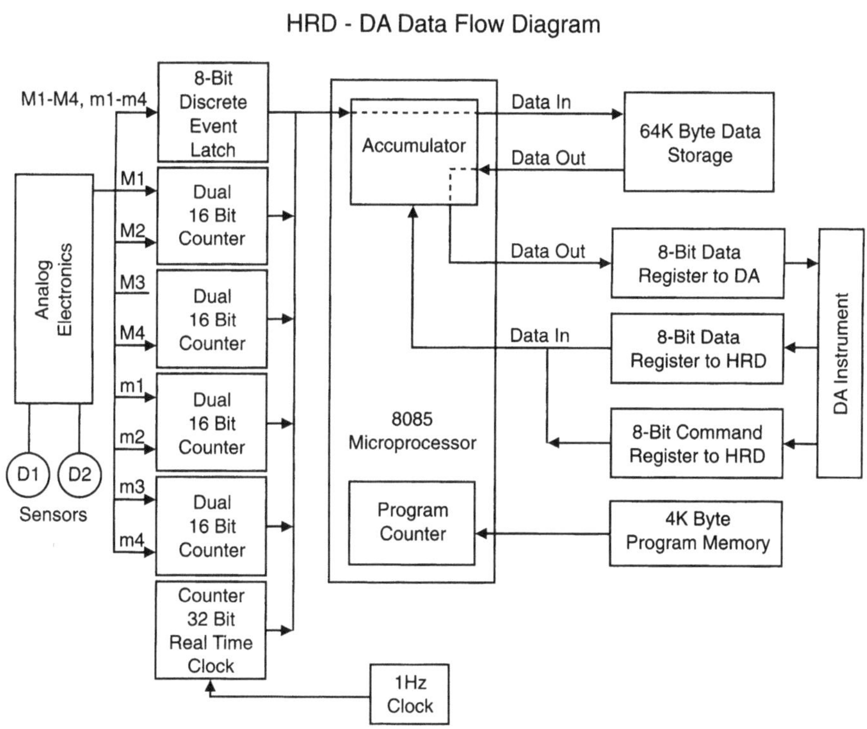

Figure 24. Diagram of the HRD-DA data flow.

for sensor #1, or the m1, m2, m3, m4 thresholds for sensor #2. As a contingency, the M1–M4, and/or m1–m4 electronic thresholds may be increased by a fixed factor (~10) by ground command, which activates a sensor mass threshold switch for either or both sensors (Figure 23).

3.4.5.2. Digital Electronics and Data. Figure 24 shows a block diagram of the important elements of the HRD digital electronics and illustrates the HRD-DA data flow. If the lowest threshold of either sensor is fired, the threshold firings will increment the corresponding 16-bit counters, and the following items (discrete event data) are stored in HRD memory when HRD is in NORMAL MODE (CRUISE):

(a) particle impact time (21 low-order bits from the HRD 32-bit clock, with 1 s resolution);
(b) CDA pointing (sector) information from DA (3 bits), and;
(c) the threshold firings for each sensor – M1, M2, M3, M4, m1, m2, m3, m4 (8 bits).

After storage of 256 discrete events in HRD memory, the next discrete event will also store in HRD memory HEADER data, which includes the contents of the eight 16-bit counters, as illustrated in Figure 25. The HEADER data also includes the HRD STATUS (8 bits), which indicates the operating mode of the HRD, and the status of the two sensor mass threshold relays, as well as the temperature (8 bits) of the interior of the HRD electronics box (TEMP #1).

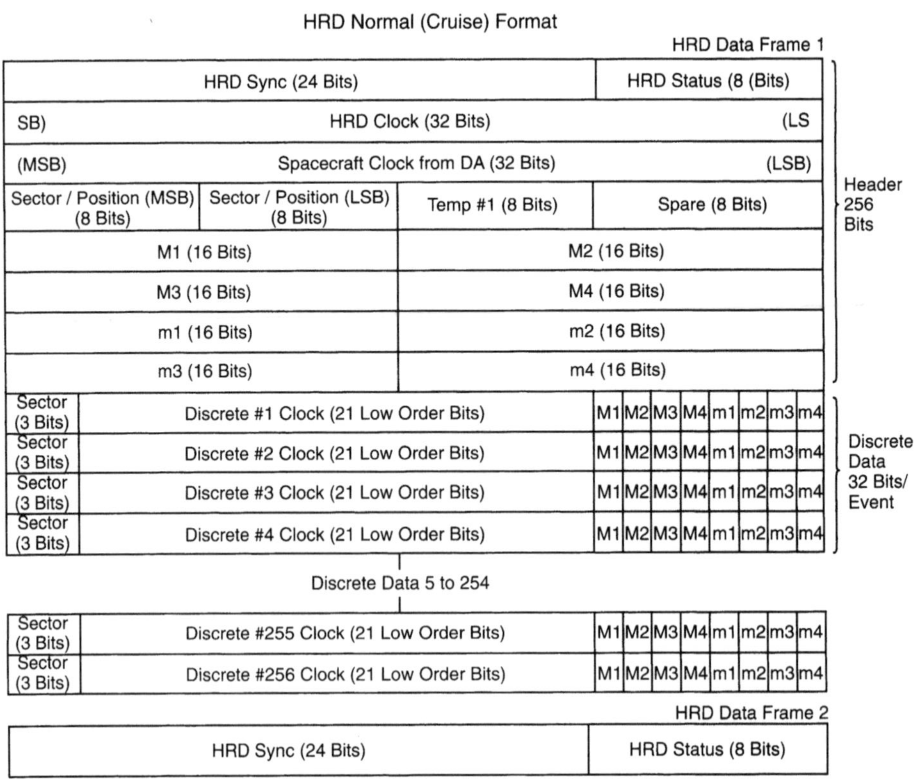

Figure 25. HRD data format for NORMAL MODE (CRUISE).

HRD "Encounter" Format

HRD Sync (24 Bits)	HRD Status (8 (Bits)
Sector (5 LS Bits)	HRD Clock (27 Bits)
M1 (16 Bits)	M2 (16 Bits)
M3 (16 Bits)	M4 (16 Bits)
m1 (16 Bits)	m2 (16 Bits)
m3 (16 Bits)	m4 (16 Bits)

Figure 26. HRD data format for ENCOUNTER MODE (ring plane crossings).

When the HRD is in FAST (ENCOUNTER) MODE, the HRD will record and store in HRD memory each Δt seconds the integral counts for the eight 16-bit counters, as shown in Figure 26, as well as HRD status, pointing and HRD clock information.

3.4.5.3. HRD Commands. When the CDA main electronics supplies power to the HRD, the HRD turn on in NORMAL (CRUISE) MODE (Figure 25), with a sensor mass threshold relays in the "LOW MASS" positions. HRD commands include:

(a) ENCOUNTER MODE: Selects Δt at one value in the range $\Delta t = 0.1, 0.2, \ldots$, 0.9, 1.0 s and runs the HRD in ENCOUNTER MODE (Figure 26).
(b) SET RELAY: Selects the threshold relay switches for each sensor to LOW MASS RANGE or HIGH MASS RANGE (see Table 2).
(c) IN-FLIGHT CALIBRATE (IFC): Initiates a sequence of electronic pulses of varying amplitude within the HRD electronics (Figure 23) which permits assessment of the stability of the HRD electronics.

3.4.6. *PVDF Sensor Calibrations*

Particle calibrations of HRD-type PVDF sensors (28 and 6 μm thick) were carried out at the Heidelberg and Munich dust accelerator facilities, as summarized in Table 10 (Tuzzolino, 1996). Our results are summarized in Figures 27 and 28 for 6 μm thick PVDF sensors and in Figures 29 and 30 for 28 μm thick PVDF sensors, where *N*(e) is the output signal amplitude from the shapers expressed in units of number of electron charges at the input to the electronics, and *m* (g) and *v* (km/s) are the mass and impact velocity of the dust particle. The calibration data show that the signal amplitude *N* may be fit by a power-law dependence on particle mass and velocity of the form $m^a v^b$, with values for the mass index *a* and velocity index *b*, as given in the expressions included.

Form these data, and an assumed ring plane crossing particle impact velocity of 15 km/s the electronic thresholds set for the HRD flight unit will result in the particle mass thresholds given in Table 11. The upper half of Table 11 lists the electronic and particle mass thresholds for the case where the sensor mass threshold switches

TABLE X

Summary of PVDF dust sensor calibrations at Heidelberg (August 1983, June 1984, January 1994) and Munich (October 1987, April 1989, January 1994).

Facility	Property	6 μm thick PVDF	28 μm thick PVDF
MPI-K Heidelberg (Fe particles)	Velocity (km/s)	1.0–12.0	1.7–11.4
	Mass (g)	3.8×10^{-13} to 1.7×10^{-10}	2.8×10^{-13} to 1.9×10^{-10}
	Diameter (μm)	0.45–3.3	0.4–3.4
Munich/Garching (glass particles)	Velocity (km/s)	1.8–15.9	2.0–11.4
	Mass (g)	2.4×10^{-9} to 2.0×10^{-6}	5.4×10^{-9} to 3.0×10^{-6}
	Diameter (μm)	12–115	16–130
Range of measured sensor signal amplitudes	Smallest	8.3×10^4 (Fe)	2.7×10^4 (Fe)
(number of electron charges e)	Largest	8.5×10^9 (glass)	4.2×10^{10} (glass)

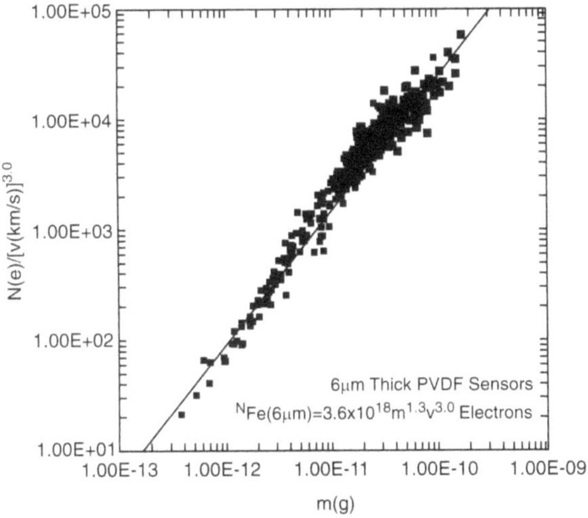

Figure 27. Calibration data for the 6 μm thick PVDF sensor (Heidelberg, Fe particles).

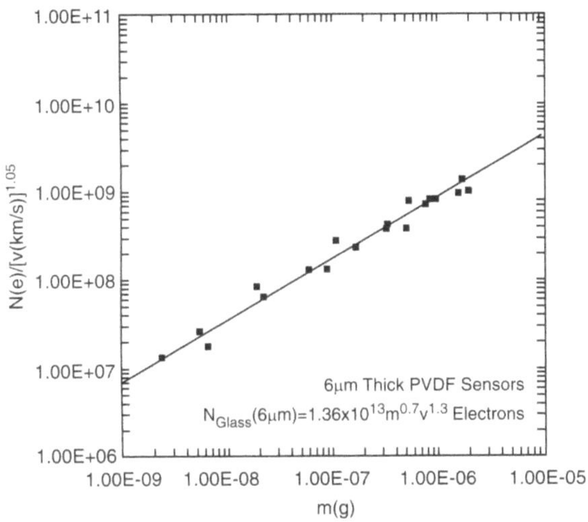

Figure 28. Calibration data for the 6 μm thick PVDF sensor (Munich, glass particles).

are set for "LOW MASS," and the lower half of Table 11 for the case where the switches are set for "HIGH MASS" (by ground command).

3.5. CDA Measurement Ranges and Payload Parameters

The impact rate will be measured from 3×10^{-7} to 1 s^{-1} (IID) and up to 10^4 s^{-1} (HRD) in order to cover both the impact rates expected in interplanetary space and during high activities such as ring plane crossings at Saturn. The measurable

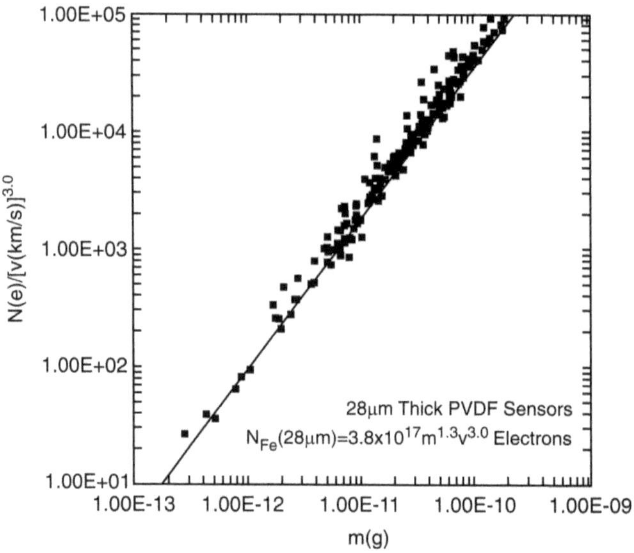

Figure 29. Calibration data for the 28 μm thick PVDF sensor (Heidelberg, Fe particles).

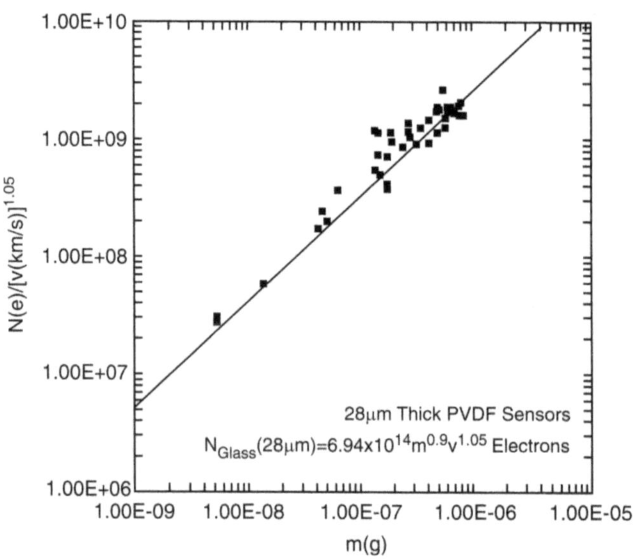

Figure 30. Calibration data for the 28 μm thick PVDF sensor (Munich, glass particles).

particle mass ranges from 5×10^{-19} to 10^{-13} kg (40 km s^{-1} impact speed) and from 10^{-16} to 10^{-10} kg (5 km s^{-1} impact speed). The detectable impact speed is greater than 1 km s^{-1}. There is no upper speed limit for the detection of particles, but the speed determination will be difficult beyond 80 km s^{-1}. The particle charges are measured from 10^{-15} to 5×10^{-13} C for both, negative and

TABLE XI

HRD PVDF sensors – electronic thresholds and corresponding particle mass thresholds for 15 km/s impact speed (upper half: "Low Mass," lower half: "High Mass").

Electronic threshold (e)	Particle mass thresholds (g)	Particle diameter[a] (μm)	Electronic threshold (e)	Particle mass threshold (g)	Particle diameter (μm)
#1 PVDF Sensor #1; area = 50 cm^2; thickness = 28 μm			PVDF Sensor #2; area = 10 cm^2; thickness = 6 μm		
LM1 = 2.1×10^6	4.0×10^{-12}	1.5	m1 = 2.0×10^6	7.5×10^{-13}	1.5
LM2 = 1.9×10^7	2.2×10^{-11}	2.6	m2 = 1.9×10^7	4.0×10^{-12}	1.5
LM3 = 4.1×10^8	5.1×10^{-9}	15.7	m3 = 4.1×10^8	2.3×10^{-9}	12.1
LM4 = 5.2×10^9	8.3×10^{-8}	40	m4 = 5.2×10^9	8.5×10^{-8}	40.2
HM1 = 1.8×10^7	2.1×10^{-11}	2.5	m1 = 2.1×10^7	4.4×10^{-12}	0.8
HM2 = 1.6×10^8	1.7×10^{-9}	10.9	m2 = 2.0×10^8	9.0×10^{-10}	8.8
HM3 = 3.5×10^9	5.5×10^{-8}	35	m3 = 4.3×10^9	6.5×10^{-8}	36.7
HM4 = 4.4×10^{10}	9.3×10^{-7}	89	m4 = 5.4×10^{10}	2.5×10^{-6}	124

[a]Assuming impacting particles with density 2.5 g/cm^3.

positive charges. The trajectory of charged particles (above 10^{-14} C) can be measured with an accuracy of 5° in one plane. The mass resolution of the time-of-flight mass spectrometer lies between 10 (lower masses) and 50 (higher masses) which is certainly sufficient to distinguish between water ice, organic or silicate particles.

The maximum angle of incidence for particles with respect to the sensor boresight is 45 ° (IIT), 28 ° (CAT) and 88 ° for the HRD. Table 12 gives an overview of the CDA properties.

3.6. Instrument Operations and Data Products.

Following the launch of Cassini-Huygens on October 15, 1997, the CDA instrument was switched on for the first time on October 31, 1997. On that day the platform was unlatched and the instrument articulated to its 270° position. On November 13, 1997, the cover was successfully jettisoned by firing the two redundant pyro devices. Since then the aperture of the instrument was open. The first decontamination of the chemical analyzer target occurred one day later. During this procedure the target was heated up to 90 °C in order to eliminate possible contamination by hydrocarbons. This target must be kept clean to obtain undisturbed mass spectra. In order to maintain the mechanics of the articulation mechanism, the instrument participated in a periodic instrument maintenance every 3 months in the first 2 years of the mission. During this process the instrument performed a 225 °

TABLE XII
Properties of the CDA instrument.

Parameter	Value	Unit
Mass total	17.151	kg
Mass Dust Analyzer	4.642	kg
Mass Main Electronics	6.801	kg
Mass Amplifier Box	3.753	kg
High Rate Detector	1.049	kg
Thermal blankets	0.905	kg
Power sleep	10.15	W
Power with HRD on	11.70	W
Power during articulation	13–17.3	W
Replacement heater	11.25	W@30 V
Decontamination	8.85	W@30 V
Supply voltage	30 (+1.5/−2.5)	V
Operating temperature		
Main Electronics Box	−15 to +40	°C
High Rate Detector	−25 to +40	°C
Radiation hardness	100	krad
Data bus S/C interface	1553 Bus	
Data rate	19 bps HK; 524 bps Science	
Articulation about 1 axis	0–270	degree
Dimensions		
Impact Ionization Target	Ø 0.41	m
Chemical Analyzer Target	Ø 0.16	m

turn. This total activity took approximately 2 h. The CDA instrument check-out was in December 1998 and took approximately 3 weeks. During the first part of the checkout, the instrument health was checked and the high voltages were stepped up to their normal values. This time was also used to verify the execution of some commands. Then a 5-day period followed during which the trigger thresholds of all channels were adjusted. The goal was to find an instrument state which can detect small particle impacts by using low thresholds but which was not too disturbed by noise events. In the following weeks the first dust impacts were awaited on all subsystems: the IID, the CAT and the HRD. Since the sensitive area of the CAT is rather small, many weeks are necessary to ensure the detection of a mass spectrum.

The planned data products are manifold. The final result is a data base containing the most significant particle parameters such as impact speed, mass, charge, direction, composition, and impact time. The software for analysis and archiving is still in development and their description will be given elsewhere.

Acknowledgements

The development, construction, implementation and testing of this instrument and its data processing system was only achieved by the efforts of a great number of individuals from various departments of the Max-Planck-Institut für Kernphysik and from outside institutions and companies. The authors are especially indebted to many coworkers who supported the team at MPI-K, DLR-Berlin, UK Canterbury, ESTEC and RAL. We are grateful to G. Pahl who designed the instrument and to A. Kaiser and S. Helfert who developed significant parts of the onboard software. This project is supported by the Max-Planck-Institut für Kernphysik and by the DLR under the grant 50OH91019.

References

Anders, E. and Grevesse, E.: 1989, *Cosmochim. Acta* **53**, 197–214.
Arndt, P., Bohsung, J., Maetz, M., and Jessberger, E. K.: 1996, *Meteoritics Planet. Sci.* **31**, 817–833.
Auer, S.: 1975, *Rev. Sci. Instrum.* **46**(2), 127–235.
Auer, S., Grün, E., Srama, R., Kempf, S., and Auer, R.: *Plan. Space Sci.* **50**(7/8), 773–779.
Baggaley, W. J. and Galligan, D. P.: 1997, *Plan. Space Sci.* **45**(7), 865–868.
Bradley, J. G., Grün, E., and Srama, R.: 1996, *The Cosmic Dust Analyzer for Cassini*, Vol. 2803, 5–6 August, Proceedings of SPIE, Denver.
Berg, O. E. and Richardson, F. F.: 1968, *Rev. Sci. Instrum.* **40**, 1333.
Bradley, J. P.: 1988, *Cosmochim. Acta* **52**, 889–900.
Brownlee, D. E.: 1985, *Ann. Rev. Earth Planet. Sci.* **13**, 147–173.
Burns, J. A., Lamy, Ph. L., and Soter S.: 1979, *Icarus* **40**, 1–48.
Bus, S. J., Bowell, E., Harris, A. W., and Hewitt, A. V.: 1989, *Icarus* **77**, 223–238.
Colwell, J. E., Horanyi, M., and Grün, E.: 1998, *Am. Inst. Phys. Conf. Proc.* **446**, 299–306.
Cuzzi, J. N. and Estrada, P. R.: 1998, *Icarus* **132**, 1–35.
Dietzel, H., Eichhorn, G., Fechtig, H., Grün, E., Hoffman, H.-J., and Kissel, J.: 1973, *J. Phys. (E) Sci. Instrum.* **6**, 209–217.
Dikarev, V. V.: 1999, *Astron. Astrophys.* **346**, 1011–1019.
Dikarev, V. and Grün, E.: 2002, *Astron. Astrophys.* **383**(1), 302–308.
Dikarev, V. V. and Krivov, A. V.: 1998, *Solar Syst. Res.* **32**(2), 128–143.
Goertz, C. K. and Morfill, G.: 1983, *Icarus* **53**, 219–229.
Goertz, C. K. and Morfill, G. E.: 1988, *Icarus* **74**, 3815.
Goldsworthy, B. J., Burchell, M. J., Cole, M. J., Green, S. F., Leese, M. R., McBride, N., McDonnell, J. A. M., Müller, M., Grün, E., Srama, R., Armes, S. P., and Khan, M. A.: 2002, *Adv. Space Res.* **29**(8), 1139–1144.
Göller, J. R. and Grün, E.: 1989, *Planet. Space Sci.* **37**(10), 1197–1206.
Graps, A. L., Grün, E., Svedhem, H., Krüger, H., Horányi, M., Heck, A., and Lammers, S.: 2000, *Nature* **405**, 48–50.
Grün, E., Baguhl, M., Divine, N., Fechtig, H., Hamilton, D. P., Hanner, M. S., Kissel, J., Lindblad, B.-A., Linkert, D., Linkert, G., Mann, I., McDonnell, J. A. M., Morfill, G. E., Polanskey, C., Riemann, R., Schwehm, G., Siddique, N., Staubach, P., and Zook, H. A.: 1995, *Planet. Space Sci.* **43**(8), 971–999.
Grün, E., Baguhl, M., Hamilton, D. P., Riemann, R., Zook, H. A., Dermott, S., Fechtig, H., Gustafson, B. A., Hanner, M. S., Horanyi, M., Khurana, K. K., Kissel, J., Kivelson, M., Lindblad, B.-A.,

Linkert, D., Linkert, G., Mann, I., McDonnell, J. A. M., Morfill, G. E., Polanskey, C., Schwehm, G., and Srama, R.: 1996, *Nature* **381**, 395–398.

Grün, E., Fechtig, H., Giese, R. H., Kissel, J., Linkert, D., Maas, D., McDonnell, J. A. M., Morfill, G. E., Schwehm, G., and Zook, H. A.: 1992, *Astron. Astrophys. Suppl. Ser.* **92**, 411–423.

Grün, E., Fechtig, H., Giese, R. H., Kissel, J., Linkert, D., McDonnell, J. A. M., Morfill, G. E., Schwehm, G., and Zook, H. A.: 1983, in: Wenzel, K. P., Marsdon, R. G., and Battrick, B. (eds.), *The ISPM Dust Experiment in The International Solar Polar Mission – Its Scientific Investigations*, ESA SP-1050, pp. 227–241.

Grün, E., Fechtig, H., Hanner, M. S., Kissel, J., Lindblad, B.-A., Linkert, D., Linkert, G., Morfill, G. E., and Zook, H.: 1992, *Space Sci. Rev.* **60**, 317–340.

Grün, E., Hamilton, D. P., Riemann, R., Dermott, S., Fechtig, H., Gustafson, B. A., Hanner, M. S., Heck, A., Horanyi, M., Kissel, J., Krüger, H., Lindblad, B.-A., Linkert, D., Linkert, G., Mann, I., McDonnell, J. A. M., Morfill, G. E., Polanskey, C., Schwehm, G., Srama, R., and Zook, H. A.: 1996, *Science* **274**, 399–401.

Grün, E., Krüger, H., Dermott, S., Fechtig, H., Graps, A., Gustafson, B. A., Hamilton, D. P., Hanner, M. S., Heck, A., Horanyi, M., Kissel, J., Lindblad, B.-A., Linkert, D., Linkert, G., Mann, I., McDonnell, J. A. M., Morfill, G. E., Polanskey, C., Schwehm, G., Srama, R., and Zook, H. A.: 1997, **24**, 2171–2174.

Grün, E., Krüger, H., Graps, A., Hamilton, D. P., Heck, A., Linkert, G., Zook, H. A., Dermott, S., Fechtig, H., Gustafson, B. A., Hanner, M. S., Horányi, M., Kissel, J., Lindblad, B.-A., Linkert, D., Mann, I., McDonnell, J. A. M., Morfill, G. E., Polanskey, C., Schwehm, G., and Srama, R.: 1998, *J. Geophys. Res.-Planets.* **103**(E9), 20011–20022.

Grün, E. and Landgraf, M.: 2000, *J. Geophys. Res.* **105**(A5), 10291–10297.

Grün, E., Morfill, G. E., and Mendis, D. A.: 1984, in: Greenberg, R., and Brahic, A. (eds.), *Dust-Magnetospere Interactions in "Planetary Rings"*, University of Arizona Press, Tuscon, pp. 275–332.

Grün, E., Morfill, G., Terrile, R. J., Johnson, T. V., and Schwehm, G.: 1983, *Icarus* **54**, 227–252.

Grün, E., Zook, H. A., Baguhl, M., Balogh, A., Bame, S. J., Fechtig, H., Forsyth, R., Hanner, M. S., Horanyi, M., Kissel, J., Lindblad, B.-A., Linkert, D., Linkert, G., Mann, I., McDonnell, J. A. M., Morfill, G. E., Phillips, J. L., Polanskey, C., Schwehm, G., Siddique, N., Staubach, P., Svestka, J., and Taylor, A.: 1993, *Nature* **362**, 428–430.

Grün, E., Zook, H. A., Fechtig, H., and Giese, R. H.: 1985, *Icarus* **62**, 244–272.

Gurnett, D. A., Ansher, J. A., Kurth, W. S., and Granroth, L. J.: 1997, *Geophys. Res. Lett.* **24**, 3125ff.

Gurnett, D. A., Grün, E., Gallagher, D., Kurth, W. S., and Scarf, F. L.: 1983, *Icarus* **53**, 236–254.

Haff, P. K., Eviatar, A., and Siscoe, G. L.: 1983, *Icarus* **56**, 426.

Hamilton, D. P. and Burns, J. A.: 1993, *Nature* **364**, 695–699.

Hamilton, D. P. and Burns, J. A.: 1994, *Science* **264**, 550–553.

Hanner, M. S., Sparrow, J. G., Weinberg, J. L., and Beeson, D. E.: 1976, Pioneer 10 observations of zodiacal light brightness near the ecliptic: Changes with heliocentric distance, in: Elsasser, H., and Fechtig, H. (eds.), *Interplanetary Dust and Zodiacal Light*, Lecture Notes in Physics, Vol. 48, Springer-Verlag, New York, pp. 29–35.

Hashimov, N. M. and Tarakanov, V. L.: 1982, *Komety i Meteory* **32**, 3–9 (in Russian).

Havnes, O.: 1984, *Adv. Space Res.* **4**(9), 75–83.

Havnes, O., Goertz, C. K., Morfill, G. E., Grün, E., and Ip, W.: 1987, *J. Geophys. Res.* **92**, 2281.

Havnes, O. and Morfill, G. E.: 1984, *Adv. Space Res.* **4**, 85.

Hoffmann, H.-J., Fechtig, H., Grün, E., and Kissel, J.: 1975, *Plant. Space Sci.* **23**, 985–991.

Horanyi, M.: 2000, *Phys. Plasmas* **7**, 3847–3850.

Horanyi, M., Morfill, G. E., and Grün, E.: 1993, *Nature* **363**, 144–146.

Horanyi, M., Morfill, G. E., and Grün, E.: 1993, The dusty ballerina skirt of jupitor. *Jour. Geophys. Res.* **98**, 21245–21251.

Howard, J. and Horanyi, M.: 2001, *Geophys. Res. Lett.* **28**, 1907–1911.

Humes, D. H.: 1980, *J. Geophys. Res.* **85**, 5841–5852.

Humes, D. H., Alvarez, J. M., O'Neal, R. L., Kinard, and W. H.: 1974, *J. Geophys. Res.* **79**(25), 3677–3684.

Jackson, A. A. and Zook, H. A.: 1989, *Nature* **337**, 629–631.

Jaffe, L. D. and Herrell, L. M.: 1997, *J. Spacecraft Rockets* **34**(4), 509.

Jessberger, E. K.: 1999, On the elemental, isotopic and mineralogical ingredients of ROCKY cometary particulates, in: Altwegg, K., Ehrenfreund, P., Geiss, J., and Huebner, W. (eds.), *Proceedings of the Origin and Composition of Cometary Materials*, Space Science Series of ISSI, Kluwer Academic Publishers, pp. 91–97.

Jessberger, E. K., Bohsung, J., Chakaveh, S., and Traxel, K.: 1992, *Earth Planet. Sci. Lett.* **112**, 91–99.

Jessberger, E. K., Christoforidis, A., and Kissel, J.: 1988, *Nature* **332**, 691–695.

Jessberger, E. K., Stephan, T., Rost, D., Arndt, P., Maetz, M., Stadermann, F. J., Brownlee, D. E., Bradley, J., and Kurat, G.: 2001, Properties of interplanetary dust: Information from collected samples, in: Dermott, S. F., Fechtig, H., Grün, E., and Gustafson, B. A. S. (eds.), *Interplanetary Dust*, Springer, Heidelberg, pp. 253–294.

Jessberger, E. K. and Wallenwein, R.: 1986, *Adv. Space Res.* **6**, 5–8.

Johnson, R. E., Pospieszalska, M. K., Sittler, E. C., Jr., Cheng, A. F., Lanzerotti, L. J., and Sievka, E. M.: 1989, *Icarus* **77**, 311.

Jyoti, G., Gupta, S. C., Ahrens, T. J., Kossakovski, D., and Beauchamp, J. L.: 1999, in: Furnish, M. D. (ed.), *Mass Spectrometer Calibration of Cosmic Dust Analyzer, Shock Compression of Condensed Matter.*

Jyoti, G., Gupta, S. C., Ahrens, T. J., Kossakovski, D., and Beauchamp, J. L.: 1999, *Int. J. Impact Eng.* **23**, 401–408.

Kissel, J., Fechtig, H., Grün, E., Igenbergs, E. B., Jessberger, E. K., McDonnell, J. A. M., Morfill, G. E., Schwehm, G. H., and Zook, H. A.: 1986, *Nature* **321**, 336–338.

Krivov, A. V. and Banaszkiewicz, M.: 2001, *Plan. Space Sci.* **49**(13), 1265–1279.

Krivov, A. V., Dikarev, V. V., Sremcevic, M., and Kholshevnikov, K. V.: 2001, *Dust Clouds of Saturnian Outer Satellites: Guidelines for Cassini Dust Instrument*, white paper.

Krüger, H., Krivov, A. V., and Grün, E.: 2000, *Plant. Space Sci.* **48**(15), 1457–1471.

Krivov, A. V., Krüger, H., and Grün, E.: 2002, *J. Geophys. Res.* **107**(E1), 2-1–13.

Lamy, P. L., Lefevre, J., Millet, J., and Lafon, J. P.: 1985, Electrostatic charge of interplanetary dust grains: New results, in: Giese, R. H., and Lamy, P. (eds.), *Properties and Interactions of Interplanetary Dust*, D. Reidel Publishing Co., Dordrecht, pp. 335–339.

Landgraf, M., Augustsson, K., Grün, E., and Gustafson, B. A. S.: 1999, *Science* **286**, 5448, 2319–2322.

Landgraf, M., Baggeley, W. J., Grün, E., Krüger, H., and Linkert, G.: 2000, *J. Geophys. Res.* **105**(A5), 10343–10352.

Landgraf, M., Liou, J.-C., Zook, H. A., and Grün, E.: 2002, *Astron. J.* **123** (5), 2857–2861.

Leinert, C. and Grün, E.: 1990, in: Schwenn, R., and Marsch, E. (eds.), *Interplanetary Dust in Physics and Chemistry in Space*, Springer, Heidelberg, pp. 207–275.

Leinert, C., Richter, I., Pitz, E., and Planck, B.: 1981, *Astron. Astrophys.* **103**, 177–188.

Low, F. J., Beintema, D. A., Gautier, T. N., Gillette, F. C., Beichman, C. A., Neugebauer, G., Young, E., Aumann, H. H., Boggess, N., Emerson, J. P., Habing, H. J., Hauser, M. G., Houck, J. R., Rowan-Robinson, M., Soifer, B. T., Walker, R. G., and Wesselius, P. R.: 1984, *Astrophys. J.* **278**, L19–L22.

McDonnell, J. A. M.: 1978, Microparticle studies for space instrumentation, in: McDonnell, J. A. M. (ed.), *Cosmic Dust*, Wiley and Sons, New York, p. 337.

McDonnell, J. A. M., Berg, O. E., and Richardson, F. F.: 1975, *Planet. Space Sci.* **23**, 205–214.

Melandso, F. and Havnes, O.: 1991, *J. Geophys. Res.* **96**(A4), 5837–5845.

Mendis, D. A., Hill, I. R., Houpis, H. L. F., and Whipple, E. C., 1981, *Astrophys. J.* **249**, 787–797.
Mendis, D. A., Hill, I. R., Ip, W.-H., Goertz, C. K., and Grün, E.: 1984, Electrodynamic processes in the ring system of saturn, in: Gehrels, T., and Matthews, M. S. (eds.), *Saturn*, University of Arizona Press, Tuscon, p. 546.
Morfill, G. E., Grün, E., and Leinert, C.: 1986, The interaction of solid particles with the interplanetary medium, in: Marsden, R. G. (ed.), *The Sun and the Heliosphere in Three Dimensions*, D. Reidel Publishing Co., Dordrecht, pp. 455–474.
Morfill, G. E., Havnes, O., and Goertz, C. K.: 1990(in preparation), Origin and maintenance of the oxygen torus in Saturn's magnetosphere.
Mukai, T., Fechtig, H., Grün, E., and Giese, R. H.: 1989, *Icarus* **80**, 254–266.
Perkins, M. A., Simpson, J. A., and Tuzzolino, A. J.: 1985, *Nucl. Instrum. Method* **A239**, 310.
Ratcliff, P., Gogu, F., Grün, E., and Srama, R.: 1996, *Adv. Space Res.* **17**(12), 111–115.
Reach, W. T., Franz, B. A., Weiland, J. L., Hauser, M. G., Kelsall, T. N., Wright, E. L., Rawley, G., Stemwedel, S. W., and Spiesmann, W. J.: 1995, *Nature* **374**, 521–523.
Schaffer, L. and Burns, J. A.: 1987, *J. Geophys. Res.* **92**, 2264.
Showalter, M. R., Burns, J. A., Cuzzi, J. N., and Pollack, J. B.: 1985, *Nature* **316**, 526–528.
Simpson, J. A., Rabinowitz, D., and Tuzzolino, A. J.: 1989, *Nucl. Instrum. Methods* **A279**, 611.
Simpson, J. A. and Tuzzolino, A. J.: 1985, *Nucl. Instrum. Methods* **A236**, 187.
Simpson, J. A. and Tuzzolino, A. J.: 1989, *Nucl. Instrum. Methods* **A279**, 625.
Spahn, F., Thiessenhusen, K.-U., Colwel, J. E., and Srama, R.: 1999, *J. Geophys. Res.* **104**, 24111–24120.
Srama, R. and Grün, E.: 1997, *Adv. Space Res.* **20**(8), 1467–1470.
Stübig, M., Schäfer, G., Ho, T. M., Srama, R., and Grün, E.: 2001, *Planet. Space Sci.* **49**(8), 853–858.
Sutton, S. R. and Flynn, G. J.: 1988, 'Stratospheric particles: Synchrotron X-ray fluorescence determination of trace element contents', in: *Proceedings of the XVIII Lunar Planetary Science Conference*, Cambridge University Press, Cambridge, pp. 607–613.
Sykes, M. V. and Greenberg, R.: 1986, *Icarus* **65**, 51–69.
Thiessenhusen, K. U., Krüger, H., Spahn, F., and Grün, E.: 2000, *Icarus* **144**(1), 89–91.
Tiersch, H. and Notni, P.: 1989, *Astron. Nachrich.* **310**(1), 67–78.
Tuzzolino, A. J.: 1996, *Adv. Space Res.* **17**(12), 123–132.
Whipple, F. L.: 1967, 'On maintaining the meteoritic complex', in: Weinberg, J. L. (ed.), *NASA SP-150: The Zodiacal Light and the Interplanetary Medium*, U.S. Government Printing Office, Washington, DC, pp. 409–426.
Whipple, F. L.: 1975, 'Sources of interplanetary dust', in: Elsässer, H., and Fechtig, H. (eds.), *Interplanetary Dust and Zodiacal Light*, Lecture Notes in Physics, Vol. 48, Springer-Verlag, New York, pp. 403–415.
Wyatt, S. P.: 1969, *Planet. Space Sci.* **17**, 155–171.
Zook, H. A.: 1975, *Planet. Space Sci.* **23**, 1391–1397.
Zook, H.: 1980, 'Evidence for ice meteoroids beyond 2 AU', in: Halliday, I., and McIntosh, B. A. (eds.), *Solid Particles in the Solar System*, D. Reidel, Dordrecht, Holland, pp. 375–380.
Zook, H. A. and Berg, O. E.: 1975, *Planet. Space Sci.* **23**, 183–203.
Zook, H. A., Grün, E., Baguhl, M., Hamilton, D. P., Linkert, G., Liou, J.-C., Forsyth, R., and Phillips, J. L.: 1996, *Science* **274**, 1501–1503.

MIX
Papier aus verantwortungsvollen Quellen
Paper from responsible sources
FSC® C105338

If you have any concerns about our products,
you can contact us on
ProductSafety@springernature.com

In case Publisher is established outside the EU,
the EU authorized representative is:
Springer Nature Customer Service Center GmbH
Europaplatz 3, 69115 Heidelberg, Germany

Printed by Libri Plureos GmbH
in Hamburg, Germany